PROCEEDINGS

International Conference on Rough Sets, Fuzzy Sets and Soft Computing

November 5-7, 2009

in collaboration with

IRSS and ISFUMIP

R.N. Bhaumik

Conference Chair

Department of Mathematics
Tripura University
Suryamaninagar, Tripura (W),
India-799130

PROCEEDINGS

International Conference on Rough Sets, Fuzzy Sets and Soft Computing

November 5–7, 2009

ORGANISED BY

Department of Mathematics

Tripura University

in Collaboration with

INTERNATIONAL ROUGH SETS SOCIETY

Indian Society for Fuzzy Mathematics and Information Processing

Serials Publications
New Delhi (India)

First Published-2011

ISBN: 978-81-8387-417-5

Published by

SERIALS PUBLICATIONS
4830/24, Prahlad Street, Ansari Road
Darya Ganj, New Delhi-110002 (India)
Phone: 23245225, Fax: 91-11-23272135
E-mail: serials@satyam.net.in

GENERAL CHAIR

Prof. A. Saha

CONFERENCE CHAIR

R.N. Bhaumik, India

INTERNATIONAL ADVISORY COMMITTEE

1. K.T. Atanassov, Bulgaria
2. G. Cattaneo, Italy
3. D. Dubois, France
4. M. Inuiguchi, Japan
5. S. Jafari, Denmark
6. E.E. Kerre, Belgium
7. D. Datta Majumdar, India
8. J.F. Peters, Canada
9. S. Ramanna, Canada
10. S.K. Pal, India
11. Andrzej Skowron, Poland
12. A.K. Srivastava, India
13. E. Turunen, Finland
14. Y.Y. Yao, Canada
15. T. Thrivikraman, Kerala
16. G. Wang, China
17. Prof. W. Ziarko

PROGRAM CHAIRS

Dominik Slezak, Poland
M.K. Chakraborty, India

FINANCE CHAIR

S. Bhattacharya (Halder)

TECHNICAL SUPPORT COMMITTEE

Reception and Accomodation

Prof. A. Mukherjee

Fooding

Prof. A. De

Arrangement

S. Bhowmik and Dr. A. Hossain

Registration

Dr. D. Bhattacharya

Conveyance

Dr. H. Chakraborty

Inauguration

Dr. Sunil De and S. Debbarma

Publicity

Azad Hossain and S.B. Chakraborty

Prof. A. Saha
Vice Chancellor

Tripura University
(A Central University)
P. O. Suryamaninagar
Tripura- 799130, INDIA
✆ : 0381 237 4801, 0381 251 4824 (R)
Fax : 0381-2374802
e-mail: arunodaysaha@rediffmail.com

No. F. TU/VCS/CORRS/01/2009 Date: 5th October, 2009

MESSAGE

I am glad to know that the Department of Mathematics, Tripura University, is going to organise an International Conference on Rough sets, Fuzzy sets and Soft computing during November 5-7, 2009. In this 3-days conference, young researchers, students & teachers of Tripura will get an opportunity for in-depth discussions on the latest trends and their applications in the frontier of knowledge. The invited lectures by the leading Mathematicians & Scientists on different aspects with trust on modern developments will motivate our young generation.

The proceedings of the International Conference on Rough sets, Fuzzy sets and Soft computing to be published in book form (Serial Publication, New Delhi) will, I believe, be a meaningful venture. I hope that the research materials to be contained in the proceedings will inspire the researchers in planning their future course of research.

(A. Saha)

ISBN: 978-81-8387-417-5

PREFACE

Mathematics, well known in dealing with precise or exact concepts, are not useful in modeling the present real life situation which consists of inexactness or impreciseness or linguistic variables.

Fuzzy sets were introduced in 1965 by L. Zadeh with a view to reconcile mathematical modeling and human knowledge in the engineering sciences. Since then, a considerable literature has blossomed around the concepts of fuzzy sets in wide range of areas, from mathematics to engineering methodologies. Applications are found in many contexts from medicine to finance, from human factors to consumer products, from vehicle control to linguistics and so on. Fuzzy logic is now currently used in the industrial practice of advanced I.T.

Rough Sets, introduced by Z.Pawlak in 1981, is expressed by a boundary region of a set . It is also an approach to vagueness. Thus Fuzzy sets and Rough sets are two different approaches to vagueness or impreciseness of the real life problems.

In the North-Eastern Region (NER) of India, research work on the theory of fuzzy sets and rough sets and their applications were not considerably cultivated. Though Tripura, geographically isolated state in India, is a backward place in NER, more than 25 persons of Tripura are working in these area. So the Department of Mathematics, Tripura University is organizing an International Conference on Rough Sets, Fuzzy sets and Soft Computing on November 5 – 7, 2009. The aim of the conference is to bring the researchers, all over the world, working in Rough Sets, Fuzzy Logic, Fuzzy Mathematics, Soft Computing and related areas to enable exchange of ideas and interaction between them regarding the advancement of imprecise conceptual phenomena with their wide variety of applications in Science and Technology. The 3-day Conference will include introductory tutorials and panel discussions by the external experts for the benefit of the scholars of this region.

It is great pleasure to welcome you all to this International Conference on Rough sets, Fuzzy sets and Soft Computing. The conference is organized in collaboration with International Rough Sets Society and ISFUMIP. The conference covers Rough Set Theory and its Applications, Fuzzy Logic and its applications, Fuzzy Topology, Fuzzy Algebra, Fuzzy Linear Programming and Soft Computing.

We are grateful to the collaborating societies, especially to Prof. J.F. Peters, Prof. S. Ramanna, Prof. G. Wang, Prof. T.Y. Lin, Prof. S. Dominik and Prof. M.K. Chakraborty.

We would like to express our sincere thanks to the members of the organizing committee for their whole hearted support, special mention must be made of the Program Chairs Prof. Slezak Dominik and Prof. M.K.Chakraborty for their initiative, co-operation and leading roles in organizing the conference.

We gratefully acknowledge the financial support provided by different organizations, especially to DST, New Delhi. Lastly we express our thanks to Mr. Vijoy Kumar of Serial Publications for his support in bringing out the proceedings on time.

Prof. R.N. Bhaumik
Conference Chair

Prof. A. Mukherjee
Head, Dept. of Mathematics

A Word from Program Co-chairs' Desk

In the North-Eastern corner of India at the University of Tripura a few enthusiasts have been engaged in the research of fuzzy set theory and rough set theory for quite a long time. Besides contributing to the theories by their work they have organized several national level conferences. This is the first time that under the leadership of Professor R.N. Bhaumik the Department of Mathematics has come forward to hold an international one. Since there is a group of people here who are interested in soft computing and since within its mathematical foundation fall both the above mentioned theories, it has been justifiably planned to put all the three areas together. Presence of some pioneers of the fields from various countries will undoubtedly enhance to a considerable degree researches in these fields in this North-Eastern zone of India in particular and India in general... It may be mentioned on this occasion that the year 2009 has been declared as the rough set year of India by the international community and this conference has been considered as one of the nodal events. I feel greatly honored and happy to be associated with the organizing team and hope this conference would be very successful.

Dominik Slezak, Poland
M.K. Chakraborty, India

INVITED SPEAKERS

1. Ahsanullah, Saudi Arabia
2. K.T. Atanassov, Bulgaria
3. M. Banerjee, IIT-Kanpur, India
4. G. Cattaneo, Italy
5. M.K. Chakraborty, India
6. D. Dubois, France
7. D. Datta Majumdar, India
8. A. Deshpande, Pune, India
9. T.K. Datta, CU, India
10. M. Inuiguchi, Japan
11. S. Jafari, Denmark
12. E.E. Kerre, Belgium
13. M.N. Mukherjee, CU, India
14. J.F. Peters, Canada
15. S. Ramanna, Canada
16. S.K. Pal, India
17. Dominik Slezak, Poland
18. S.K. Samanta, India
19. T. Som, BHU, India
20. Andrzej Skowron, Poland
21. A.K. Srivastava, India
22. E. Turunen, Finland
23. T. Thrivikraman, Kerala
24. Y.Y. Yao, Canada
25. G. Wang, China
26. W. Ziarko

REFEREES

1. D. Slezak (PC), Poland
2. M.K. Chakraborty (PC), India
3. S. Jafari, Denmark (Adv. Com), Denmark
4. Ahsanullah, Saudi Arabia
5. A. Deshpande, Pune, India
6. T.Thrivikraman, Kerala (Adv. Com), India
7. A.K. Srivastava, BHU (Adv. Com), India
8. S.K. Samanta, Visva Bharati, India
9. T. Som, BHU, India
10. R.N. Bhaumik, CC, India
11. Pawan Jha, Nepal
12. M.N. Mukherjee, CU, India
13. M. Banerjee, IIT-Kanpur, India
14. S. Mukherjee, Burdwan, India
15. P. Datta, Visva Bharati, India
16. T.K. Datta, CU, India
17. A. Mukherjee, TU, India
18. B.B. Pal, KU, India
19. G.V. Navalagi, Karnataka, India
20. S. Chatterjee, ISM, India
21. B.C. Giri, JU, India

LIST OF PARTICIPANTS (OUTSIDE INDIA)

S.No.	Name	Place	E-mail Address
1.	G. Cattaneo	Italy	cattang@disco.unimib.it
2.	Esko Turunen	Finland	esko.turunen@tut.fi
3.	Vilem Novak	University of Ostrava, Czech Republic	
4.	Irina PERFILJEV	University of Ostrava, Czech Republic	
5.	Piero Pagliani	Rome	p.pagliani@agora.it
6.	Hossam A. Nabwey	Menofia University Egypt	eng_hossam21@yahoo.com
7.	Thanat	Rome, Italy	

LIST OF PARTICIPANTS (INDIA, OUTSIDE TRIPURA)

S.No.	Name	Place	E-mail Address
1.	Javid Ali,	UP	
2.	Mahua Banerjee	IIT-Kanpur	mohua@iitk.ac.in
3.	Stuti Borgohain	IASST, Guwahati	tripathybc@yahoo.com
4.	P. Biswas	WB	papunbiswas@yahoo.com
5.	M.K. Chakraborty	C.U.	mihirc99@vsnl.com (09831145979)
6.	D. Chakraborty	Kalyani WB	elebjani_333@yahoo.co.in
7.	D. Datta Majumdar	ISI, Kolkata	ddmdr@hotmail.com (033 25777107)
8.	Ashok Deshpande	Bioinformatics, Pune	deshpande@bioinfo.ernet.in ashokwd@yahoo.com,
9.	T. Datta	CU	
10.	D. Datta,	Maharastra	ddatta@barc.gov.in
11.	Oshmita Dey	IIT, Kharagpur	oshmi_kgp@yahoo.co.uk
12.	Mrs. M. Dey (Maji)	Eco, Asansol	
13.	Soma Dutta	Kolkata	somadutta77@yahoo.co.in
14.	Somsubhra Gupta	WB	gsomsubhra@gmail.com
15.	Debashree Guha	IIT, Kharagpur	deb17116@yahoo.co.in
16.	Debajit Hajarika	IIT, Kharagpur	debajit@tezu.ernet.in
17.	D.S. Jamwal	DOM, Univ. of Jammu, Jammu-180006, India	dalipsj@yahoo.com

Contd.

LIST OF PARTICIPANTS (INDIA, OUTSIDE TRIPURA) (*Contd.*)

18.	Mousumi Kumar	University of North Bengal	
19.	Sanjeev Kumar	Agra	sanjeevibs@yahoo.co.in
20.	S.R. Kannan,	Gandhigram Rural Univ., 624302, T.N, India	s.r.kannan@ruraluniv.ac.in
21.	Nilaratna Kalia	Mayurbhanj Orissa	nilaratna_math@rediffmail.com
22.	Nirmal Laha	WB	pkverma123@gmail.com
23.	M.N. Mukherjee	Cal. Univ.*	mnmukherjee@yahoo.co.in
24.	D.K. Mitra	Tezpur Univ.	dkmitra@tezu.ernet.in
25.	A. Mani	Kolkata Calcutta Math Society	a_mani_sc_gs@yahoo.co.in
26.	Pabitra Kr. Maji	Calcutta University,	pabitra_maji@yahoo.com
27.	Debadutta Mohanty	S. Mohavidyalaya, Orissa, 757086	debadutta.mohanty@rediffmail.com
28.	B.B. Pal	DOM, Kal. Univ. Kalyani-WB	bbpal18@hotmail.com
29.	Chandra Kanta Phukan	DOM, Dib. Univ.	Chandra_kanta25@yahoo.com
30.	Swapan Raha	Viswa Bharati	raha_swapan@rediffmail.com
31.	Shakeel (Ahmed) Raina	Govt. Deg. Coll., Thanna Mandi, Rajouri, J. and K. India. 185212	shakeelar2002@yahoo.co.in
32.	T. Som	BHU*	som_tanmoy@yahoo.co.in 09236916279
33.	S. Sen	WB	ssenk@yahoo.co.in
34.	A. Sathya	Gandhigram Rural Univ., 624302, T.N., India	s.r.kannan@ruraluniv.ac.in
35.	Syamal Kumar Samanta	Viswa Bharati*	syamal_123@yahoo.co.in
36.	A.K. Srivastava	BHU	arunksrivastava@gmail.com 09433045266
37.	Shashi Bajaj	Jadavpur University, WB	
38.	B.C. Tripathy	IASST, Guwahati	tripathybc@yahoo.com tripathybc@rediffmail.com
39.	Renu	Jammu University	renu_ju@yahoo.com
40.	S.P. Tiwari	ISMU, Dhanbad	
41.	S. Coumaresrane	Pondicherry University	

LIST OF PARTICIPANTS (TRIPURA)

1. Mr. Debashis Bhattacharya (NIT, Agartala)
2. Mrs. Baby Bhattacharya (Agartala)
3. Mr. Ashish Bhattacharjee (Asst. Professor, DDM College, Khowai)
4. Dr. Debasish Bhattacharya (Asst. Professor, NIT, Agartala)
5. Dr. Baby Bhattacharya (Lecturer, Dept. of Math, NIT, Agartala)
6. Dr. Joydip Bhattacharya (Asst. Professor, Belonia College, Belonia)
7. Dr. Sharmistha Bhattacharya (Halder) (Lecturer, Tripura University)
8. Mr. Shouvik Bhattacharya (RS, TU)
9. Mr. Mrinal Kanit Bhowmik (Lecturer, CSE Dept. T.U. Dhaleswar, Agartala)
10. Prof. R.N. Bhaumik (Professor of Mathematics, T.U.)
11. Mr. Subrata Bhowmik (Lecturer, Dept. of Mathematics, T.U.)
12. Ms. Sayanta Chakraborty (RS, NIT, Agartala)
13. Mr. Snigdhendu Bikash Chakraborty (PGT)
14. Mrs. Jhilly Choudhury
15. Dr. Sutapa Das (PGT, Banividya Pith Class XII School, Ramnagar, Road No.-4, Agartala)
16. Mr. Shyamal Debnath (Ass. Professor, N.S. Mahavidyalaya)
17. Mr. Kamal Krishna Debnath (PGT)
18. Mr. Bishnupada Debnath (RS, T.U.)
19. Mr. Pranab Kumar Deb (Asst. Professor, Ramkrishna Mahavidyalaya, North Tripura)
20. Mr. Dipankar De (Asst. Professor, Ramthakur College, Badharghat)
21. Mr. Runu Dhar (Asst. Prof., BBM College, Agartala)
22. Dr. Dulal Dey (Asst. Prof., N.S. Mahavidyalaya, Udaipur)
23. Mrs. Mousumi Gupta (Sengupta) (RS, NIT, Agartala)
24. Mrs. Chhaya Gangwal (RS, T.U.)
25. Md. Azad Hussain (Lecturer, Women's Polytechnic, Hapania)
26. Mr. Goutam Majumder (Research Assistant, Dept. of CS and Eng. T.U.)
27. Prof. Anjan Mukherjee (Professor of Mathematics, T.U.)
28. Mr. Pratap Kumar Panigrahi (Asst. Professor, DDM College, Khowai)
29. Mr. Gour Pal (Asst. Professor, Women's College)
30. Mr. Debasish Pal (P.G. Teacher, Sudhanwa D/B Memorial H.S. School)
31. Ms. Susmita Roy (RS, Tripura University)
32. Mr. Prasenjit Roy (Asst. Teacher, B.K.N, Vidyabhaban)
33. Mr. Rupak Sarkar (Con. Lecturer, NIT, Agartala)
34. Mr. Subrata Saha (Lecturer, Bhavan's Tripura College)

35. Mr. Biswajit Saha (Lecturer Sr. Scale, TIT, Narsingarh)
36. Mr. Apu Kumar Saha (Lecturer, NIT, Agartala)
37. Mr. Kankan Saha (Research Assistant, Dept. of CS and Eng., T.U.)
38. Mrs. Jayanti Tripathi (RS, T.U.)

STUDENTS DELEGATES (TRIPURA)

1. Pinki Majumder
2. Manidipha Debnath
3. Payel Aine
4. Sudipa Poddar
5. Kalyani Debnath
6. Piyali Debnath
7. Bhaskar Mallik
8. Mitthun Datta
9. Jayanta Debnath
10. Banti Datta
11. Monmon Sarkar
12. Atikul Islam
13. Biplab Rudrapal
14. Kaushik Deb
15. Abhijit Debnath
16. Sharmistha Debnath
17. Sunny Biswas
18. Aditi Bhatacharjee
19. Ajoy Debnath
20. Sudipa Chaudhary
21. Sanghamitra Datta
22. Payel Baishab
23. Ajoy Kanti Das
24. Kanistha Sinha
25. Jaysri Pal
26. Suparna Biswas
27. Rajib Datta
28. Karna Debnath
29. Gouri Biswas
30. Jeheda Khatun
31. Debaro Das
32. Raju Debnath
33. Paramita Debnath

CONTENTS

KEYNOTE ADDRESS

A. INVITED TALK

B. PAPERS

Rough Sets and its Applications

Fuzzy Logic and its Applications

Fuzzy Topology

1

FUZZY SETS, NEAR SETS, AND ROUGH SETS, SETS IN THE COMPUTATIONAL INTELLIGENCE SPECTRUM

James F. Peters

Computational Intelligence Laboratory
Department of Electrical & Computer Engineering,
University of Manitoba Winnipeg,
Manitoba R3T 5V6 Canada
E-mail: jfpeters@ee.umanitoba.ca

Abstract: *This keynote talk considers how one might utilize fuzzy sets, near sets, and rough sets, taken separately or taken together in hybridizations in solving a variety of problems commonly faced in science and engineering. These technologies offer set theoretic approaches to solving problems such as classifying sensor output, image retrieval and image correspondence. Fuzzy sets result from the introduction of a membership function that generalizes the traditional characteristic function. The notion of a fuzzy set was introduced by L. Zadeh in 1965. Fifteen years later, rough sets were introduced by Z. Pawlak in 1981. A set is considered rough the boundary between its lower and upper approximation is non-empty. Of the three forms of sets, a near set is newest, introduced in 2007 by J.F. Peters in a perception-based approach to the study of perceptual objects. Perceptual systems provide stepping stones leading to nearness relations and properties of near sets. This work has been motivated by an interest in finding a solution to the problem of discovering perceptual granules that are, in some sense, near each other. Near set theory provides a description-based approach to observing, comparing and classifying perceptual granules. Near sets result from the introduction of a description-based approach to perceptual objects and a generalization of the traditional rough set approach to granulation that is independent of the notion of the boundary of a set approximation. Near set theory has strength by virtue of the strength it gains from rough set theory, starting with a L_2 norm-based perceptual indiscernibility relation, a new extension of the traditional indiscernibility equivalence relation. This keynote talk highlights a context for three forms of sets that* *are now part of the computational intelligence spectrum of tools useful in pattern recognition. By way of introduction to near sets, we consider various perceptual nearness relations that define partitions of sets of perceptual objects that are near each other. Every perceptual granule is represented by a set of perceptual objects that have their origin in the physical world. Objects that have the same appearance are considered perceptually near each other, i.e., objects with matching descriptions. Pixels, pixel windows, and segmentations of digital images are given by way of illustration of sample near sets. The contribution of this paper is an overview of the links between fuzzy sets, near sets and rough sets as well as the relation between these sets and the original notion of a set introduced by Cantor in 1883.*

Keywords: *Description, fuzzy set, near set, perceptual system, rough set.*

Near To

How near to the bark of a tree are drifting snowflakes, swirling gently round, down from winter skies?

How near to the ground are icicles, slowly forming on window ledges?

–Fragment of a Philosophical Poem.

–Z. Pawlak and J.F. Peters, 2002.

1. INTRODUCTION

This paper briefly presents the notion of a Cantor set. From the definition of a Cantor set, it is pointed out that fuzzy sets, near sets and rough sets are special forms of Cantor sets. In addition, this paper points to links between the three types of sets that are part of the computational intelligence spectrum. Probe functions in near set theory provide a link between fuzzy sets and near sets, since every fuzzy membership function is a particular form of probe function. Probe functions are realvalued functions introduced by M. Pavel in 1993 as part of a study of image registration and a topology of images [6]. Z. Pawlak originally thought of a rough set as a new form of fuzzy set [7]. It has been shown that every rough set is a near set (this is Theorem 4.8 in [11]) but not every near set is a rough set. For this reason, near sets are considered a generalization of rough sets. The contribution of this paper is an overview of the links between fuzzy sets, near sets and rough sets as well as the relation between these sets and the original notion of a set introduced by Cantor in 1883 [3].

2. CANTOR SET

By a 'manifold' or 'set' I understand any multiplicity, which can be thought of as one, i.e., any aggregate [inbegri f f] of determinate elements which, can be united into a whole by some law.

– Foundations of a General Theory of Manifolds,

– G. Cantor, 1883.

In this mature interpretation of the notion of a set, G. Cantor points to a property or law that determines elementhood in a set and "unites [the elements] into a whole" [3], elaborated in [2], and commented on in [4]. In 1851, Bolzano [1] writes that "an aggregate so conceived that is indifferent to the arrangement of its members I call a *set*". At that time, the idea that a set could contain just one element or no elements (null set) was not contemplated. This is important in the current conception of a near set, since such a set must contain pairs of perceptual objects with similar descriptions and such a set is never null. That is, a set is a perceptual near set if, and only if it is never empty and it contains pairs of perceived objects that have descriptions. that are within some tolerance of each other (see Def. 1).

3. NEAR SETS

The basic idea in the near set approach to object recognition is to compare object descriptions. Sets of objects X, Y are considered near each other if the sets contain objects with at least partial matching descriptions.

– Near sets. General theory about nearness of objects,

– J.F. Peters, 2007.

In a more recent interpretation of the notion of a near set, the nearness of sets is considered in the context of a perceptual system [15]. The trivial case is excluded. That is, an element $x \in X$ is not considered near itself. In addition, the empty set is excluded from near sets, since the empty set is never something that we perceive, i.e., a set of perceived objects is never empty. In the case where one set X is near another set Y, this leads to the realization that there is a third set containing pairs of elements $x, y \in X \times Y$ with similar descriptions. The key to an understanding of near sets is the notion of a description. Each perceived object is represented by a vector of feature values and each feature is represented by what is known as a probe function that maps an object to a real value. Since our main interest is in detecting similarities between seemingly quite disjoint sets such as subimages in an image or pairs of classes in coverings on a pair of images, a near set is defined in context of a tolerance space.

Table 1: *Nomenclature*

Symbol	*Interpretation*
O, X	Set of perceptual objects, $X, Y \subseteq O$,
$\mathbb{F}, \mathcal{B}$	Sets of probe functions, $\mathcal{B} \subseteq \mathbb{F}$,
ε	$\varepsilon \in \mathfrak{R}$ (reals) such that $\varepsilon \geq 0$,
$\phi_i(x)$	ith probe function $\phi_i : X \to \mathfrak{R}$ representing feature of x,
$\phi(x)$	$(\phi_1(x), \phi_2(x), ..., \phi_i(x), ..., \phi_l(x))$, $\phi_i \in \mathbb{F}, x \in O$, description of x,
$\cong_{\mathcal{B}, \varepsilon}$	$\{(x, y) \in O \times O : \lVert\phi(x) - \phi(y)\rVert_2 \leq \varepsilon\}$, tolerance relation
$x/\cong_{\mathcal{B}, \varepsilon}$	$\{y \in X \mid x \cong_{\mathcal{B}, \varepsilon} y\}$, tolerance class,
$X \bowtie_{\mathcal{B}, \varepsilon} Y$	X resembles (is near) $Y \Leftrightarrow X \cong_{\mathcal{B}, \varepsilon} Y$.

3.1 Definition 1: Tolerance Near Sets [12]

Let $\langle O, \mathbb{F}\rangle$ be a perceptual system, where O is a set of images (sets of points) and $\mathcal{B} \subseteq \mathbb{F}$ be a set of probe functions representing image features. Let $\mathcal{B}$ contains probe functions used to measure features of subimages in $X, Y \subset O$. A set X is perceptually near a Y *within* the perceptual system $\langle O, \mathbb{F}\rangle$ (i.e., $(X \bowtie_{\mathbb{F}, \varepsilon} Y)$) iff there are $x \in X$ and $y \in Y$ and there is $\mathcal{B} \subseteq \mathbb{F}$ such that $x \cong_{\mathcal{B}, \varepsilon} y$.

In effect, a Cantor set is a near set contains pairs of objects that satisfy the nearness description property enunciated in Def. 1.

4. FUZZY SET

A fuzzy set is a class of objects with a continuum of grades of membership.

– Fuzzy sets, Information and Control 8

– L.A. Zadeh, 1965.

> *... A fuzzy set is characterized by a membership function which assigns to each object its grade of membership (a number lying between 0 and 1) in the fuzzy set.*
>
> –A new view of system theory
>
> –L.A. Zadeh, 20-21 April 1965.

The notion of a fuzzy set was introduced by L.A. Zadeh in 1965 [16]. In effect, a Cantor set is a fuzzy set if, and only if every element of the set has a grade of membership assigned to it by a specified membership function. Notice that a membership function $\phi: X \rightarrow [0, 1]$ is a special case of what is known as a probe function in near set theory. A fuzzy set X is a near set relative to a set Y if the grade of membership of the objects in sets X, Y is assigned to each object by the same membership function ϕ and there is a least one pair of objects $x, y \in X \times Y$ such that $\| \phi(x) - \phi(y) \|_2 \leq \varepsilon \}$, i.e., the description of x is similar to the description y within some ε.

5. ROUGH SET

> *A new approach to classification, based on information systems theory, given in this paper.... This approach leads to a new formulation of the notion of fuzzy sets (called here the rough sets). The axioms for such sets are given, which are the same as the axioms of topological closure and interior.*
>
> –Classification of objects by means of attributes.
>
> –Z. Pawlak, 1981.

Table 2: *Pawlak Indiscernibility Relation and Partition Symbols*

Symbol	*Interpretation*
$\sim_{\mathcal{B}}$	$\{(x, y) \mid f(x) = f(y)\ \forall f \in \mathcal{B}\}$, indiscernibility relation, *cf.* [7],
$x/\sim_{\mathcal{B}}$	$x/\sim_{\mathcal{B}} = \{y \in X \mid y \sim_{\mathcal{B}} x\}$, elementary set (class),
$O/\sim_{\mathcal{B}}$	$O/\sim_{\mathcal{B}} = \{x/\sim_{\mathcal{B}} \mid x \in O\}$, quotient set.

In effect, in its original conception, Z. Pawlak thought of a rough set as a new formulation of the notion of a fuzzy set [7]. In a rough set approach to classifying sets of objects X, one considers the size of the boundary region in the approximation of X. By contrast, in a near set approach to classification, one does not consider the boundary region of a set. In particular, assume that X is a non-empty set belonging to a universe U and that $\mathbb{F}$ is a set of features defined either by total or partial functions. The lower approximation of X relative to $\mathcal{B} \subseteq \mathbb{F}$ is denoted by $\mathcal{B}_*(X)$ and the upper approximation of X is denoted by $\mathcal{B}^*(X)$, where

$$\mathcal{B}_*(X) = \bigcup_{x/\sim_{\mathcal{B}} \subseteq X} x/\sim_{\mathcal{B}}'$$

$$\mathcal{B}^*(X) = \bigcup_{x/\sim_{\mathcal{B}} \cap X \neq \emptyset} x/\sim_{\mathcal{B}}$$

The $\mathcal{B}$-boundary region of an approximation of a set X is denoted by $Bnd_{\mathcal{B}}(X)$, where

$$Bnd_{\mathcal{B}}(X) = \mathcal{B}^*(X) \setminus \mathcal{B}_*(X) = \{x \mid x \in \mathcal{B}^*(X) \text{ and } x \notin \mathcal{B}_*(X)\}.$$

A set X is roughly classified whenever $Bnd_{\mathcal{B}}(X)$ is not empty. In other words, X is a rough set whenever the boundary region $Bnd_{\mathcal{B}}(X) \neq \emptyset$. In sum, a Cantor set is a rough set if, and only if its approximation boundary is non-empty. It should also be noted that rough sets differ from near sets, since near sets are defined without reference to an approximation boundary region. This means, for example, with near sets the image correspondence problem can be solved without resorting to set approximation.

6. CONCLUSION

From the beginning, the near set approach to perception has had direct links to rough sets in its approach to the perception of objects [7, 5] and the classification of objects [7, 10, 9, 8]. This is evident in the early work on nearness of objects and the extension of the approximation space model (see, e.g., [13, 14]). Unlike the focus on the approximation boundary of a set, the study of near sets focuses on the discovery of affinities between perceptual granules such as digital images viewed as sets of points. In the context of near sets, the term *affinity* means *close relationship relationship between perceptual granules (particularly images) based on common description.* Affinities are discovered by comparing the descriptions of perceptual granules, e.g., descriptions of objects contained in classes found in coverings defined by the tolerance relation $\cong_{\mathbb{F},\varepsilon}$. In sum, fuzzy sets, near sets and rough sets are particular forms of Cantor sets. In addition, each of these sets in the computational intelligence spectrum offer very useful approaches in classifying objects.

ACKNOWLEDGEMENTS

The insights and suggestions by Mihir Chakraborty, Homa Fashandi, Amir H. Meghdadi, Christopher Henry, Dan Lockery, Leszek Puzio, Andrzej Skowron, Piotr Wasilewski, and Sheela Ramanna concerning topics in this article are gratefully acknowledged. This research has been supported by the Natural Sciences and Engineering Research Council of Canada (NSERC) grant 185986, Canadian Arthritis Network grant SRI-BIO-05, Manitoba Hydro grant T277.

REFERENCES

[1] Bolzano B., "Paradoxien des Unendlichen (Paradoxes of the Infinite), Trans. by D.A. Steele". Routledge and Kegan Paul, London, (1959).

[2] Cantor G., "Gesammelte Abhandlungen Mathematischen and Philosophischen Inhalts", Ed. E. Zermelo. Springer, Berlin, (1932).

[3] Cantor G., Über Unendliche, Lineare Punktmannigfaltigkeiten. Mathematische Annalen", **201**, 72–81, (2008).

[4] Lavine S., "Understanding the Infinite". Harward University Press, Cambridge, MA, (1994).

[5] Orłowska E., "Semantics of Vague Concepts. Applications of Rough Sets". Polish Academy of Sciences, **469**, (1982). In G. Dorn, P. Weingartner (Eds.), "Foundations of Logic and Linguistics. Problems and Solutions", Plenum Press, London/NY, 1985, 465–482.

[6] Pavel M., "Fundamentals of Pattern Recognition", 2nd edn. Marcel Dekker, Inc., N.Y., U.S.A., (1993).

[7] Pawlak Z., "Classification of Objects by Means of Attributes". Polish Academy of Sciences, **429**, (1981).

[8] Pawlak Z., Skowron A., "Rough Sets and Boolean Reasoning". Information Sciences, **177**, 41–73, (2007).

[9] Pawlak Z., Skowron A., "Rough Sets: Some Extensions". Information Sciences, **177**, 28–40, (2007).

[10] Pawlak Z., Skowron A., "Rudiments of Rough Sets", Information Sciences, **177**, 3–27, (2007).

[11] Peters J., Near Sets General Theory About Nearness of Objects". Applied Mathematical Sciences, **1 (53)**, 2609–2029, (2007).

[12] Peters J., "Tolerance Near Sets and Image Correspondence". *Int. J. of Bio-Inspired Computation*, **4 (1)**, 239–445, (2009).

[13] Peters J., Henry C., "Reinforcement Learning with Approximation Spaces". *Fundamenta Informaticae*, **71**, 323–349, (2006).

[14] Peters J., Skowron A., Stepaniuk J., "Nearness of Objects: Extension of Approximation Space Model". *Fundamenta Informaticae*, **79 (3-4)**, 497–512, (2007).

[15] Peters J., Wasilewski P., "Foundations of Near Sets". *Information Sciences. An International Journal*, **26**, 1–35 (2009). *Digital object identifier: doi:10.1016/j.ins.2009.04.018, in press.*

[16] Zadeh L., "Fuzzy Sets. Information and Control", **201**, 72–81, (1965).

2

THE LATTICE TOPOLOGICAL APPROACH TO ROUGHNESS THEORY BY APPROXIMATION SPACES*

Gianpiero Cattaneo

Dipartimento di Informatica, Sistemistica e Comunicazione Università di Milano–Bicocca Viale Sarca 336–U14, I–20126 Milano (Italia)
E-mail: {cattang, ciucci}@disco.unimib.it

Abstract: *The notion of abstract approximation space for roughness theory is introduced, and its relationship with Tarski interior and closure operations in lattice context is studied. Their categorical isomorphism is proved and the role of the Tarski interior and closure with an algebraic semantic of a S4–like model of modal logic is widely investigated.*

A hierarchy of closure operators on the abstract context of lattice structures is then investigated, comparing their role with respect to the abstract approach to rough approximation spaces. In particular, besides the Tarski closure (and listed from the weaker situation to the stronger one), the Kuratowski closure and the Halmos closure (or existential quantifier) are considered.

This hierarchy of closure operations is exemplified by three particular models based on a concrete universe: covering spaces, topological spaces (by open basis), and partition spaces.

1. INTRODUCTION

The abstract notion of *approximation space* in the context of roughness theory has been introduced in [Cat 98] (see also [CC04])

* The author's work has been supported by MIUR\PRIN project "Mathematical aspects and forthcoming applications of automata and formal languages".

in order to describe imprecise, uncertain, or vague concepts which can be approximated from the bottom and the top by crisp (sharp) ones. The aim was to give an *abstract axiomatization* to the Pawlak rough set theory based on a partition space, i.e., a concrete set as the *universe* of the discourse equipped with an *equivalence relation of indistinguishability* between objects [Paw81, Paw82], [Paw92]. In the formal treatment of abstract approximation spaces it is possible to consider them as a "merge" of the two structures of a *lower approximation space* and an *upper approximation space*, which furnish the lower crisp and upper crisp approximations of any approximable element, respectively.

On the other hand, another theoretical approach to rough sets is the topological one (see for instance [Zak83, Pom87, Wiw89, Yao96, Yao98b, Pol02]). A closure (resp., interior) operator gives an upper (resp., lower) approximation of a generic element through a closed (resp., open) one. In particular, one can consider Tarski closure operators in the abstract context of de Morgan lattices (and hence in particular also on Boolean algebras) and it is possible to show that these structures are categorically equivalent to abstract upper approximation spaces [CC06a,CC06b]. The de Morgan complementation allows one to define by duality the corresponding interior operator or, equivalently, the corresponding lower approximation space.

Concrete models of this abstract notion of approximation space (or equivalently of Tarski closure on de Morgan lattices) are given either by partitions or by coverings of a given universe arising from complete or incomplete Information Systems, which, as well known, are able to generate equivalence (reflexive, symmetric, and transitive) relations or *similarity* (reflexive and symmetric, in general non transitive) relations, respectively.

2. ABSTRACT APPROXIMATION SPACES

The notion of *Approximation Space* is introduced in this section as an abstraction of the Pawlak concrete approach to rough set theory. In the description of this abstract formalism, *imprecise concepts*, with their associated *lower* and *upper knowledge*, are mathematically realized by *points* of an abstract lattice, which are said to be the *approximable elements*, contrary to the corresponding concrete model in which they are realized by *subsets* of a given universe.

In this abstract context some criteria must be given in order to "approximate" any imprecise concept by a pair consisting of a unique lower crisp concept and a unique upper crisp concept. Since we want that these approximations are the best possible inside the classes of corresponding crisp concepts, it is necessary to have also a criterium to state how an approximation is sufficiently good. This is realized by a partial order relation $\leq$ on the set of all approximable elements which mathematically describes the fact that an element a (resp. b) is a *lower* (resp., an *upper*) *approximation* of the element b (resp., a), written $a \leq b$. This partial ordering must be such that an algebraic structure of lattice is induced by it. Then, a "lower approximation map" and an "upper approximation map" furnish, respectively, the best approximations from the bottom and from the top of any element with respect to the involved partial ordering and according to the following definition.

Definition 1: *An abstract* approximation space (*according to [Cat98], [CC04]*) *is a system* $\mathfrak{R} := \langle \Sigma, \mathbb{L}(\Sigma), \mathbb{U}(\Sigma) \rangle$, *where:*

1. The set Σ has a structure $\langle \Sigma, \wedge, \vee, 0, 1 \rangle$ of a lattice with respect to the partial order relation $a \leq b$ iff $a = a \wedge b$ (or, equivalently, $b = a \vee b$).

   This lattice is bounded by the least element 0 ($\forall a \in \Sigma$, $0 \leq a$) and the greatest element 1 ($\forall a \in \Sigma$, $a \leq 1$).

   Elements from Σ are interpreted as concepts, data, etc. and are called the elements which can be approximated (approximable elements);

2. $\mathbb{L}(\Sigma)$ is a sub-poset of Σ (i.e., a subset of Σ equipped with the partial order relation $\leq$ of Σ) containing the element $1 \in \mathbb{L}(\Sigma)$, whose elements are said to be lower (also, inner) crisp;

3. $\mathbb{U}(\Sigma)$ is a sub-poset of Σ (i.e., a subset of Σ equipped with the partial order relation $\leq$ of Σ) containing the element $0 \in \mathbb{U}(\Sigma)$, whose elements are said to be upper (also, closure) crisp.

The structure satisfies the following conditions.

($A \times 1$) For any element $a \in \Sigma$ which can be approximated, there exists (at least) one element $l(a)$ such that:

(In 1) $l(a) \leq a$, i.e. $l(a)$ is a lower approximation of a;

(In 2) $l(a) \in \mathbb{L}(\Sigma)$, i.e. $l(a)$ is a lower crisp element;

(In 3) $\forall \beta \in \mathbb{L}(\Sigma), \beta \leq a \Rightarrow \beta \leq l(a)$, i.e. $l(a)$ is the best lower approximation of a from the bottom by upper crisp elements.

($A \times 2$) For any element $a \in \Sigma$ which can be approximated, there exists (at least) one element $u(a)$ such that:

(Up 1) $a \leq u(a)$, i.e. $u(a)$ ia an upper approximation of a;

(Up 2) $u(a) \in \mathbb{U}(\Sigma)$, i.e. $u(a)$ is an upper crisp element;

(Up 3) $\forall \gamma \in \mathbb{U}(\Sigma)$, $a \leq \gamma \Rightarrow u(a) \leq \gamma$, i.e. $u(a)$ is the best upper approximation of a from the top by upper crisp elements.

Let us stress that $\mathbb{L}(\Sigma)$ is the collection of *lower crisp* elements and $\mathbb{U}(\Sigma)$ the collection of *upper crisp* elements of an approximation space based on the set of approximable elements described by the lattice Σ. In general these two sub-posets of Σ are different between them. As shown later (see corollary 3) in any approximation space the sub-poset $\mathbb{L}(\Sigma)$ (resp., $\mathbb{U}(\Sigma)$) turns out to be a join (resp., meet) sub semi-lattice of Σ; that is, for all $a, b \in \mathbb{L}(\Sigma)$ (resp., $a, b \in \mathbb{U}(\Sigma)$) the corresponding lattice join (resp., meet) $a \vee b$ (resp., $a \wedge b$) in Σ is such that $a \vee b \in \mathbb{L}(\Sigma)$ (resp., $a \wedge b \in \mathbb{U}(\Sigma)$).

Remark 1: In the standard literature on rough sets, usually it is preferred the term *definable* element instead of the here adopted term of *crisp* element. However, the term *definable* has not a clear semantic and it refers to the ideas of intension and extension which are not relevant here (see [Yao07] for a discussion about this argument). On the other hand, the term "crisp" is more coherent with our approach, where the lower and upper approximations give a valuation inside two "a priori" singled out classes of elements which are interpreted as support of some exact knowledge about approximable elements.

The meaning of ($A \times 1$) (resp., ($A \times 2$)) is that $l(a)$ (resp., $u(a)$) furnishes the best approximation of the "imprecise" approximable element a from the bottom (resp., top) by lower (resp., upper) crisp elements. In any approximation space it is easy to prove that for any approximable element $a \in \Sigma$, the lower $l(a) \in \mathbb{L}(\Sigma)$ and the upper $u(a) \in \mathbb{U}(\Sigma)$ crisp elements, whose existence is assured by ($A \times 1$) and ($A \times 2$), are unique. For this reason one introduces the following.

Definition 2: *For any approximable element* $a \in \Sigma$, the element $l\,(a)$ *is said to be the* lower (inner) crisp approximation *of a and the element* $u\,(a)$ *the* upper (outer) crisp approximation of a.

Thus, it is possible to introduce in an *equivalent way* an approximation space as a structure $\mathfrak{R} = \langle \Sigma, \mathbb{L}\,(\Sigma), \mathbb{U}\,(\Sigma), l, u \rangle$, consisting of a bounded distributive lattice Σ and two of its sub-posets $\mathbb{L}\,(\Sigma)$ and $\mathbb{U}\,(\Sigma)$, under the assumption of the *existence* of a *lower approximation mapping* $l\colon \Sigma \mapsto \mathbb{L}\,(\Sigma)$ and an *upper approximation mapping* $u\colon \Sigma \mapsto \mathbb{U}\,(\Sigma)$, given for any arbitrary $a \in \Sigma$ respectively by the laws:

$$l\,(a) := \max\,\{\beta \in \mathbb{L}\,(\Sigma) : \beta \le a\} \qquad \text{... (1a)}$$

$$u\,(a) := \min\,\{\gamma \in \mathbb{U}\,(\Sigma) : a \le \gamma\}. \qquad \text{... (1b)}$$

A brief comment about this formulation, taking into account the first condition (1a) (the second one can be discussed in a dual way). Of course, for any element $a \in \Sigma$ it is possible to collect all its crisp lower bounds $\mathbb{L}\,(a) := \{\beta \in \mathbb{L}\,(\Sigma) : \beta \le a\}$, but the condition that $\mathbb{L}\,(\Sigma)$ is a sub-poset of Σ does not assure that the greatest lower bound (*g.l.b.*) of this family exists. The meaning of $(A \times 1)$ is that in order to have an approximation space it is required that the system $\mathfrak{R}$ must satisfy the condition that for any element of Σ this g.l.b. exists and, further, that it is an element of $\mathbb{L}\,(\Sigma)$.

In any approximation space the *rough approximation* of any element $a \in \Sigma$ is then defined as the lower–upper crisp pair

$$r\,(a) := (l\,(a), u\,(a)) \in \mathbb{L}\,(\Sigma) \times \mathbb{U}\,(\Sigma), \text{ with}$$

$$l\,(a) \le a \le u\,(a) \qquad \text{... (2)}$$

which is the image of the element a under the *rough approximation mapping* $r\colon \Sigma \mapsto \mathbb{L}\,(\Sigma) \times \mathbb{U}\,(\Sigma)$ described by the diagram:

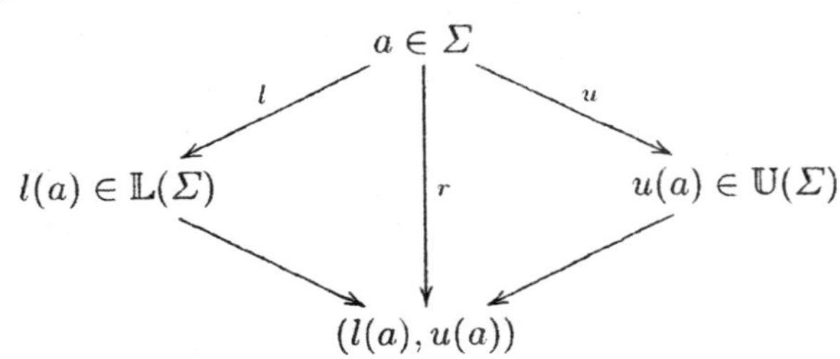

It is possible to characterize lower crisp and upper crisp elements according to the following result.

Proposition 1: Let $\mathfrak{R}$ be an approximation space. Then,

(*i*) the element $a \in \Sigma$ is lower crisp, i.e. $a \in \mathbb{L}(\Sigma)$, iff $a = l(a)$;

(*ii*) the element $a \in \Sigma$ is upper crisp, i.e. a $\in \mathbb{L}(\Sigma)$, iff $a = u(a)$.

Proof: (*i*) If $a = l(a)$ from (In 2) it follows that $a \in \mathbb{L}(\Sigma)$. Vice versa, if $a \in \mathbb{L}(\Sigma)$ then, applying the (In3) to the element $\beta = a \in \mathbb{L}(\Sigma)$, the relationship $a \leq a$ implies $a \leq l(a)$; but from the (In 1) it follows that $a = l(a)$.

The point (*ii*) is proved by duality.

Following [Cat98], an element e of Σ is said to be *crisp* (also *exact*, *sharp*) if and only if its lower and upper approximations coincide: $l(e) = u(e)$, equivalently, iff its rough approximation is the trivial one $r(e) = (e, e)$ (and this implies a fortiori that $l(e) = e = u(e)$). Therefore, the collection Σ_c of all crisp elements from Σ consists of those elements which are simultaneously lower-upper crisp, formally $\Sigma_c := \mathbb{L}(\Sigma) \cap \mathbb{U}(\Sigma)$ (sometimes denoted also by $\mathbb{LU}(\Sigma)$), which is not empty owing to the following result.

Proposition 2: In any approximation space $\mathfrak{R}$ both $0, 1 \in \mathbb{LU}(\Sigma)$ (i.e., they are crisp elements). In particular, $0 \in \mathbb{L}(\Sigma)$ and $1 \in \mathbb{U}(\Sigma)$.

Proof: The (In 1) applied to the element 0 gives $l(0) \leq 0$, and so necessarily $l(0) = 0$, but (In 2) says that $0 = l(0) \in \mathbb{L}(\Sigma)$. Applying the (In 3) to the element $a = 1$ we get that for every $\beta \in \mathbb{L}(\Sigma)$ the condition $\beta \leq 1$ implies $\beta \leq l(1)$; so, since for point (2) of definition 1 the element $1 \in \mathbb{L}(\Sigma)$, we have in particular that $1 \leq 1$ implies $1 \leq l(1)$, i.e. $1 = l(1) \in \mathbb{L}(\Sigma)$. In a dual way we also obtain that $0, 1 \in \mathbb{U}(\Sigma)$ using the three conditions (Up1)–(Up3).

Let us stress that from a formal point of view, an approximation space $\mathfrak{R}$ can be considered as the "merge" of the following two substructures.

(LAS) The *lower approximation space* $\mathfrak{L}\mathfrak{R} = \langle \Sigma, \mathbb{L}(\Sigma), l \rangle$, consisting of the lattice Σ of *approximable* elements, the join sub semi-lattice $\mathbb{L}(\Sigma)$ of *lower crisp* elements, and the *lower approximation mapping* (1a).

(UAS) The *upper approximation space* $\mathfrak{U}\mathfrak{R} = \langle \Sigma, \mathbb{U}(\Sigma), u \rangle$, consisting of the lattice Σ of *approximable* elements, the meet sub semi-lattice $\mathbb{U}(\Sigma)$ of *upper crisp* elements, and the *upper approximation mapping* (1b).

Of course in some approximation space it may happen that the two sets of lower crisp and upper crisp sets coincide, formally, $\mathbb{L}(\Sigma) = \mathbb{U}(\Sigma)$. In this case, this common lattice is just the collection of crisp elements Σ_c, and the corresponding approximation space will be simply denoted by the structure $\mathfrak{R} = \langle \Sigma, \Sigma_c, l, u \rangle$, with $l: \Sigma \mapsto \Sigma_c$ and $u: \Sigma \mapsto \Sigma_c$ defined according to the above Eq. (1). In this case the lower approximation space is the triple $\mathfrak{L}\mathfrak{R} = \langle \Sigma, \Sigma_c, l \rangle$ and the upper approximation space is $\mathfrak{U}\mathfrak{R} = \langle \Sigma, \Sigma_c, u \rangle$.

2.1 Complete Approximation Spaces

Let us note that in presence of a complete approximation space, which we are going to define now, the existence of the two approximation mappings according to Eq. (1) is assured, without any further requirement.

Definition 3: A complete abstract approximation space *for rough theories is a triple* $\mathfrak{C}\mathfrak{R} = \langle \Sigma, \mathbb{L}(\Sigma), \mathbb{U}(\Sigma) \rangle$, *where*

(CR1) Σ is a complete lattice with respect to a partial order relation $\leq$, in which the *g.l.b.* (resp., *l.u.b.*) of any family of its elements $\{a_m \in \Sigma : m \in M\}$ exists and is denoted by $\wedge_{m \in M} a_m$ (resp., $\vee_{m \in M} a_m$).

As any complete lattice, the meet $0 := \wedge_{a \in \Sigma} a$ and the join $1 := \vee_{a \in \Sigma} a$ are the least and the greatest element of the lattice;

(CR2) $\mathbb{L}(\Sigma)$ is a complete join semi-lattice (i.e., for any family $\{l_\alpha\}$ of elements from $\mathbb{L}(\Sigma)$ the lattice join $\vee l_\alpha$ on Σ is an element of $\mathbb{L}(\Sigma)$) containing both 0 and 1);

(CR3) $\mathbb{U}(\Sigma)$ is a complete meet semi-lattice (i.e., for any family $\{u_\beta\}$ of elements from $\mathbb{U}(\Sigma)$ the lattice meet $\wedge u_\beta$ on Σ is an element of $\mathbb{U}(\Sigma)$) containing both 0 and 1.

As usual, the elements from the lattice Σ are approximable, *whereas the elements from the join semi-lattice $\mathbb{L}(\Sigma)$ are the* lower crisp *and the element from the meet semi-lattice $\mathbb{U}(\Sigma)$ are the* upper crisp.

Complete approximation spaces will be the frame of study when discussing of concrete examples based on the power set of a universe. Let us recall that, according to [Moo10] (and see also [Bir67, pp. 111]), the condition (CR3) defines $\mathbb{U}(\Sigma)$ as a *Moore family* or, following [Coh65], the structure $\mathfrak{C}\mathfrak{S} := \langle \Sigma, \mathbb{U}(\Sigma), 1 \rangle$ as a *closure system*, satisfying the further condition of containing the least element 0 of the lattice. By duality, we can say that the structure $\mathfrak{IS} := \langle \Sigma, \mathbb{L}(\Sigma), 0\rangle$ is an *interior system*. The main difference with respect to Moore families is that $\mathbb{U}(\Sigma)$ (resp., $\mathbb{L}(\Sigma)$) is a closure (resp., an interior) system containing the least (resp., greatest) element 0 (resp., 1). This latter condition in general is not required to hold in a generic interior (resp., closure) system.

Example 1: Let us consider the (finite) collection of rational numbers $\Sigma = \left\{0, \frac{1}{k}, \frac{2}{k}, \ldots, \frac{k-1}{k}, 1\right\}$, for a fixed integer $k > 0$, totally ordered with respect to the standard ordering on numbers. The sublattice $\mathbb{L}(\Sigma) = \left\{0, \frac{1}{k}, \frac{2}{k}, \ldots, \frac{k-1}{k}\right\}$ is an interior (resp., a closure) system without the 1 (resp., 0) element.

In any complete approximation space, for any approximable element $a \in \Sigma$ it is possible to associate the following two elements, which owing to the conditions (CR2) and (CR3) exist and are elements of $\mathbb{L}(\Sigma)$ and $\mathbb{U}(\Sigma)$, respectively:

$$l(a) := \vee \{l \in \mathbb{L}(\Sigma) : l \le a\} \qquad \text{... (3a)}$$

$$u(a) := \wedge \{u \in \mathbb{U}(\Sigma) : a \le u\} \qquad \text{... (3b)}$$

In this way we have constructed the two mappings $l: \Sigma \mapsto \mathbb{L}(\Sigma)$, $a \to l(a)$ and $u: \Sigma \mapsto \mathbb{U}(\Sigma)$, $a \to u(a)$ which satisfy all the conditions defining an *abstract approximation space*.

3. LOWER AND UPPER APPROXIMATION SPACES: THEIR DUALITY IN THE CONTEXT OF DE MORGAN LATTICES

A particular case of abstract approximation space is the one in which the lattice of approximable elements Σ is equipped with a *de Morgan complementation mapping*.

Definition 4: *A* de Morgan lattice *is a system* $\langle \Sigma, \wedge, \vee, ', 0, 1\rangle$ where:

1. the substructure $\langle \Sigma, \wedge, \vee, 0; 1\rangle$ is a lattice, bounded by the least element 0 and the greatest element 1;
2. the mapping $': \Sigma \mapsto \Sigma$ is a unary operation on Σ, called de Morgan complementation, assigning to any element $a \in \Sigma$ the

corresponding complement $a' \in \Sigma$ under the conditions that for arbitrary $a, b \in \Sigma$ the following hold:

(dM1) $a = a''$ (double negation law)

(dM2) $(a \vee b)' = a' \wedge b'$ (de Morgan law)

Remark 2: The notion of de Morgan lattice just given is based on a non necessarily distributive lattice. This choice is made also by Kalman in [Kal58] where de Morgan lattices are called *i-lattices*, as distinguished by *distributive i-lattices*. Quoting [Cig65]: "This notion [of de Morgan distributive lattice] has been introduced by Gr. Moisil [Moi35, pp. 91] and studied by J. Kalman [Kal58] under the name of *distributive i-lattice*".

On the contrary, some other authors consider *de Morgan lattice* as a *distributive lattice* equipped with a de Morgan complementation. From the terminological point of view we can quote [Cig65]: "If Σ has the last element 1, we shall say that Σ is a *de Morgan Algebra*. This notion has been studied by A. Bialynicki – Birula and H. Rasiowa [BB57, BBR57] under the name of *quasi – Boolean algebras*. In this case, $0 = 1'$ is the first element of Σ." For some more recent contributions see for instance [Cig75, Dun86].

The choice of considering non distributive lattices is due to the desire of being more general as possible, since there exist de Morgan lattices which are not distributive. For example, let us consider the de Morgan lattice of Fig. 1.

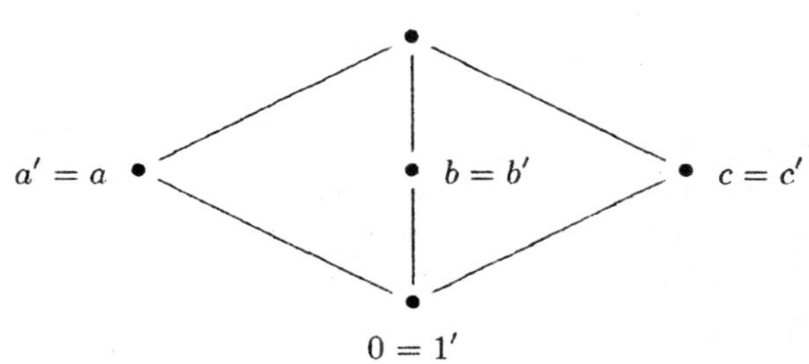

Fig. 1: *An Example of de Morgan Lattice which is not Distributive*

This lattice is not distributive, since $a \vee (b \wedge c) = a \neq 1 = (a \vee b) \wedge (a \vee c)$.

The following interesting result holds.

Lemma 1: Under condition (dM1), the following properties are mutually equivalent among them:

(dM2) $(a \vee b)' = a' \wedge b'$ (*first de Morgan law*)

(dM2a) $(a \wedge b)' = a' \vee b'$ (*second de Morgan law*)

(dM2b) $a \leq b$ implies $b' \leq a'$ (*contraposition law*)

(dM2c) $b' \leq a'$ implies $a \leq b$ (*dual contraposition law*)

Proof: For the first three statements see proposition 3.1 of [CM88]. Property (dM2c) is clearly equivalent to (dM2b) due to double negation (dM1).

Note that in any de Morgan lattice one has that $1 = 0'$ and $0 = 1'$, since from the trivial inequality $0 \leq 1'$, by contraposition and double negation law, one obtains $1 = 1'' \leq 0'$. Similarly, from $0' \leq 1$ we get $1' \leq 0'' = 0$.

As an immediate consequence of the lemma 1 we have the following.

Corollary 1: Let us denote the de Morgan complementation mapping in the form $\phi: \Sigma \mapsto \Sigma$, defined by the correspondence $a \in \Sigma \to \phi(a) := a'$, then this map is a poset anti-isomorphism, i.e. the following hold:

(aiso-1) ϕ is a bijective (one-to-one and onto) mapping of Σ into itself;

(aiso-2) $a \leq b$ implies $\phi(b) \leq \phi(a)$ (first poset anti-morphism condition).

(aiso-3) $\phi(b) \leq \phi(a)$ implies $a \leq b$ (second poset anti-morphism condition).

Proof: The condition $\phi(a) = \phi(b)$ means that $a' = b'$, from which $a'' = b''$ follows, and in virtue of (dM1) one gets $a = b$, i.e. the injectivity of ϕ. On the other hand, for every $b \in \Sigma$, there exists $b' \in \Sigma$ such that $\phi(b') = b'' = b$, i.e. the surjective condition is verified. (aiso-2) and (aiso-3) are nothing else than the (dM2b) and (dM2c).

In general for the de Morgan negation neither the *noncontradiction* law, $\forall a \in \Sigma, a \wedge a' = 0$, nor the *excluded middle* law, $\forall a \in \Sigma, a \vee a' = 1$ are required to hold. Let us recall that a Boolean lattice is a de Morgan lattice for which either the noncontradiction and the excluded middle laws hold. So any Boolean lattice is a particular case of de Morgan lattice.

In order to distinguish these two categories of lattice structures we introduce the following sufficient condition. A lattice with de Morgan complementation is said to be *genuine* iff there exists an element $\frac{1}{2}$ ($\neq 0, 1$), called the *half element*, such that $\left(\frac{1}{2}\right)' = \frac{1}{2}$. Indeed, for this element $\frac{1}{2} \wedge \left(\frac{1}{2}\right)' = \frac{1}{2}$ ($\neq 0$) and $\frac{1}{2} \vee \left(\frac{1}{2}\right)' = \frac{1}{2}$ ($\neq 1$) and so any de Morgan lattice with half element cannot be Boolean. The de Morgan lattice of figure 1 is an example of a lattice with more than one half element; precisely, it has the three half elements a, b, and c.

However, there could be de Morgan lattices with no half element and which are not Boolean. So, the existence of half elements is only a sufficient condition for assuring the non Boolean behavior of the lattice.

Example 2: Let us consider the de Morgan lattice of Fig. 2.

This lattice is not Boolean since the de Morgan complementation does not satisfy the noncontradiction law ($d \wedge d' = b \neq 0$) and the excluded middle law ($a \vee a' = a' \neq 1$). But this lattice has no half elements.

A strengthening of the de Morgan lattice structure, intermediate with respect to the Boolean one, are *Kleene lattices*, i.e. de Morgan lattices for which the further Kleene condition holds (*distributive* Kleene lattices are called by Kalman in [Kal58] *normal i-lattices*):

(K) for any pair of elements $a, b \in \Sigma$, it is $a \wedge a' \leq b \vee b'$.

Proposition 3: Under conditions ($dM1$) and ($dM2$), property (K) is equivalent to the following (pure poset, i.e. non-equational) condition:

$$a \leq a' \text{ and } b' \leq b \text{ implies } a \leq b. \qquad \text{... (4)}$$

Proof: Indeed, let us suppose that (K) holds; if $a \leq a'$ and $b' \leq b$ then $a = a \wedge a' \leq b \vee b' = b$. On the other hand, for all a, b it holds $a \wedge a' \leq a \vee a' = (a' \wedge a)'$ and $b \wedge b' = (b \vee b')' \leq b \vee b'$. Thus by Eq. (4), $a \vee a' \leq b \vee b'$.

Corollary 2: In a Kleene lattice the half element, if it exists, is unique.

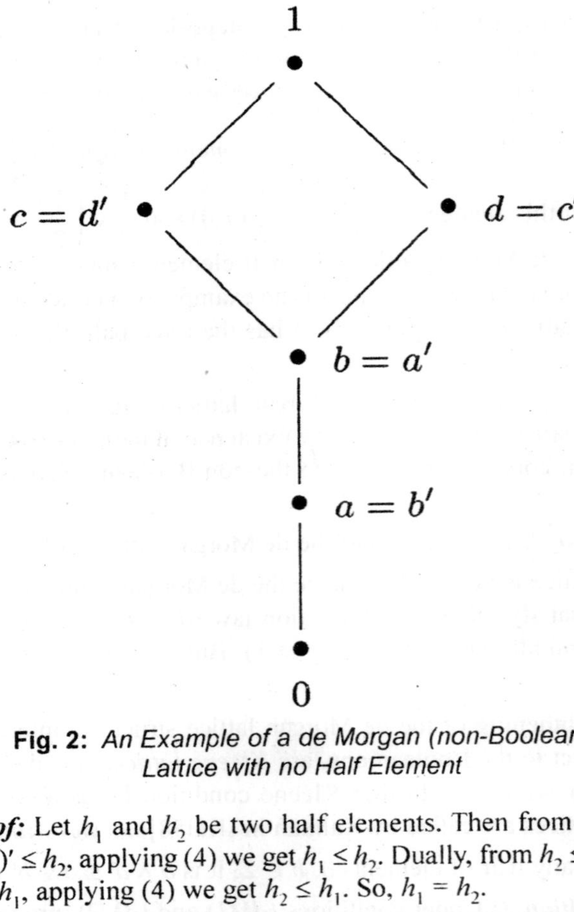

Fig. 2: *An Example of a de Morgan (non-Boolean) Lattice with no Half Element*

Proof: Let h_1 and h_2 be two half elements. Then from $h_1 \leq (h_1)'$ and $(h_2)' \leq h_2$, applying (4) we get $h_1 \leq h_2$. Dually, from $h_2 \leq (h_2)'$ and $(h_1)' \leq h_1$, applying (4) we get $h_2 \leq h_1$. So, $h_1 = h_2$.

The two categories of de Morgan and Kleene lattice are distinct between them since there exist de Morgan lattices which are not Kleene, as can be seen in the following example.

Example 3: The de Morgan lattice of Fig. 3, denoted by **T** in [Cig75], is the smallest de Morgan lattice which is not Kleene since $a \wedge a' = a \nleq b = b \vee b'$.

Any *Boolean lattice* is a Kleene lattice since the *non-contradiction* and the *excluded middle* laws implies that $a \wedge a' = 0 \leq 1 = b \vee b'$ for pair of elements a, b. Thus these two laws substitute (and imply) the weaker condition (K).

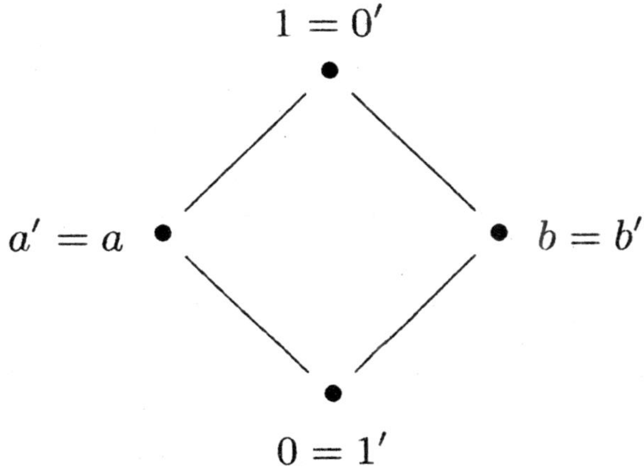

Fig. 3: *An Example of de Morgan Lattice which is not a Kleene Lattice*

Coming back to the more general context, we have the following result.

Proposition 4: Let Σ be a de Morgan lattice.

(D1) Given a lower approximation space $\mathfrak{LR} = \langle \Sigma, \mathbb{L}(\Sigma), l \rangle$ and introduced the collection

$$\mathbb{U}_d(\Sigma) := \{\gamma \in \Sigma : \exists \beta \in \mathbb{L}(\Sigma) \ s.t.\ \gamma = \beta'\}$$

and the mapping defined for any $a \in \Sigma$ by the law

$$u_d(a) := (l\,(a'))' \qquad \text{... (5)}$$

then the triplet $\mathfrak{UR}_d = \langle \Sigma, \mathbb{U}_d(\Sigma), u_d \rangle$ is an upper approximation space, dual of the original lower approximation space $\langle \Sigma, \mathbb{L}(\Sigma), l \rangle$ in the sense that $\mathbb{U}_d(\Sigma)$ is naturally a lattice of upper crisp elements, the dual of $\mathbb{L}(\Sigma)$, and $u_d : \Sigma \mapsto \mathbb{U}(\Sigma)$ is an upper approximation map, the dual of l (i.e., it satisfies all the above conditions (Up1)–(Up3)).

Similarly,

(D2) Given an upper approximation space $\mathfrak{UR} = \langle \Sigma, \mathbb{U}(\Sigma), u \}$ and introduced the collection

$$\mathbb{L}_d(\Sigma) := \{\beta \in \Sigma : \exists \gamma \in \mathbb{U}(\Sigma) \text{ s.t. } \beta = \gamma'\} \qquad \text{... (6)}$$

and the mapping defined for any $a \in \Sigma$ by the law

$$l_d(a) := (u(a'))' \qquad \text{... (7)}$$

then the triplet $\mathfrak{LR}_d = \langle \Sigma, \mathbb{L}_d(\Sigma), l_d \rangle$ is a lower approximation space, dual of the original upper approximation space $\langle \Sigma, \mathbb{U}(\Sigma), u \rangle$ in the sense that $\mathbb{L}_d(\Sigma)$ is naturally a lattice of upper crisp elements, the dual of $\mathbb{U}(\Sigma)$, and $l_d : \Sigma \mapsto \mathbb{L}(\Sigma)$ is a lower approximation map, the dual of u (i.e., it satisfies all the above conditions (In1)–(In3)).

Proof: We shall prove the point ($D1$) only, since the other can be proved in a similar way by duality. First of all, for any pair of elements $\gamma_1, \gamma_2 \in \mathbb{U}_d(\Sigma)$ we have that there exist two corresponding elements $\beta_1, \beta_2 \in \mathbb{L}(\Sigma)$ such that $\gamma_1 = (\beta_1)'$ and $\gamma_2 = (\beta_2)'$. Moreover $\beta_1 \wedge \beta_2 \in \mathbb{L}(\Sigma)$, since $\mathbb{L}(\Sigma)$ is a meet semi-lattice, from which it follows $\gamma_1 \vee \gamma_2 = (\beta_1)' \vee (\beta_2)' = (\text{dM2a}) = (\beta_1 \wedge \beta_2)'$, with $(\beta_1 \wedge \beta_2)' \in \mathbb{U}_d(\Sigma)$, and so $\mathbb{U}_d(\Sigma)$ is a join sub semi-lattice of Σ.

On the other hand, from (In1) we have that for any a in Σ one has $l(a) \leq a$, which applied to a' leads to $l(a') \leq a'$; from this result, using (dM1) and (dM2b), one obtains $a \leq (l(a'))' = u_d(a)$.

Moreover, for any a of Σ its lower approximation $l(a) \in \mathbb{L}(\Sigma)$, and applying this property to the particular case of a', as element of Σ, we have that also $l(a') \in \mathbb{L}(\Sigma)$, from this it follows that $u_d(a) = (l(a'))' \in \mathbb{U}_d(\Sigma)$, since there exists the element $\beta = l(a') \in \mathbb{L}(\Sigma)$.

Finally, let an element $\gamma \in \mathbb{U}_d(\Sigma)$ (i.e., for which there exists $\beta \in \mathbb{L}(\Sigma)$ such that $\gamma = \beta'$) be such that $a \leq \gamma$, then by (dM2b) we get $\beta = (\text{dM1}) = \beta'' = \gamma' \leq a'$; making use of (In3) applied to the element a' we have that $\beta \leq l(a')$ and so, by (dM2b), $(l(a'))' \leq \beta' = \gamma$, i.e. we have proved the (Up3).

In terms of the involutive anti-morphism $\phi : \Sigma \mapsto \Sigma$ introduced in corollary 1, recalling the self-invertibility condition $\phi^{-1} = \phi$, the two situations described by ($D1$) and ($D2$) in proposition 4 can be depicted by the two commutative diagrams which describe their inter-connections by de Morgan complementation:

Hence the de Morgan complementation determines a duality between the lower and upper approximations of any element a according to Eqs (5) and (7). Such a relationship can be represented by the diagram described in Fig. 5, where for simplicity we continue to use the symbol $l(a)$ (resp., $u(a)$) instead of $l_d(a)$ (resp., $u_d(a)$), if no confusion is likely.

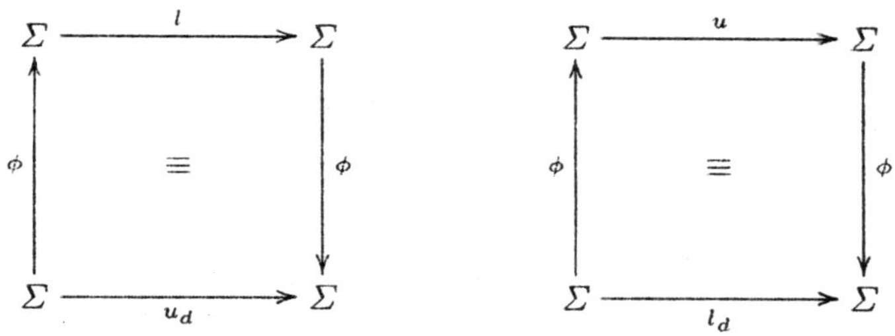

Fig. 4: *Lower and Upper Approximations Interconnection*

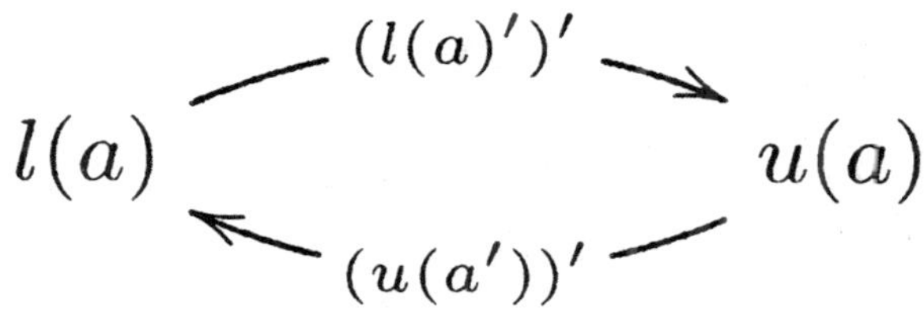

Fig. 5: *de Morgan Duality between Lower and Upper Approximations*

In this de Morgan context, two new notions involving the elements of Σ can be now introduced: given $a \in \Sigma$ its *exterior* is the lower crisp element $e(a) := u(a)' \in \mathbb{L}(\Sigma)$, i.e. the complement of its upper approximation; moreover, its *boundary* is the upper crisp element $b(a) := u(a) \wedge l(a)' \in \mathbb{U}(\Sigma)$, i.e. the relative complement of its interior with respect to the closure. For any approximable element $a \in \Sigma$ the triplet $\pi(a) := \{l(a), b(a), e(a)\}$ consists of mutually orthogonal elements (two elements a, b of a complemented lattice Σ are said to be *orthogonal*, written $a \perp b$, iff $a \leq b'$ or equivalently iff $b \leq a'$). Moreover, if the de Morgan lattice Σ is distributive they *orthogonally decompose* the whole Σ in the sense that $l(a) \vee b(a) \vee e(a) = 1$.

Making use of the exterior of a generic element, we can consider the *ortho-rough approximation* of any element $a \in \Sigma$ as the *lower-exterior* pair

$$r_{\perp}(a) := (l(a), e(a)) \in \mathbb{L}(\Sigma) \times \mathbb{L}(\Sigma), \text{ with } l(a) \perp e(a) \quad \ldots (8)$$

which is the image of the element a under the *ortho-rough approximation mapping* $r\perp : \Sigma \mapsto \mathbb{L}(\Sigma) \times \mathbb{L}(\Sigma)$ described by the diagram:

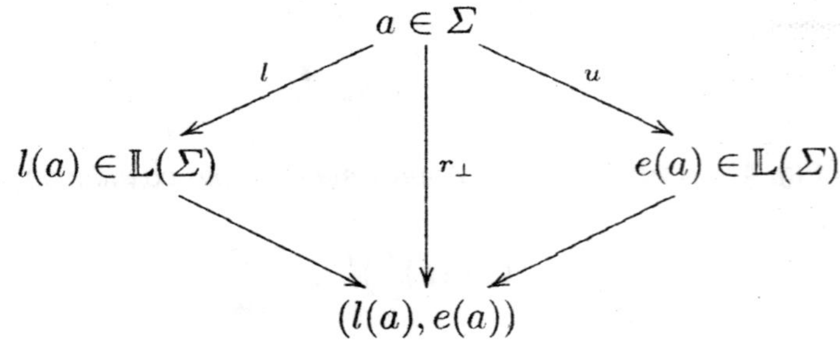

In the sequel we also (or almost always, if not explicitly declared) consider approximation spaces according to the following definition.

Definition 5: *A* bi-approximation space *is an approximation space* $\mathfrak{R} = \langle \Sigma, \mathbb{L}_d(\Sigma), \mathbb{U}(\Sigma), l_d, u\rangle$, where

(b-RAS1) The substructure $\mathfrak{U}\mathfrak{R} = \langle \Sigma, \mathbb{U}(\Sigma), u\rangle$ is an upper approximation space, according to the (UAS) of Section 2, based on a de Morgan lattice $\langle \Sigma, \wedge, \vee, ', 0, 1\rangle$;

(b-RAS2) $\mathbb{L}_d(\Sigma)$ is the dual lattice of $\mathbb{U}(\Sigma)$ consisting of the lower crisp elements according to Eq. (6);

(b-RAS3) l_d: $\Sigma \mapsto \mathbb{L}_d(\Sigma)$ is the dual lower approximation map relatively to $u: \Sigma \mapsto \mathbb{U}(\Sigma)$ according to Eq. (7).

Let us observe that in the abstract treatment of approximation spaces, and in particular in the use of one of the two dualities proved in proposition 4, the de Morgan conditions (dM1) and (dM2) are the only necessary conditions, i.e. the minimal requirements on the complementation associated to a generic lattice in order to be able to prove all the required results. Of course, this does not exclude that in some particular applicative situation the involved lattice satisfies the stronger condition of being Kleene or Boolean complemented, as in the case of the concrete models treated in the forthcoming section, but from a pure theoretical point of view the only de Morgan conditions are sufficient to start from a lower (resp., upper) approximation space in order to construct by duality the corresponding upper (resp., lower) of a global bi-approximation space. Clearly, this holds in the present formalization of abstract approximation spaces. Things could be very different if weaker properties than the ones axiomatized in definition 1 are taken into account (see for instance [Ciu08]). But in this case the reasons of these weakening must be clearly declared and semantically explained.

4. MODELS OF ABSTRACT APPROXIMATION SPACES: SOME INTRODUCTORY CONSIDERATIONS

In the forthcoming subsections three concrete *models* of the abstract notion of approximation space introduced by definition 1 are widely discussed. These models are based on a (non necessarily finite) nonempty set X, considered as the *universe* of the discourse, with corresponding *power set* $\mathcal{P}(X)$, collection of all its possible subsets. In all these models the elements of the abstract lattice $a \in \Sigma$ are *realized* by concrete subsets A of the universe X, i.e. by $A \in \mathcal{P}(X)$. Formally,

$$a \in \Sigma \xrightarrow[\text{Realized as}]{\text{Concretely}} A \in \mathcal{P}(X)$$

Let us recall that with respect to the set theoretic inclusion between subsets $A \subseteq B$, with corresponding lattice operations of meet as set theoretical inter-section $A \cap B$ and join as set theoretical union $A \cup B$, and the set theoretical complementation $A^c = X \setminus A$, the family $\mathcal{P}(X)$ is an atomic, complete Boolean lattice whose atoms are the singletons $\{x\}$, for x running in X.

The operations of the lattice structure Σ are realized by the usual set theoretic operations according to the following table:

Let us recall that in general a structure $\langle \mathcal{P}(X), ...\rangle$ based on the power set of a concrete universe is a model of the abstract theory $\langle \Sigma, ...\rangle$ iff any AXIOM of the abstract theory $\langle \Sigma, ...\rangle$ turns out to be a THEOREM in the concrete model $\langle \mathcal{P}(X), ...\rangle$.

4.1 Incomplete Information Systems

In this subsection we introduce a model of abstract approximation space based on incomplete information systems, which are generalization of Pawlak (complete) information systems. Let us give now the formal definition.

ABSTRACT		CONCRETE
		Universe X
$a \in \Sigma$	$\xrightarrow[Realized\ as]{Concretely}$	$A \in \mathcal{P}(X)$
$a \leq b$	$\xrightarrow[Realized\ as]{Concretely}$	$A \subseteq B$
$a \wedge b$	$\xrightarrow[Realized\ as]{Concretely}$	$A \cap B$
$a \vee b$	$\xrightarrow[Realized\ as]{Concretely}$	$A \cup B$
a'	$\xrightarrow[Realized\ as]{Concretely}$	$A^c = X \setminus A$

Definition 6: An (Incomplete) Information System (IIS) is a structure

$$\mathcal{K} = \langle X, Att, Val, F \rangle$$

where:

- X *(called the* universe) *is a nonempty set of* objects (situations, entities, states);

- *Att is a nonempty set of* attributes, *which assume values for the objects belonging to the set X*;
- *V al is the set of all* possible values *that can be observed for an attribute a from Att in the case of an object x from X*;
- *F* (called the information map) *is a mapping* $F : (X \times Att)_p \rightarrow V\,al$ *partially* defined on a subset $(X \times Att)_p \subset X \times Att$, *which assigns only to some object-attribute pair, precisely the ones* $(x, a) \in (X \times Att)_p$, *the value* $F(x, a) \in V\,al$ *assumed by a for the object x.*

The information system is thus *incomplete* iff there exists at least an object-attribute pair $(x, a) \in X \times Att$ for which the information function is not defined. This corresponds to a situation in which an attribute value for an object is unknown (missing, null).

Before starting with the formal treatment of incomplete information systems, let us quote from [Zia01]:

A tough problem waiting for the definite solution is the treatment of data records with missing or unknown data item values. Although a number of strategies have been developed in the past [...] none of the proposed solutions seems to address the problem of missing values satisfactorily.

So, let $\mathcal{K}$ be an incomplete information system. In order to denote the fact that the value possessed by an object x with respect to the attribute a is unknown, or missing, we introduce a *null value* * and write $F(x_i, a_j) = *$. This is the approach introduced by Kryszkiewicz in [Kry98], which according to [ST99] has the advantage that null values "corresponds to the idea that such values are just missing, but they do exist. In other words, it is our imperfect knowledge that obliges us to work with a partial information table."

Let us clarify this concept of incomplete information system by an example.

Example 4: Let us consider the information Table 1 involving as object a finite number of flats, with respect to which some interesting attributes are valuated by a sale agency.

Unfortunately, in each of these °ats some information is missing. For instance for the flat f_3 it has not been possible to know its price, whereas for the flat f_2 it is the number of rooms which is unknown.

Table 1: *Flats Incomplete Information System*

Flat	*Price*	*Rooms*	*Down-Town*	*Furniture*
f_1	high	2	yes	*
f_2	high	*	yes	no
f_3	*	2	yes	no
f_4	low	*	no	no
f_5	low	1	*	no
f_6	*	1	yes	*

In this incomplete context of information systems, a natural binary relation on objects is the one for which two objects are *indistinguishable* if they possesses the same values for all the attributes for which the value is known. In a more formal way let $\mathcal{A}$ be a subset of attributes, then $\mathcal{S}_A \subseteq X \times X$ is defined for any pair of objects $x, y \in X$ in the following way:

$x\mathcal{S}_A y$ if and only if

$$\forall a_i \in \mathcal{A}, \text{ either } F(x, a_i) = F(y, a_i) \text{ or } F(x, a_i) = * \text{ or } F(y, a_i) = * \qquad \text{... (9)}$$

We introduce now the above relation of indistinguishability between objects in an equivalent, and in some sense more compact, way. To this purpose, for any fixed object $x \in X$ we define as *attribute domain* generated by x the following collection:

$$\mathcal{A}(x) := \{a \in Att: (x, a) \in (X \times Att)_p\}$$

and the set of *possible values* associated with the attribute $a \in Att$

$$val(a) := \{\lambda \in Val: \exists x \in X s.t. (x, a) \in (X \times Att)_p \text{ and } F(x, a) = \lambda\}$$

Then the attribute a is represented by the mapping $f_a: X(a) \to val(a)$ defined by the correspondence:

$$x \in X(a) \to f_a(x) := F(x, a) \in val(a)$$

Thus the above definition of indiscernibility relation according to Eq. (9) can be reformulated in the following manner:

$$x\mathcal{S}_A y \quad \text{iff} \quad \forall a \in \mathcal{A}(x) \cap \mathcal{A}(y), f_a(x) = f_a(y) \qquad \text{... (10)}$$

From the formal point of view, such a binary relation is *reflexive, symmetric*, but in general the *transitivity condition is not satisfied.* Relations of this kind (i.e., reflexive and symmetric) are called

- *similarity* relation (after Poincaré [Poi93]) in the context of Kripke semantics of modal logic (see [Che88, p. 83]), or
- *tolerance* relation (after Zeeman [Zee62]) in the context of incomplete information systems [Zak83,WS89].

In the roughness context the generalization of the basic definitions to the case of non-equivalence binary relations can be found in [Pom87, Yao98a].

Example 5: In the case of example 4 for the choice of the three attributes $\mathcal{A} = \{P, R, D\}$, where P stands for "Price", R for "Rooms", and D for "Down-town", we have that

$$f_4 \mathcal{S}_{\mathcal{A}} f_5 \quad \text{and} \quad f_5 \mathcal{S}_{\mathcal{A}} f_6, \quad \text{but it is not } f_4 \mathcal{S}_{\mathcal{A}} f_6$$

Similarity spaces: Thus we have seen that any incomplete information system generates a similarity relation of indistinguishability according to (9) or (10). This is a particular case of a more general situation according to the following definition.

Definition 7: A similarity space, *also* tolerance space, *is a structure* $\mathfrak{S} = (X, \mathcal{S})$, *where* X *(called the* universe *of the space) is a nonempty set of objects and* $\mathcal{S}$ *(called the* similarity, *or* tolerance, *relation of the space)* is a reflexive and symmetric binary relation defined on X. In other words, $\mathcal{S}$ satisfies the following two conditions (for any $x, y \in X$):

(*Si*1)	$x \mathcal{S} x$	(*reflexivity*)
(*Si*2)	$x \mathcal{S} y$ *implies* $y \mathcal{S} x$	(*symmetry*)

Given a similarity space $\mathfrak{S} = (X, \mathcal{S})$, the *similarity* (*tolerance*) *class* generated by the object $x \in X$ is the collection of all objects similar to x, i.e.

$$S(x) := \{y \in X : x\mathcal{S}y\}$$

In other words, the similarity class $S(x)$ generated by x is the maximal set of objects which are possibly indiscernible with x with respect to the similarity relation $\mathcal{S}$. In this way this class constitute a *granule of knowledge* about x by similarity and is also called the *similarity granule* generated by x.

Trivially, being $\mathcal{S}$ a reflexive relation, it holds that $x \in S(x)$ and so no similarity class is empty. Now, it is possible to introduce the collection of all the similarity class generated by the similarity relation S for x running on X,

$$\gamma(X) := \{S(x) : x \in X\}, \qquad \text{... (11)}$$

The following result is now straightforward.

Lemma 2: Let $\mathcal{S} = (X, \mathcal{S})$ be a similarity space, for instance induced from an incomplete information system. Then the collection $\gamma(X)$ of all information granules is a covering of the universe X, i.e.,

$$X = \bigcup_{x \in X} S(x) \qquad \text{... (12)}$$

Let us note that, as a consequence of the non-transitivity of the relation, the similarity classes of the above covering are not necessarily disjoint, i.e. there may exist x, y such that $S(x) \neq S(y)$ and $S(x) \cap S(y) \neq \emptyset$.

Example 6: Making reference to the incomplete information system of Example 4, if one considers the set of all attributes (i.e. $\mathcal{A} = Att$) and the induced similarity relation according to (9), we have that the similarity classes are the following ones:

$$S(f_1) = \{f_1, f_2, f_3\}, S(f_2) = \{f_1, f_2, f_3, f_6\}, S(f_3) = \{f_1, f_2, f_3\}$$
$$S(f_4) = \{f_4, f_5\}, S(f_5) = \{f_4, f_5, f_6\}, S(f_6) = \{f_2, f_5, f_6\}$$

As the remark about the general nonempty intersection of distinct similarity classes, in this example we have for instance that $S(f_2) \cap S(f_5) = \{f_6\}$, i.e. the intersection of these two similarity classes is nonempty.

Let us note that some of these classes may be equal (for instance $S(f_4) = S(f_5)$), and so the covering of the universe is

$$\gamma(X) = \{\{f_4, f_5\}, \{f_1, f_2, f_3\}, \{f_2, f_5, f_6\}, \{f_4, f_5, f_6\}, \{f_1, f_2, f_3, f_6\}\}$$

Of course, there may be incomplete information systems for which any pair of distinct similarity classes are disjoint (and so the covering is more correctly a partition of the universe).

Example 7: Let us consider the information table based on the universe of objects $X = \{x, y, z, u, w\}$ and the set of attributes $Att = \{a, b, c\}$.

	a	b	c
x	0	h	*
y	0	*	m
z	*	k	n
u	1	*	m
w	1	h	*

The corresponding similarity classes generated by objects are

$$S(x) = \{x, y\} = S(y),\ S(z) = \{z\},\ S(u) = \{u, w\} = S(w)$$

whose induced covering of X is

$$\gamma(X) = \{\{x, y\}, \{z\}, \{u, w\}\}$$

which more correctly is a partition of X, since they constitute a covering of X by mutually disjoint nonempty subsets.

4.2 Covering Approximation Spaces

In the previous discussion we have seen that any similarity space generates (and so it is a particular case of) a covering of the universe. In this section we investigate covering spaces from a general point of view, that is independently from any possible generation of their obtained on the basis of some stronger structure. Let us recall that the treatment of roughness theory in the covering context has been treated in the seminal papers [Pom87, WS89, Yao98c]. Formally, we have the following.

Definition 8: A covering of X is a family $\gamma = \{C_i \in \mathcal{P}(X) : i \in I\}$ (indexed by the index set I) of nonempty subsets of X which satisfies the unique

(C) covering condition:

$$X = \cup \{C_i \in \gamma : i \in I\}$$

Any subset C_i of the covering γ is also called a pseudo-open granule. A pair (X, γ) consisting of a universe X and one of its covering γ is called a covering space.

Let us stress that from the theoretical point of view no requirement of finiteness about the universe X is done. All the (in the following) obtained results are true in the general context. Given a covering γ of X, a lower (pseudo-open) *crisp* set is, for definition, either the empty set $\emptyset$ or any set theoretic union of subsets from the covering. Denoting by $\mathcal{L}_\gamma(X)$ their collection, we have that for any nonempty subset $L \neq \emptyset$ of X

$$L \in \mathcal{L}_\gamma(X) \quad \text{iff} \quad \exists \{C_{i_r} \in \gamma : r \in R\} \ s.\ t.\ L = \bigcup_{r \in R} C_{i_r}$$

Of course, $\gamma \subseteq \mathcal{L}_\gamma(X)$ and so this latter is a fortiori also a covering of X, i.e. $X = \cup\{L : L \in \mathcal{L}_\gamma(X)\}$. Moreover, the following properties of a pseudo-open family hold:

(PO1) $\mathcal{L}_\gamma(X)$ contains either the empty set and the whole universe, and

(PO2) it is closed with respect to arbitrary set theoretic union.

Remark 3: The name "pseudo-open" family rises from the fact that in order to have a real family of open set for a topology of X it must be satisfied the further condition of closure with respect to finite intersection. Condition which in general is not verified by the family $\mathcal{L}_\gamma(X)$ induced from a generic covering.

Example 8: Considering the covering of the universe $X = \{f_1, f_2, f_3, f_4, f_5, f_6\}$ induced in the Example 6 from the incomplete information table introduced in the Example 4,

$$\gamma = \{\{f_4, f_5\}, \{f_1, f_2, f_3\}, \{f_2, f_5, f_6\}, \{f_4, f_5, f_6\}, \{f_1, f_2, f_3, f_6 g\}\}$$

The corresponding family of pseudo-open sets is

$$\mathcal{L}_\gamma(X) = \{\emptyset, \{f_4, f_5\}, \{f_1, f_2, f_3\}, \{f_2, f_5, f_6\}, \{f_4, f_5, f_6\}, \{f_1, f_2, f_3, f_6\}, \{f_2, f_4, f_5, f_6\}, \{f_1, f_2, f_3, f_4, f_5\}, \{f_1, f_2, f_3, f_5, f_6\}, X\}$$

This is a family of pseudo-open sets since it is closed with respect to any set theoretic union, but it is not closed with respect to finite intersection. For instance, $\{f_2, f_4, f_5, f_6\} \cap \{f_1, f_2, f_3, f_4, f_5\} = \{f_2, f_4, f_5\} \notin \mathcal{L}_\gamma(X)$.

In this covering framework one can consider the structure of *complete lower approximation space* induced from the topological space by an open base $\tau = (X, \mathcal{B})$ as follows

$$\mathfrak{LR}_\gamma = \langle \mathcal{P}(X), \mathcal{L}_\gamma(X), l_\gamma \rangle$$

where

(LC1) the role of the lattice of approximable elements Σ is played by the power set $\mathcal{P}(X)$ of the universe X, which (as stressed before) is an atomic Boolean complete lattice.

(LC2) from the above outlined properties of pseudo-open sets it immediately follows that the role of family of lower crisp elements $\mathbb{L}(\Sigma)$ is played by $\mathcal{L}_\gamma(X)$, since in particular (PO2) assures that it is a complete meet sub semi-lattice of $\mathcal{P}(X)$;

(LC3) as usual in any complete meet semi-lattice, for any approximable subset A of X it is possible to introduce, according to the (1a), its lower approximation:

$$l_\gamma(A) := \cup \{L \in \mathcal{L}_\gamma(X) : L \subseteq A\} \qquad \dots (13)$$

which exists and, as set theoretic union of pseudo-open sets, is a pseudo-open set: $l(A) \in \mathcal{L}_\gamma(X)$.

In other words, the lower approximation of the subset A (the approximable element) is its interior, i.e. the greatest pseudo-open set contained in A, usually denoted by $A^o := l_\gamma(A)$.

Since the power set $\mathcal{P}(X)$ of X is a Boolean lattice, and so in particular a de Morgan lattice in which the complementation mapping is the set theoretic complementation $A \in \mathcal{P}(X) \to A^c = X \setminus A \in \mathcal{P}(X)$, we can applied to the present case the part (D1) of Proposition 4 of Section 3 and construct the *dual anti-covering* of the covering $\gamma = \{C_i : i \in I\}$ which is, for definition, the family $\gamma_d := \{D \in \mathcal{P}(X) : \exists C \in \gamma \; s.t \; D = C^c\}$ of pseudo-closed granules.

This dual anti-covering can be characterized by the following.

(*D*) *disjointness* condition:

$$\emptyset = \cap \{D : D \in \gamma_d\}$$

An *upper (pseudo-closed) crisp* set is then the set theoretic intersection of elements from the dual covering. The collection of all *upper (pseudo-closed) crisp* sets will be denoted by $\mathcal{U}_\gamma(X)$ and so for any nontrivial subset $U \neq X$

$$U \in \mathcal{U}_\gamma(X) \quad \text{iff} \quad \exists \{D_s \in \gamma d : s \in S\} \text{ s.t. } U = \bigcap_{s \in S} D_s$$

Of course, the upper crisp set can be characterized by the condition

$$U \in \mathcal{U}_\gamma(X) \quad \text{iff} \quad \exists L \in \mathcal{L}_\gamma(X) \text{ s.t. } U = L^c$$

Thus, as a consequence of the duality, the family of crisp (pseudo-closed) sets satisfies the following properties:

(PC1) $\mathcal{U}_\gamma(X)$ contains either the empty set and the whole universe, and

(PC2) it is closed with respect to arbitrary set theoretic intersection.

These properties qualify $\mathcal{U}_\gamma(X)$ as a particular *Moore family* of subsets of X (see [Bir67, pp. 111], and the footnote in that page), which further on contains the empty set (property which is not required to hold in the general definition of Moore family).

The corresponding dual *upper (pseudo-closed) approximation space* with respect to the covering γ is then the structure

$$\mathfrak{U}\mathfrak{R}_\gamma = \langle \mathcal{P}(X), \mathcal{U}_\gamma(X), u_\gamma \rangle$$

where

(UT1) the role of the lattice of approximable elements Σ is played by the power set $\mathcal{P}(X)$ of the universe X, which is an atomic Boolean complete lattice.

(UT2) from the above outlined properties of pseudo-closed sets it immediately follows that the role of lattice of upper crisp elements $\mathbb{U}(\Sigma)$ is played by $\mathcal{U}(X)$, since in particular (PC2) assures that this latter is a complete meet sub semi-lattice of $\mathcal{P}(X)$;

(UT3) as usual in any complete meet semi-lattice, for any approximable subset A of X it is possible to introduce, according to the (1b), the corresponding upper approximation:

$$u_\gamma(A) := \cap \{C \in \mathcal{C}(X) : A \subseteq C\} \qquad \text{... (14)}$$

which exists and, as set theoretic intersection of pseudo-closed sets, is a pseudo-closed set: $u_\gamma(A) \in \mathcal{C}(X)$.

In other words, the upper approximation of the subset A (the approximable element) is its *pseudo-closure*, i.e. the smallest pseudo-closed set containing A, usually denoted by $A^* := u_\gamma(A)$.

The *covering approximation space*, complete according to the definition discussed in Section 2.1, is then the structure

$$\mathfrak{R}_\gamma = \langle \mathcal{P}(X), \mathcal{L}_\gamma(X), \mathcal{U}_\gamma(X), l_\gamma, u_\gamma \rangle$$

The *covering rough approximation mapping* $r_\gamma : \mathcal{P}(X) \mapsto \mathcal{L}_\gamma(X) \times \mathcal{U}_\gamma(X)$ is the mapping which assigns to any approximable subset A of the universe X the pair $r_\gamma(A) = (A^o, A^*)$, with $A^o \subseteq A \subseteq A^*$, consisting of its interior (pseudo-open subset $A^o \in \mathcal{L}_\gamma(X)$) and its closure (pseudo-closed subset $A^* \in \mathcal{U}_\gamma(X)$). Trivially, a subset E of a topological space X is crisp iff it is *pseudo-clopen* (in particular the empty set and the whole space are pseudo-clopen, and so crisp).

Pseudo-closed lower and pseudo-open upper approximations in a covering space. In the present covering context let us stress the property (PC2) which assures that set theoretic intersection of an arbitrary family of pseudo-closed sets is pseudo-closed and the property (PO2) which assures that set theoretic union of an arbitrary family of pseudo-open sets is pseudo-open.

So we can introduce a new pair of lower-upper approximations of A according to the following definitions:

$$l_\gamma^{(cl)}(A) = \cap \{U \in \mathcal{U}_\gamma(X) : U \cup A \neq X\} \qquad \text{... (15a)}$$

$$u_\gamma^{(op)}(A) = \cup \{L \in \mathcal{L}_\gamma(X) : L \cap A \neq \emptyset\} \qquad \text{... (15b)}$$

Let us note that $l_\gamma^{(cl)}(A) \in \mathcal{U}_\gamma(X)$ is a pseudo-closed set (as set theoretic intersection of pseudo-closed sets), whereas $l_\gamma(A) = A^o \in \mathcal{L}_\gamma(X)$ is a pseudo-open set (as set theoretic union of pseudo-open sets). From a terminological point of view, $l_\gamma(A)$ can be called the pseudo-open upper approximation and $l_\gamma^{(cl)}(A)$ the pseudo-closed upper approximation of A. Similarly, $u_\gamma^{(op)}(A) \in \mathcal{L}_\gamma(X)$ is a pseudo-open set (as set theoretic union of pseudo-open sets), whereas $u_\gamma(A) = A^* \in \mathcal{U}_\gamma(X)$ is a pseudo-closed set (as set theoretic intersection of pseudo-closed sets). From a terminological point of view, $u_\gamma(A)$ can be called the *pseudo-closed upper approximation*, and $u_\gamma^{(op)}(A)$ the *pseudo-open upper approximation* of A.

Moreover, the operators $l_\gamma^{(cl)}$ and $u_\gamma^{(op)}$ are in duality since:

$$[u_\gamma^{(op)}(A^c)]^c = l_\gamma^{(cl)}(A) \quad \text{... (16)}$$

In general l_γ and $l_\gamma^{(cl)}$ (and dually u_γ and $u_\gamma^{(op)}$ applied to the same approximable set A give different results.

Example 9: Let us consider the finite universe $X = \{1, 2, 3, 4, 5, 6\}$, with associated lower (pseudo-open) base $\gamma = \{\{1, 6\}, \{2, 3\}, \{1, 2, 3, 4\}, \{3, 4, 5, 6\}\}$. The corresponding collection of lower (pseudo-open) crisp sets is

$$\mathcal{L}_\gamma(X) = \{\emptyset, \{1, 6\}, \{2, 3\}, \{1, 2, 3, 4\}, \{3, 4, 5, 6\}, \{1, 2, 3, 6\}, \{1, 2, 3, 4, 6\}, \{1, 3, 4, 5, 6\}, X\}$$

The dual upper (pseudo-closed) base is $\gamma_d = \{\{1, 2\}, \{5, 6\}, \{2, 3, 4, 5\}, \{1, 4, 5, 6\}\}$, with collection of upper (pseudo-closed) crisp sets

$$\mathcal{U}_\gamma(X) = \{\emptyset, \{1, 2\}, \{5, 6\}, \{1, 4, 5, 6\}, \{2, 3, 4, 5\} \{1\}, \{2\}, \{5\}, \{4, 5\}, X\}$$

Now, the set $A = \{2, 3, 4\}$ is such that $l_\gamma^{(cl)}(A) = \emptyset, \subset \{2, 3\} = l_\gamma(A)$, and $u_\gamma(A) = \{2, 3, 4, 5\} \subset X = u_\gamma^{(op)}(A)$. So one has the rough approximation of $A = (\{2, 3, 4\})$ by the pseudo-open–pseudo-closed pair $r_\gamma(\{2, 3, 4\}) = (\{2, 3\}, \{2, 3, 4, 5\})$ and the rough approximation of the same A by the pseudo-closed–pseudo-open pair $r_\gamma^{(co)}(\{2, 3, 4\}) = (\emptyset, X)$.

In this example it happens that the pseudo-open upper approximation is always greater than or equal to the corresponding pseudo-closed lower version and, dually, the pseudo-closed upper approximation is always less or equal than the corresponding pseudo-open upper version. This indeed is a general behavior according to the following result.

Proposition 5: In any covering space the following inclusions between approximations hold:

$$\forall A \in \mathcal{P}(X),\ l_\gamma^{(cl)} A \subseteq l_\gamma(A) \subseteq A \subseteq u_\gamma(A) \subseteq u_\gamma^{(op)}(A)$$

Proof: We prove the inequality for the upper approximations, the one for the lower approximation can be derived by duality. Let $x \in u_\gamma(A)$. Then, $\forall U \in \mathcal{U}(X)$, $[A \subseteq U \to x \in U]$. Since $U = L^c$ for some $L \in \mathcal{L}(X)$, we have $\forall L \in \mathcal{L}(X)$, $[L \subseteq A^c \to x \in L^c]$, that is $\forall L \in \mathcal{L}(X)$, $[L \cap A = \emptyset \to x \notin L]$ or $\forall L \in \mathcal{L}(X)$, $[\neg(L \cap A = \emptyset) \vee x \notin L]$ and so $\forall L \in \mathcal{L}(X)$, $[(x \notin L) \vee (L \cap A \neq \emptyset)]$. Since, by the covering property, $\exists L_0$: $x \in L_0$ this last result assures that $L_0 \cap A \neq \emptyset$, i.e. $x \in u_\gamma^{(op)}(A)$.

So the "pseudo-open – pseudo-closed" rough approximation of any subset A, i.e. the pair $r_\gamma(A) = (l_\gamma(A), u_\gamma(A))$, is "better" than the "pseud-closed – pseudo-open" rough approximation of the same A, defined as $r_\gamma^{(co)}(A) := (l_\gamma^{(cl)}(A), u_\gamma^{(op)}(A))$. This can be formally written as

$$r_\gamma(A) = (l_\gamma(A), u_\gamma(A)) \sqsubseteq (l_\gamma^{(cl)}(A), u_\gamma^{(op)}(A)) = r_\gamma^{(co)}(A)$$

where we have denoted by $\sqsubseteq$ the partial order relation on the collection of all pairs of subsets from X, $(\mathcal{P}(X) \times \mathcal{P}(X))_{\subseteq} := \{(A_1, A_p) \in \mathcal{P}(X) \times \mathcal{P}(X) : A_1 \subseteq A_p\}$, defined as

$$(A_1, A_p) \sqsubseteq (B_1, B_p) \text{ iff } B_1 \subseteq A_1 \subseteq A_p \subseteq B_p$$

Remark 4: This definition is similar to the one which formalizes the approximation by real numbers. In this numerical case, the pair of real numbers (r_l, r_u), with $r_l \leq r_u$, furnishes a better approximation with respect to the pair (s_l, s_u), with $s_l \leq s_u$, iff the following chain of inequalities is verified: $s_l \leq r_l \leq r_u \leq s_u$. This condition is denoted by $(r_l, r_u) \sqsubseteq (s_l, s_u)$. For instance, in approximating the (irrational) real number $\sqrt{2}$ it is trivial that the pair (1.414, 1.415) furnishes a better approximation with respect to the pair (1, 3) owing to the fact that $1 < 1.414 < \sqrt{2} < 1.415 < 3$.

Let us note that the pseudo-open – pseudo-open approximation $r_\gamma^{(op)}(A) = (l_\gamma(A), u_\gamma^{(op)}(A))$ of the generic subset A has been studied also by other authors, for instance in [Pom87] for the case of coverings. Of course, also in this case $r_\gamma(A) \sqsubseteq r_\gamma^{(op)}(A)$. In particular, Pomykala in [Pom87] addresses the problem of studying under which conditions the two above approximations, based on coverings, are *Kuratowski topological* interior and closure. In the next section, we will shaw that they are indeed a *Tarski* interior and closure, which is not (Kuratowski) topological.

Yao lower and upper approximations for coverings. Further, Yao in [Yao98c] gives another possibility for the case of neighborhood systems (at least under the reflexivity condition of a binary relation on a universe), which we call "local" since it involves the *neighborhood* of a generic point $x \in X$ relatively to the covering γ as the pseudo-open set $\gamma(x) := \cup \{C \in \gamma : x \in C\} \in \mathcal{L}_\gamma(X)$. Using this notion, the lower and upper approximations of a subset A are defined as:

$$l_\gamma^{(l)}(A) = \{x \in X : \gamma(x) \subseteq A\} \text{ and}$$
$$u_\gamma^{(l)}(A) = \{x \in X : \gamma(x) \cap A \neq \emptyset \qquad \text{... (17)}$$

Trivially, these two maps give rise to a different approximation of the generic approximable subset A of the universe X according to the inequalities:

$$l_\gamma^{(l)}(A) \subseteq l_\gamma(A) \subseteq A \subseteq u_\gamma(A) \subseteq u_\gamma^{(l)}(A)$$

with the "global" rough approximation $r_\gamma(A) := (l_\gamma(A), u_\gamma(A))$ better than the "local" one $r_\gamma^{(l)}(A) := (l_\gamma^{(l)}(A), u_\gamma^{(l)}(A))$, i.e. $r_\gamma(A) \sqsubseteq r_\gamma^{(l)}(A)$.

Let us illustrate this situation with one example.

Example 10: In the universe $X = \{1, 2, 3, 4, 5\}$ let us consider the covering $\gamma = \{\{1, 2\}, \{2, 3, 4\}, \{4, 5\}\}$. The corresponding family of pseudo-open sets is $\mathcal{L}_\gamma(X) = \{\emptyset, \{1, 2\}, \{2, 3, 4\}, \{4, 5\}, \{1, 2, 4, 5\}, \{1, 2, 3, 4\}, \{2, 3, 4, 5\}, X\}$.

The pseudo-open neighborhood of the points of the universe are then:

$$\gamma(1) = \{1, 2\} \; \gamma(2) = \{1, 2, 3, 4\} \; \gamma(3) = \{2, 3, 4\}$$
$$\gamma(4) = \{2, 3, 4, 5\} \; \gamma(5) = \{4, 5\}$$

Notice that the family of pseud-closed sets is $\mathcal{U}_\gamma(X) = \{\emptyset, \{3, 4, 5\}, \{1, 5\}, \{1, 2, 3\}, \{3\}, \{5\}, \{1\}, X\}$.

Let us consider the subset $A = \{3, 4, 5\}$ of the universe, then

$$l_\gamma^{(l)}(A) = \{5\} \text{ and } l_\gamma(A) = \{4, 5\} \text{ with } l_\gamma^{(l)}(A) \subset l_\gamma(A)$$
$$u_\gamma(A) = \{3, 4, 5\} \text{ and } u_\gamma^{(l)}(A) = \{2, 3, 4, 5\} \text{ with}$$
$$u_\gamma(A) \subset u_\gamma^{(l)}(A)$$

4.3 Topological Approximation Spaces

Another concrete model of approximation space is based on the notion of topological space, in this context called *topological approximation space*. To this purpose, let us recall that a topological space can be defined as a pair $\tau = (X, \mathcal{B})$ consisting of a set X and a base for a topology (simply *open base*). An open base is a collection $\mathcal{B} = \{B_i \in \mathcal{P}(X) : i \in I\}$ of subsets of X, each of which is an *open granule* or (adopting the topological terminology, see [Sim63, pp. 99]) a *basic open set*, such that the following hold:

(C) *covering* condition:

$$X = \cup \{B_i \in \mathcal{B} : i \in I\}$$

(OB) *open base* condition: for any pair B_i, B_j of subsets of the base $\mathcal{B}$ for which $B_i \cap B_j \neq \emptyset$ a collection $\{\hat{B}_{ih} \in \mathcal{B} : h \in H\}$ of elements from the base exists such that

$$B_i \cap B_j = \cup \{\hat{B}_{ih} \in \mathcal{B} : h \in H\}$$

An open base is then a particular covering of the universe X for which the further condition (OB) holds. So we can apply all the procedures considered in the previous covering case. In particular, the family of *open sets* $\mathcal{O}(X)$ is obtained by arbitrary set theoretic unions of elements from the open base $\mathcal{B}$ plus the empty set $\emptyset$. In this way we have induced a topological space as the pair $(X, \mathcal{O}(X))$ consisting of a nonempty set X equipped with a family of *open subsets* $\mathcal{O}(X)$, satisfying the following conditions:

(O1) the empty set $\emptyset$ and the whole space X are open;

(O2) the family $\mathcal{O}(X)$ is closed with respect to arbitrary unions;

(O3) the family $\mathcal{O}(X)$ is closed with respect to finite intersections.

In particular any element of the open basis is an open set, that is $\mathcal{B} \subseteq \mathcal{O}(X)$.

As in the covering case, also in this topological framework one can consider the structure of *complete lower approximation space* induced from the topological space by an open base $\tau = (X, \mathcal{B})$ as follows $\mathcal{LR}_\tau = \langle \mathcal{P}(X), \mathcal{O}(X), I \rangle$ where as usual the role of lattice of approximable elements Σ is played by the power set $\mathcal{P}(X)$ of the universe X, which is a complete lattice, the role of lower (open) crisp sets is played by elements from $\mathcal{O}(X)$.

The lower (open) approximation of an approximable subset A is then its *topological interior* A^o formally identified with the open set

$$l(A) := \cup \{O \in \mathcal{O}(X) : O \subseteq A\} \qquad \text{... (18)}$$

The *dual base* for a topology of *closed granules* is the collection $\mathcal{K} := \{K \in \mathcal{P}(X) : \exists B \in \mathcal{B} \text{ s.t. } K = B^c\}$. This dual base is a base for a topology of closed sets (simply, closed base for a topology, see [Sim63, pp. 112]), in the sense that the following hold:

(DB) *disjointness* condition

$$\emptyset = \cap \{K \in \mathcal{P}(X) : K \in \mathcal{K}\}$$

(CB) *closed base* condition: for any pair K_i, K_j of subsets of the closed base $\mathcal{K}$ for which $K_i \cup K_j \neq X$ a collection $\{\hat{K}_{i_v} \in \mathcal{K} : v \in V\}$ of elements from the closed base exists such that

$$K_i \cup K_j = \cap \{\hat{K}_{i_v} \in \mathcal{K} : v \in V\}$$

A *closed* set is either the set theoretic intersection of elements from the closed base or the whole space X and a subset of X is *closed* iff it is the set theoretic complement of an open set:

$$C \in \mathcal{C}(X) \text{ iff } \exists O \in \mathcal{O}(X) \text{ s.t. } C = O^c$$

Therefore, the collection $\mathcal{C}(X)$ of all closed subsets of X satisfies the following conditions:

(C1) both $\emptyset$ and X are closed;

(C2) the family $\mathcal{C}(X)$ is closed with respect to finite unions;

(C3) the family $\mathcal{C}(X)$ is closed with respect to arbitrary intersections.

All the elements of the closed base are also closed, that is $\mathcal{K} \subseteq \mathcal{C}(X)$.

In this topological framework of closed sets one can introduce the structure of *complete upper approximation space* induced by the topological space $\tau = (X, \mathcal{B})$

$$\mathfrak{UR}_\tau = \langle \mathcal{P}(X), \mathcal{C}(X), u \rangle$$

where as said before the role of lattice of approximable elements Σ is played by the power set $\mathcal{P}(X)$ of the universe X, which is a complete lattice. Moreover, $\mathcal{C}(X)$ plays the role of lattice of upper (closed) crisp elements and for any approximable subset A of X it is possible to introduce its upper approximation:

$$u(A) := \cap \{C \in \mathcal{C}(X) : A \subseteq C\} \quad \text{... (19)}$$

which is the *topological closure* A^* of A.

The (complete) *topological approximation space* is then the structure

$$\mathfrak{R}_T = \langle \mathcal{P}(X), \mathcal{O}(X), \mathcal{C}(X), l, u \rangle$$

The *topological rough approximation mapping* $r_T : \mathcal{P}(X) \mapsto \mathcal{O}(X) \times \mathcal{C}(X)$ is the mapping which assigns to any subset A of the topological space X the pair $r_T(A) = (A^o, A^*)$, with $A^o \subseteq A \subseteq A^*$, consisting of its interior (open subset $A^o \in \mathcal{O}(X)$) and its closure (closed subset $A^* \in \mathcal{C}(X)$). Similarly to the covering case, a subset E of a topological space X is crisp iff it is *clopen*, i.e. iff $E \in \mathcal{C}(X) \cap \mathcal{O}(X)$ (in particular the empty set and the whole space are clopen, and so crisp). The collection of all clopen (crisp) subsets of X is denoted as $\mathcal{E}(X) := \mathcal{C}(X) \cap \mathcal{O}(X)$.

Example 11: Let us consider the universe $X = \{1, 2, 3, 4, 5, 6\}$ and the open base $\mathcal{B} = \{\{2\}, \{4\}, \{3, 4\}, \{1, 3, 4\}, \{4, 5, 6\}, \{3, 4, 5, 6\}\}$, which trivially is not a partition of the universe X and is more than a simple covering. These are the open granules of the topology on X given by the following open sets

$$\begin{aligned}\mathcal{O}(X) = \{&\emptyset \{2\}, \{4\}, \{2, 4\}, \{3, 4\}, \{1, 3, 4\}, \{2, 3, 4\},\\ &\{4, 5, 6\}, \{1, 2, 3, 4\}, \{2, 4, 5, 6\}, \{3, 4, 5, 6\},\\ &\{1, 3, 4, 5, 6\}, \{2, 3, 4, 5, 6\}, X\}\end{aligned}$$

The corresponding dual base of closed granules is $\mathcal{K} = \{\{1, 2\}, \{1, 2, 3\}, \{2, 5, 6\}, \{1, 2, 5, 6\}, \{1, 3, 4, 5, 6\}, \{1, 2, 3, 5, 6\}\}$. Note that in particular, $\{2, 5, 6\} \cap \{1, 2, 3\} \cap \{1, 3, 4, 5, 6\} = \emptyset$, but there is no pair of different closed granules $K_i \neq K_j$ such that $K_i \cap K_j = \emptyset$. The induced topology of closed sets is

$$\begin{aligned}\mathcal{C}(X) = \{&\emptyset, \{1\}, \{2\}, \{1, 2\}, \{1, 3\}, \{5, 6\}, \{1, 2, 3\},\\ &\{1, 5, 6\}, \{2, 5, 6\}, \{1, 2, 5, 6\}, \{1, 3, 5, 6\},\\ &\{1, 3, 4, 5, 6\}, \{1, 2, 3, 5, 6\}, X\}\end{aligned}$$

Let us consider now the approximable subset $A = \{1, 2, 5\}$, then

$$l(A) = A^o = \{2\} \text{ and } u(A) = A^* = \{1, 2, 5, 6\}$$

On the contrary, the open upper approximation of A according to (15b) is

$$u^{(op)}(\{1, 2, 5\}) = \{1, 3, 4\} \cup \{3, 4, 5, 6\} \cup \{4, 5, 6\} \cup \{2\} = X$$

On the other hand, computing the approximations of the set $B = \{1, 2, 4\}$ we have $l(B) = \{2, 4\}$ and $u(B) = X$, whereas the closed lower approximation of B according to (15a) is:

$$l^{(cl)}(B) = \{1, 2\} \cap \{1, 2, 3\} \cap \{2, 5, 6\} \cup \{1, 2, 5, 6\}$$
$$= \{2\}$$

Let us remark that in this example we have seen a particular subset A of a concrete topological space such that $u(A) \subset u^{(op)}(A)$ and a particular subset B such that $l^{(cl)}(B) \subset l(B)$. These are indeed two particular cases of the general property shown in the previous section by Proposition 5.

4.4 The Partition Approach to Rough Set Theory

The usual Pawlak approach to rough set theory as introduced in the papers [Paw81, Paw82, Paw92] is formally based on a partition of a concrete (non necessarily finite) universe. Formally,

Definition 9: A partition of X is a family $\pi = \{G_i \in \mathcal{P}(X) : i \in I\}$ (indexed by the index set I) of nonempty subsets of X such that the following hold:

(*C*) covering *condition:*

$$X = \cup \{G_i \in \pi : i \in I\}$$

(*D*) disjointness *condition:* for any pair $G_h \neq G_k$ of different subsets of the partition π, it is $G_h \cap G_k = \emptyset$.

Any subset G_i of the partition π is also called an elementary sets, considered as a knowledge granule. A pair (X, π) consisting of a universe X and one of its partition π is called partition space.

As usual in any concrete model, subsets of X are the approximable elements. Let us stress again that also in this partition case no requirement of finiteness about the universe X is done.

Any partition π can be characterized by the induced equivalence relation $\mathcal{R}_\pi \subseteq X \times X$, defined as

$(x, y) \in R_\pi$ iff $\exists G \in \pi$ s.t. $x, y \in G$ ... (20)

In this case the two objects x, y are said to be *indistinguishable* with respect to π and the equivalence relation $\mathcal{R}_\pi$ is called an *indistinguishability* relation. In this indistinguishability context the partition π is considered as the support of some knowledge available on the objects of the universe and so any equivalence class (i.e., elementary set) G is interpreted as a *granule* (or *atom*) of knowledge contained in (or supported by) π. For any object $x \in X$ we shall denote by $gr(x)$ the (unique) equivalence class which contains x ($gr(x) = G$ iff $x \in G$), called the *granule generated* by x.

Of course, any partition is in particular an open base for a topology on X since condition (D) about partitions implies condition (OB) about open basis.

Thus a partition space (X, π) generates a topological space whose open base is just the family π, i.e. $\mathcal{B} = \pi$. Applying in this partition context the usual approach to topology from open basis, the corresponding closed base consists of subsets of X which are the set complement of an equivalence class. As an immediate consequence, an element of the closed base is just a set union of equivalence classes. But it is not true that a generic union of equivalence classes generates an element of $\mathcal{K}$.

Example 12: Let us consider the partition $\pi = \{\{1, 3\}, \{2\}, \{4\}, \{5\}\}$ of the universe $X = \{1, 2, 3, 4, 5\}$. The corresponding closed base is $\mathcal{K} = \{\{x_2, x_4, x_5\}, \{x_2, x_3, x_4, x_5\}, \{x_1, x_2, x_3, x_5\}, \{x_1, x_2, x_3, x_4\}\}$ which for instance does not contain the set $\{x_2, x_4\}$, union of two open granules.

If as usual in the topological context we construct an open set as union of subsets of the open base (i.e., equivalence classes from the partition π) according to $E \in \mathcal{O}_\pi(X)$ iff $\exists \{G_j \in \pi : j \in J \subseteq I\}$ s.t. $E = \cup \{G_j \in \pi : j \in J \subseteq I\}$, then the set complement E^c of any open set is also a set theoretic union of equivalence classes from the open base π, i.e. an open set. This means that $E = (E^c)^c$ as set theoretic complement of the open set E^c is, from the topological point of view, also closed, i.e. a *clopen* set.

From this it follows that a partition gives rise to a topology of clopen: formally $\mathcal{O}_\pi(X) = \mathcal{C}_\pi(X)$. This unique family will be denoted by $\mathcal{E}_\pi(X)$ in the sequel, formally $\mathcal{O}(X) = \mathcal{C}(X) = \mathcal{E}_\pi(X)$, and it is the collection of all crisp sets induced from the partition space. In particular any granule G_i of the partition π is crisp. The following hold:

(*i*) The collection of all crisp sets is a Boolean algebra $\langle \mathcal{E}_\pi(X), \cap, \cup, c, \emptyset, X \rangle$ with respect to set theoretic intersection, union, and complementation. In particular for any crisp set $E \in \mathcal{E}_\pi(X)$ its set theoretic complement E^c is crisp too.

(*ii*) From the topological point of view $\mathcal{E}_\pi(X)$ contains both the empty set and the whole space, moreover it is closed with respect to any arbitrary set theoretic union and intersection, i.e. it is a family of clopen (simultaneously closed and open) subsets for a topology on X.

The following is now straightforward.

Theorem 1: *The concrete* approximation space *generated by the partition* π *of the universe* X, *or* partition approximation space, *consists of the system*

$$\mathfrak{R}_\pi := \langle \mathcal{P}(X), \mathcal{E}_\pi(X), l_\pi, u_\pi \rangle$$

where:

(P1) the set of approximable *elements* Σ is the Boolean atomic (*complete*) *lattice* $\langle \mathcal{P}(X), \cap, \cup, ^c, \emptyset, X \rangle$ of *all* approximable *subsets* A *of the universe* X;

(P2) the two sets of lower crisp elements $\mathbb{L}(\Sigma)$ and upper crisp elements $\mathbb{U}(\Sigma)$ are the same and coincide with the Boolean atomic (complete) lattice $\mathcal{E}_\pi(X)$. Whose atoms are the equivalence classes (granules) G of the partition $\pi(X)$. Formally, in this model $\mathbb{L}(\Sigma) = \mathbb{U}(\Sigma) = \mathcal{E}_\pi(X)$ and so the set Σ_c of crisp elements is just $\mathcal{E}_\pi(X)$;

(P3) the map $l_\pi : \mathcal{P}(X) \mapsto \mathcal{E}_\pi(X)$ associating with any subset A of X its lower approximation defined by the (clopen) crisp set

$$l_\pi(A) := \cup \{E \in \mathcal{E}_\pi(X) : E \subseteq A\} \qquad \text{... (21)}$$

satisfies the conditions (In1)–(In3) of *Ax* 1, i.e. it is a lower approximation map;

(P4) the map $u_\pi : \mathcal{P}(X) \mapsto \mathcal{E}_\in(X)$ associating with any subset A of X its upper approximation defined by the (clopen) crisp set

$$u_{\pi}(A) := \cap \{F \in \mathcal{E}_{\pi}(X) : A \subseteq F\} \qquad \text{... (22)}$$

satisfies the conditions (Up1)–(Up3) of *Ax* 2, i.e. it is an upper approximation map.

The *rough approximation* of a subset A of X is then the clopen pair

$$r_{\pi}(A) := (l_{\pi}(A), u_{\pi}(A)), \text{ with } l_{\pi}(A) \subseteq A \subseteq u_{\pi}(A) \qquad \text{... (23)}$$

Let us stress the following result, useful for a successive discussion.

Proposition 6: Let (X, π) be a partition space, with associated approximation space $\mathfrak{R}_{\pi}$. Then, making reference to the granules from π instead of the collection of crisp sets $\mathcal{E}_{\pi}(X)$, the inner and the upper approximations of a subset A of X can be equivalently expressed as

$$l_{\pi}(A) = \cup \{G \in \pi : G \subseteq A\} \qquad \text{... (24a)}$$

$$u_{\pi}(A) = \cup \{H \in \pi : H \cap A \neq \emptyset \qquad \text{... (24b)}$$

Note that for any subset A, the universe turns out to be the set theoretic union of the mutually disjoint sets $X = l_{\pi}(A) \cup b_{\pi}(A) \cup e_{\pi}(A)$, where $b_{\pi}(A) := u_{\pi}(A) \setminus l_{\pi}(A)$ is the *boundary* and $e_{\pi}(A) := X \setminus u_{\pi}(A)$ is the *exterior* of A. The triplet of subsets $\pi(A) := \{l_{\pi}(A), b_{\pi}(A), e_{\pi}(A)\}$ is a new (*local*, i.e. depending from A) partition of X induced by A from the original partition π.

Let us stress that in the case of a topological approximation space, as generalization of the partition approximation space, the Eqs (18) and (19) (in which are involved open sets from $\mathcal{O}(X)$ and closed sets from $\mathcal{C}(X)$ respectively) are generalizations of the corresponding partition versions (21) and (22), in which are involved crisp sets from $\mathcal{E}_{\pi}(X)$. Then, one can try to generalize to the topological case also the equivalent formulation (24a) and (24b) expressed in terms of the partition π.

Partition Approximation Spaces from (Complete) Information Systems. In this section we show as the usual Pawlak approach to roughness by (complete) information system (IS) gives rise to a partition approximation space (we recall the already quoted seminal papers [Paw81, Paw82, Paw92]).

Definition 10: A structure $\mathcal{IF}(X) = \langle X, Att, Val, F\rangle$, *is an* Information System *iff*:

- X (*called the* universe) *is a non empty set of* objects (situations, entities, states);
- *Att* is a non empty set of attributes, which assume values for the objects belonging to the set X;
- $V\,al$ is the set of all possible values that can be assigned to objects from X for attributes ranging on *Att*;
- F (called the information function) is a mapping $F: X \times Att \rightarrow V\,al$ which associates with any pair (x, a), consisting of an object $x \times X$ and an attribute $a \in Att$, the corresponding value $F(x, a) \in V\,al$.

In general in the sequel we assume that an information system satisfies the following conditions of coherence:

1. The information mapping F must be surjective; this means that if there exists a value $v \in V\,al$ which is not the result of the application of the information map F to at least a pair $(x, a) \in X \times Att$, then this value has no interest with respect to the knowledge stored in the information system and can be eliminated.
2. For any attribute $a \in Att$ at least two different objects $x_1 \neq x_2$ must exist such that $F(x_1, a) \neq F(x_2, a)$, otherwise this attribute does not supply any knowledge and can be suppressed by a reduct operation.

The *indiscernibility* relation that may hold between the objects from the universe under investigation represents the mathematical basis for the rough set approach. Suppose an information system $\mathcal{IS}(X) = \langle X, Att, V\,al, F\rangle$, and let $\mathcal{A}$ be any family of attributes (a convenient subset of *Att*). Then, $\mathcal{A}$ determines a binary relation $I_{\mathcal{A}}$ on the set of objects X, called the *indiscernibility relation* generated by $\mathcal{A}$, and de¯ned as follows:

$$(x, y) \in \mathcal{I}_{\mathcal{A}} \text{ iff } \exists a \in \mathcal{A}, F(x, a) = F(y, a) \qquad \dots (25)$$

In other words, two objects x and y are $\mathcal{I}_{\mathcal{A}}$-indiscernible iff they assume the same values for *all* the attributes belonging to the family $\mathcal{A}$. The relation $\mathcal{I}_{\mathcal{A}}$ turns out to be an *equivalence relation* whose equivalence classes are the *granules* of knowledge $gr_{\mathcal{A}}(x) := \{y \in X : (x, y) \in \mathcal{I}_{\mathcal{A}}\}$, i.e. $gr_{\mathcal{A}}(x)$ consisting of all the objects $y \in X$ that share with x the same values for all the attributes belonging to A (see the Eq. (25)). We shall denote by τ_A the collection of all such granules (equivalence classes) and by $\mathcal{E}_A(X)$ the induced collection of crisp sets.

On the basis of the partition space (X, π_A) generated by the family of attributes $\mathcal{A}$, all the results analyzed in Section 4.4 can be applied. In this information system context the lower approximation of a subset A of the universe will be denoted by $l_A(A)$ and the corresponding upper approximation by $u_A(A)$.

The rough approximation of A is then the pair

$$r_A(A) = (l_A(A), u_A(A)) \text{ with } l_A(A) \subseteq A \subseteq u_A(A)$$

Example 13: Let us consider an information system represented by the information Table 2 involving a universe X of 5 flats with attributes "P = Price", "R = Rooms", "F = Furniture", and "PF = Prestigious Flat".

The partition generated by the set of attributes $\mathcal{A}_3$ = {Price, Rooms, Furniture} is the following:

$$\pi_{A3} = \{\{x_1, x_3\}, \{x_2\}, \{x_4\}, \{x_5\}\}.$$

With respect to the attribute "PF = Prestigious flat", interpreted as a *decision attribute*, we can single out two *decision classes*: $[PF = y]$ consisting of all prestigious flats, and $[PF = n]$ consisting of the not prestigious flats:

$$[PF = y] = \{x_1, x_2, x_4\} \text{ and } [PF = n] = \{x_3, x_5\}.$$

Table 2: *A Simple Information Table Involving a Universe X of 5 Flats*

X/Att	*Price*	*Rooms*	*Furniture*	*Prestigious Flat*
x_1	high	1	no	yes
x_2	high	3	yes	yes
x_3	high	1	no	no
x_4	high	2	no	yes
x_5	low	2	yes	no

The collection $\pi_{PF} = \{[PF = y], [PF = n]\}$ gives rise to a different partition of the universe X, called the *decision-partition* induced by the decision attribute *Prestigious flat PF*.

The collection of prestigious flats $[PF = y]$, considered as a subset of the universe X, is lower and upper approximated relatively to the collection of attributes $\mathcal{A}_3$ in the following way

$$l_{A3}[PF = y] = \{x_2, x_4\} = [P = h, R = 3, F = y] \cup [P = h, R = 2, F = n]$$

$$u_{A3}[PF = y] = \{x_1, x_2, x_3, x_4\} = [P = h, R = 3, F = y] \cup [P = h, R = 2, F = n] \cup [P = h, R = 1, F = n]$$

That is, in this information system

1. a flat is *necessarily* prestigious iff its price is high and it consists of exactly 2 or 3 rooms; the role of furniture is irrelevant;
2. a flat is *possibly* prestigious iff the price is high; in this case not only the role of furniture continues to be irrelevant, but also irrelevant is the number of rooms.

5. THE TARSKI CLOSURE AND INTERIOR ABSTRACT APPROACH TO APPROXIMATION SPACES

Summarizing what previously discussed, the abstract approach to roughness introduced in Section 2 is based on a lattice of "approximable" concepts, with associated two well defined sublattices describing "lower" and "upper" crisp concepts respectively, with respect to which it is possible to construct "inner" and "outer" approximations of any element. In order to introduce the equivalence (isomorphism) between this abstract approaches to roughness theory and another one based on topological notions of closure and interior operator, let us give the following definition.

Definition 11: A structure $\langle \Sigma, \wedge, \vee, *, 0, 1\rangle$ is a closure lattice iff $\langle \Sigma, \wedge, \vee, 0, 1\rangle$ is a bounded lattice and the mapping $* : \Sigma \to \Sigma$, that associates with any element a from Σ its closure $a^* \in \Sigma$, is a closure operation, that is, it satisfies the properties:

(C1) $0^* = 0$ (*normalized*)

(C2) $a \leq a^*$ (*increasing*)

(C3) $a^* = a^{**}$ (*idempotent*)

(C4) $a^* \vee b^* \leq (a \vee b)^*$ (*sub-additive*)

Note that the just given definition of closure operation is equational. The first interesting result is that the sub-additive property (C4) is equivalent to the following (non-equational) *monotonicity* condition, which allows one to introduce a notion of closure in the more general context of a poset, which is not necessarily a lattice (see for instance [MR42]).

Lemma 3: In any lattice structure, the sub-additive property (C4) is equivalent to the monotonicity condition:

(C4a) $a \leq b$ implies $a^* \leq b^*$

Proof: Let (C4) be true. Then, from $a \leq b$ we get $b^* = (a \vee b)^* \geq a^* \vee b^* \geq a^*$. Vice versa, since in a lattice we have $a, b \leq a \vee b$, by the monotonicity condition (C4a) it follows that $a^*, b^* \leq (a \vee b)^*$, and so the l.u.b. condition of the lattice implies $a^* \leq b^* \leq (a \vee b)^*$.

In a closure lattice, owing to the condition (C2), it is possible to select the subset of *closed elements* defined as the collection of elements which are equal to their closure. Formally,

$$C(\Sigma) := \{a \in \Sigma : a = a^*\}$$

The set $C(\Sigma)$ is not empty since the element 0 is closed as consequence of (C1). Further, in general it does not coincide with the set of all elements Σ, i.e. there may exist elements which are not closed. Condition (C3) says that for any element $a \in \Sigma$ the corresponding a^* is closed; for this reason in definition 11 this element has been called the *closure* of *a*.

This notion has been developed by Tarski during the '30 s in his study of *consequences* in logic [Tar30], but it can be considered as the most fascinating applications of closure operator on the power set of a given universe. Also if Tarski did not point out the importance of his approach in relation with the closure theory developed in the same period by others, the co-paternity of the notion of closure system can been attributed, besides to Tarski, also to Monteiro-Ribeiro [MR42] (in the abstract poset context), Ward [War42] (in the abstract lattice context), Ore [Ore42] (in the concrete power set context) and, in connection with Galois theory, Ore [Ore44] and Everett [Eve44] (both these latter with a poset version of the closure). For this reason, in the sequel we also refer to this notion with the term of *Tarski closure operation*.

Proposition 7: The set $\mathcal{C}(\Sigma)$ of all closed element of a Tarski closure lattice satisfies the following (finite) Moore conditions:

(M1) the greatest element is closed: $1 \in \mathcal{C}(\Sigma)$;

(M2) for any pair of closed elements $a, b \in \mathcal{C}(\Sigma)$, their meet is closed too: $a \wedge b \in \mathcal{C}(\Sigma)$.

Proof: Applying (C2) to the element 1 we get $1 \le 1^*$, and so $1 = 1^*$. Moreover, from $a \wedge b \le \{a, b\}$ and the monotonicity (C4a) we obtain $(a \wedge b)^* \le \{a^*, b^*\} = \{a, b\}$, where the last equality follows from the closure of a and b. So $(a \wedge b)^*$ is a lower bound of the pair a, b and from this we get that $(a \wedge b)^* \le a \wedge b$; but from (C2) it is $a \wedge b \le (a \wedge b)^*$.

These considerations can be extended to any arbitrary family if Σ is a complete lattice, with the following characterization.

Proposition 8: Let Σ be a complete lattice. Then,

(*i*) Any Tarski closure operator on Σ determines the closure system $\mathcal{C}(\Sigma)$ satisfying the Moore conditions:

(M1) the greatest element is closed: $1 \in \mathcal{C}(\Sigma)$;

(M2) for any collection of closed elements $\{c_\alpha\} \subseteq \mathcal{C}(\Sigma)$ their meet is closed too: $\wedge c_\alpha \in \mathcal{C}(\Sigma)$.

(*ii*) Conversely, every closure system $\mathcal{C}(\Sigma)$ (also Moore family according to [Bir67, pp. 111]) satisfying conditions (M1)–(M2) on the complete lattice Σ determines an operator*: $\Sigma \mapsto \Sigma$ defined as follows:

$$a^* := \wedge \{c \in \mathcal{C}(\Sigma) : a \le c\} \quad \text{... (26)}$$

satisfying properties (C2)–(C4), but in general not the (C1).

Proof: The part (*i*) can be proved with a slight modification of the proof of the previous Proposition 7. For the part (*ii*), if $\mathcal{C}(\Sigma)$ is a Moore family then for any element $a \in \Sigma$ let us consider the lattice meet $a^* := \wedge \{c \in \mathcal{C}(\Sigma) : a \le c\}$, which exists owing to the completeness of the lattice Σ and the fact that the involved family is not empty since $a \le 1$ with $1 \in \mathcal{C}(\Sigma)$. Trivially, $a \le a^*$ since any c of the family is such that $a \le c$, and so the (C2) is true. Applying the just proved (C2) to the element a^* we get $a^* \le (a^*)^*$; moreover, from the particular case $a^* \le a^*$ it follows that $(a^*)^* = \wedge \{d \in \mathcal{C}(\Sigma): a^* \le d\} \le a^*$, i.e. the (C3). Finally, if $a \le b$ then $\{c \in \mathcal{C}(\Sigma): b \le c\} \subseteq \{d \in \mathcal{C}(\Sigma): a \le d\}$, which trivially leads to the monotonicity condition $a^* \le b^*$ equivalent to (C4).

As stated in the previous proposition, a Moore family generates an operator which satisfies condition (C2)–(C4). In general, the normalized condition (C1) does not hold, as can be seen in the following example.

Example 14: Let us consider the set $\left\{0, \frac{1}{k}, \frac{2}{k}, \dots \frac{k-1}{k}, 1\right\}$ with the induced complete lattice structure given by the usual order relation on numbers, previously investigated in Example 1. As seen before, the subset $\left\{\frac{1}{k}, \frac{2}{k}, \dots \frac{k-1}{k}, 1\right\}$ is a Moore family (i.e., closure system), but according to Eq. (26), $0^* = \frac{1}{k} \neq 0$.

Another important notion is the dual (with respect to closure) one of interior lattice.

Definition 12: An interior lattice is a system $\langle \Sigma, \wedge, \vee, o, 0, 1\rangle$ where the mapping $^o: \Sigma \mapsto \Sigma$, that associates to any element a from Σ its interior $a^o \in \Sigma$, is an interior operation, i.e. it satisfies the followings:

(I1)	$1^o = 1$	(*normalized*)
(I2)	$a^o \le a$	(*decreasing*)
(I3)	$a^o = a^{oo}$	(*idempotent*)
(I4)	$(a \wedge b)^o \le a^o \wedge b^o$	(*sub-multiplicative*)

Also in this interior case by duality the following (non{equational and pure poset) result holds.

Lemma 4: In any lattice the sub-multiplicative property (I4) is equivalent to the monotonicity condition:

$$a \le b \textit{ implies } a^o \le b^o$$

Given an interior operator, the subset of *open elements* is defined as the collection of elements which are equal to their interior. Formally,

$$\mathcal{O}(\Sigma) = \{a\,\Sigma : a = a^o\}$$

Also the set $\mathcal{O}(\Sigma)$ is not empty since 1 is open as consequence of (I1); moreover, in a dual way as respect to the proof of condition (M1) in Proposition 7, one has that also 0 is open. Further, in general it does not coincide with the set of all elements Σ, since there may exist elements which are not open. Condition (I3) says that for every $a \in \Sigma$ the element a^o is open, for this reason called in Definition 12 the *interior* of a.

In general, the two subsets of open and closed elements do not coincide, neither one is a subset of the other. Their intersection is the set of *clopen* elements:

$\mathcal{CO}(\Sigma) := \mathcal{C}(\Sigma) \cap \mathcal{O}(\Sigma)$, which is not empty because it contains the least element 0 and the greatest element 1 of the lattice Σ.

The equivalence (isomorphism) between approximation spaces for rough theory (as introduced in definition 1) and lattices with interior and closure is given by the following result.

Theorem 2:

(*i*) Suppose an approximation space $\mathcal{A} = \langle \Sigma, \mathbb{L}(\Sigma), \mathbb{U}(\Sigma) \rangle$ and for arbitrary $a \in \Sigma$ let us define

$$a^o := l(a) \text{ and } a^* := u(a).$$

Then, $\mathcal{A}^\blacktriangle := \langle \Sigma, {}^o, * \rangle$ is a lattice equipped with a Tarski interior operation ^{o} and a Tarski closure operation $*$ such that

$$\mathcal{O}(\Sigma) = \mathbb{L}(\Sigma) \text{ and } \mathcal{C}(\Sigma) = \mathbb{U}(\Sigma)$$

(*ii*) Suppose a lattice equipped with a Tarski interior and a Tarski closure operations $\mathcal{A} = \langle \Sigma, {}^o, * \rangle$ and let us define

$$\mathbb{L}(\Sigma) := \mathcal{O}(\Sigma) \text{ and } \mathbb{U}(\Sigma) := \mathcal{C}(\Sigma).$$

Then, $A^\blacktriangledown := \langle \Sigma, \mathbb{L}(\Sigma), \mathbb{U}(\Sigma) \rangle$ is an approximation space in which for arbitrary $a \in \Sigma$ it is

$$l(a) := a^o \text{ and } u(a) := a^*$$

(*iii*) Let $\mathcal{A} = \langle \Sigma, \mathbb{L}(\Sigma), \mathbb{U}(\Sigma) \rangle$ be an approximation space. Then: $(\mathcal{A}^\blacktriangle)^\blacktriangledown = \mathcal{A}$.

(*iv*) Let $\mathcal{A} = \langle \Sigma, {}^o, * \rangle$ be a lattice equipped with an interior and a closure Tarski operator. Then: $(\mathcal{A}^\blacktriangledown)^\blacktriangle = \mathcal{A}$.

Proof: (*i*)–the upper/closure case (the lower/interior case can be proved in a dual way).

(C1) From (Up1) we have that $0 \leq u(0) = 0^*$; but owing to condition (3) in definition 1 it is $0 \in \mathbb{U}(\Sigma)$, and so by (Up3) applied to $\gamma = 0$ one has that condition $0 \leq 0$ implies $0^* = u(0) \leq 0$, i.e. $0^* = 0$.

(C2) is nothing else that the (Up1) using $a^* = u(a)$.

(C3) Applying (Up1) to a^* we get $a^* \leq (a^*)^*$. On the other hand, from (Up2), it is $a^* = u(a) \in \mathbb{U}(\Sigma)$; so this condition and $a^* \leq a^*$, owing to (Up3), lead to $(a^*)^* = u(a^*) \leq a^*$.

(C4) By (Up1) $a \vee b \leq (a \vee \mathrm{b})^*$, with $(a \vee b)^* \in \mathbb{U}(\Sigma)$ as a consequence of (Up2). Moreover, the order chain $a, b \leq a \vee b \leq (a \vee b)^*$ and (Up3) applied to $\gamma = (a \vee b)^*$ imply $a^*, b^* \leq (a \vee b)^*$, an so for the definition of l.u.b $a^* \vee b^* \leq (a \vee b)^*$.

The identity $\mathcal{C}(\Sigma) = \mathbb{U}(\Sigma)$ is nothing else that Proposition 1.

(*ii*)–Also in this case we prove the closure/upper case since the interior/lower is dual. Then, considering the set $\mathcal{O}(\Sigma)$ as the collection $\mathbb{U}(\Sigma)$ of upper crisp elements and setting $u(a) := a^*$, the (Up1) is a reformulation of (C2). Moreover, the (C3) implies that $a^* \in \mathcal{O}(\Sigma)$, i.e. the (Up2): $u(a)\, \Sigma\, \mathbb{U}(\Sigma)$. Finally, applying the (C4a) to the case $b = \gamma \in \mathcal{O}(\Sigma)$ we get $a \leq \gamma$ implies $a^* \leq \gamma^* = \gamma$, i.e. the (Up3).

Points (*iii*) and (*iv*) are trivial to prove.

In this way we have shown the isomorphism between the structure $\langle \Sigma, \mathbb{L}(\Sigma), \mathbb{U}(\Sigma) \rangle$ of approximation space based on the lattice Σ and satisfying axioms ($A \times 1$) and ($A \times 2$), and the structure $\langle \Sigma, {}^o, * \rangle$ based on the same lattice Σ and equipped with an interior and a closure operation, satisfying conditions (I1)–(I4) and (C1)–(C4) respectively. Clearly, the set of *crisp* elements $\Sigma_c = \mathbb{L}\mathbb{U}(\Sigma)$ of an approximation space coincide with the set of clopen elements $\mathcal{CO}(\Sigma)$ of the Tarski topological approach.

Let us note that as a consequence of the point (*i*) of this theorem, in any approximation space the lower and upper approximation mappings l and u can be considered as Tarski interior and closure operations and so the proved results about these operations can be applied. In particular, the point (M2) of Proposition 7 which lead to the following.

Corollary 3: The sub-poset $\mathbb{L}(\Sigma)$ (resp., $\mathbb{U}(\Sigma)$) of any approximation space $\mathfrak{R}$ is a join (resp., meet) sub semi-lattice of the lattice Σ.

Definition 13: A structure $\langle \Sigma, \wedge, \vee, ', *, 0, 1 \rangle$ is a Tarski closure lattice (instead of a de Morgan lattice with a Tarski closure operation) iff the following hold:

(CdM1) The substructure $\langle \Sigma, \wedge, \vee, ' 0, 1 \rangle$ is a de Morgan lattice.

(CdM2) The mapping $*: \Sigma \to \Sigma$ is a Tarski closure operation that associates with any element a from Σ its closure $a^* \in \Sigma$.

A dual notion is the one of interior de Morgan lattice.

Definition 14: A Tarski interior lattice is a system $\langle \Sigma, \wedge, \vee, ', {}^o, 0, 1 \rangle$ where:

(IdM1) The substructure $\langle \Sigma, \wedge, \vee, ', 0, 1 \rangle$ is a de Morgan lattice.

(IdM2) The mapping $o: \Sigma \to \Sigma$ is a Tarski interior operation that associates with any element a from Σ its interior $a^o \in \Sigma$.

Clearly, the notions of Tarski interior lattice and Tarski closure lattice are dual, in the sense that in any Tarski interior lattice it is possible to define a Tarski closure operator and vice versa obtaining an equivalence between the two structures.

Theorem 3:

(*i*) Suppose a Tarski closure lattice $\mathcal{T} = \langle \Sigma, \wedge, \vee, ', *, 0, 1 \rangle$. Define the map ${}^o: \Sigma \to \Sigma$ by the law

$$\exists a \in \Sigma, a^o := ((a')^*)'$$

Such a map is a Tarski interior operator and $\mathcal{T}^i = \langle \Sigma, \wedge, \vee, ', {}^o, 0, 1 \rangle$ is a Tarski interior lattice.

(*ii*) Suppose a Tarski interior lattice $\mathcal{T} = \langle \Sigma, \wedge, \vee, ', {}^o, 0, 1 \rangle$. Define the map $*: \Sigma \to \Sigma$ by the law

$$\exists a \in \Sigma, a^* := ((a')^o)'$$

Such a map is a Tarski closure operator and $\mathcal{T}^c = \langle \Sigma, \wedge, \vee, ', *, 0, 1 \rangle$ is a Tarski closure lattice.

(*iii*) Let $\mathcal{T} = \langle \Sigma, \wedge, \vee, ', *, 0, 1 \rangle$ be a Tarski closure lattice. Then $(\mathcal{T}^i)^c = \mathcal{T}$.

(*iv*) Let $\mathcal{T} = \langle \Sigma, \wedge, \vee, ', {}^o, 0, 1 \rangle$ be a Tarski interior lattice. Then $(\mathcal{T}^c)^i = \mathcal{T}$.

Proof: (*i*) We have to prove that the map ^{o} satisfies conditions (I1)–(I4) of Definition 12.

(I1) Since $1 = 0'$, we obtain, by definition, $1^{o} = (0')'^{*\prime}$. Thus, by double negation and normalization (C1), we can conclude $1^{o} = 1$.

(I2) By increase property (C2) applied to a' we have $a' \leq a'^{*}$. Thus, by contra-position and double negation: $a^{o} := a'^{*\prime} \leq a'' = a$.

(I3) By definition, $a^{oo} = a'^{*\prime\prime*\prime}$. Thus, by double negation and idempotent property (C3), we obtain $a^{oo} = a'^{*\prime} = a^{o}$.

(I4) Suppose $a \leq b$. By contraposition and monotonicity (C4a), $b'^{*} \leq a'^{*}$. Thus, again by contraposition, $a^{o} := a'^{*\prime} \leq b'^{*\prime} = b^{o}$.

(*ii*) By duality. Finally, (*iii*) and (*iv*) are trivial.

Hence the de Morgan complementation determines a duality relation between the Tarski closure and interior of any element a according to:

$$a^{o} = a'^{*\prime} \text{ and } a^{*} = a'^{o\prime}$$

Such a relationship can be represented by the diagram described in Fig. 6.

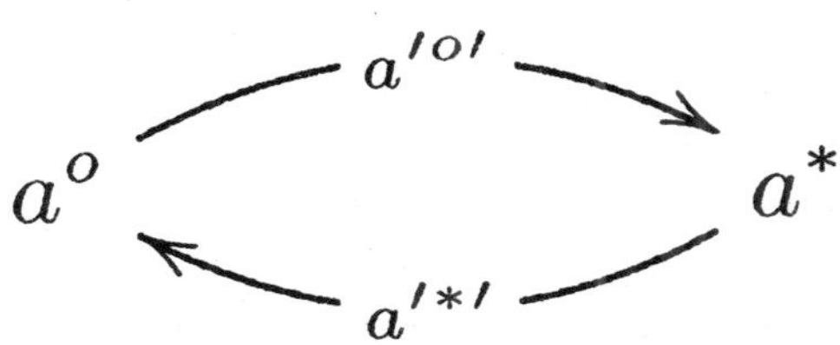

Fig. 6: *de Morgan Duality between Tarski Interior and Closure*

If in the same structure we are able to define both the interior and closure operators it is meaningful and interesting to consider the relation between closed and open elements. It is trivial to prove that an element a is closed iff a' is open, and vice versa an element a is open iff a' is closed. In this way there is a duality between open and closed elements determined by the de Morgan negation:

$$\mathcal{O}(\Sigma) = \{e \in \Sigma : e' \in \mathcal{C}(\Sigma)\} \text{ and } \mathcal{C}(\Sigma) = \{\{f \in \Sigma f' \in \mathcal{O}(\Sigma)\}$$

In this abstract context we are also interested to the so-called *exterior* of an element $a \in \Sigma$ defined as the open element $e(a) := (a^*)' \in \mathcal{O}(\Sigma)$.

Recalling that a rough approximation space $\mathfrak{R} = \langle \Sigma, \mathbb{L}(\Sigma), \mathbb{U}(\Sigma), l, u \rangle$ can be considered (see Section 2) as the "pasting" of *a lower approximation space* $\mathfrak{LR} := \langle \Sigma, \mathbb{L}(\Sigma), l \rangle$ and an *upper approximation space* $\mathfrak{UR} := \langle \Sigma, \mathbb{U}(\Sigma), u \rangle$, we restate theorem 2 in the following more precise way.

1. Any lower approximation space $\langle \Sigma, \mathbb{L}(\Sigma), l \rangle$ satisfying $(A \times 1)$ is equivalent to a Tarski interior space $\langle \Sigma, o \rangle$ satisfying conditions (I1)–(I4).
2. Any upper approximation space $\langle \Sigma, \mathbb{U}(\Sigma), u \rangle$ satisfying $(A \times 2)$ is equivalent to a Tarski closure space $\langle \Sigma, * \rangle$ satisfying conditions (C1)–(C4).

Some remarks on the equivalence between rough approximation spaces and Tarski closure (interior) lattices.

(*i*) The non-equational equivalent conditions expressed in Definition 1 capture the intuitive aspects of an expected approximation space. Indeed, as stressed, the semantical content of conditions (In1)–(In3) (resp., (Up1)–(Up3)) is that the lower (resp., upper) approximations $l(a)$ (resp., $u(a)$) is the best approximation of the approximable element a from the bottom (resp., top) by lower (resp., upper) crisp elements, that in the Tarski topological approach are just open (resp., closed) elements.

(*ii*) Theorem 2 can be summarized by the statement that the category of de Morgan lattice with closure (and induced interior) operator and the one of de Morgan lattice with upper (and induced lower) approximation map are equivalent (isomorphic) between them, according to the general category theory [Lan71].

Indeed, as seen in points (*iii*) and (*iv*) of Theorem 2, starting with a Tarski closure lattice, forming its upper approximation space version and then coming back to the corresponding Tarski closure structure, one recovers the original starting structure. Vice versa, starting from an upper approximation space, forming its Tarski closure lattice version and then coming back to the corresponding upper approximation space structure, one recovers the original starting structure.

(*iii*) This means that if one works in the context of a generalization of the closure operator conditions, for instance weakening (or suppressing) some of the above conditions (C1)–(C4) of Definition 11, then some of the conditions expressed by (Up1)–(Up3) are correspondingly lost, or modified. A choice which, as stressed before, needs a justification from the interpretative point of view.

In Section 4.2 we have seen that given a covering space (X, γ) it is possible to construct the structure $\mathfrak{R}_\gamma = \langle \mathcal{P}(X), \mathcal{L}_\gamma(X), \mathcal{U}_\gamma(X), l_\gamma, u_\gamma \rangle$ which is a concrete model of approximation space. Making use of Theorems 2, 3 and the specific properties of the involved models, one obtains the following.

Proposition 9: Let (X, γ) be a covering space. Then, on the Boolean lattice $\langle \mathcal{P}(X), \cap, \cup, ^c, \emptyset, X \rangle$, one has that

(Cov1) the mapping* : $\mathcal{P}(X) \mapsto \mathcal{P}(X)$ associating with any subset A of X its upper pseudo-closed approximation

$$A^* = u_\gamma(A) = \cap \{U \in \mathcal{U}_\gamma(X) : A \subseteq U\}$$

is a Tarski closure operation.

(Cov1) the mapping ^{o}: $\mathcal{P}(X) \mapsto \mathcal{P}(X)$ associating with any subset A of X its lower pseudo-open approximation

$$A^o = l_\gamma(A) = \cup \{L \in \mathcal{L}_\gamma(X) : L \subseteq A\}$$

is a Tarski interior operation.

Example 15: In the covering $\gamma = \{\{1, 6\}, \{2, 3\}. \{1, 2, 3, 4\}, \{3, 4, 5, 6\}\}$ of the universe $X = \{1, 2, 3, 4, 5, 6\}$ discussed in Example 9, the two subsets $A = \{4\}$ and $B = \{6\}$ are such that the strict inclusion $\{4\}^* \cup \{6\}^* = \{4, 5, 6\} \subset \{1, 4, 5, 6\} = \{4, 6\}^*$ holds. So it is a concrete Tarski closure lattice.

Similarly, the two sets $C = \{1, 2, 3, 5, 6\}$ and $D = \{1, 2, 3, 4, 5\}$ are such that the corresponding interiors satisfy the strict inclusion $(C \cap D)^o = \{1, 2, 3\} \subset \{1, 2, 3, 4\} = C^o \cap D^o$, i.e. against to what asserted by Eq. (6) in [Zak83] this is an example of a Tarski interior generated by a covering which is not topological, in the sense that the stronger conditions $(a \wedge b)^* = a^* \vee b^*$ and $(a \wedge b)^o = a^o \wedge b^o$ which, respectively, characterize the Kuratowski topological interior and closure with respect to the Tarski ones, do not hold.

5.1 The Lattice-Algebraic Semantic of S4-Like Modal Logic Induced by Tarski Closure

In this section we want to analyze in the Tarski lattice context the *algebraic semantic* of some modal-logic system. We denote by $\neg a$ the de Morgan complement a' and by $i(a)$ (resp., $u(a)$) the Tarski interior a^o (resp., closure a^*) of a in order to capture in a *algebraic model* the formalism of usual propositional logic of a modal language based on the negation Not $\neg$, and on the standard connectives AND $\wedge$ and OR $\vee$. The lower (interior) l is interpreted as a *necessity* operator and the upper (closure) u as a *possibility* operator (L and M in a standard modal language $\mathcal{L}$ whose sentences are denoted by Greek letters α, β, γ, ...).

Let us discuss now how the Tarski interior (and dually the closure) operator satisfies the algebraic versions of axioms and rules of some well known modal principles. As usual "when interpreting [complemented lattice] structures from the viewpoint of logic, it is customary to interpret the lattice elements [a, b, ...] as *propositions*, the meet operation [$\wedge$] as *conjunction* [and], the join operation [$\vee$] as *disjunction* [or], and the orthocomplement [$\neg$] as *negation* [not]" [Har81]. To this purpose, let us quote [Che88, pp. 212] "intuitively, the points in Σ may be thought of as propositions–including 'truth' and 'falsity' (1 and 0)–closed under propositional analogues of negation, conjunction, disjunction, and necessitation ($\neg$, $\wedge$, $\vee$, and l). [...] It should be emphasized in this connection that the points in an algebraic model are not possible worlds, nor even, necessarily, sets of possible worlds."

In this lattice context "it is generally agreed that the partial ordering $\leq$ of a lattice [...] can be logically interpreted as a *relation* of implication, more specifically *semantic entailment* (see van Frassen [vF71]). [...] The *implication relation*, which is fundamental to logic, must not however be confused with an implication or conditional *operation*, which is a logical *connective* which, like conjunction and disjunction, forms propositions out of propositions remaining at the same linguistic level" [Har79]. In this context, the partial order relation $a \leq b$ in Σ is rather the algebraic counterpart of the statement "$\alpha \supset \beta$

is true" with respect to some implication connective $\supset$ involving sentences α and β of the object language $\mathcal{L}$. In this sense the *implication relation* $a \leq b$ in the lattice Σ corresponds to *semantic entailment* among formulas from the language $\mathcal{L}$.

Hence, the statement "α semantically entails β" is not a formula in the object language, but occurs in the *metalanguage*.

Thus, in order to express the lattice conditions in a formal way nearer to the usual logical formalism, it is required a binary operation $\rightarrow$ on the lattice Σ playing the role of algebraic counterpart of the logical implication connective $\supset$ in the language $\mathcal{L}$, and which assigns to each pair of lattice elements $a, b \in \Sigma$ another lattice element $a \rightarrow b \in \Sigma$. "Since not just any binary lattice operation should qualify as a material implication, we must determine what criteria must (should) be satisfied by a lattice operation in order to be regarded as a material implication. First, it seems plausible to require that every implication operation $\rightarrow$ be related to the implication relation ($\leq$) in such a way that if a proposition a implies [is less or equal to] a proposition b, then the conditional proposition $a \rightarrow b$ is universally true, and conversely. [...] Translating this into the general lattice context, we obtain

$$a \rightarrow b = 1 \quad \text{iff} \quad a \leq b. \qquad \text{... (27)}$$

Here 1 is the lattice unit element, which corresponds to the *universally true proposition*" [Har81]. In this just quoted paper, this condition is assumed as one of the *minimal implicative conditions*.

Let us enter in some formal detail supposing that in our Tarski lattice structure there exists an implication operator $\rightarrow$ on Σ for which at least the minimal condition (27) is required to hold. Let us define as a *truth functional tautology* (also *universally valid sentence*) any well formed formula based on a Tarski interior lattice which is equal to the universally true element 1. It is now easy to prove that the l and u, interpreted as necessity and possibility operators, satisfy the algebraic versions of axioms and rules of some celebrated modal principles.

First of all, from the logic-algebraic semantic point of view the Tarski interior lattice rules can be interpreted as follows:

(mod – 1n)

$$l(1) = 1.$$

That is: if a sentence is true, then also its necessity is true (necessitation rule or according to [Che88, pp. 20] the *N* principle).

(mod–2n)

$$l(a) \leq a.$$

In other words: necessity implies actuality (the characteristic *T* principle whatever is necessary is so: $L(\alpha) \supset \alpha$; in other words, if necessarily α, then α [Che88, pp. 6]).

(mod–3n)

$$a \leq b \text{ implies } l(a) \leq l(b)$$

if a conditional and its antecedent are both necessary, then so is the consequent (which is the characteristic *K* principle of modal logic *L* $(\alpha \supset \beta) \supset (L(\alpha) \supset L(\beta))$, see [Che88, pp. 7]).

(mod–4n)

$$l(a) = l(l(a)).$$

Whatever is necessary is necessarily necessary (the characteristic 4–principle $L(\alpha) \leftrightarrow LL(\alpha)$, see [Che88, pp. 18]).

(mod–5n)

$$u(a) = \neg(l(\neg a))$$

which "embodies the idea that what is possible is just what is not–necessarily–not. [...] possibility is always expressible in terms of necessity and negation, and so is theoretically superflous" (the DF ◊ principle $M(\alpha) \leftrightarrow \neg L \neg(\alpha)$, see [Che88, pp. 7]).

Let us enter in some details.

(*N*) The above algebraic version of the *N* principle can assume also the form "$a = 1$ implies $l(a) = 1$" corresponding at the language level $\mathcal{L}$ to the *necessitation transformation rule* of [HC84, pp. 4] "$\vdash \alpha$ implies $\vdash L(\alpha)$", i.e. if α is true so is $L(\alpha)$. In other words, according to [Che88, pp. 7], "this corresponds to a rule of inference (*RN*, the rule of necessitation). It states that the necessitation of a valid sentence is itself always valid." In symbols: $\dfrac{\alpha}{L(\alpha)}$.

(*T*) According to the above minimal requirement (27) about an implication connective one has that (mod-2n) can be expressed as $i(a) \to a = 1$, the algebraic version of the *T* modal tautology $L(\alpha) \supset \alpha$.

(*K*) As a consequence of "$a \leq b$ implies $l(a) \leq l(b)$", under the axiom (mod-2n) we now prove that "if the conditional and its antecedent are both necessarily true, then the consequent is necessarily true" [Che88, pp. 7], i.e. formally, that

$$l(a \to b) = 1 \text{ and } l(a) = 1 \text{ imply } l(b) = 1. \qquad \text{... (28)}$$

Indeed, both the two conditions $l(a \to b) = 1$ and the (mod-2n), in the form $l(a \to b) \leq a \to b$, imply $a \to b = 1$, from which $a \leq b$ follows by (27). So, if also $l(a) = 1$ (i.e., if the conditional and its antecedent are both necessarily true), then from the (mod-2n) principle of the necessity operator applied to $a \leq b$ one obtains that also $l(b) = 1$ (i.e., then the consequent is necessarily true).

Let us show that property (28) is equivalent to

$$l(a \to b) \to (l(a) \to l(b)) = 1. \qquad \text{... (29)}$$

Let the (28) be true. Under the hypothesis $l(a \to b) = 1$, i.e. $a \leq b$, the fact that "$l(a) = 1$ implies $l(b) = 1$" leads to the identity $l(a) = l(b) = 1$, written in particular as $l(a) \leq l(b)$, which by the (27) means that $l(a) \to l(b) = 1$. Therefore, both conditions $l(a \to b) = 1$ and $l(a) = 1$ imply that also $l(a) \to l(b) = 1$, and so under these conditions one gets $l(a \to b) = l(a) \to l(b) = 1$, which in particular can be written as $l(a \to b) \leq l(a) \to l(b)$, i.e. $l(a \to b) \to (l(a) \to l(b)) = 1$. Conversely, let $l(a \to b) \to (l(a) \to l(b)) = 1$, i.e. $l(a \to b) \leq (l(a) \to l(b))$. If $l(a \to b) = 1$, then also $l(a) \to l(b) = 1$, i.e. $l(a) \leq l(b)$, and so condition $l(a) = 1$ implies $l(b) = 1$.

Finally, let us prove that (29) implies condition (mod-3n). Under the hypothesis $a \leq b$, i.e. $a \to b = 1$, the Eq. (29) (taking into account (mod-1n)) assumes the form $1 \to (l(a) \to l(b)) = 1$, i.e. $1 \leq (l(a) \to l(b))$ which leads to $(l(a) \to l(b)) = 1$, i.e. $l(a) \leq l(b)$.

Thus, we have proved the equivalence of (mod-3n) with (29), the algebraic version of the characteristic *K* principle formalized in a modal language $\mathcal{L}$ as the tautology $L(\alpha \supset \beta) \supset (L(\alpha) \supset L(\beta))$.

(5) Is nothing else than the point (*ii*) of Theorem 3.

A particular consequence of lemma 4 is that the monotonicity condition of the interior is equivalent to the sub-multiplicative rule $l(a \wedge b) \leq l(a) \wedge l(b)$, which according to (27) can be written as the "tautology" $l(a \wedge b) \to l(a) \wedge l(b) = 1$, algebraic version of the modal principle M expressed as $L(\alpha \wedge \beta) \rightarrow (L\alpha \wedge L\beta)$, see [Che88, pp. 20]). That is the algebraic version of the modal principle K is equivalent to the algebraic version of the modal principle M:

$$l(a \wedge b) \leq l(a) \wedge l(b) \qquad M \text{ principle}$$

In general the following property of modality is ***not*** valid:

$$l(a) \wedge l(b) \leq l(a \wedge b) \qquad C \text{ principle}$$

as shown in the following example.

Example 16: Let us consider the Tarski distributive lattice given in the Hasse diagram of Fig. 7. Then, we have that $l(b \wedge c) = l(a) = 0 < a = l(b) \wedge l(c)$.

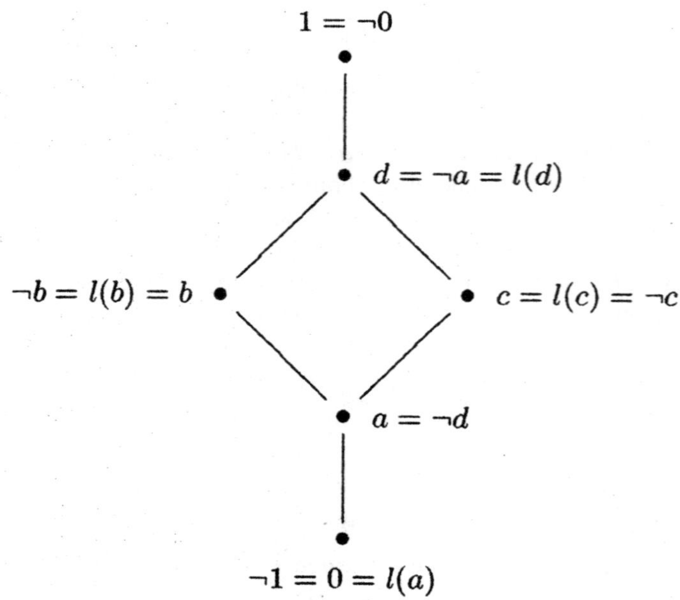

Fig. 7.

These results can be compared with the axiomatization of system $S4$ in [Che88], where the underlying basic structure of any algebraic model is the Boolean one. But what we have proved here is that, in order to obtain the algebraic versions of characteristic T, K, 4, and $Df\Diamond$ axioms plus the N rule of inference, it is sufficient to consider a de Morgan lattice. To be precise we can state the following.

(AM_T) Let $\langle \Sigma, \wedge, \vee, \neg, l, 0, 1 \rangle$ be a Tarski interior lattice. Then this structure is a modal algebra of a S4 – like logical system with respect to the necessity connective l, where the term "like" stresses that in general it is based on a de Morgan lattice instead of on a Boolean algebra, as assumed in any modal algebra (see [Che88, pp. 212]). Precisely,

(a) the modal principles N, T, K, 4, and $DF\ \Diamond$ are satisfied, and so also the equivalent to K modal principle M, but not the "dual" modal principle C (on the contrary accepted in the standard $S4$ Lewis modal system).

(b) From $DF\ \Diamond$ and T it follows the further property $l(a) \leq u(a)$, i.e. according to (27) that $l(a) \to u(a) = 1$, algebraic version of the modal principle D that "whatever is necessary is possible" $L(\alpha) \supset M(\alpha)$, see [Che88, pp. 16]).

So, summarizing, the in general non Boolean structure of the lattice and the non validity of principle C are the main differences between the standard $S4$ modal system and our $S4$ – like Tarski modal algebra

In general, besides the refused modal principle C, also the following properties of modalities are not valid:

$a \leq l(u(a)$	B principle
$l(a) = u(l(a)),\ u(a) = l(u(a))$	5 principle

This is shown by the following example.

Example 17: Let us consider the Tarski non-distributive (and so non – Boolean, but Kleene) lattice given in the Hasse diagram of Fig. 8.

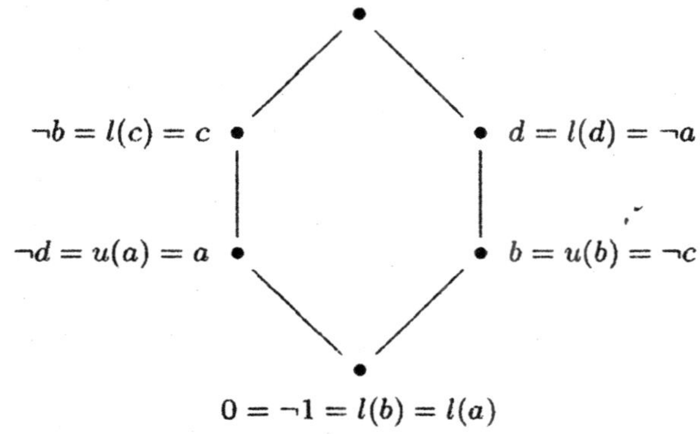

Fig. 8

Then, we have that b is a closed element, i.e. $b = u(b)$ and $l(u(b)) = l(b) = 0$. So, $u(b) \neq l(u(b))$ and $l(u(b)) < b$. Further, c is an open element, i.e. $l(c) = c$ and $u(l(c)) = 1$. So $l(c) \neq u(l(c))$. On the contrary, in this lattice the C principle holds.

6. A HIERARCHY OF CLOSURE OPERATORS

Once stated the categorical equivalence between closure lattices and upper approximation spaces, with the minimal intuitive interpretation of these latter, starting from the original Tarski closure version given in Definition 11 it is possible to introduce a hierarchy of closure operators, each of which is a stronger version of the previous one. As it happens in many situations, the equivalence between two categories of structures does not prevent the deep investigation of one of them, formally more suitable to perform an accurate analysis of the involved description.

This is just the case we are investigating with. The notion of approximation spaces introduced in Definition 1 gives as an intuitive (and adherent to reality) description of a procedure in which generic objects (the approximable elements) can be approximated by the knowledge furnishes from other well known (in the sense "given a priori") categories of objects, precisely lower and upper crisp elements. The equational formalization of the Tarski closure (and, in a de Morgan lattice context, induced interior) operation gives the hint to explore some stronger axiomatic versions of the condition (C1)–(C4) (dually (I1)–(I4)) in order to fit the richness of situations which can be described by them. This is the object of the further investigation we perform now.

6.1 The Kuratowski Closure on de Morgan Lattices

Let us start with a closure operation which is stronger than the above introduced Tarski one.

Definition 15: A closure operation on a de Morgan lattice is said to be topological iff the sub–additive condition (C4) is substituted by the stronger additive property:

(C4T) $a^* \vee b^* = (a \vee b)^*$

About this argument we can cite from Rasiowa–Sikorski [RS70]: "The closure operation * satisfies the conditions (C1)–(C3) and

(C4T). These Axioms are due to Kuratowski [Kur22]. For a detailed exposition of the theory of topological spaces see, e.g. Kelley [Kel55], Kuratowski [Kur33, Kur61]." In the introduction of his paper, Nakamura [Nak41] says: "The concept of closure was axiomatized by F. Riesz and Kuratowski [Kur33] on the field of sets, and Terasaka [Ter37] generalized it onto abstract Boolean algebras. The object of this note is to extend it onto general lattices. Incidentally, 'combinations of topologies' of G. Birkhoff [Bir36] are teated from more general point of view." Quoting M. Ward [War42]: "These axioms are satisfied by Kuratowski's closure operator over a Boolean algebra with points [Kur33]." For a more recent treatment see also the Birkhoff book [Bir67, pp. 116]. Finally, according to Halmos "a closure operator is a normalized [(C1)], increasing [(C2)], idempotent [(C3)], and additive [(C4T)] mapping [...]. The first systematic investigation of the algebraic properties of closure operators was carried out by McKinsey and Tarski [MT44]." In agreement with all these quotations, in the sequel we shall call Kuratowski closure operation any topological closure on a de Morgan lattice. A de Morgan lattice equipped with a Kuratowski closure operations is called a Kuratowski closure lattice.

Note that under conditions (C2) and (C3) the additive property (C4T) is equivalent to the two conditions:

(C4a) $a \leq b$ implies $a^* \leq b^*$ (monotonicity).

(C4b) $a = a^*$ and $b = b^*$ imply $(a \vee b)^* = a \vee b$ (topological closure).

Given a Kuratowski (i.e., topological) closure operator, the *Kuratowski* (i.e. topological) *interior operator* formally defined by duality according to point (*i*) of Theorem 3 satisfies the multiplicative property:

(I4T) $a^o \wedge b^o = (a \wedge b)^o$.

Moreover, under conditions (I2)–(I3) the multiplicative property (I4) is equivalent to the two conditions:

(I4a) $a \leq b$ implies $a^o \leq b^o$ (monotonicity).

(I4b) $a = a^o$ and $b = b^o$ imply $(a \wedge b)^o = a \wedge b$ (topological interior).

Extending to the present Kuratowski case the results of Proposition 9 which say that covering spaces, with their lower and upper approximation maps, are models of Tarski interior and closure operations, and taking into consideration the terminology introduced in Section 4.3, it is possible to state the following.

Proposition 10: Let $\mathcal{B}$ be an open base for a topology on X. Then, on the Boolean lattice $\langle \mathcal{P}(X), \cap, \cup, {}^c, \emptyset, X\rangle$, one has that

(Top1) the mapping $*: \mathcal{P}(X) \mapsto \mathcal{P}(X)$ associating with any subset A of X its upper approximation

$$A^* = \cap\{C \in \mathcal{C}(X) : A \subseteq C\}$$

is a Kuratowski closure operation.

(Top1) the mapping ${}^o: \mathcal{P}(X) \mapsto \mathcal{P}(X)$ associating with any subset A of X its lower approximation

$$A^o = \cup\{O \in \mathcal{O}(X) : O \subseteq A\}$$

is a Kuratowski interior operation.

So topological spaces are models of abstract Kuratowski lattices.

With respect to the algebraic semantic of modal logic based on a Tarski lattice (AMT) discussed in Section 5.1, in the present Kuratowski case the multiplicative property (I4T) says that besides the M (or equivalently K) principle, also the C principle Holds. So we can assert the following.

(AMK) Let $\langle \Sigma, \wedge, \vee, \neg, l, 0, 1\rangle$ be a Kuratowski interior lattice. Then this structure is a modal algebra of a $S4$–quasi logical system with respect to the necessity connective l, where the term "quasi" stresses that in general it is based on a de Morgan lattice instead of on a Boolean algebra, but for which the following principles of the "real" $S4$ system are verified:

(*a*) all the modal principles N, T, M, C, 4, and $DF\Diamond$ are satisfied, and so also the equivalent to M modal principle K (according to the standard $S4$ Lewis modal system).

(*b*) From $DF\Diamond$ and T it follows the further algebraic version of the modal principle D: $l(a) \leq u(a)$, i.e. $l(a) \to u(a) = 1$.

6.2 The Halmos Closure in a de Morgan Lattice

Now, we investigate a further strengthening of the Kuratowski closure operator.

Definition 16: *A* Halmos closure operation on a de Morgan lattice Σ is a mapping $*: \Sigma \mapsto \Sigma$ which satisfies the following conditions for arbitrary $a, b \in \Sigma$:

(C1) $0^* = 0$ (normalized)

(C2) $a \leq a^*$ (increasing)

(sC3) $a^{*\prime *} = a^{*\prime}$ (interconnection)

(C4T) $a^* \vee b^* = (a \vee b)^*$ (additive)

A Halmos closure lattice *is any de Morgan lattice equipped with a Halmos closure operation.*

Remark 5: Making use of the interior operator, the condition (sC3) in the form $a^* = (a^*)'^{*\prime}$, can be equivalently formalized as

(sC3a) $a^* = a^{*\,o}$

This definition has been introduced by Halmos in [Hal56] (collected in [Hal62]) on the base of a Boolean algebra structure as the algebraic formalization of the *existential quantifier*, where it is pointed out that "The concept of existential quantifier occurs implicitly in a brief announcement of some related work of Tarski and Thompson [TT52]". In the context of the abstract approach to roughness theory we prefer to generalize this structure to the more general case of de Morgan lattices, keeping the terminology of closure instead of the one of quantifier.

The following result assures that the Halmos closure is a Kuratowski closure too.

Proposition 11: Let Σ be a de Morgan lattice with an operation $*: \Sigma \mapsto \Sigma$ which satisfies conditions $(C1)$, $(C2)$, and $(C4T)$. Then the following implication holds:

$$\text{(sC3)} \quad \forall a \in \Sigma, a^{*\prime *} = a^{*\prime} \text{ implies } (C3)\ \forall a \in \Sigma, a^* = a^{**}$$

In other words, any Halmos closure operation is a Kuratowski closure operation.

If as usual we denote by $\mathcal{C}(\Sigma)$ (resp., $\mathcal{O}(\Sigma)$) the collection of all closed (resp., open) elements, condition (sC3) can also be expressed saying that for any $a \in \Sigma$ the element $a^{*\prime}$ 0 is closed. The following property characterizes the category of de Morgan lattices with Halmos closure relatively to the category of de Morgan lattices with Kuratowski closure.

Proposition 12: Let Σ be a de Morgan lattice with a topological (i.e., Kuratowski) closure operation. Then the following are equivalent:

1. the topological closure operation satisfies condition (sC3), i.e. it is Halmos;
2. the collection of closed elements and the collection of open elements coincide:

$$\mathcal{C}(\Sigma) = \mathcal{O}(\Sigma).$$

Taking into account the point (2) of Theorem 1 in Section 4.4, i.e. that partition approximation spaces are characterized by the fact that the collections of lower and of upper crisp sets coincide, this proposition leads to the following improvement of the Proposition 10.

Proposition 13: Let (X, π) be a partition space with $\mathcal{E}_\pi(X)$ the (unique) collection of lower crisp and upper crisp subsets of X. Then, on the Boolean lattice $\langle \mathcal{P}(X), \cap, \cup, {}^c, \emptyset, X\rangle$, one has that

(Par1) the mapping $* : \mathcal{P}(X) \mapsto \mathcal{P}(X)$ associating with any subset A of X its upper approximation

$$A^* = \cap \{C \in \mathcal{E}_\pi(X) : A \subseteq C\}$$

is a Halmos closure operation.

(Par1) the mapping ${}^o : \mathcal{P}(X) \mapsto \mathcal{P}(X)$ associating with any subset A of X its lower approximation

$$A^o = \cup \{O \in \mathcal{E}_\pi(X) : O \subseteq A\}$$

is a Halmos interior operation.

So partition spaces are models of abstract Halmos lattices.

With respect to the algebraic semantic of modal logic based the present Halmos case is a model of a S5 modal logic.

(AMP) Let $\langle \Sigma, \wedge, \vee, \neg, I, 0, 1\rangle$ be a Halmos interior lattice. Then this structure is a modal algebra of a $S5-$quasi logical system with respect to the necessity connective ***i***, where as usual the term "quasi" stresses that in general it is based on a de Morgan lattice instead of on a Boolean algebra,

(a) all the modal principles N, T, M (equivalently K), C, 5, and $DF\Diamond$ are satisfied; let us stress that from these principle it is possible to deduce principles $S4$ and B (i.e., $a \leq l\ (u\ (a))$) of modal logic.

(b) From $DF\Diamond$ and T it follows the further property $l\ (a) \leq \boldsymbol{u}\ (a)$, i.e. $l\ (a) \rightarrow \boldsymbol{u}\ (a) = 1$, algebraic version of the modal principle D that "whatever is necessary is possible".

7. CONCLUSIONS

A collection of results about closure and approximation spaces in the abstract context of lattice structures has been given, with complete proofs. In particular, the categorical isomorphism between Tarski closure and Abstract Rough Approximation Spaces is analyzed in detail. We underline that the former one has an equational definition whereas the last one does not, but it grasps the intuitive aspects of a rough approximation.

As concrete models of the abstract lattice approach, we analyzed a hierarchy of situations based on the Boolean lattice $\mathcal{P}\ (X)$, power set of a universe X: partition spaces, topological spaces by open basis and covering spaces. These are all models of de Morgan lattices with Tarski closure. However, stronger structures are needed to capture the different behavior of these models. In particular, different notions of closure in the lattice context has been considered and besides the Tarski closure, we have introduced the Kuratowski and the Halmos ones, whose concrete models are just the before mentioned covering, topological and partition spaces on X, respectively.

All the obtained results can be summarized in the Table 3, in which all the specific models are based on the power set $\mathcal{P}\ (X)$ of a universe X and the realization of an element a of the abstract lattice Σ is represented by a subset A of the concrete lattice $\mathcal{P}\ (X)$.

Table 3: *Summary of Closure Lattices with Corresponding Models on the Power Set P (X) of the Universe X*

Abstract lattice Σ	$\xrightarrow{Realization}$	Concrete model on $\mathcal{P}(X)$
Tarski closure lattice	$\xrightarrow{Realization}$	Covering space
$\Uparrow$		$\Uparrow$
Kuratowski closure lattice	$\xrightarrow{Realization}$	Topological space
$\Uparrow$		$\Uparrow$
Halmos closure lattice	$\xrightarrow{Realization}$	Partition space

REFERENCES

[BB57] A. Bialynicki-Birula, "Remarks on Quasi–Boolean Algebras", *Bull. Acad. Pol. Sci.*, Cl III, **5**, (1957), 615–619.

[BBR57] A. Bialynicki-Birula and H. Rasiowa, "On the Representation of Quasi–Boolean Algebras", *Bull. Acad. Pol. Sci.*, Cl III, **5**, (1957), 259–261.

[Bir36] G. Birkhoff, "On the Combinations of Topologies", *Fundamenta Mathematicae*, **26**, (1936), 156–166.

[Bir67] ...., "Lattice Theory", third ed., *American Mathematical Society Colloquium Publication*, **XXV**, American Mathematical Society, Providence, Rhode Island, 1967, (first edition 1940, second (revisited) edition 1948).

[Cat98] G. Cattaneo, "Abstract Approximation Spaces for Rough Theories", *Rough Sets in Knowledge Discovery 1* (L. Polkowski and A. Skowron, eds.), Physica–Verlag, Heidelberg, New York, 1998, pp. 59–98.

[CC04] G. Cattaneo and D. Ciucci, "Investigation About Time Monotonicity of Similarity and Preclusive Rough Approximations in Incomplete Information Systems", *Lecture Notes in Artificial Intelligence*, **3066**, Springer, 2004, pp. 38–48.

[CC06a] ...., "Some Methodological Remarks About Categorical Equivalence in the Abstract Approach to Roughness. Part I", *Lecture Notes in Artificial Intelligence*, **4062**, Springer, 2006, pp. 277–283.

[CC06b] ...., "Some Methodological Remarks About Categorical Equivalence in the Abstract Approach to Roughness. Part II", *Lecture Notes in Artificial Intelligence*, **4062**, Springer, 2006. pp. 284–289.

[CC08] ...., "A Hierarchical Lattice Closure Approach to Abstract Approximation Spaces". *Lecture Notes in Artificial Intelligence*, **5009**, 2008, pp. 363–370.

[Che88] B.F. Chellas, "Modal Logic, an Introduction", Cambridge University Press, Cambridge, MA, 1988.

[Cig65] R. Cignoli, "Boolean Elements in Łukasiewicz Algebras". *I, Proceedings of the Japan Academy*, **41**, (1965), 670–675.

[Cig75] ...., "Injective de Morgan and Kleene Algebras", *Proceedings of the American Mathematical Society*, 47, (1975), 269–278.

[Ciu08] D. Ciucci, "A Unifying Abstract Approach for Rough Models", *Lecture Notes in Artificial Intelligence*, **5009**, 2008, pp. 371–378.

[CM88] G. Cattaneo and G. Marino, "Non-usual Orthocomplementations on Partially Ordered Sets and Fuzziness", *Fuzzy Sets and Systems*, **25**, (1988), 107–123.

[Coh65] P.M. Cohn, "Universal Algebra". *Harper and Row*, 1965.

[Dun86] J.M. Dunn, "Relevance Logic and Entailment". *Handbook of Philosophical Logic*, (D. Gabbay and F. Guenther, eds.), **3**, Kluwer, 1986, pp. 117–224.

[Eve44] C.J. Everett, "Closure Operators and Galois Theory in Lattices", *Transaction of the American Mathematical Society*. **55**, (1944), 514–525.

[Hal56] P.R. Halmos, "The Basic Concepts of Algebraic Logic", *American Mathematical Monthly*, **53**, (1956), 363–387.

[Hal62] ...., "Algebraic Logic", *Chelsea Pub. Co.*, New York, 1962.

[Har79] G.M. Hardegree, "The Conditional in Abstract and Concrete Quantum Logic", *Logico - Algebraic Approach to Quantum Mechanics II*, (C. A. Hooker, ed.), D. Reidel, Dordrecht, Holland, 1979, pp. 49–108.

[Har81] ...., "Material Implication in Orthomodular (and Boolean) Lattices", *Notre Dame Journal of Modal Logic*, **22**, (1981), 163–182.

[HC84] G.E. Hughes and M.J. Cresswell, "A Companion to Modal Logic", Methuen, London, 1984.

[Kal58] J.A. Kalman, "Lattices with Involution", *Transactions of the American Mathematica Society*, **87**, (1958), 485–491.

[Kel55] J.L. Kelley, "General Topology", Springer - Verlag, New York, 1955, (second edition 1957 by Van Nostrand).

[Kri63] S.A. Kripke. "Semantical Analysis of Modal Logic I. Normal Modes Propositional Calculi". *Zeitschrift für Mathematische Logik und Grundlagen der Mathematik*, **9**, (1963), 67–96.

[Kry98] M. Kryszkiewicz, "Rough Set Approach to Incomplete Information Systems". *Information Sciences*, **112**, (1998), 39–49.

[Kur22] C. Kuratowski, "Sur l'opération $\bar{A}$ de l'analysis Situs", *Fundamenta Mathematicae*, **3**, (1922), 182–199.

[Kur33] ...., "Topologie I", *Monografie Matematyczne*, Warszawa, 1933, (Other editions 1948 and 1958).

[Kur61] ...., "Topologie II". Monografie Matematyczne, Warszawa, 1961.

[Lan71] S. MacLane, "Categories for the Working Mathematicians", *Graduate Text in Mathematics*, **5**, Springer–Verlag, Berlin, Heidelberg, 1971.

[Moi35] G.C. Moisil, "Recherches sur l'algebres de la Logiques", *Annales Sc. Univ. Jassy*, **22**, (1935), 1–117.

[Moo10] E.H. Moore, "Introduction to a Form of General Analysis", AMS Colloq. Publ., **2**, *American Mathematical Society*, Providence, Rhode Island, 1910.

[MR42] A. Monteiro and H. Ribeiro, "L'operation de Fermeture et ses Invariants dans les Systemes Partiellement Ordennes", *Portugaliae Mathematica*, **3**, (1942), 171–184.

[MT44] J.C.C. McKinsey and A. Tarski, "The Algebra of Topology", *Annals of Mathematics*, **45**, (1944), 141–191.

[Nak41] M. Nakamura, "Closure in General Lattices", *Proceedings of the Imperial Academy*, **XVII**, (1941), 5-6.

[Ore42] O. Ore, "Combinations of Closure Relations", *Annals of Mathematics*, **44**, (1942), 514–533.

[Ore44] ...., "Galois Connexions", *Transactions of the American Mathematical Society*, **55**, (1944), 493–513.

[Paw81] Z. Pawlak, "Information Systems–Theoretical Foundations", *Information Systems*, **6**, (1981), 205–218.

[Paw82] ...., "Rough Sets", *Int. J. of Computer and Information Sciences*, **11**, (1982), 341–356.

[Paw92] ...., "Rough Sets: A New Approach to Vagueness", *Fuzzy Logic for the Management of Uncertainty*, (L.A. Zadeh and J. Kacprzyc, eds.), J. Wiley and Sons, New York, 1992, pp. 105–118.

[Poi93] H. Poincaré, "Le Continu Mathématique", *Revue de Métaphysique et de Morale I*, (1893), 26–34, (Reprinted in [Poi03] as Chapter II).

[Poi03] ...., "La Science et l'hypothµese", *Flammarion*, Paris, 1903, (English translation as Science and Hypothesis, Dover, New York, 1952).

[Pol02] L. Polkowski, "Rough Sets Mathematical Foundations", Physica Verlag, Heidelberg, 2002.

[Pom87] J.A. Pomykala, "Aproximation Operations in Approximation Space", *Bulletin of the Polish Academy of Sciences – Mathematics*, **35**, (1987), 653 – 662.

[PSZ94] L. Polkowski, A. Skowron, and J. Zytkow, "Tolerance Based Rough Sets", *Third International Workshop on Rough Sets and Soft Computing* (University of San Jose, California) (T.Y. Lin and A.M. Wildberger, eds.), 1994, pp. 55 – 58.

[RS70] H. Rasiowa and R. Sikorski, The Mathematics of Metamathematics", third ed., *Monografie Matematyczne*, **41**, Polish Scientific Publishers, Warszawa, 1970.

[Sim63] G.F. Simmons, "Topology and Modern Analysis", McGraw-Hill Book Company, Inc., NY, 1963.

[ST99] J. Stefanowki and A. Tsoukiμas, "On the Extension of Rough Sets Under Incomplete Information", New Directions in Rough Sets, Data Mining, and Granular-Soft Computing, 7th International Workshop, RSFDGrC '99, Yamaguchi, Japan (Heidelberg) (Setsuo Ohsuga Ning Zhong, Andrzej Skowron, ed.), *Lecture Notes in Computer Science*, **1711**, Springer, 1999, pp. 73 – 81.

[Tar30] A. Tarski, "Fundamentale Begriffe der Methodologie der Deduktiven Wissennschaften I", *Monathshefte fur Mathematik und Physik*, **37**, (1930), 361 – 404, (English version in [Tar83]).

[Tar83] .... "Logic, Semantics, Metamathematics", Hackett, Indianapolis, 1983, (Second Edition – First Edition, 1956 by Oxford).

[Ter37] Terasaka, "Theorie der Topologischen Verbände", *Proc.*, 13, (1937).

[TT52] A. Tarski and F.B. Thompson, "Some General Properties of Cylindric Algebras", *Bulletin of the American Mathematical Society*, **58**, (1952), 65, (Abstracts of the Meeting in Pasadena, December 1, 1951).

[vF71] B.C. van Frassen, "Formal Semantic and Logic", MacMillan, New York, 1971.

[War42] M. Ward, "The Closure Operator on Lattice", *Annals of Mathematics*, **43**, (1942), 191 – 196.

[Wiw89] A. Wiweger, "On Topological Rough Sets", *Bullettin of the Polish Academy of Sciences – Mathematics*, **37**, (1989), 89 – 93.

[WS89] U. Wybraniec-Skardowska, "On a Generalization of Approximation Space", *Bulletin of the Polish Academy of Sciences – Mathematics*, **37** (1989), 51 – 61.

[Yao96] Y.Y. Yao, "Two Views of the Theory of Rough Sets in Finite Universes, *International Journal of Approximate Reasoning*, **15**, (1996), 291 – 317.

[Yao98a] ...., "Constructive and Algebraic Methods of the Theory of Rough Sets", *Journal of Information Sciences*, **109**, (1998), 21–47.

[Yao98b] ...., "On Generalizing Pawlak Approximation Operators", *Lecture Notes in Artificial Intelligence*", **1424**, Springer-Verlag, 1998, pp. 298–307.

[Yao98c] ...., "Relational Interpretations of Neighborhood Operators and Rough Set Approximation Operators", *Information Sciences*, **111**, (1998), 239–259.

[Yao07] ...., A Note on Definability and Approximations", *Lecture Notes in Computer Science: Transactions on Rough Sets VII*, **4400**, Springer-Verlag, 2007, pp. 274–282.

[Zak83] W. Zakowski, "Approximations in the Space (U, π)", *Demonstration Mathematica*, **XVI**, (1983), 761–769.

[Zee62] E.C. Zeeman, "The Topology of the Brain and Visual Perception, Topology of 3-manifolds and Related Topics, Prentice-Hall, Englewood Cliffs, 1962, pp. 240–256.

[Zia01] W. Ziarko, "Rough Sets: Trends, Challenges, and Prospects". 2001, pp. 1–7.

3

TOWARDS DESCRIBING RIVER WATER QUALITY STRAIGHTWAY IN LINGUISTIC TERMS–IS IT ACHIEVABLE?–A PARADIGM SHIFT

Ashok Deshpande

Berkeley Initiative in Soft Computing (BISC)-Special Interest Group (SIG)-Environment Management Systems (EMS), and Bioinformatics Center, University of Pune, India

E-mail: ashok_deshpande@hotmail.com

Abstract: *One of the most disturbing prospects facing water quality scientists is how to turn often very complex water quality data into information which is understandable and usable by non-scientists, managers, and public at large. The present practice of communicating the status of river water quality to policy makers and the end users is with Water Quality Index (WQI). In spite of the development of numerous water quality indices over the last four decades, there has been no universal acceptance.*

What do we understand by the term water quality, and more particularly excellent river water quality for bathing? Different people mean differently as water quality has no precise meaning-it is fuzzy term! Why should water quality index value to be worked out firstly on an interval scale, and then describe its meaning in linguistic terms? Is it not possible to straightway describe river water quality in linguistic terms with some degree of certainty attached to each linguistic term, and is it achievable? The answer is: yes. A paradigm shift in describing fuzzy terms (such as water quality, environmental quality, and alike) is the departure from the conventional approach. In this sequel, we have also compared the results obtained using WQI approach with fuzzy logic based formalism.

Keywords: *Water quality index, dissolved oxygen, Bootstrapping, fuzzy logic, degree of match.*

1. INTRODUCTION

Water quality index is in essence, a function used to simplify large quantities of data into a more useful form in order to convey an image of overall quantity to a variety of different users. The most critical deficiency of the selected indices is the lack of dealing with uncertainty and subjectivity present in this complex environmental problem (Chang et al., 2001; Mpimpas et al., 2001; Silvert, 2000).

What do we understand by the term *water quality*, and more particularly *excellent river water quality for bathing*? Different people mean differently as *water quality* has no precise meaning-it is fuzzy term! Why should water quality index value to be worked out firstly on an interval scale, and then describe its meaning in linguistic terms? Is it not possible to straightway describe river water quality in linguistic terms with some degree of certainty attached to each linguistic term, and is it achievable? The answer is: yes. A paradigm shift in describing fuzzy terms (such as water quality, environmental quality, and alike) is the departure from the conventional approach. In this sequel, fuzzy logic based formalism for straightway describing water quality in linguistic terms with some degree of certainty for the specific usage is presented, and its utility over the conventional water quality indexing system is demonstrated.

2. WATER QUALITY CLASSIFICATION

The efforts on the development of Water Quality Indices (WQI) started in 1990s and are considered a subset of indices used to measure environmental quality on an interval scale. Some of wisely used indexing systems are discussed in brief.

2.1 Conventional Methods

In general five different types of water quality indices have been considered for the calculation and comparison of the water quality. However, the two commonly used water quality indices are considered in the present study. The first WQI is an index proposed by Horton hereafter called as the arithmetic water quality index (WQI_A), which is denoted as

$$WQI_A = \sum_{i=1}^{n} w_i q_i \qquad \text{... (1)}$$

Where n is the number of variables, w_i is the relative weight of the ith parameter such that $\sum_{i=1}^{n} w_i = 1$ and q_i is the quantity rating of ith parameters. The second WQI is a multiplicative form of index proposed by Brown et. al. Multiplicative water quality index (MQI_M) defined as followed has been used

$$WQI_M = \prod_{i=1}^{n} q_i^{W_i} \qquad \text{... (2)}$$

WQI is classified in linguistic terms as: (0-25)-Not acceptable; (26-50)-Just acceptable; (51-75) and Acceptable; (76-100)-Highly acceptable.

2.2 Fuzzy Logic Based Method

In spite the development of numerous water quality indices over the last three decades, there has been no universal acceptance. There exists aleatory uncertainty due to randomness in water quality parametric data, and epistemic uncertainty in the perception of experts about rating these parameters (partial belief). The problem of water quality classification can, therefore, be approached using fuzzy logic based approach. *In this sequel, we have followed the conceptual framework of the model presented by* Deshpande, in his paper (1994) and made an attempt to compare the conventional and fuzzy logic based approach in a case study. It is important to understand some of the important terms used in the paper.

2.2.1 Matching Between Fuzzy Values

The fuzzy number for field data (A) on parameters and the fuzzy number characterizing linguistic terms (A') are matched together to arrive a measure called a Degree of Match (DM) defined by

$$DM\,(A, A') = \frac{\int \mu_{A \cap A'}(x)\,dx}{\int \mu_A(x)\,dx}, x \in X \qquad \text{... (3)}$$

In which X denotes the universe, and $\mu_{A \cap A'}(x)$ is membership grade for $A \cap A'$. The degree of match between A and A' present in Fig. 1 shows the extent to the assertion is included in the antecedent part of rule. Furthermore, if A and A' is the discrete possibility distribution the measure is defined as:

$$DM(A, A') = \frac{\sum \mu_{A \cap A'}(x)}{\sum \mu_A(x)}, x \in X \qquad \dots (4)$$

In the present study, we consider relevant water quality parameters. One such parameter in the context of bathing in river water is Most Probable Number (MPN expressed as MPN/100 ml of water sample).

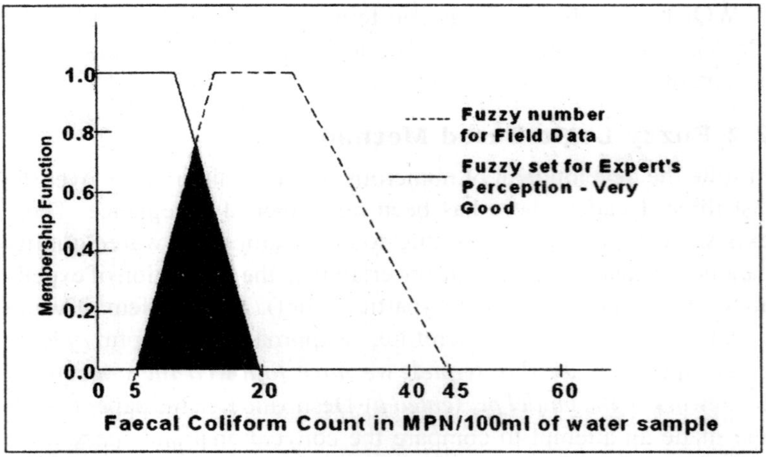

Fig. 1: *Fuzzy Matching Between A and A'*

2.2.2 Fuzzy Rule Based System

A set of rule is constructed for classifying water quality as highly acceptable, just acceptable, not acceptable (rejected) in order to aggregate the set of attributes. Each rule has antecedent propositions connected together using AND operator, resulting in some consequences. The assertions related to its antecedent part are obtained from the users, which are imprecise or fuzzy. Therefore, a fuzzy rule based system can be developed for the knowledge representation or reasoning process. Here the partial matching is allowed and the analyst can estimate the extent to which the assertion satisfies the antecedent part of the rule contrary to the rule-base system which examines as to whether the antecedent part is satisfied or not

(Lee, 1992). Figure 2 presents a hierarchical structure for water classification in a set rule.

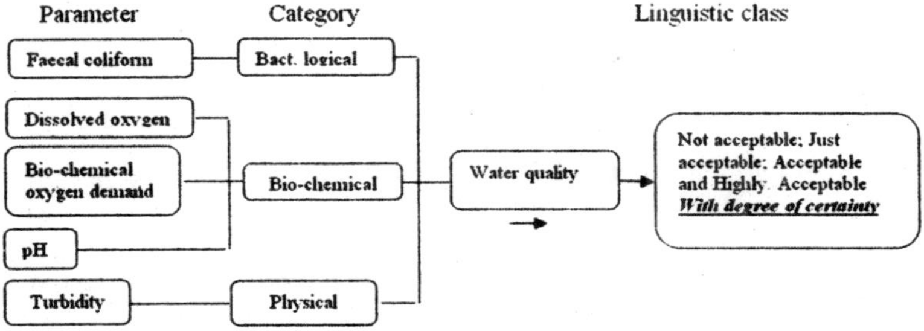

Fig. 2: *Hierarchical Structure for Water Quality Classification*

The chemical status of water is judged in the first hierarchical level of knowledge. While the second level characterizes bacteriological, chemical and physical status of water to arrive at the ultimate acceptable strategy of water quality for bathing purpose. Following are examples rules stored at two different hierarchical levels of the knowledge base:

If **DO** is *<fair>* and **BOD** is *<good>* and **pH** is *<very good>* Then chemical status of water is *<good>*. The rule at the next level could be If **Bacteriological** status of water is *<fair>* and **Chemical** status of water is *<good>* and **Physical** status of water is *<fair>* Then water quality for bathing is *<just acceptable>* with certain degree of certainty.

3. RELEVANCE OF BOOTSTRAP

It was observed that high pollution level persists in Indrayani River in different seasons, and there was no significant variation in the concentration of the pollution indicating parameters. Collection of several samples for the analysis in all the seasons would have been a drain on the financial resources. It was, therefore, considered appropriate to use bootstrapping procedure to arrive at the conclusion on the average parametric concentration.

As an allusion to lifting oneself up by one's own bootstraps, the term ***bootstrapping*** means using a special process to perform a task that one would be unable to do in general. Furthermore, Bootstrap sampling procedures has adopted for resampling by increasing sample size up to (n = 1000) with replacement procedure without any loss of generosity of mother sample. The boot estimation was done by using the formula;

$$\bar{x} = \frac{1}{n}\sum_{i=1}^{n} Xi \quad \text{for the bootstrap mean} \qquad \text{... (5)}$$

$$\bar{X}_b = \frac{1}{n}\sum_{i=1}^{n} X(i), \qquad \text{... (6)}$$

All parametric mean values obtained from bootstrap technique have been used for computation of convention water quality index approach and fuzzy logic based formalism. It could be stated that bootstrapping has increased the data precision without loss of the generosity of the mother sample. It could be argued that the relevance of this re-sampling technique will be helpful for solving environmental issues based on very limited data sets.

4. CASE STUDY

Using WQI formalism, it can be stated that the WQI values of both the sampling locations (S_1 and S_2) are almost same revealing same pollution status of the river at both locations, and linguistically described as ***fair*** category. The computational details of WQIs at both locations in Indrayani River are presented in the Table 1.

Table 1: *Water Quality Index Using Bootstrap Mean Values*

Parameters	*Bootstrap Mean*		*Weight (w_i)*	*WQI at Station (S_1)*		*WQI at Station (S_2)*	
	S_1	S_2		Q_1	Q_1^{W1}	Q_2	Q_2W_2
FC (MPH/100 ml)	742	709	0.26	22	2.23	25	2.31
DO (mg/l)	4	4.29	0.21	30	2.04	32	2.07
BOD (mg/l)	15.2	14.15	0.14	18	1.50	20	1.52
pH	7.56	7.66	0.18	95	2.27	95	2.27
Turbidity (ppm)	16.16	16.54	0.11	65	1.58	63	1.58
Temperature (deg. C)	19.9	19.9	0.10	5	1.17	5	1.17
$\prod_{i=1}^{n} Qi^{wi}$			1		28.63		30.49

Limited water samples were collected in different seasons and the boostrap mean values were worked out (Table 1). Indrayani river water quality for bathing remains almost the same in rainy season and fair seasons and could be linguistically described as ***fair for bathing with no degree of certainty attached to the description.*** It is very hard to communicate such information to the public or end users with confidence.

The problem of river water quality classification was, therefore, approached using combination of degree of match (DM) and the fuzzy rule-based system (Deshpande et. al., 1994). In a fuzzy rule based system, the experts represent their knowledge concerning the classification of the object (*water quality*) in the form of rules. Each rule has a set of antecedent propositions comprising to parameters names for example: pH, Dissolved Oxygen, etc. and attributes values or linguistic description like *very good, good, etc.* These linguistic descriptions are invariably fuzzy or vague keeping in view the inadequate information on the health implications of each parameter on the users and the integrated effect of all the parameters on human health. In order to describe the aggregated effect of water quality a hierarchical structure for the overall water quality classification straightway in linguistic terms is developed.

Table 2: *Degree of Match with Fuzzy Values for Each Linguistic Class*

Sampling Stations	*Parameters*	*Linguistic Class*			
		Very Good	*Good*	*Fair*	*Poor*
S_1	FC	0	0	0	1
	DO	0.2	1	0.98	0.1
	BOD	0	0	0	1
	pH	1	0.28	0	0
	Turbidity	0.02	1	0.03	0
S_2	FC	0	0	0	1
	DO	0.6	1	0.8	0
	BOD	0	0	0	1
	pH	1	0.7	0.001	0
	Turbidity	0.016	1	0.33	0

Table 2 presents the degree of match for each parameter for both the sampling stations (ghats). The dominating parameters namely, FC and BOD exhibit *poor* linguistic class description with degree of certainty 1. But the linguistic value of parameters pH and DO is in class *very good* and *good* meaning thereby that these parameters might not be responsible for the deterioration of for Indrayani River water quality for bathing purpose.

Table 3: *Fuzzy Description of River Water Quality for Bathing with Degree of Certainty*

Water Quality Description	*Degree of Certainty*
Highly acceptable/very good	0
Acceptable/good	0
Just acceptable/fair	0
Not acceptable/poor	1

Table 3 presents fuzzy description of river water quality for bathing with degree of certainty. The parameters FC and BOD contribute towards worsening water quality status for bathing in Indrayani. The water quality is *not acceptable/poor* for bathing with degree of certainty 1.00 at two ghats of the Indrayani River, near Pune, India (Table 3).

The calculated values of QWI_M are 28.63 and 30.49 at the stations S_1 and S_2 in fare season, respectively (Table 1) indicating water quality as *fair for bathing* at these two stations. The results obtained from two different approaches are contradictory. More importantly, the computational procedure for indexing system does not attach any degree of certainty to the statement *water quality is fair for the specific usage-bathing*. Fuzzy logic based approach, straightway describes water quality at the ghat as poor or rejected for bathing with degree of certainty 1.00.

5. CONCLUDING REMARKS

The fuzzy logic based describes water quality straightway in linguistic terms with degree of certainty which, in turn, would help the decision makers. This is what is expected of modelling exercises and this is how human brain works. In the real world, no one works out first a number, and then say that the water quality is *good /bad* for bathing. Fuzzy logic based method with a case study delineated in the paper is the complete departure from indexing system, and have received severe criticism in the past. There is an urgent need to exploit the potential of fuzzy logic based approaches in many areas of *S* and *T* in general and environmental management system in particular. *The biggest room in this world is the room for improvement*– so we would continue to carry out further research.

REFERENCES

[1] APHA (1992), "Standard Methods for the Examination of Water and Waste Water 18th Edition". *American Public Health Association, American Water Works Association, Water Environment Federation*, pp. 949.

[2] Wiener V., Cyrus D.P., Vermeulen L.A., and Wade P., (2006). "Development of a Water Quality Index for Estuarine Water Quality Management in South Africa". *WRC Report No. 1163/1/06. www.Wrc.org.za/download.*

[3] Chang N., Chen H.W., and Ning K.S., (2001). "Identification of River Water Quality Using the Fuzzy Synthetic Evaluation Approach". *Journal of environmental management*, **63**, 293–305.

[4] Silvert W. (2000). "Fuzzy Indices of Environmental Conditions". *Ecol Model*, **130**, 111–119.

[5] Deshpande A.W., Raje D.V., and Khanna P., (1994). "Fuzzy Description of River Water Quality", *Paper for International conference-EUFIT*.

[6] Efron B., 1981. "Censored Data and Bootstrap". *Journal of American Statistical Association*, **76**, 312–319.

[7] Miller R., and Turkey J. (1974). "The Jackknife–A Review". *Biometrika*, **61**, 1–15.

[8] Rijal K., Deshpande A., and Ghole V., (2008). "Bathing in Polluted Rivers, Water Borne Diseases, and Fuzzy Measure: A Case Study in India". *International journal of environment and Waste Management (IJEWM)*. In press-manuscript number-2809.

[9] Ross T.J., (2004). "Fuzzy Logic with Engineering Application". John Wiley and Sons.

4

MEAN-VARIANCE-SKEWNESS PORTFOLIO SELECTION WITH FUZZY INTERVAL COEFFICIENTS

Rupak Bhattacharyya[1], **Samarjit Kar**[2] **and Dwijesh Dutta Majumder**[3]

[1, 2]Department of Mathematics, National Institute of Technology, Durgapur-713209, India
E-mail: mathsrup@gmail.com[1], kar_s_k@yahoo.com[2],

[3]Electronics and Communication Science Unit, Indian Statistical Institute, Kolkata 108, India
E-mail: ddmdr@hotmail.com[3]

Abstract: *Portfolio returns are in general asymmetric in nature. So in addition to the general concept of considering mean and variance, the third moment skewness should be considered for a better decision making in portfolio selection problem. In a fuzzy environment, the expected return, risk and skewness can not be predicted accurately. Based on this fact, a fuzzy model under the assumption that the expected return, risk and skewness are interval numbers is proposed. Three mathematical models have been proposed by defining the future financial market optimistically, pessimistically and in the combined form. Finally, numerical examples extracted from Indian Stock Market are used to illustrate the method.*

Keywords: *Fuzzy interval number, Portfolio selection problem, Transaction cost, Mean-varianceskewness model.*

1. INTRODUCTION

Most of the reasonable works on portfolio selection has been done based on only the first two moments of return distribution. The first order moment about the origin, i.e. the mean, quantify the return and the second order moment about the mean, i.e. the variance, quantify the risk. Now the third order moment about the mean of a return distribution i.e. skewness measures the asymmetry of the distribution. There will be three goals, maximizing the mean and the skewness and minimizing the variance. One interested in considering skewness prefers a portfolio with a higher probability of large payoffs when mean and variance remain same. Consideration of skewness in portfolio selection was started by Lai (1991) and then continued by Konno et. al. (1993, 1995), Chunhachinda et. al. (1997), Liu et. al. (2003), Briec et. al. (2007), etc.

In the decision making field an approach, based on interval analysis, allows handling imprecise input data. This approach consists in assuming that the data of a decision making problem are not well defined but may vary in given interval. Hence not only vagueness (lack of sharp class boundaries), but also a feature of uncertainty (lack of information) can be addressed intuitively. Lai et. al. (2002) extends the Markowitz's model to an interval programming model by quantifying the expected return and the covariance as intervals. Ida (2003) solved a multi-objective portfolio selection problem with interval coefficients. Giove et. al. (2006) formulate a minimax regret portfolio selection problem in which the prices of the securities are considered as interval variables. Though a considerable number of research papers have been published for portfolio selection problem in fuzzy environment, none has considered interval valued mean-variance- skewness model for fuzzy portfolio selection.

In this paper, fuzzy expected return, covariance and central co-moment are considered as interval numbers. Mean, variance and skewness in fuzzy intervals are considered to model portfolio selection model in Section 2. Three different models are constructed here for fuzzy portfolio optimization. In Section 3 share price data from Indian stock market (BSE index) are used to illustrate the effectiveness of the algorithm and finally in Section 4 some conclusions are specified.

2. INTERVAL MEAN-VARIANCE-SKEWNESS MODEL

Let us consider a financial market with n risky assets offering random rates of return. An investor allocates his capital among the risky assets.

Let

x_i = the proportion invested in the risky asset i;

r_i = the random rate of return of the risky asset i;

R_i = $E(r_i)$, the expected rate of return of the risky asset i;

σ_{ij} = cov (r_i, r_j), the covariance between r_i and r_j and $(\sigma_{ij})_{n \times n}$ is semi-positive definite;

c_i = the transaction cost of the risky asset i;

k_i = the constant cost per change in a proportion of the risky asset i, $k_i \geq 0$; $i, j = 1, 2, 3, \ldots, n$.

The total transaction cost between a given portfolio $x^0 = (x_1^0, x_2^0, \ldots, x_n^0)$ and a new portfolio $x = (x_1, x_2, \ldots, x_n)$ of risky asset i can be expressed as

$$\sum_{i=1}^{n} c_i = \sum_{i=1}^{n} k_i \left| x_i - x_i^0 \right|.$$

The expected return, variance and skewness of portfolio $x = (x_1, x_2, \ldots, x_n)$ respectively are

$$R(x) = \sum_{i=1}^{n} R_i x_i - \sum_{i=1}^{n} k_i \left| x_i - x_i^0 \right|,$$

$$\sigma^2(x) = \sum_{i=1}^{n} \sum_{j=1}^{n} \sigma_{ij} x_i x_j,$$

$$S(x) = \frac{\sum_{i=1}^{n} \sum_{j=1}^{n} \sum_{k=1}^{n} E\left(R_i R_j R_k\right) x_i x_j x_k}{\left(\sum_{i=1}^{n} \sum_{j=1}^{n} \sigma_{ij} x_i x_j \right)^{3/2}}$$

For a new investor it can be assumed that $x_i^0 = 0$, $i = 1, 2, \ldots, n$.

Since in a fuzzy environment, the expected return, risk and skewness can not be predicted accurately, so the expected return, risk and skewness can be considered as interval numbers. Let the fuzzy expected return, covariance and third order co-moment are denoted by the following interval numbers:

$$\tilde{R}_i = [R_i - \delta_{il}, R_i + \delta_{ir}],$$
$$\tilde{\sigma}_{ij} = [\sigma_{ij} - \delta_{ijl}, \sigma_{ij} + \delta_{ijr}],$$
$$\tilde{\psi}_{ijk} = [E(R_i R_j R_k) - \delta_{ijkl}, E(R_i R_j R_k) + \delta_{ijkr}].$$

Then the fuzzy expected return, risk and skewness are respectively defined by

$$\tilde{R}(x) = \sum_{i=1}^{n} \tilde{R}_i x_i - \sum_{i=1}^{n} k_i \left| x_i - x_i^0 \right|,$$

$$\tilde{\sigma}^2(x) = \sum_{i=1}^{n} \sum_{j=1}^{n} \tilde{\sigma}_{ij} x_i x_j,$$

$$\tilde{S}(x) = \frac{\sum_{i=1}^{n} \sum_{j=1}^{n} \sum_{k=1}^{n} \tilde{\psi}_{ijk} x_i x_j x_k}{\left(\sum_{i=1}^{n} \sum_{j=1}^{n} \tilde{\sigma}_{ij} x_i x_j \right)^{3/2}}$$

Since $x_i \geq 0$, we have fuzzy numbers

$$\tilde{R}(x) = [R(x) - \delta_{RL}(x), R(x) + \delta_{RR}(x)],$$
$$\tilde{\sigma}^2(x) = [\sigma^2(x) - \delta_{VL}(x), \sigma^2(x) + \delta_{VR}(x)],$$
$$\tilde{S}(x) = [S(x) - \delta_{SL}(x), S(x) + \delta_{SR}(x)].$$

Here

$$R(x) - \delta_{RL}(x) = \sum_{i=1}^{n} (R_i - \delta_{il}) x_i - \sum_{i=1}^{n} k_i \left| x_i - x_i^0 \right|,$$

$$R(x) + \delta_{RR}(x) = \sum_{i=1}^{n} (R_i + \delta_{ir}) x_i - \sum_{i=1}^{n} k_i \left| x_i - x_i^0 \right|,$$

$$\sigma^2(x) - \delta_{VL}(x) = \sum_{i=1}^{n} \sum_{j=1}^{n} (\sigma_{ij} - \delta_{ijl}) x_i x_j,$$

$$\sigma^2(x) - \delta_{VR}(x) = \sum_{i=1}^{n} \sum_{j=1}^{n} (\sigma_{ij} + \delta_{ijr}) x_i x_j,$$

$$S(x) - \delta_{SL}(x) = \frac{\sum_{i=1}^{n} \sum_{j=1}^{n} \sum_{k=1}^{n} \left(E(R_i R_j R_k) - \delta_{ijkl} \right) x_i x_j x_k}{\left(\sum_{i=1}^{n} \sum_{j=1}^{n} (\sigma_{ij} - \delta_{ijl}) x_i x_j \right)^{3/2}},$$

$$S(x) + \delta_{SR}(x) = \frac{\sum_{i=1}^{n} \sum_{j=1}^{n} \sum_{k=1}^{n} \left(E(R_i R_j R_k) + \delta_{ijkr} \right) x_i x_j x_k}{\left(\sum_{i=1}^{n} \sum_{j=1}^{n} (\sigma_{ij} - \delta_{ijr}) x_i x_j \right)^{3/2}},$$

Corresponding to the optimization problem, the fuzzy optimization investment problem becomes

$$Min\left(\alpha \cdot \sigma^2(x) - \beta \cdot \tilde{R}(x) - \gamma \cdot \tilde{S}(x)\right)$$

subject to

$$\sum_{i=1}^{n} x_i = 1$$

$$x_i \geq 0,\ i = 1, 2, \ldots, n.$$

Denote

$$F_l(x) = \alpha \cdot (\sigma^2(x) - \delta_{VL}(x)) - \beta \cdot (R(x) + \delta_{RR}(x)) - \gamma \cdot (S(x) + \delta_{SR}(x))$$

$$F_r(x) = \alpha \cdot (\sigma^2(x) + \delta_{VR}(x)) - \beta \cdot (R(x) - \delta_{RL}(x)) - \gamma \cdot (S(x) - \delta_{SL}(x))$$

$$F(x) = \alpha \cdot \sigma^2(x) - \beta \cdot R(x) - \gamma \cdot S(x)$$

Then clearly,

$$\alpha \cdot \tilde{\sigma}^2(x) - \beta \cdot \tilde{R}(x) - \gamma \cdot \tilde{S}(x) = \left[F_l(x), F_r(x)\right].$$

From the above, let us construct the following three models:

Model I

$$\min F_l(x)$$

subject to

$$\sum_{i=1}^{n} x_i = 1$$

$$x_i \geq 0,\ i = 1, 2, \ldots, n.$$

From the objective function $F_l(x)$, we can observe that the investor estimates the return, skewness and risk optimistically, and aims to optimize the total objective in this case.

Model II

$$\min F_r(x)$$

subject to

$$\sum_{i=1}^{n} x_i = 1$$

$$x_i \geq 0,\ i = 1, 2, \ldots, n.$$

In contrast to Model I, the investor estimates the return, skewness and risk of risky assets pessimistically in Model II, and the investor aims to optimize the total objective based on his assumption.

Model III

$$\min F(x) = \lambda \cdot F_r(x) + (1 - \lambda) \cdot F_l(x)$$

subject to

$$\sum_{i=1}^{n} x_i = 1$$

$$x_i \geq 0, i = 1, 2, ..., n, \lambda \in (0, 1).$$

This model covers the scenario where the investor makes his portfolio selection neither too optimistically nor too pessimistically; to some extend optimistic; but to some extend he is pessimistic, i.e., this model covers a balanced one.

The investor selects a model on the basis of different estimations of return, risk and skewness.

3. CASE STUDY FROM BSE

The Here in the portfolio selection process, initially monthly share price data for 60 months of 30 companies, which are included in BSE (Bombay Stock Exchange) index, are taken. In order to shortlist 5 companies among them, the average monthly return of the 30 companies are calculated. On the basis of return generated by them, 5 best performing companies are short listed here.

The vector of their expected rates of return is

$$(R_1, R_2, R_3, R_4, R_5) = (0.032306, 0.048434, 0.043003, 0.030623, 0.034704).$$

The covariance matrix of the return rates of these risky assets are

$$\begin{bmatrix} 0.02379169 & 0.0113415 & 0.00856463 & 0.00718686 & 0.00626257 \\ 0.0113415 & 0.01838454 & 0.00793816 & 0.00746377 & 0.00539194 \\ 0.00856463 & 0.00793816 & 0.01709313 & 0.00736477 & 0.00497023 \\ 0.00718686 & 0.00746377 & 0.00736477 & 0.01832598 & 0.01841618 \\ 0.00626257 & 0.00539194 & 0.00497023 & 0.01841618 & 0.01261857 \end{bmatrix}$$

The central co-moments are computed as

	R_1	R_2	R_3	R_4	R_5
R_1R_1	0.2058	0.0655	0.0159	0.0199	−0.0245
R_2R_2	0.0902	−0.029	0.0254	0.0051	−0.0128
R_2R_3	0.0068	0.0254	0.0187	−0.0418	−0.0527
R_2R_4	−0.0326	0.0051	−0.0418	−0.0261	−0.0279
R_2R_5	−0.0585	−0.0128	−0.0527	−0.0279	−0.0381
R_3R_3	−0.0168	0.0187	−0.1088	0.0053	−0.0042
R_3R_4	−0.0361	−0.0418	0.0053	−0.0261	−0.0379
R_3R_5	−0.0643	−0.0527	−0.0042	−0.0379	−0.0416
R_4R4	−0.0045	−0.0261	−0.0261	−0.0309	−0.0159
R_4R_5	−0.0354	−0.0279	−0.0379	−0.0159	−0.0204
R_5R_5	−0.0204	−0.0381	−0.0416	−0.0204	−0.0168

Now we consider the fuzzy case that the return, risk and skewness are given in the form of interval numbers as follows:

$$(\widetilde{R}_1, \widetilde{R}_2, \widetilde{R}_3, \widetilde{R}_4, \widetilde{R}_5) = ([0.0323, 0.0324], [0.0484, 0.0485], [0.0430, 0.0431], [0.0306, 0.0307], [0.0347, 0.0348]),$$

$$\begin{bmatrix} [0.02379, 0.02380] & [0.01134, 0.01135] & [0.00856, 0.00857] & [0.00718, 0.00719] & [0.00626, 0.00627] \\ [0.01134, 0.01135] & [0.01838, 0.01839] & [0.00793, 0.00794] & [0.00746, 0.00747] & [0.00539, 0.00540] \\ [0.00856, 0.00857] & [0.00793, 0.00794] & [0.01709, 0.01710] & [0.00736, 0.00737] & [0.00497, 0.00498] \\ [0.00718, 0.00719] & [0.00746, 0.00747] & [0.00736, 0.00737] & [0.01832, 0.01833] & [0.01841, 0.01841] \\ [0.00626, 0.00627] & [0.00539, 0.00540] & [0.00497, 0.00498] & [0.01841, 0.01841] & [0.01261, 0.01262] \end{bmatrix}$$

	$\widetilde{R}_1$	$\widetilde{R}_2$	$\widetilde{R}_3$	$\widetilde{R}_4$	$\widetilde{R}_5$
$\widetilde{R}_1\widetilde{R}_1$	[0.205, 0.206]	[0.065, 0.066]	[0.015, 0.016]	[0.019, 0.020]	[−0.025, −0.24]
$\widetilde{R}_2\widetilde{R}_2$	[0.090, 0.091]	[−0.030, −0.029]	[0.025, 0.026]	[0.005, 0.006]	[−0.013, −0.012]
$\widetilde{R}_2\widetilde{R}_3$	[0.006, 0.007]	[0.025, 0.026]	[0.018, 0.019]	[−0.042, −0.041]	[−0.053, −0.052]
$\widetilde{R}_2\widetilde{R}4$	[−0.033, −0.032]	[0.005, 0.006]	[−0.042, −0.041]	[−0.027, −0.026]	[−0.028, −0.027]
$\widetilde{R}_2\widetilde{R}_5$	[−0.059, −0.058]	[−0.013, −0.012]	[−0.053, −0.052]	[−0.028, −0.027]	[−0.039, −0.038]

$\tilde{R}_3\tilde{R}_3$	[−0.017, −0.016]	[0.018, 0.019]	[−0.109, −0.108]	[0.005, 0.006]	[−0.005, −0.004]
$\tilde{R}_3\tilde{R}_4$	[−0.037, −0.036]	[−0.042, −0.041]	[0.005, 0.006]	[−0.027, −0.026]	[−0.038, −0.037]
$\tilde{R}_3\tilde{R}_5$	[−0.065, −0.064]	[−0.053, −0.052]	[−0.005, −0.004]	[−0.038, −0.037]	[−0.042, −0.041]
$\tilde{R}_4\tilde{R}4$	[−0.005, −0.004]	[−0.027, −0.026]	[−0.027, −0.026]	[−0.031, −0.030]	[−0.016, −0.015]
$\tilde{R}_4\tilde{R}_5$	[−0.036, −0.035]	[−0.028, −0.027]	[−0.038, −0.037]	[−0.016, −0.015]	[−0.021, −0.020]
$\tilde{R}_5\tilde{R}5$	[−0.021, −0.020]	[−0.039, −0.038]	[−0.042, −0.041]	[−0.021, −0.020]	[−0.017, −0.016]

respectively.

For the above sets of data Model I take the form:

$$\begin{aligned}
\text{Min } F_1(x) = {} & \alpha\cdot(0.02379x_1^2 + 0.01838x_2^2 + 0.01709x_3^2 + \\
& 0.01832x_4^2 + 0.01261x_5^2 + 0.02268x_1x_2 + \\
& 0.01612x_1x_3 + 0.01436x_1x_4 + 0.01252x_1x_5 + \\
& 0.01586x_2x_3 + 0.01492x_2x_4 + 0.01078x_2x_5 + \\
& 0.01472x_3x_4 + 0.00994x_3x_5 + 0.03682x_4x_5) - \\
& \beta\cdot(0.0314x_1 + 0.0475x_2 + 0.0421x_3 + 0.0297x_4 \\
& + 0.0338x_5) - \gamma\cdot(0.206x_1^3 - 0.029x_2^3 - 0.108x_3^3 - \\
& 0.030x_4^3 - 0.016x_5^3 + 0.198x_1^2x_2 + 0.048x_1^2x_3 + \\
& 0.060x_1^2x_4 - 0.072x_1^2x_5 + 0.272x_1x_2^2 + 0.078x_2^2x_3 + \\
& 0.018x_2^2x_4 - 0.036x_2^2x_5 - 0.048x_1x_3^2 + 0.057x_2x_3^2 + \\
& 0.018x_3^2x_4 - 0.012x_3^2x_5 - 0.012x_1x_4^2 - 0.078x_2x_4^2 - \\
& 0.078x_3x_4^2 - 0.045x_4^2x_5 - 0.060x_1x_5^2 - 0.114x_2x_5^2 - \\
& 0.123x_3x_5^2 - 0.60x_4x_5^2 + 0.042x_1x_2x_3 - 0.192x_1x_2x_4 \\
& - 0.348x_1x_2x_5 - 0.216x_1x_3x_4 - 0.384x_1x_3x_5 - \\
& 0.210x_1x_4x_5 - 0.246x_2x_3x_4 - 0.312x_2x_3x_5 - \\
& 0.162x_2x_4x_5 - 0.222x_3x_4x_5)/(0.02380x_1^2 + \\
& 0.01839x_2^2 + 0.01710x_3^2 + 0.01833x_4^2 + 0.01262x_5^2 \\
& + 0.02270x_1x_2 + 0.01614x_1x_3 + 0.01438x_1x_4 + \\
& 0.01254x_1x_5 + 0.01588x_2x_3 + 0.01494x_2x_4 + \\
& 0.01080x_2x_5 + 0.01474x_3x_4 + 0.00996x_3x_5 + \\
& 0.03684x_4x_5)^{3/2}
\end{aligned}$$

Subject to

$$x_1 + x_2 + x_3 + x_4 + x_5 = 1,$$

$$x_i \geq 0, i = 1, 2, 3, 4, 5.$$

Similarly Models II and III can be formulated.

The portfolio corresponding to Model I is obtained as:

x_1	x_2	x_3	x_4	x_5	Return	Risk	Skewness
0.277	0.278	0.212	0.152	0.081	0.0381	0.011	0.072

4. CONCLUSION

This paper has introduced a new framework of fuzzy MVS portfolio selection using the concept of interval-valued fuzzy numbers. Since skewness is incorporated in the portfolio selection model, the model has become more sensible. Again keeping in view the vagueness and uncertainty of historical data, the interval fuzzy number is used in place of crisp numbers. BSE data is used to show the effectiveness of the method. For symmetries all the elements of co-central moments need not to be computed. The dimension of the variance – covariance matrix is $n \times n$ but only $n(n+1)/2$ elements are to compute. Skewness-co skewness matrix has dimension $(n^2 \times n)$ but only $n(n+1)(n+2)/6$ elements are independent. In our example of five portfolios we have to compute 15 and 35 independent elements for the variance-covariance and skewness-co skewness matrices respectively (whereas these matrices have respectively 25, 125 terms). The efficiency of the portfolios are evaluated by looking for risk contraction on one hand and expected return and skewness augmentation on the other hand. An empirical application has served to illustrate the computational tractability of the approach and the effectiveness of the algorithm. The methodology presented here can be extended to the portfolio selection problems in hybrid and general uncertain environments.

REFERENCES

[1] Lai T., "Portfolio Selection with Skewness: A Multiple–Objective Approach", *Review of the Quantitative Finance and Accounting*, **1**, 293–305, (1991).

[2] Konno H., Shirakawa H., "A Mean-Absolute Deviation-Skewness Portfolio Optimization Model", *Annals of Operations Research*, **45**, 205–220, (1993).

[3] Konno H., Suzuki K., "A Mean-Variance-Skewness Optimization Model", *Journal of the Operations Research Society of Japan*, **38**, 137–187, (1995).

[4] Chunhachinda P., Dandapani P., Hamid S., Prakash A.J., "Portfolio Selection and Skewness: Evidence from International Stock Markets", *Journal of Banking and Finance*, **21**, 143–167, (1997).

[5] Liu S.C., Wang S.Y., Qiu W.H., "A Mean-Variance-Skewness Model for Portfolio Selection with Transaction Costs", *International Journal of System Science*, **34**, 255–262, (2003).

[6] Briec W., Kerstens K., Jokung O., "Mean-Variance-Skewness Portfolio Performance Gauging: A General Shortage Function and Dual Approach", *Management Science*, **53**, 135–149, (2007).

[7] Lai K.K., Wang S.Y., Xu J.P., Zhu S.S., Y. Fang, "A Class of Linear Interval Programming Problems and its Application to Portfolio Selection", *IEEE Transactions on Fuzzy Systems*, **10**, 698–704, (2002).

[8] Ida M., "Solutions for the Portfolio Selection Problem with Interval and Fuzzy Coefficients", *Reliable Computing*, **10**, 389–400, (2004).

[9] Giove S., Funari S., Nardelli C., "An Interval Portfolio Selection Problems Based on Regret Function", *European Journal of Operational Research*, **170**, 253–264, (2006).

[10] Fang Y., Lai K.K., Wang S., "Fuzzy Portfolio Optimization", Springer (2007).

5

NEW FORMS OF FUZZY MINIMAL-CLOSED FUNCTIONS

Anjan Mukherjee

Department of Mathematics, Tripura Universitry, Surjyamaninagar, Agartala-799130, Tripura, India

E-mail: anjan2002_m@yahoo.co.in

Abstract: *Fuzzy minimal structure and fuzzy minimal space were introduced by M. Alimohammady and M. Roohi in 2006. Fuzzy semiopen sets, fuzzy preopen sets, fuzzy α-open sets, fuzzy β-open sets play an important role in the researches of generalizations of closed functions in fuzzy topological space. By using these sets many authors introduced and studied various types of modifications of fuzzy closed functions. The analogy in their definitions and results suggests to unify several characterizations and properties of some kind of modifications of fuzzy closed functions. The purpose of this paper is to unify several characterizations and properties of some kind of modifications of fuzzy closed functions. We obtain some characterizations and several properties of fuzzy minimal closed function defined between sets satisfying some minimal conditions. We recall some modifications of fuzzy open sets and point out the possibility of new forms of fuzzy minimal closed functions. The notion of fuzzy m-regular space and fuzzy m-normal space were introduced by the author. Some characterization theorems are studied when the fuzzy minimal space satisfies the property B. Lastly preservations of fuzzy m-regularity and fuzzy normality between two fuzzy m-spaces are shown.*

Keyword: *Fuzzy minimal structures, fuzzy m-open (closed) set, fuzzy m-open (closed) function, fuzzy m-continuity, fuzzy bitopological space, fuzzy m-regularity, fuzzy m-normality.*

AMS Subject Classification: *54 A 40.*

1. INTRODUCTION

Fuzzy minimal structures and fuzzy minimal space were introduced by M. Alimohammady and M. Roohi in [1, 2] in the following way:

A family m_X of fuzzy sets in X is called a fuzzy minimal structure (briefly m-structure) on X if 0_X and I_X belong to m_X and (X, m_X) is a fuzzy minimal space. The members of m_X are called fuzzy m-open sets and their complements are called fuzzy m-closed sets.Using fuzzy semiopen sets [3], fuzzy preopen sets [15], fuzzy α-open sets [16], fuzzy semi preopen sets [10, 17] and fuzzy δ-open sets [5] several authors introduced and studied various types of modifications of fuzzy open functions in fuzzy topological spaces and fuzzy bitopological spaces. S. Sampath Kumar [13] introduced the fuzzy semiopen sets, fuzzy semi continuity and fuzzy semi open mappings in fuzzy bitopological space.

In Section 2 of the present paper we unify several characterizations and properties of some kind of modifications of fuzzy closed functions and also obtain some characterizations and several properties of fuzzy m-closed function between fuzzy minimal spaces. We also recall some modifications of fuzzy open sets and point out the possibility of new forms of fuzzy m-closed functions. The notion of fuzzy quasi m-closed function is also introduced. In [8] we introduce the notion of fuzzy m-regular space and fuzzy m-normal space. Now some characterization theorems are studied when the fuzzy minimal space satisfies the property B. Lastly preservations of fuzzy m-regularity and fuzzy normality between two fuzzy m-spaces are shown.

Let (X, τ) be a fuzzy topological space and λ a fuzzy subset of X. The fuzzy closure of λ and fuzzy interior of λ are denoted by F Cl (λ) and F Int (λ) respectively. Throughout the present paper (X, τ), (Y, Δ) always denote fuzzy topological spaces and (X, τ_1, τ_2), (Y, Δ_1, Δ_2) denote fuzzy bitopological spaces [6].

We denote the family of all fuzzy semiopen, fuzzy preopen, fuzzy α-open, fuzzy β-open and fuzzy semi preopen, fuzzy δ-open subsets of (X, τ) by $FSO(X)$, $FPO(X)$, $F\alpha(X)$, $F\beta(X)$, $FSPO(X)$, $F\delta(X)$ respectively. Similarly we denote the family of all (i, j) fuzzy semi open, (i, j) fuzzy pre open, (i, j) fuzzy α-open, (i, j) fuzzy β-open, (i, j) fuzzy semi pre open subsets of (X, τ_1, τ_2) by (i, j) $FSO(X)$, (i, j) $FPO(X)$, (i, j) $F\alpha(X)$, (i, j) $F\beta(X)$, (i, j) $FSPO(X)$ respectively. The fuzzy closurer and fuzzy interior of the fuzzy subsets of λ with respect to τ_i are denoted by τ_i-Cl (λ) and τ_i-Int (λ) respectively for $i = 1, 2$.

Let (X, τ) be a fuzzy topological space. Then the family τ, $FSO(X)$, $FPO(X)$, $F\alpha(X)$, $F\beta(X)$, $FSPO(X)$ are all fuzzy minimal structures on X.

A fuzzy point [7] with singleton support $x \in X$ and the value p $(0 < p \leq 1)$ is denoted by x_p. For any fuzzy point x_p and fuzzy set μ we write $x_p \in \mu$ if and only if $p \leq \mu(x)$.

Definition 1.1: [1, 2] Let (X, m_X) be a fuzzy minimal space and λ be a fuzzy subset of X, the fuzzy m_X-closure of λ and the fuzzy m_X-interior of λ are defined as follow:

$$m_X\text{-Int}(\lambda) = \vee\{\alpha : \alpha \leq \lambda, \alpha \in m_X\}$$

$$mX\text{-Cl}(\lambda) = \wedge\{\gamma : \lambda \leq \gamma, \gamma^C \in m_X\}.$$

Proposition 1.2: [1, 2] For two fuzzy sets α and λ

(*a*) m_X-Int $(\alpha) \leq \alpha$, and m_X-Int $(\alpha) = \alpha$ if α is a fuzzy m-open set.

(*b*) $\alpha \leq mX$-Cl (α) and m_X-Cl $(\alpha) = \alpha$ if α is a fuzzy m-closed set.

(*c*) m_X-Int $(\alpha) \leq m_X$-Int (λ) and m_X-Cl $(\alpha) \leq m_X$-Cl (λ) if $\alpha \leq \lambda$.

(*d*) m_X-Int $(\alpha \wedge \lambda) = (m_X\text{-Int}(\alpha)) \wedge (m_X\text{-Int}(\lambda))$ and
$(mX\text{-Int}(\alpha)) \vee (m_X\text{-Int}(\lambda)) \leq m_X\text{-Int}(\alpha \vee \lambda)$.

(*e*) m_X-Cl $(\alpha \vee \lambda) = (m_X\text{-Cl}(\alpha)) \vee (m_X\text{-Cl}(\lambda))$ and
$m_X\text{-Cl}(\alpha \wedge \lambda) \leq (m_X\text{-Cl}(\alpha)) \wedge (m_X\text{-Cl}(\lambda))$.

(*f*) m_X-Int $(m_X\text{-Int}(\alpha)) = m_X$-Int (α) and
$m_X\text{-Cl}(m_X\text{-Cl}(\lambda)) = m_X\text{-Cl}(\lambda)$.

(*g*) $(m_X\text{-Cl}(\alpha))^C = m_X\text{-Int}(\alpha^C)$ and $(m_X\text{-Int}(\alpha))^C = m_X\text{-Cl}(\alpha^C)$.

Definition 1.3: [2] A fuzzy minimal space (X, m_X) is said to have the property 'B' if arbitrary union of fuzzy m-open sets is fuzzy m-open sets.

Proposition1.4: [1, 2] For a fuzzy minimal structure m_X on a set X the following are equivalent:

(*a*) m_X has the property *B*.

(*b*) If m_X-Int $(\alpha) = \alpha$ then $\alpha \in m_X$

(*c*) If m_X-Cl $(\lambda) = \lambda$ then $\lambda^C \in m_X$.

Definition 1.5: [2] Let (X, m_X) and (Y, m_Y) be two fuzzy minimal space then a function $f: (X, m_X) \to (Y, m_Y)$ is fuzzy minimal continuous (briefly fuzzy m-continuous) if $f^{-1}(\lambda) \in m_Y$ for every $\lambda \in m_Y$.

Definition 1.6: [2] Suppose (X, m_X) and (Y, m_Y) be two fuzzy minimal space. A function $f: (X, m_X) \to (Y, m_Y)$ is said to be

(*a*) Fuzzy m-open if $f(\alpha) \in m_Y$ for any fuzzy m-open set α in X.

(*b*) Fuzzy m-closed if $f(\lambda)$ is a fuzzy m-closed set in (Y, m_Y) for any fuzzy m-closed λ in (X, m_X), or equivalently for two fuzzy m-closed sets α_1, α_2 of (Y, m_Y) there exists $v_1, v_2 \in m_Y$ such that $\alpha_1 \le v_1, \alpha_2 \le v_2$ and $f^{-1}(v_1) \le \sigma_1$ and $f^{-1}(v_2) \le \sigma_2$, where $\sigma_1, \sigma_2 \in m_X$.

Definition 1.7: A fuzzy subset λ of a fuzzy bitopological space (X, τ_1, τ_2) is said to be

(*a*) [13] (i, j) fuzzy semiopen if $\lambda \le \tau_j - \text{Cl}\,(\tau_i - \text{Int}\,\lambda)$ where $i \ne j$, $i, j = 1, 2$.

(*b*) [15] (i, j) fuzzy preopen if $\lambda \le \tau_i - \text{Int}\,(\tau_j - \text{Cl}\,\lambda)$ where $i \ne j$, $i, j = 1, 2$.

(*c*) [14] (i, j) fuzzy α-open if $\lambda \le \tau_i - \text{Int}\,(\tau_j - \text{Cl}\,(\tau_i - \text{Int}\,\lambda))$ where $i \ne j$, $i, j = 1, 2$.

(*d*) [11] (i, j) fuzzy semi pre open if then exists an (i, j) fuzzy preopen set α such that $\alpha \le \lambda \le \tau_j - \text{Cl}\,(\alpha)$ where $i \ne j$, $i, j = 1, 2$.

2. NEW FORMS OF MINIMAL CLOSED FUNCTIONS

In this section we obtain several characterizations of fuzzy m-closed functions [1, 2]. We call some modifications of fuzzy open sets and point out the possibility of new forms of fuzzy m-closed functions. M. Alimohammady and M.Roohi [1, 2], introduced the notion of m-closed between two fuzzy minimal spaces in the following way: A function $(X, m_X) \to (Y, m_Y)$ is said to be fuzzy m-closed if $f(\lambda)$ is a fuzzy m-closed set in (Y, m_Y) for any fuzzy m-closed set λ in (X, m_X).

Lemma 2.1: Let (X, τ) and (Y, α) be fuzzy topological spaces and $f: (X, m_X) \to (Y, m_Y)$ be a fuzzy m-closed function then

(*a*) If $m_X = \tau$ and $m_Y = \sigma$ (resp *FSO* (Y), *FPO* (Y), $F\alpha\,(Y)$, $F\beta\,(Y)$, $F\delta\,(Y)$) then f is fuzzy closed (resp. fuzzy semi closed, fuzzy preclosed, fuzzy α-closed, fuzzy β-closed and fuzzy star closed) function.

(*b*) If $m_X = FSO(X)$, (resp. $FPO(X)$, $F\alpha(X)$, $F\beta(X)$, $F\delta(X)$) and if $m_Y = FSO(Y)$, (resp. $FPO(Y)$, $F\alpha(Y)$, $F\beta(Y)$, $F\delta(Y)$ then f is fuzzy pre semiclosed (resp. fuzzy m-preclosed, fuzzy strongly α-closed, fuzy pre β-closed and fuzzy δ-closed) function.

(*c*) If $m_X = F\alpha(X)$ (FRO (X)) and $m_Y = \sigma$ then f is fuzzy quasi α-closed (resp. fuzzy almost closed or fuzzy regular closed) function.

Theorem 2.2: For a one-one function $f: (X, m_X) \to (Y, m_Y)$, where m_Y has property B, the following properties are equivalent:

(*i*) f is fuzzy m-closed function.

(*ii*) For each fuzzy subset β of Y and each $\alpha \in m_X$ with $f^{-1}(\beta) \leq \alpha$, there exists $\lambda \in m_Y$ such that $\beta \leq \lambda$ and $f^{-1}(\lambda) \leq \alpha$.

(*iii*) For each fuzzy point $y_t \in Y$ and each $\alpha \in m_X$ with $f^{-1}(y_t) \leq \alpha$, the exists $\lambda \in m_Y$ containing y_t such that $f^{-1}(\lambda) \leq \alpha$.

Proof: (i) → (ii): Let f be a fuzzy m-closed function. Let β be any fuzzy subset of Y and $\alpha \in m_X$ with $f^{-1}(\beta) \leq \alpha$. We put $\lambda = I_Y - f(I_X\text{-}\alpha)$. Then $f(I_X\text{-}\alpha)$ is fuzzy m-closed in (Y, m_Y), hence λ is fuzzy m-open in (Y, m_Y). Now using definition 1.6 (b) $[f^{-1}(\beta) \leq \alpha \Rightarrow \beta \leq f(\alpha) \leq I_Y - f(I_X\text{-}\alpha)$. Thus $\beta \leq \lambda$ and $f^{-1}(\lambda) \leq \alpha$.

(ii) → (iii): Obvious.

(iii) → (i): Let γ be a fuzzy m-closed set of (X, m_X) and $y_t \in I_Y\text{-}f(\gamma) = f(I_X\text{-}\gamma)$. Since $f^{-1}(y_t) \leq I_X\text{-}\gamma \in m_X$ there exists $\lambda \in m_Y$ with $y_t \in \lambda$ and $f^{-1}(\lambda) \leq I_X\text{-}\gamma$. Therefore $\lambda \wedge f(\gamma) = 0_X$ then $y_t \in I_Y\text{-}m_Y\text{-cl}\,(f(\gamma))$. Hence $m_Y\text{-cl}\,(f(\gamma)) = f(\gamma)$ and hence $f(\gamma)$ is m-closed in (Y, m_Y). This implies f is a fuzzy m-closed function.

Theorem 2.3: A function $f: (X, m_X)\ (Y, m_Y)$, where m_X has property B is fuzzy m-closed if and only if $m_Y\text{-cl}\,(f(\alpha)) \leq f(m_X\text{-cl}\,(\alpha))$ for every fuzzy subset α of X.

Proof: Suppose that f is a fuzzy m-closed function and let α be any fuzzy subset of X. Since m_X has property 'B', $m_X\text{-cl}\,(\alpha)$ is a fuzzy m-closed set in $(X \cdot m_X)$. So $f(m_X\text{-cl}\,(\alpha))$ is fuzzy m-closed in (Y, m_Y). Now $f(\alpha) \leq f(m_X\text{-cl}\,(\alpha))$ and hence $m_Y\text{-cl}\,(f(\alpha)) \leq m_Y\text{-cl}\,(f(m_X\text{-cl}\,(\alpha)) = f(m_X\text{-cl}\,(\alpha))$. So we obtain $m_Y\text{-cl}\,(f(\alpha)) \leq f(m_X\text{-cl}\,(\alpha))$. On the other hand, let γ be any fuzzy m-closed set of X. Then by the hypothesis m_Y-cl $(f(\alpha)) \leq f(m_X$-cl $(\alpha)) = f(\alpha)$. Therefore we have m_Y-cl $(f(\alpha)) = f(\alpha)$ and hence $f(\gamma)$ is fuzy m-closed in (Y, m_Y). So, f is a fuzzy m-closed function.

Theorem 2.4: For a one-one function f: $(X, m_X) \rightarrow (Y, m_Y)$ where m_Y has property B, the following properties are equivalent.

(*i*) f is a fuzzy m-closed function.

(*ii*) If $\alpha \in m_X$ the set $G = \{\gamma \in Y: f^{-1}(\gamma) \leq \alpha\}$ is fuzzy m-open in (Y, m_Y).

(*iii*) If β is fuzzy m-closed in (X, m_X) then the set $M = \{\gamma \in Y: f^{-1}(\gamma) \wedge \beta \neq 0_X\}$ is fuzzy m-closed in (Y, m_Y).

Proof: (i) → (ii): Let $\alpha \in m_X$ and $\gamma \in G$. By the Theorem 2.2, there exists $\lambda \in m_Y$ containing γ such that $f^{-1}(\lambda) \leq \alpha$. Thus $\lambda \leq G$ and $G = m_Y$-int G. So G is fuzzy m_Y-open.

(ii) → (i): Let β be any fuzzy m-closed set in (X, m_X). Then 1_X-β is fuzzy open in (X, m_X) and by (*ii*) $G = \{\gamma \in Y: f^{-1}(\gamma)\ 1_X$-$\beta\}$ is fuzzy m-open in (Y, m_Y). Since $f(\beta) = 1_Y$-G, it follows that $f(\beta)$ is fuzzy m-closed in (Y, m_Y). Therefore, f is a fuzzy m-closed function.

(i) → (iii): This is obvious.

Definition 2.5: A function f: $(X, m_X) \rightarrow (Y, \sigma)$ is said to be fuzzy quasi m-closed if $f(\beta)$ is fuzzy closed in (Y, σ) for every fuzzy m-closed set β of (X, m_X).

Theorem 2.6: For a function f: $(X, m_X) \rightarrow (Y, \sigma)$ the following properties are equivalent:

(*i*) f is a fuzzy quasi m-closed function.

(*ii*) for each fuzzy subset β of Y and each $\alpha \in m_X$ with $f^{-1}(\beta) \leq \alpha$, there exists a fuzzy open set λ of Y such that $\beta \leq \lambda$ and $f^{-1}(\lambda) \leq \alpha$.

(*iii*) For each fuzzy set $\gamma \in Y$ and each $\alpha \in m_X$ with $f^{-1}(\gamma) \leq \alpha$, there is a fuzzy open set λ of Y containing γ such that $f^{-1}(\lambda) \leq \alpha$.

Proof: This follows immediately from Theorem 2.2.

Lemma 2.7: A function f: $(X, m_X) \rightarrow (Y, \sigma)$ where m_X has property 'B', is a fuzzy quasi m-closed if and only if cl $(f(\alpha)) \leq f(m_X$-cl $(\alpha))$ for every fuzy subset α of X.

Proof: This follows immediately from Theorem 2.3.

Now we recall some modifications of fuzzy open sets defined recently.

Definition 2.8: A fuzzy subset α of a fts (X, τ) is said to be fuzzy δ-semiopen [9], (fuzzy δ-preopen [4], fuzzy δ-βopen [9]) if $\alpha \leq$ Fcl (Fint$_\delta$ (α)) [resp. $\alpha \leq$ Fint (Fcl$_\delta$ (α)), Fcl (Fint (Fcl$_\delta$ (α)))].

By $F\delta SO$ (X) (resp. $F\delta PO$ (X), $F\delta\beta$ (X)) we denote the collection of all fuzzy δ-semiopen (resp. fuzzy δ-preopen, fuzzy δ-β open) sets of a fuzzy topological space (X, τ). These three collections are all fuzzy m-structure with property 'B'. Then the following implication holds:

Fδ-open → F open → F-preopen → Fuzzy δ-pre open

↓ ↓ ↓ ↓

Fuzzy δ-semi open → F semi open → F β-open → F $\delta\beta$-open

Diagram-1

Definition 2.9: A fuzzy subset β of a fts (X, τ) is said to be fuzzy δ-semi closed [9] (resp. fuzzy δ-pre closed [4], fuzzy δ-β closed [9]) if the complement of β is fuzzy δ-semiopen (resp. fuzzy δ preopen, fuzzy δ-β open).

Definition 2.10: A function f: $(X \cdot \tau) \rightarrow (Y, \sigma)$ is said to be

(*i*) fuzzy δ-semiclosed if $f(\beta)$ is fuzzy δ-semiclosed in Y for every fuzzy closed set β of X.

(*ii*) fuzzy δ-preclosed if $f(\beta)$ is fuzzy δ-preclosed in Y for every fuzzy closed set of X.

(*iii*) fuzzy δ-β closed if $f(\beta)$ is fuzzy δ-β closed in Y for every fuzzy closed set β of X.

Remark 2.11: Let f: (X, τ) (Y, σ) be a function. If f: $(X, m_X) \rightarrow (Y, m_Y)$ is a fuzzy m-closed function, $m_X = \tau$ and $mY = F\delta SO$ (Y) (resp. $F\delta\beta O$ (Y), F$\delta\beta$ (Y)), then f: $(X, \tau) \rightarrow (Y, \sigma)$ is fuzzy δ semiclosed (resp. fuzzy δ-preclosed, fuzzy δ-β closed) function.

Lemma 2.12: Let m_X^1 and m_X^2 be two fuzzy m-structures on a nonempty set X. If $m_X^1 \subset m_Y^2$ and a function f: (X, m_X^1) (Y, m_Y) is fuzzy m-closed then f: $(X, m_X{}^2) \rightarrow (Y, m_Y)$ is also a fuzzy m-closed function.

Proof: Straight forward.

Lemma 2.13: Let m_Y^1 and m_Y^2 be two fuzzy m-structures on a nonempty set Y. If $m_Y^1 \leq m_Y^2$ and a function f: $(X, m_X) \rightarrow (Y, m_Y{}^1)$ is fuzzy m-closed then f: (X, m_X) $(Y, m_Y{}^2)$ is also a fuzzy m-closed function.

Proof: Straight forward.

By diagram 1 and lemma 2.12 we obtain the following

Fuzzy Star closed → fuzzy closed → fuzzy pre closed → fuzzy δ preclosed

↓ ↓ ↓ ↓

Fuzzy δ semi closed → fuzzy semi closed → fuzzy β-closed → fuzzy δβ-closed

Diagram-2

So if as m_X and m_Y, we take the collection of fuzzy δ-open sets, fuzzy δ-preopen sets or fuzzy δ-β open sets, we can define new kind of function and use the results already established in Section 2. In [8] we introduce the notion of fuzzy m-regular space and fuzzy m-normal space. Now some characterization theorems are studied when the fuzzy minimal space satisfies the property B. Lastly preservations of fuzzy m-regularity and fuzzy normality between two fuzzy m-spaces are shown.

Definition 2.14: [8] A fuzzy m-space (X, m_X) is said to be fuzzy m-regular if for each fuzzy point x_p in X and each fuzzy m-closed set λ such that $x_p \in \lambda^c$ there exist fuzzy m-open sets α and ν such that $x_p \in \alpha$, $\lambda \leq \nu$ and $\alpha \leq \nu^c$.

Remark 2.15: Let (X, τ) be a fuzzy topological space and $m_x = \tau$ (resp. $FSO(X)$, $FPO(X)$, $F\beta(X)$). Then fuzzy m-regularity coincides with fuzzy regularity (resp. fuzzy semi regularity, fuzzy pre regularity, fuzzy semi pre-regularity).

Theorem 2.16: Let (X, m_X) be a fuzzy m-space where m_x satisfies property B. Then the following are equivalent:

(*i*) (X, m_X) is fuzzy m-regular.

(*ii*) For each $x_p \in X$ and each fuzzy m-open set λ containing x_p there exists a fuzzy m-open set α such that $x_p \in \alpha \leq m_X$-Cl (α) $\leq \lambda$.

(*iii*) For every fuzzy m-closed set λ in X and any fuzzy point $x_p \in \lambda^c$ there exist fuzzy m-open sets μ_1 and μ_2 such that $x_p \in \mu_1$, $\lambda \leq \mu_2$ and m_X-Cl $\mu_1 \leq (m_X\text{-Cl } \mu_2)^c$.

***Proof:* (*i*) → (*ii*):** Let x_p be any fuzzy point in X and let λ be a fuzzy m-open set in X such that $x_p \in \lambda$. Since $x_p \in \lambda$, $x_p \in (\lambda^c)^c$ and λ^c is a fuzzy m-closed set in X. Then by fuzzy m-regularity there exist fuzzy m-open sets α and ν such that $x_p \in \alpha$, $\lambda^c \leq \nu$ and $\alpha \leq \nu^c$. Since $\lambda^c \leq \nu$ we have $\nu^c \leq \lambda$ and $\alpha \leq \nu^c \leq \lambda$. Moreover, $\alpha \leq m_X$-Cl $\alpha \leq m_X$-Cl $\nu^c = \nu^c \leq \lambda$. Hence $x_p \in \alpha \leq m_X$-Cl $(\alpha) \leq \lambda$.

(ii) → (iii): Let λ be a fuzzy m-closed in X and x_p be any fuzzy point in X such that $x_p \in \lambda^c$. Using (*ii*) there exists a fuzzy m-open set σ_1 such that $x_p \in \sigma_1 \leq m_X$-Cl $(\sigma_1) \leq \lambda^c$. Applying (*ii*) again there exists a fuzzy m-open set σ_2 such that $x_p \in \sigma_2 \leq m_X$-Cl $(\sigma_2) \leq \sigma_1 \leq m_X$-Cl $(\sigma_1) \leq \lambda^c$. The proof is completed by taking $\mu_1 = \sigma_2$ and $\mu_2 = m_X$-Cl $(\sigma_1)^c$ because $x_p \in \sigma_2$, $\lambda \leq (m_X$-Cl $(\sigma_1))^c = \mu_2$ and m_X-Cl $(\sigma_2) \leq [m_X$-Cl $(m_X$-Cl $(m_X$-Cl $(\sigma_1)^c)]^c$. Indeed, $\sigma_1 \leq m_X$-Cl $(\sigma_1) \Rightarrow m_X$-Int $(\sigma_1) \leq m_X$-Int $(m_X$-Cl $(\sigma_1))$. Thus we get, $\sigma_1 \leq [m_X$-Cl $(m$-Cl $(\sigma_1)^c]^c$, since m_X-Cl $(\sigma_2) \leq \sigma_1$. We have, m_X-Cl $(\sigma_2) \leq [m_X$-Cl $(m_X$-Cl$\sigma_1)^c]^c$, i.e, m_X-Cl $(\mu_1) \leq (m_X$-Cl$\mu_2)^c$.

(iii) → (i): Let λ be a fuzzy m-closed in X and x_p be any fuzzy point in X such that $x_p \in \lambda^c$. Using (*iii*) there exists fuzzy m-open sets sets μ_1 and μ_2 such that m_X-Cl $\mu_1 \leq (m_X$-Cl $\mu_2)^c$. Since $\mu_1 \leq m_X$-Cl μ_1 and $\mu_2 \leq m_X$-Cl μ_2, we have $\mu_1 \leq (\mu_2)^c$.

Theorem 2.17: If f: $(X, m_X) \rightarrow (Y, m_Y)$ where m_X and m_Y have property B be a fuzzy m-continuous, fuzzy m-open and fuzzy m-closed surjection and (X, m_X) is fuzzy m-regular then (Y, m_Y) is also fuzzy m-regular.

Proof: Let $y_t \in Y$ and λ be any fuzzy m-open set of Y containing y_t. Let $x_p \in f^{-1}(y_t)$.

Since f is fuzzy m-continuous we have $x_p \in f^{-1}(\lambda) \in m_X$. Since (X, m_X) is fuzzy m-regular, there exists $\alpha \in m_X$ such that $x_p \in \alpha$ mX-Cl $\alpha \leq f^{-1}(\lambda)$. Then we have $y_t \in \leq f(\alpha) \leq . f(m_X$-Cl $\alpha) \leq \lambda$. Since f is fuzzy m-open, $f(\alpha)$ is fuzzy m-open in Y. Now m_X has property B and m_X-Cl α is fuzzy m-closed set, hence f $(m_X$-Cl $\alpha)$ is fuzzy m-closed in Y. Hence we obtain $y_t \in \leq f(\alpha) \leq m_X$-Cl $(f(\alpha)) \leq \lambda$. It follows that (Y, m_Y) is also fuzzy m-regular.

Corollary 2.18: If f: $(X, \tau) \rightarrow (Y, \Delta)$ be a fuzzy continuous, fuzzy open and fuzzy closed surjection and (X, τ) is fuzzy regular then (Y, Δ) is fuzzy regular.

Theorem 2.19: If f: $(X, m_X) \to (Y, m_Y)$ where m_X has property B is a fuzzy m-continuous and fuzzy m-closed injection and (Y, m_Y) is fuzzy m-regular then (X, m_X) is also fuzzy m-regular.

Proof: Let α be any fuzzy m-closed set of X and $x_p \in \alpha^c = I_X\text{-}\alpha$. Since f is fuzzy m-closed, $f(\alpha)$ is fuzzy m-closed in Y and $f(x_p) \in f(I_X - \alpha) = I_Y - f(\alpha) = (f(\alpha))^c$. Since (Y, m_Y) is fuzzy m-regular, there exist fuzzy m-open sets λ and ν in Y such that $f(x_p) \in \lambda$ and $f(\alpha) \le \nu$ and $\lambda \le I_Y - \nu = \nu^c$. Now f is fuzzy m-continuous hence $f^{-1}(\lambda)$ and $f^{-1}(\nu)$ are fuzzy m-open sets in X. Thus we obtain $x_p \in f^{-1}(\lambda)$, $\alpha \le f^{-1}(\nu)$ and $f^{-1}(\lambda) \le f^{-1}(I_Y - \nu) = I_X - f^{-1}(\nu) = (f^{-1}(\lambda))^c$. This shows that (X, m_X) is fuzzy m-regular.

Definition 2.20: [8] A fuzzy m-space (X, m_X) is said to be fuzzy m-normal if for each pair of fuzzy m-closed sets α_1 and α_2 of X with $\alpha_1 \le \alpha_2$ there exists μ_1 and $\mu_2 \in m_X$ such that $\alpha_1 \le \mu_1$, $\alpha_2 \le \mu_2$ and $\mu_1 \le \alpha_2^c$.

Remark 2.21: Let (X, τ) be a fuzzy topological space and $m_X = \tau$ (resp. *FSO* (X), *FPO* (X), $F\beta$ (X)). Then fuzzy m-normality coincide with fuzzy normality (resp. fuzzy semi normality, fuzzy pre normality, fuzzy β-normality).

Theorem 2.22: Let (X, m_X) be a fuzzy m-space, m_X satisfies the property 'B'. Then the following are equivalent

(*i*) (X, m_X) is fuzzy m-normal

(*ii*) For every pair of fuzzy m-closed sets σ and fuzzy m-open set λ with $\sigma \le \lambda$ there exists a fuzzy open set μ such that $\sigma \le \mu \le m_X\text{-Cl}\ \mu \le \lambda$.

(*iii*) For every pair of fuzzy m-closed sets σ and λ such that $\lambda \le \sigma^C$ there exists fuzzy m-open sets μ_1 and μ_2 such that $\sigma \le \mu_2$ and $\lambda \le \mu_1$ and $m_X\text{-Cl}\ \mu_1 \le (m_X\text{-Cl}\ \mu_2)^C$.

(*iv*) For every fuzzy m-closed set λ and for every fuzzy m-open set μ such that $\lambda \le \mu$ there exists a fuzzy set σ such that $\lambda \le m_X\text{-Int}\ \sigma \le m_X\text{-Cl}\ \sigma \le \mu$.

Proof: (*i*) $\to$ (*ii*): Let σ be fuzzy m-closed set and λ be a fuzzy m-open set in X such that $\sigma \le \lambda$. By (*i*) there exists fuzzy m-open sets v and μ such that $\sigma \le \mu$; $\lambda^C \le v$ and $\mu \le v^C$.

Since $\sigma \le \mu \le v^C \le \lambda$ and $\sigma \le m_X\text{-Cl}\ \mu \le m_X\text{-Cl}\ v^C = v^C$.

Therefore we have $\sigma \le \mu \le m_X\text{-Cl}\ \mu \le \lambda$.

***(ii)* → *(iii)*:** Let λ and μ be fuzzy m-closed sets in X such that $\mu \le \lambda^C$. Using (*ii*) there exists a fuzzy m-open set σ_1 such that $\mu \le \sigma_1 \le m_X$-Cl $\sigma_1 \le \lambda^C$. Applying (*ii*) again there exists a fuzzy m-open set σ_2 such that $\mu \le \sigma_2 \le m_X$-Cl $\sigma_2 \le \sigma_1 \le m_X$-Cl $\sigma_1 \le \lambda^C$. The proof is completed by taking $\mu_1 = \sigma_2$ and $\mu_2 = (m_X\text{-Cl } \sigma_1)^C$ because $\mu \le \sigma_2$, $\lambda \le \mu_2$ and $\sigma_2 \le [m_X\text{-Cl } (m_X\text{-Cl } \sigma_1)^C]^C$. Indeed $\sigma_1 \le m_X$-Cl σ_1 => m_X-Int $\sigma_1 \le m_X$-Int m_X-Cl $\sigma_1 = [m_X\text{-Cl } (m_X\text{-Cl } \sigma_1)^C]^C$. Thus we get $\sigma_1 \le [m_X\text{-Cl } (m_X\text{-Cl } \sigma_1)^C]^C$ since m_X-Cl $\sigma_2 \le \sigma_1$ we have m_X-Cl $\sigma_2 \le [m_X\text{-Cl } (m_X\text{-Cl } \sigma_1)^C]^C$ i.e. m_X-Cl $\mu_1 \le (m_X\text{-Cl } \mu_2)^C$.

***(iii)* → *(i)*:** Let λ and μ be a fuzzy m-closed set in X such that $\mu \le \lambda^C$. Using (*iii*) there exists fuzzy m-open sets μ_1 and μ_2 such that m_X-Cl $\mu_1 \le (m_X\text{-Cl } \mu_2)^C$. Since $\mu_1 \le m_X$-Cl $\mu_1 \le (m_X\text{-Cl } \mu_2)^C = m_X$-Int $\mu_2^C \le \mu_2^C$.

***(ii)* → *(iv)*:** It is immediate since m_X-Int $\sigma \le \sigma$.

***(iv)* → *(ii)*:** Let λ be a fuzzy m-closed set and μ be a fuzzy m-open sets in X such that $\lambda \le \mu$. Using (*iv*) there exists a fuzzy set σ such that $\lambda \le m_X$-Int $\sigma \le m_X$-Cl $\sigma \le \mu$. Let $v = m_X$-Int σ. Then v is a fuzzy m-open set $v = m_X$-Int $\sigma \le \sigma$.=> m_X-Cl $v \le m_X$-Cl $\sigma \le \mu$. Consequently v is a fuzzy m-open set satisfying $\lambda \le v \le m_X$-Cl $v \le \mu$.

Theorem 2.23: If f: $(X, m_X) \to (Y, m_Y)$ where m_X and m_Y have property B is a fuzzy m-closed, fuzzy m-continuous surjection and (X, m_X) is fuzzy m-normal space then (Y, m_Y) is also fuzzy m-normal.

Proof: Let α_1, α_2 be any fuzzy m-closed sets of (Y, m_Y) with $\alpha_1 \le \alpha_2^C$. Since f is fuzzy m-continuous, $f^{-1}(\alpha_1)$, $f^{-1}(\alpha_2)$ are fuzzy m-closed sets of (X, m_X) with $f^{-1}(\alpha_1) \le (f^{-1}(\alpha_2))^C$. Now (X, m_X) is fuzzy m-normal; then there exists $\sigma_1, \sigma_2 \in m_X$ such that $f^{-1}(\alpha_1) \le \sigma_1$, $f^{-1}(\alpha_2) \le \sigma_2$ and $\sigma_1 \le \sigma_2^C$. Since f is fuzzy m-closed, (by the result 1.7 (b)) there exists $v_1, v_2 \in m_Y$ such that $\alpha_1 \le v_1$, $\alpha_2 \le v_2$ and $f^{-1}(v_1) \le \sigma_1$ and $f^{-1}(v_2) \le \sigma_2$. Since $\sigma_1 \le \sigma_2^C$ and f is surjection we obtain $v_1 \le v_2^C$, therefore (Y, m_Y) is fuzzy m-normal. [$v_1 \le f(\sigma_1)$, $v_2 \le f(\sigma_2)$ => $(f(\alpha_2))^C \le v_2^C$, $\sigma_1 \le \sigma_2^C$ => $f(\sigma_1) \le (f(\sigma_2))^C$ => $v_1 \le f(\sigma_1) \le v_2^C$]

Theorem 2.24: If f: $(X, m_X) \to (Y, m_Y)$ where m_X has property 'B' is a fuzzy m-continuous and fuzzy m-closed injection and (Y, m_Y) is fuzzy m-normal space then (X, m_X) is also fuzzy m-normal.

Proof: Let α_1, α_2 be any fuzzy m-closed sets of X with $\alpha_1 \le \alpha_2^C$. Since f is fuzzy m-closed injection $f(\alpha_1)$, $f(\alpha_2)$ are fuzzy m-closed sets of Y with $f.(\alpha_1) \leq (f(\alpha_2))^C$. By the m-normality of (Y, m_Y) there exists fuzzy open sets μ_1, μ_2 such that $\mu_1 \leq \mu_2^C$ and $f(\alpha_1) \leq \mu_1, f(\alpha_2) \leq \mu_2$. Now f is fuzzy m-continuous so $f^{-1}(\mu_1)$, and $f^{-1}(\mu_2)$ are fuzzy m-open sets containing α_1 and α_2 respectively with $f(\mu_1) \leq ((f(\mu_2))^C$. This shows (X, m_X) is fuzzy m-normal.

REFERENCES

[1] M. Alimohammady, M. Roohi, "Fuzzy Minimal Structure and Fuzzy Minimal Vector Spaces", *Chaos, Solitons and Fractals*, **27 (3)**, (2006), 599–605.

[2] M. Alimohammady, M. Roohi, "Compactness in Fuzzy Minimal Spaces", *Chaos, Solitons and Fractals*, **28 (4)**, (2006), 906–912.

[3] K.K. Azad, "On Fuzzy Semi Continuity, Fuzzy Almost Continuity and Fuzzy Weakly Continuity", *J. Math. Anal. Appl.*, **82**, (1981), 14–32.

[4] M. Caldas, S. Jafari and R.K. Saraf, "Fuzzy (δ, p)-T_1 Topological Spaces", *Journal Tri. Math. Soc.*, **V-9**, (2007), 1–4.

[5] S. Ganguly and S. Saha; "A Note on δ-continuity and δ-connected Sets in Fuzzy Set Theory", *Simon Stevin*, **62**, (1988), 127–141.

[6] A. Kandil, A.A. Nouth and S.A. El-Sheikh, "On Fuzzy Bitopological Soaces, Fuzzy Sets and Systems", **74**, (1995), 353–363.

[7] P. Pao-Ming, L. Ying-Ming, "Fuzzy Topology I, Neighbourhood Structure of a Fuzzy Point and Moore Smith Convergence", *J. Math. Anal. Appl.*, **76 (2)**, (1980), 571–599.

[8] Anjan Mukherjee, "Fuzzy Minimal Structures, Fuzzy *m*-regularity, Fuzzy *m*-normality and Fuzzy Bitopological Spaces". (to appear).

[9] Anjan Mukherjee and S. Debnath, "On δ-semi Open Sets in Fuzzy Settings", *Journal Tri. Math. Soc.*, **V-8**, (2006), 51–54.

[10] J.H. Park, B.Y. Lee, "Fuzzy Semi Pre Open and Fuzzy Semi Pre Continuous Maps, Fuzzy Sets and Systems", **67**, (1994), 359–364.

[11] J.H. Park, "On Fuzzy Pair Wise Semi Pre Continuity, Fuzzy Sets and Systems", **93**, (1998), 375–379.

[12] Z. Petricevic, "On Fuzzy Semi Regularization, Separation Properties and Mappings", *Indian J. Pure Appl. Math.*, **22**, (1991), 971–982.

[13] S. Sampath Kumar, "Semi Open Sets, Semi Continuity and Semi Open Mappings in Fuzzy Bitopological Spaces, Fuzzy Sets and Systems", **64**, (1994), 421–426.

[14] S. Sampath Kumar, "On Fuzzy Pair Wise α-continuity and Fuzzy Pair Wise Pre Continuity, Fuzzy Sets and Systems", **62**, (1994), 231–238.

[15] M.K. Singal and N. Prakash, "Fuzzy Pre Open Sets and Fuzzy Pre Separation Axioms, Fuzzy Sets and Systems", **44**, (1991), 345–351.

[16] M.K. Singal and Niti Rajvansi, "Fuzzy Alpha Sets and Alpha Continuous Maps", *Fuzzy Sets and Systems*, **48**, (1992), 383–390.

[17] S.S. Thakur and Surendra Singh, "On Fuzzy Semi Pre Open Sets and Fuzzy Semi Pre Continuity, Fuzzy Sets and Systems", **98**, (1998), 383–391.

[18] L.A. Zadeh, "Fuzzy Sets, Information and Control", **8**, (1965), 338–353.

6

MATHEMATICAL FUZZY LOGIC: TODAY AND TOMORROW*

Vilém Novák

University of Ostrava Institute for Research and Applications of Fuzzy Modeling 30. dubna 22, 70103, Ostrava 1, Czech Republic
E-mail: Vilem.Novak@osu.cz

Abstract: *We discuss the present state of mathematical fuzzy logic in narrow sense and recall its extension, fuzzy logic in broader sense (FLb) as a logic of natural human reasoning. We will try to answer the questions: where do we stand and what should be further directions of its development. Our main conclusion is that we should now concentrate more on FLb since there are many interesting unsolved problems and it has various practical applications.*

1. INTRODUCTION

Mathematical fuzzy logic is a special mathematical theory whose goal is to provide a working mathematical model of the vagueness phenomenon and develop a well established sound formal system that makes its applications well justified. The mathematization is based on introduction of degrees of truth taken from some scale. Many people have significantly contributed to its development, for example P. Hájek, F. Esteva, L. Godo, S. Gottwald, F. Montagna, P. Cintula, D. Mundici, G. Gerla, V. Novák, and others.

Two directions (branches) of mathematical fuzzy logic can be currently distinguished:

(*i*) Fuzzy logic in narrow sense (FLn),

(*ii*) Fuzzy logic in broader sense (FLb).

* The research was supported by project MSM 6198898701.

The goal of this paper is to give an overview of the present state of mathematical fuzzy logic. We will try to answer the questions: where do we stand and what should be further directions of its development. Our main conclusion is that we should now concentrate more on FLb since there are many interesting unsolved problems in it. Moreover, FLb has many practical applications.

2. MATHEMATICAL FUZZY LOGIC

2.1 Fuzzy Logic in Narrow Sense

Let us stress that after the seminal paper on fuzzy set theory written by L.A. Zadeh [40], the next seminal paper on fuzzy logic is [12] written by J. A. Goguen. In this paper, the residuated lattice has been proposed as a convenient structure of truth values. The paper contains a lot of solid material accompanied by detailed philosophical discussion but it is not sufficiently formal. The first, mathematically highly sophisticated paper is that of J. Pavelka [36]. Since then, FLn achieved a significant development. Further division of it can be seen in two directions:

(*i*) Fuzzy logic with traditional syntax,

(*ii*) Fuzzy logic with evaluated syntax.

Fuzzy logic with traditional syntax has been elaborated in a deep and seminal way by P. Hájek in [13]. It is specific for it that formulas are dealt with classically, i.e. the language of such fuzzy logic differs from that of classical logic especially by having more connectives. On the other hand, the concepts of inference rule, theory, proof and many other ones remain classical.

The semantics is many valued and based on the concept of a commutative, integral, bounded residuated lattice $\mathcal{L}$ (shortly, residuated lattice) which is an algebra of type (2, 2, 2, 2, 0, 0)

$$\mathcal{L} = \langle L, \wedge, \vee, \otimes, \rightarrow, \mathbf{0}, \mathbf{1} \rangle, \qquad \text{... (1)}$$

where $\langle L, \wedge, \vee, \mathbf{0}, \mathbf{1} \rangle$ is a lattice with the bottom element $\mathbf{0}$ and the top element $\mathbf{1}$. The operation $\otimes$ is called *multiplication* and $\langle L, \otimes, \mathbf{1} \rangle$ is a commutative monoid. The operation $\rightarrow$ is the *residuation* operation fulfilling the adjointness property: for all $a, b, c \in L$ it holds that

$$a \otimes b \leq c \quad \text{iff} \quad a \leq b \rightarrow c. \qquad \text{... (2)}$$

A BL-algebra is a residuated lattice fulfilling, moreover, *prelinearity* ($(a \rightarrow b) \vee (b \rightarrow a) = \mathbf{1}$) and *divisibility* ($a \otimes (a \Rightarrow b) = a \wedge b$).

The most prominent is the standard Łukasiewicz MV-algebra

$$\mathcal{L}_Ł = \langle [0,1], \vee, \wedge, \otimes, \rightarrow, 0, 1 \rangle$$

where

(*i*) $a \otimes b = \max(0, a + b - 1)$, (Łukasiewicz conjunction)

(*ii*) $a \rightarrow b = \min(1, 1 - a + b)$. (Łukasiewicz implication)

At present, there are many formal systems of it both in propositional as well as predicate version. The prominent among them are MTL, IMTL, BL, product, Łukasiewicz logics (cf. [28]). All these logics are now embraced in a wide class of the, so called, *core fuzzy logics* [3, 14].

All these logics enjoy completeness in the form

$$\vdash A \quad \text{iff} \quad \models A \qquad \text{... (3)}$$

where $\models A$ means that $\mathcal{M}(A) = 1$ for every interpretation $\mathcal{M}$ of the language of the given core fuzzy logic.

FLn is in detail studied especially from the algebraic point of view. One can hardly estimate, how far this work can still continue. It seems that FLn is in the position of a theory promising a deeply justified technique for modeling various manifestations of the vagueness phenomenon and for various other applications. There are not, however, so many results in the latter. One of exceptions is the program of formalization of fuzzy set theory using fuzzy logic as a metatheory that was announced by P. Cintula and L. Běhounek in [1, 2]. Several paper have already been published and the program seems to be very promising since it encompasses and quite often even generalizes most (if not all) of the results in fuzzy set theory obtained so far.

A special role in this direction is played by higher-order fuzzy logics, the so called *fuzzy type theories* (FTT). These logics have been initiated by V. Novák in [24] where the first formal system of IMTL-FTT has been introduced. Further systems of FTT are BL-FTT and Łukasiewicz-FTT. Quite recently, a new special algebra has been introduced in which the essential operation is that of *fuzzy equality*. More precisely, it is an algebra of type (2, 2, 2, 0)

$$\mathcal{E} = \langle E, \wedge, \otimes, \sim, 1 \rangle, \quad \text{... (4)}$$

where for all a, $b, c, d \in E$:

(*E*1) $\langle E, \wedge, \mathbf{1} \rangle$ is a commutative idempotent monoid (i.e. $\wedge$-semilattice with top element **1**),

(*E*2) $\langle E, \otimes, \mathbf{1} \rangle$ is a commutative monoid and $\otimes$ is isotone w.r.t. $\leq$ (with $a \leq b$ defined as $a \wedge b = a$),

(*E*3) $a \sim a = \mathbf{1}$,

(*E*4) $((a \wedge b) \sim c) \otimes (d \sim a) \leq c \sim (d \wedge b)$,

(*E*5) $(a \sim b) \otimes (c \sim d) \leq (a \sim c) \sim (b \sim d)$,

(*E*6) $(a \wedge b \wedge c) \sim a \leq (a \wedge b) \sim a$,

(*E*7) $(a \wedge b) \sim a \leq (a \wedge b \wedge c) \sim (a \sim c)$,

(*E*8) $a \otimes b \leq a \sim b$.

The implication operation is derived using the formula

$$a \to b = a \wedge b \sim a.$$

This algebra gives rise to a new kind of fuzzy type theory (EQ-FTT) (see [21]) which is not identical with any of the above mentioned ones.

Fuzzy logic with evaluated syntax ($Ev_{\text{Ł}}$) is specific by introducing formulas that are evaluated also on syntactic level. Namely, we deal with evaluated formulas which have the form a/A where A is a formula and $a \in L$ is its syntactic evaluation. This has nice interpretation since it allows to consider axioms which need not be fully satisfactory and so, their initial truth value can be lower than 1.

This logic has been in propositional version established by J. Pavelka in [36] and extended to predicate version by V. Novák in [22] and especially in the book [35]. $Ev_{\text{Ł}}$ has been established as a generalization of classical mathematical logic with clearly distinguished syntax and semantics. The syntax consists of precise definitions of evaluated formula, evaluated proof, fuzzy theory, model, evaluated provability, etc. and the semantics is formed by the standard

Łukasiewicz algebra only. The reason for choosing the latter follows from [36] where it has been proved that completeness of the syntax w.r.t. semantics can be kept providing that the corresponding algebra fulfils the following four equations:

$$\bigvee_{i \in I} (a \to b_i) = a \to \bigvee_{i \in I} b_i, \quad \bigwedge_{i \in I} (a \to b_i) = a \to \bigwedge_{i \in I} (b_i) \quad \ldots (5)$$

$$\bigvee_{i \in I} (a_i \to b) = \bigwedge_{i \in I} a_i \to b, \quad \bigwedge_{i \in I} (a_i \to b) = \bigvee_{i \in I} a_i \to b \quad \ldots (6)$$

These equations are in [0, 1] equivalent to continuity of $\to$, which is fulfilled only by Łukasiewicz implication and its isomorphs. The resulting logic is quite strong since it is complete with respect to the generalized syntax, in which all formulas are evaluated by elements from the underlying algebra and the completeness thus takes the form of generalization of the Gödel completeness:

$$T \vdash_a A \text{ iff } T \models_a A, \quad a \in L,$$

for all all formulas $A \in F_{J(T)}$ and *fuzzy theories* T where the latter are determined by fuzzy sets of axioms. The system of $Ev_Ł$ is open to extension by new connectives. The fact that it is the only one (up to isomorphism), however, makes many researchers unsatisfied with it.

2.2 Fuzzy Logic in Broader Sense

The concept of fuzzy logic in broader sense was proposed by V. Novák in 1995 in [23] as a program for extension of FLn, which aims at developing a *formal theory of natural human reasoning that would include mathematical models of special expressions in natural language with regard to their vagueness*. This program overlaps with two other paradigms proposed in the literature, namely *commonsense reasoning* and *precisiated natural language* (PNL).

The idea of commonsense reasoning has been proposed by J. McCarthy in [20] as a part of the program of logic-based artificial intelligence. Its paradigm is to develop formal commonsense theories and systems using mathematical logic that exhibit commonsense behavior. The reason is that commonsense reasoning is a central part of human behavior and no real intelligence is possible without it.

The main drawback of the up-to-date formalizations of commonsense reasoning, in our opinion, is that it neglects vagueness present in the meaning of natural language expressions (cf. [4] and the citations therein).

PNL, on the other hand, is based on two main premises:

(*a*) Much of the world knowledge is perception based,

(*b*) Perception based information is intrinsically fuzzy.

It is important to stress that the term *precisiated natural language* means "a reasonable working formalization of semantics of natural language without pretension to capture it in detail and fineness". Its goal is to provide an acceptable and applicable technical solution. It should also be noted that the term *perception* is not considered here as a psychological term but rather as a result of human, intrinsically imprecise measurement.

PNL methodology requires the presence of *World Knowledge Database* and Multiagent, *Modular Deduction Database*. The former contains all the necessary information including perception based propositions describing the knowledge acquired by direct human experience, which can be used in the deduction process. The latter contains various rules of deduction. No exact formalization of PNL, however, has been developed until recently, and so it should be taken mainly as a reasonable methodology.

The concept of FLb is a glue between both paradigms that should consider the best of each. During the years, it has been slowly developed and so far, it consists of the following theories:

(*a*) Formal theory of evaluative linguistic expressions, which is in detail explained in [29].

(*b*) Formal theory of fuzzy **IF-THEN** rules which is presented in [7, 33].

(*c*) Formal theory of perception-based logical deduction, which is presented in [5, 34, 26, 25],

(*d*) Formal theory of intermediate and generalized quantifiers presented in [16, 6, 27, 30].

The mathematical basis for all these theories was originally provided by $Ev_{\text{Ł}}$. An essential constituent of FLb, though, is related to natural language semantics and so, a more convenient formal system as the basis for further development of FLb seems to be *fuzzy type theory*. One of the reasons supporting the latter is argumentation of many logicians and linguists (cf. [18, 19, 39]) that the first order logic is not sufficient for capturing semantics of natural language. Although $Ev_{Ł}$ is more general and so, it is more powerful than classical logic, still formalization of items (*i*) and (*ii*) using $Ev_{Ł}$ as presented in [35, Chapter 6] is not fully satisfactory.

3. WHERE SHOULD WE GO?

In FLn, many formal (residuated lattice-based) logical systems have been studied in detail including their interrelations. Therefore, structure of FLn is now already known quite well. For example, in the paper [3], 60 various fuzzy logics are analyzed together. We still do not know which logic is the most convenient to solve problems related to models of vagueness and their applications. We have a noble mathematical theory and a deep metamathematical knowledge. Therefore, we think that it is already time to turn the attention to the proclaimed goals of fuzzy logic so that its power could be fully acknowledged!

As can be seen from the presentation of FLb, some steps have already been done. We think that there are good reasons to continue the development of FLb and move it further towards human (i.e., commonsense) reasoning, following the methodology of PNL and results obtained in the AI *theory of commonsense reasoning*. Another promising direction is to include also uncertainty in FLb in the sense that has been nicely established by T. Flaminio and F. Montagna in [10]. In their paper, fuzzy logic is joined with probability theory.

There are already convincing formal solutions of the sorites (falakros) paradox which is core of the vagueness phenomenon. The solutions can be found in [15, 35].

Another, more general solution of the sorites paradox using fuzzy type theory is presented in [29] where also a solution with respect to the meaning of basic evaluative expressions "small" and "big" is presented. In [31], a model of higher order vagueness has also been proposed.

A list of some of the possible future directions of FLb is the following:

(*a*) Extend the repertoire of natural language expressions, for which a reasonable working mathematical model can be elaborated.

(*b*) Find a reasonable class of "intended models" for the theory of intermediate and generalized quantifiers.

(*c*) Extend the theory of generalized quantifiers (cf. [17, 37]) using the formalism of FLn. Some steps have already been done (cf. [6, 11, 16]).

(*d*) Study various forms of commonsense human reasoning and search for a reasonable formalization of them (cf. [8, 32]). As a subgoal, extend the list of generalized syllogisms with intermediate quantifiers (cf. [38, 30]).

(*e*) Develop a reasonable formalization of FLb on the basis of which the above mentioned constituents of PNL methodology (namely the World Knowledge Database and Multiagent, Modular Deduction Database) could be formed.

(*f*) Extend the technique started in [10] to be able to include also uncertainty inside FLb.

(*g*) Study the ways, how other formal systems of FLn (besides the fuzzy type theory) could be used for solution of problems raising in FLb and develop them accordingly.

In general we propose to shift interest from metamathematics of fuzzy logic to applications of the latter. One of the problems we face is quite often the great complexity of formal means when applied. For example, in [8, 32], a methodology for commonsense reasoning on an example of a solution of a detective story is presented. The methodology is quite general; in [9] it has also been applied to analysis of economic texts. A question is raised whether it is possible to simplify the formalization and not to loose the expressive power.

As can be seen, we face a lot of various kinds of interesting problems both inside fuzzy logic as well as in its applications that must be solved in the future.

REFERENCES

[1] L. Běhounek and P. Cintula. "Fuzzy Class Theory". *Fuzzy Sets and Systems*, **154**, 34–55, 2005.

[2] L. Běhounek and P. Cintula. "From Fuzzy Logic to Fuzzy Mathematics: A Methodological Manifesto". *Fuzzy Sets and Systems*, **157 (5)**, 642–646, 2006.

[3] P. Cintula, P. Hájek, and R. Horčík. "Formal Systems of Fuzzy Logic and their Fragments. *Annals of Pure and Applied Logic*. **150**, 40–65, 2007.

[4] E. Davis and L. Morgenstern. "Introduction: Progress in Formal Commonsense Reasoning". *Artifical Intelligence*. **153**, 1 - 12, 2004.

[5] A. Dvořák and V. Novák. "Fuzzy Logic Deduction with Crisp Observations". *Soft Computing*. **8**, 256–263, 2004.

[6] A. Dvořák and M. Holčapek. "L-fuzzy Quantifiers of the Type $\langle 1 \rangle$ Determined by Measures". *Fuzzy Sets and Systems*. 2009 (to appear).

[7] A. Dvořák and V. Novák. "Formal Theories and Linguistic Descriptions". *Fuzzy Sets and Systems*. **143**, 169–188, 2004.

[8] A. Dvořák and V. Novák. "Fuzzy Logic Model of Detective Reasoning". *In Proc. 11th Int. Conf. IPMU, Paris*, July 2006, **2**, pp. 1890–1897, Les Cordeliers, Paris, 2006. Éditions EDK.

[9] A. Dvořák and V. Novák. "Towards Automatic Modeling of Economic Texts". *Mathware and Soft Computing*, **14 (3)**, 217–231, 2007.

[10] T. Flaminio and F. Montagna. "Mv-algebras with Internal States and Probabilistic Fuzzy Logics. *International Journal of Approximate Reasoning*, 2009.

[11] I. Glöckner. "Fuzzy Quantifiers: A Computational Theory". Springer, Berlin, 2006.

[12] J.A. Goguen. "The Logic of Inexact Concepts. *Synthese*, **19**, 325–373, 1968–69.

[13] P. Hájek. "Metamathematics of Fuzzy Logic". Kluwer, Dordrecht, 1998.

[14] P. Hájek and P. Cintula. "On Theories and Models in Fuzzy Predicate Logics. *Journal of Symbolic Logic*, **71 (3)**, 863–880, 2006.

[15] P. Hájek and V. Novák. "The Sorites Paradox and Fuzzy Logic" *International Journal of General Systems*, **32**, 373–383, 2003.

[16] M. Holčapek. "Monadic *l*-fuzzy Quantifiers of the Type $\langle 1^n, 1 \rangle$. Fuzzy Sets and Systems, 159, 1811–1835, 2008.

[17] E.L. Keenan and D. Westerstå hl. "Quantifiers in Formal and Natural Languages. In J. van Benthem and A. ter Meulen, editors, *Handbook of Logic and Language*, pp. 837–893. Elsevier, Amsterdam, 1997.

[18] P. Materna. "*Concepts and Objects*". Acta Philosophica Fennica 63, Helsinki, 1998.

[19] P. Materna. "*Conceptual Systems*". Logos Verlag, Berlin, 2004.

[20] J. McCarthy. "Programs with Common Sense". *In Proc. of the Teddington Conf. on Mechanization of Thought Processes*, pp. 75–91, Her Majesty's Stationary Office, London, 1959.

[21] V. Novák. EQ-algebra-based Fuzzy Type Theory and its Extensions". *Logic Journal of the IGPL*, page (submitted)..

[22] V. Novák. "On the Syntactico-semantical Completeness of First-order Fuzzy Logic I, II. *Kybernetika*, **26**, 47–66. 134–154, 1990.

[23] V. Novák. "Towards Formalized Integrated Theory of Fuzzy Logic. In Z. Bien and K.C. Min, editors, "Fuzzy Logic and its Applications to Engineering", *Information Sciences, and Intelligent Systems*, pp. 353–363. Kluwer, Dordrecht, 1995.

[24] V. Novák. "On Fuzzy Type Theory". *Fuzzy Sets and Systems*, **149**, 235–273, 2005.

[25] V. Novák. "Perception-based Logical Deduction". In B. Reusch, editor, *Computational Intelligence, Theory and Applications*, pp. 237–250. Springer, Berlin, 2005.

[26] V. Novák. "Perception-based Logical Deduction as Alternative Approximate Reasoning Method". *In Proc. Int. Conf. FUZZ-IEEE 2005*, pp. 1032–1037, Reno, USA, May 2005, 2005.

[27] V. Novák. "Fuzzy Logic Theory of Evaluating Expressions and Comparative Quantifiers. *In Proc. 11th Int. Conf. IPMU*, Paris, July 2006, **2**, pp. 1572–1579, Les Cordeliers, Paris, 2006. Éditions EDK.

[28] V. Novák. "Which Logic is the Real Fuzzy Logic?" *Fuzzy Sets and Systems*, **157**, 635–641, 2006.

[29] V. Novák. "A Comprehensive Theory of Trichotomous Evaluative Linguistic Expressions". *Fuzzy Sets and Systems*. **159 (22)**, 2939–2969, 2008.

[30] V. Novák. "A Formal Theory of Intermediate Quantifiers". *Fuzzy Sets and Systems*, **159 (10)**, 1229–1246, 2008. , DOI: 10.1016/j.fss. 2007.12.008.

[31] V. Novák and A. Dvořák. "Fuzzy Logic: A Powerful Tool for Modeling of Vagueness". *Int. Journal of General Systems*, page (submitted).

[32] V. Novák and A. Dvořák. "Towards Commonsense Reasoning in Fuzzy Logic in Broader Sense. *Fuzzy Sets and Systems*, submitted.

[33] V. Novák and S. Lehmke. "Logical Structure of Fuzzy IF-THEN Rules". *Fuzzy Sets and Systems*, **157**, 2003–2029, 2006.

[34] V. Novák and I. Perfilieva. "On the Semantics of Perception-based Fuzzy Logic Deduction". *International Journal of Intelligent Systems*, **19**, 1007–1031, 2004.

[35] V. Novák, I. Perfilieva, and J. Močkoř. "Mathematical Principles of Fuzzy Logic". Kluwer, Boston, 1999.

[36] J. Pavelka. "On Fuzzy Logic I, II, III. Zeitschrift für Mathematische Logik und Grundlagen der Mathematik, **25**, 45–52, 119–134, 447–464, 1979.

[37] S. Peters and D. Westerståhl. "Quantifiers in Language and Logic". Claredon Press, Oxford, 2006.

[38] P.L. Peterson. "*Intermediate Quantifiers. Logic, Linguistics, and Aristotelian Semantics*. Ashgate, Aldershot, 2000.

[39] P. Sgall, E. Hajičová, and J. Panevová. "The Meaning of the Sentence in Its Syntactic and Pragmatic Aspects". D. Reidel, Dordrecht, 1986.

[40] L.A. Zadeh. "Fuzzy Sets". *Information and Control*, **8**, 338–353, 1965.

7

CONCRETE AND FORMAL PRE-TOPOLOGICAL SYSTEMS AND THEIR APPROXIMATION OPERATORS

Piero Pagliani

Research Group on Knowledge and Communication Models Via Imperia, 6. Roma, Italy

E-mail: pagliani@agora.it

Abstract: *Spaces with properties weaker than topological spaces are exploited in a number of fields, such as graph theory, data analysis, perception systems, modal logic, combinatorial chemistry, and others. In this paper we aim at linking these kinds of more general spaces with the notions of a pre-topological and a topological formal space, and investigate the relationships between the various approximation operators that they induce.*

1. INTRODUCTION

In a number of fields one is not interested in single elements of the domain of concern, *U*, but in the geometrical patterns induced by some connection relations between the elements of *U*. These connection relations are typically induced by the properties manifested by entities and perceived by a subject. Since properties hardly are exhaustive, connection relations typically cannot account for any subset of entities. However subsets of entities can be approximated on the basis of a connection relation or a set of connection relations. In some cases given a subset *X* of *U*, we can compute one or more lower approximations and one or more upper approximations. With a lower approximation we intend a map $\phi : \wp(U) \mapsto \wp(U)$ such that

$\phi(X) \subseteq X$ and $X \subseteq X'$ implies $\phi(X) \subseteq \phi(X')$. Dually, we understand an upper approximation as a map ψ: $\wp(U) \mapsto \wp(U)$ such that $X \subseteq \psi(X)$ and $X \subseteq X'$ implies $\psi(X) \subseteq \psi(X')$.

1.1 Pawlak Approximation Spaces

An intuitive setting which induces approximation operators is given by an equivalence relation $E \subseteq U \times U$. The structure $\langle U, E \rangle$ is called an *Indiscernibility Space*, and E is typically induced by a Pawlak's Information System: for all $u, u' \in U$, $\langle u, u' \rangle \in E$ if u and u' fulfill exactly the same set of properties from a given Information System.

Any binary relation $R \subseteq U \times U$ and set $X \subseteq U$ determine a subset $R(X) = \{u \in U : \exists x : x \in X \text{ and } \langle x, u \rangle \in R\}$, called the R–neighborhood of X[1]. A usual, given $u \in U$, $R(\{u\})$ will be shortened by $R(u)$. As is well-known, by exploiting the notion of an R-neighborhood, from an Indiscernibility Space one can define a pair of *approximation operators*:

$$(lE)(X) = \{E(Z) : E(Z) \subseteq X\}, (uE)(X)$$
$$= \{E(Z) : E(Z) \cap X \neq \emptyset\} \quad \text{... (1.1)}$$

$(lE)(X)$ is called the *lower approximation* of X (*via E*), while $(uE)(X)$ is called the *upper approximation of* X (*via E*) and $\langle U, (uE), (lE) \rangle$ is called a *Pawlak approximation space* or *approximation space* (cf. [Pawlak 1991]).

From now on the interested reader is addressed to [Pagliani and Chakraborty 2008].

1.2 Generalised Approximation Spaces

A number of generalisations of this approach were proposed in Rough Set literature. For instance, R could be a tolerance relations or even an arbitrary binary relations. In this case, some care must be taken about the definition of generalised lower and *generalised upper approximations*. Indeed, the general form is:

[1]Or the Right Peirce Product of R and X.

$$(lR)(X) = \bigcup\{Z : R(Z) \subseteq X\},$$

$$(uR)(X) = \bigcup\{Z : R(Z) \cap X \neq \emptyset\}, \qquad \text{... (1.2)}$$

which under particular conditions coincides with

$$(lR)(X) = \bigcup\{R(Z) : R(Z) \subseteq X\},$$

$$(uR)(X) = \bigcup\{R(Z) : R(Z) \cap X \neq \emptyset\} \qquad \text{... (1.3)}$$

Moreover, properties of generalised upper and lower approximations depend on the properties of the given relation R. We call $\langle U, (lR), (uR)\rangle$ a *generalised approximation space.*

1.3 Dynamic Approximation Spaces

However, taking into account just a single binary relation is appropriate in a "static frame-work". On the contrary, entities my be related by patterns which vary over time or which vary according to different points of view, in a "dynamic framework". In a dynamic framework, we have to take into account families of binary relations. Given a set of indexes I, we call $\langle U, \{R_i\}_{i \in I}\rangle$ a *dynamic space.*

1.4 Concrete Neighborhood Systems

Given a dynamic space, $\langle U, \{R_i\}_{i \in I}\rangle$, for any $u \in U$, $\{R_i(u)\}_{i \in I}$ is a subset of $\wp(U)$. Therefore, we can synthesise a dynamic space as a relational structure $\langle U, \wp(U), R\rangle$, where $R \subseteq U \times \wp(U)$ and for all $g \in U$ and $X \in \wp(U)$, $\langle u, X\rangle$ if and only if $X = R_i(u)$ for some index i. We call $R(u)$ a *neighborhood family* of u and denote it with $\mathcal{N}_u$, and $\{\mathcal{N}_u\}_{g \in U}$ a *concrete neighborhood system*, and denote it with $\mathcal{N}(U)$. Obviously, $\mathcal{N}_u = \{R_i(u)\}_{i \in I}$.

Differently from dynamic spaces, in concrete neighborhood systems we are interested in the system of relations as a whole. From the point of view of concrete neighborhood systems, a neighborhood of u collects the elements $R_i(u)$ that are close to u with respect to a specific criterium R_i (the R_i-neighbors of u) and the neighborhood family $R(u)$ collects the results of the application of all these criteria to u. Concrete neighborhood systems give rise to a pair of operators, G and, respectively, F called a core and a vicinity operator, respectively.

Concrete neighborhood systems may be equipped with a series of properties and the behaviour of the core and vicinity operators varies according to the accepted properties.

1.5 Property Systems

Patterns of points may be induced by a fulfillment relation between points and a set of properties M. In this case, approximation operators depends on a relation $R \subseteq U \times M$. We call $\langle U, M, R \rangle$ a *property system*. A property system gives rise to a pair of operators, *int* and, respectively, *cl*.

1.6 Formal Neighborhood Systems

Given a property system $\langle U, M, R \rangle$ if $\langle u, m \rangle \in R$, then m may be seen as a formal neighborhood of u. Intuitively it is a proxy for a subset of elements of G which fulfill a nearness relation with u, defined by means of the property m.

Therefore, we can lift the level of abstraction and consider $\langle U, M, R \rangle$ as an *abstract neighborhood basis*.

Notice that a structure $\langle U, \wp(U), R \rangle$ may be alternatively considered as a concrete neighborhood basis, when we consider the set-theoretical properties of each neighborhood family $\mathcal{N}_g$, or an abstract neighborhood basis, when we take into account just its abstract properties with no care of the elements of the neighborhoods. For this reason we shall call $\langle U, \wp(U), R \rangle$ with the intermediate term of a *formal neighborhood basis*.

Actually, on the basis of abstract neighborhood bases (alias, property systems), a new approach to *pre-topological* and *topological formal systems* has been introduced in [Sambin '87] and [Sambin '89] upon suggestions of P. Martin-Löf's Intuitionistic Type Theory.

Pre-topological formal systems are characterised by a pre-covering relation ◀ between elements of B and subsets of B. This pre-covering relation fulfills a one-one correspondence with an upper approximation operator $\mathcal{A}$ which acts on the formal side B and which is symmetric dual of the interior operator *int* acting on the concrete side U.

Moreover, ◀ may fulfill different properties which make a formal system span from pre-topology to topology.

From the above picture, some questions arises:

2. THREE QUESTIONS

We have seen that a dynamic space $\mathbf{S} = \langle U, \{R_i\}_{i \in I} \rangle$ can be synthesized into a concrete neighborhood basis $\mathbf{S'} = \langle U, \wp(U), R \rangle$. In turn the concrete neighborhood system **S'** is a property system. Therefore, we have:

(Q1) What are the relations between the approximation operators F and G of **S'** *qua* concrete neighborhood basis and the approximation operators *cl* and, respectively, *int* of **S'** *qua* property system?

Moreover, since any relational structure $\mathbf{Q} = \langle U, R \rangle$ is trivially a property system $\langle U, U, R \rangle$, the following question naturally arises:

(Q2) What are the relations between the operators *int* and *cl* of **Q** *qua* property system, and the operators (lR) and, respectively, (uR)?

Finally, we have what can be considered as the key question, because Q1 will be answered through the answer to the following:

(Q3) What are the relations between *int* and G in dependence of the properties fulfilled by a pre-covering ◀ induced by **S'** *qua* abstract neighborhood basis and those fulfilled by **S'** *qua* concrete neighborhood basis?

3. PRE-TOPOLOGICAL SPACES

Definition 3.1: Let $\langle U, \wp(U), R \rangle$ be a concrete neighborhood basis, then if $u' \in X \in R(u)$, (*i*) u' is called a concrete neighbor of u; (*ii*) X is called a *concrete neighborhood of* u; (*iii*) $R(u)$ is the concrete neighborhood family of u induced by R and (*iv*) $\mathcal{N}(U) = \{R(u): u \in U\}$ is called a concrete neighborhood system of u. In this context we shall denote $R(u)$ with $\mathcal{N}_u$. The pair $\langle U, \mathcal{N}(U)\rangle$ will be called a concrete neighborhood space. *Un* such a space:

Core: If for all $X \subseteq U$, $G(X) = \{x: X \in \mathcal{N}_x\}$, then G is called a *core map*;

Vicinity: If for all $X \subseteq U$, $F(X) = -G(-X)$, then F is called a *vicinity map*.

Concrete pre-topological properties on $\mathcal{N}(U)$. We can impose the following independent properties on a concrete neighborhood system. For any $x \in U$, and $A, N, N' \subseteq U$:

1: $U \in \mathcal{N}_x$; **0:** $\emptyset \notin= \mathcal{N}_x$; **Id:** if $x \in G(A)$ then $G(A) \in \mathcal{N}_x$; **N1:** if $N \in \mathcal{N}_x$ then $x \in N$; **N2:** if $N \in \mathcal{N}_x$ and $N \subseteq N'$, then $N' \in \mathcal{N}_x$; **N3:** if $N, N' \in \mathcal{N}_x$, then $N \cap N' \in N_x$; **N4:** there is an $N \neq \emptyset$ such that $\mathcal{N}_x = \uparrow N$ (i.e. the principal order filter generated by N).

Conditions **1**, **0**, **Id**, **N1**, **N2** and **N3** characterise topological spaces. But if we consider them separately we find that each one induces a particular property on the operators G and F:

Lemma 3.1 ([Pagliani 2003]): Let U be a set and $\mathcal{N}(U)$ a neighborhood system. Then, for any $X, Y \subseteq U$,

Condition	*Equivalent properties of G*	*Equivalent properties of F*
1	$G(U) = U$ (*normality*)	$F(\emptyset) = \emptyset$; (*normality*)
0	$G(\emptyset) = \emptyset$; (*co-normality*)	$F(U) = U$ (*co-normality*)
Id	$G(X) \subseteq G(G(X))$	$F(F(X)) \subseteq F(X)$
N1	$G(X) \subseteq X$ (*deflation*)	$X \subseteq F(X)$ (*inflation*)
N2	$X \subseteq Y \Rightarrow G(X) \subseteq G(Y)$ (*isotonicity*)	$X \subseteq Y \Rightarrow F(X) \subseteq F(Y)$
	$G(X \cap Y) \subseteq G(X) \cap G(Y)$	$F(X \cup Y) \supseteq F(X) \cup F(Y)$
N3	$G(X \cap Y) \supseteq G(X) \cap G(Y)$	$F(X \cup Y) \subseteq F(X) \cup F(Y)$

In particular, a neighborhood system in which **1**, **0**, **Id**, **N1** and **N2** hold, will be said to be of type $\mathcal{N}_{2Id}$. From the above Lemma, one can promptly deduce that in $\mathcal{N}_{2Id}$ neighborhood systems, G and F are interior and, respectively, closure operators which, however, are not topological. That is, they do not enjoy idempotency. In what follows, we shall deal only with spaces of type at least $\mathcal{N}_1$.

4. FORMAL RELATIONSHIPS BETWEEN CONCRETE POINTS AND OBSERVABLE PROPERTIES

Let us consider a pure generic relational structure $\langle A, B, R\rangle$, where $R \subseteq A \times B$, for A and B sets. The relation R induces the following functions:

- $\langle R\rangle : \wp(B) \mapsto \wp(A); \langle R\rangle(Y) = \{a \in A : \exists b(b \in Y \text{ and } \langle a, b\rangle \in R)\}$,
- $[R] : \wp(B) \mapsto \wp(A); [R](Y) = \{a \in A : \forall b(\langle a, b\rangle \in R \Rightarrow b \in Y)\}$,
- $\langle R^{\smile}\rangle : \wp(A) \mapsto \wp(B); \langle R^{\smile}\rangle(X) = \{b \in B : \exists a(a \in X \text{ and } \langle a, b\rangle \in R)\}$,
- $[R^{\smile}] : \wp(A) \mapsto \wp(B); [R^{\smile}](X) = \{b \in B : \forall a(\langle a, b\rangle \in R \Rightarrow a \in X)\}$,

From the above machinery we can define the following functions:

1. $int : \wp(A) \mapsto \wp(A); int(X) = \langle R\rangle([R^{\smile}](X))$,
2. $cl : \wp(A) \mapsto \wp(A); cl(X) = [R](\langle R^{\smile}\rangle(X))$,
3. $\mathcal{A} : \wp(B) \mapsto \wp(B); \mathcal{A}(Y) = [R^{\smile}](\langle R\rangle(Y))$,
4. $\mathcal{C} : \wp(B) \mapsto \wp(B); \mathcal{C}(Y) = \langle R^{\smile}\rangle([R](Y))$,

It is not difficult to verify that, *int* (X) and *cl* (X) generalise the usual definitions of an interior and, respectively, a closure of a set $X \subseteq A$. Symmetrically, for $Y \subseteq B$, $\mathcal{A}(Y) = Y$ expresses the fact that Y is a "formal open subset" while $\mathcal{C}(Y) = Y$ means that Y is a "formal closed subset". Moreover, since one can prove that $\langle\langle R\rangle, [R^{\smile}]\rangle$ and $\langle\langle R^{\smile}\rangle, [R]\rangle$ are adjoint pairs, one obtains that *int* and $\mathcal{C}$ are indeed interior operators and *cl* and $\mathcal{A}$ are indeed closure operators. However, they lack idempotency, thus are pre-topological. Further, from the properties of adjoint operator, one can straightforwardly prove that the following inequalities hold, for any $X \subseteq U$: *int* $(X) \subseteq X \subseteq$ *cl* (X) and *cl* (X) is a minimal set enjoying this inequality relation, while *int* (X) is maximal.

Therefore, $cl(X)$ is an upper approximation and $int(X)$ is a lower approximation of X.

Eventually, in a square property system $\langle U, U, R\rangle$, the operator *int* coincides with (lR) and *cl* coincides with (uR). It follows that if R is an equivalence relation, *cl* and *it* are the usual Pawlak approximation operators.

4.1 Formal Semi-covering Relations

Given a relational structure, called a "basic pair", $\langle A, B, R\rangle$, the operator $\mathcal{A}$ makes it possible to synthesise the formal properties of pre-topologies in logical (or, better, type theoretical) terms. From [Sambin '89], the starting point is the definition, by means of the operator $\mathcal{A}$, of a relation, denoted with the symbol $\blacktriangleleft$ (in what follows we shall write $b \blacktriangleleft b'$ instead of $b \blacktriangleleft \{b'\}$ and $\langle R\rangle(x)$ instead of $\langle R\rangle(\{x\})$.

Definition 4.1: Let $\langle A, B, R\rangle$ be a basic pair. Then for any $b \in B$ and $Y, Y' \subseteq B$, the following relation is called a *formal semi-cover*, or shortly a *semi-cover*:

(basis): $b \blacktriangleleft Y$ iff $b \in \mathcal{A}(Y)$, **(step)** $Y \blacktriangleleft Y'$ iff $\forall y \in Y, y \blacktriangleleft Y'$.

Observable properties are combinable in principle. Therefore, from the abstract point of view B is assumed to be equipped with a binary operation "$\cdot$", the *combination*, or the *fusion*, of b and b', which represents an abstraction of intersection between concrete neighborhoods.

In particular three **abstract pre-topological properties** dealing with fusion are required to be carefully analysed:

$$\frac{b \blacktriangleleft Y}{b \cdot b' \blacktriangleleft Y}(left); \frac{b \blacktriangleleft Y \quad b \blacktriangleleft Y'}{b \blacktriangleleft Y \cdot Y'}(right); \frac{b \blacktriangleleft Y \quad b' \blacktriangleleft Y'}{b \cdot b' \blacktriangleleft Y \cdot Y'}(stability)$$

Indeed, one can prove that *int* (hence *cl*) is a topological operator if and only if its formal counterpart $\mathcal{A}$ is such that the above three properties hold.

What are, if any, the relationships between the above abstract pre-topological properties and the concrete pre-topological properties we have seen before?

Proposition 4.1: Let $\langle \wp (U), \cup, U, \blacktriangleleft, \perp\rangle$ be a formal neighborhood system. *If* $\{R(x)\}_{x \in U}$ fulfills **N3**, *then (right) holds.*

But the converse does not hold.

On the contrary **N2** and (*left*) are equivalent:

Proposition 4.2: Let $\langle \wp (U), \cap, U, \blacktriangleleft, \perp\rangle$ be a formal neighborhood system. Then, $\{R(x)\}_{x \in U}$ is a neighborhood system fulfilling **N2** if and only if (left) holds.

Moreover, (*stability*), (*right*) and (*left*) are independent of **N1** and **Id**, and, more generally, there is no formal property representing **N1** and **Id**.

However, in a sense **Id** and **N1** are embedded in $\blacktriangleleft$, *via* the closure properties of $\mathcal{A}$.

If we consider the double nature, concrete and abstract, of a neighborhood basis $\langle U, \wp (U), R\rangle$ we have to understand what properties must be satisfied by the neighborhood system $\mathcal{N}(U)$ in order to make G fulfill the same properties as *int*. Trivially, the immediate answer is: "Since *int* is an interior operator, G must be an interior operator, as well. Therefore in view

of *Lemma* 3.1, $\mathcal{N}(U)$ must fullfil **N1**, **N2** and **Id**".

Anyway, we want to know if the above three properties, **N1**, **N2** and **Id**, co-operate in order to make G **coincide** with *int*.

According to the definition, $int(X) = \{a \in U: \exists X' (X' \in R(a) \; \& \; R^{\smile}(\{X'\}) \subseteq X)\}$.

Therefore our question turns into the following: does the recursive equation

$$G(X) = \{a: \exists X' (X' \in R(a) \; \& \; G(X') \subseteq X)\}$$

have a solution when $\{R(x)\}_{x \in U}$ is of type $\mathcal{N}_{2Id}$?

The answer is affirmative.

Proposition 4.3: $\langle U, \wp (U), R\rangle$ be a basic neighborhood pair and $\mathcal{N}(U) = \{R(x)\}_{x \in U}$. Then, for all $X \in \wp (U)$, $G(X) = int(X)$ if and only if $\mathcal{N}(U)$ is of type $\mathcal{N}_{2Id}$.

REFERENCES

[1] [Belmandt '93], Z. Belmandt. "Manuel de Prétopologie et ses Applications". Hermes. 1993.

[2] [Demri & Orłowska 2002], S. Demri and E. Orłowska, "Incomplete Information: Structure, Inference, Complexity". Springer, 2002.

[3] [Lin '98] T.Y. Lin: "Granular Computing on Binary Relations. I: Data Mining and Neighborhood Systems. II: Rough Set Representation and Belief Functions". In L. Polkowski & A. Skowron (Eds.) "Rough Sets in Knowledge Discovery. 1: Methodology and Applications", Physica-Verlag, 1988, pp. 107–121 and 122–140.

[4] [Pagliani and Chakraborty 2008] P. Pagliani and M. Chakraborty, "A Geometry of Approximation". Springer 2008.

[5] [Pawlak 1991] Z. Pawlak, "Rough Sets: A Theoretical Approach to Reasoning About Data". Kluwer, 1991.

[6] [Sambin '87] G. Sambin, "Intuitionistic Formal Spaces—a first communication". In D. Skordev (Ed.) "Mathematical Logic and its Applications", Plenum Press, 1987, pp. 187–204.

[7] [Sambin '89] G. Sambin: "Intuitionistic Formal Spaces and their Neighbourhood". In Ferro, Bonotto, Valentini and Zanardo (Eds.), *Logic Colloquium '88*, Elsevier (North-Holland), 1989, pp. 261–285.

8

DISCOVERING AFFINITIES BETWEEN PERCEPTUAL GRANULES, TOLERANCE NEAR SET APPROACH

James F. Peters[1] and Sheela Ramanna[1, 2]

[1]Computational Intelligence Laboratory, Department of Electrical and Computer Engineering, University of Manitoba Winnipeg, Manitoba R3T 5V6, Canada
E-mail: jfpeters@ee.umanitoba.ca

[2]Department of Applied Computer Science, University of Winnipeg, Winnipeg, Manitoba R3B 2E9,
E-mail: s.ramanna@uwinnipeg.ca

Abstract: *This paper proposes an approach to detecting chains of affinities between perceptual objects contained in perceptual granules such as tolerance classes in digital image coverings. A perceptual object is something perceptible to the senses or knowable by the mind. Perceptual objects that have similar appearance are considered perceptually near each other, i.e., perceived objects that have perceived affinities or, at least, similar descriptions. A perceptual granule is a finite, non-empty set containing sample perceptual objects with common descriptions. Perceptual granules originate from observations of the objects in the physical world. Similarities between perceptual granules are measured within the context of what is known as a tolerance near space. This form of tolerance space is inspired by C.E. Zeeman's work on visual perception and H. Poincaré's work on the contrast between mathematical continua and the physical continua in a pragmatic philosophy of science that laid the foundations for tolerance spaces. The perception of nearness or closeness that underlies tolerance near relations is rooted in M. Merleau-Ponty's work on the phenomenology of perception during the mid-1940s, and, especially, philosophical reflections on description of perceived objects and the perception of nearness. Pairs of perceptual granules such as images are considered near each other to the extent that elements of tolerance classes in digital image coverings have similar descriptions. The contribution of this article is* *twofold, namely, a proposed tolerance near set-based approach to solving the image correspondence problem and a comparison the results obtained using several tolerance spacebased image resemblance measures.*

Keywords: *Affinities, image, perceptual granule, similarity, tolerance near set.*

Our classifications are often plainly influenced by chains of affinities.

–Charles Darwin, 1859.

1. INTRODUCTION

This paper focuses on various approaches to detecting chains of affinities between perceptual objects contained in perceptual granules such as images with tolerance near preclasses. A *perceptual object* is something perceptible to the senses or knowable by the mind. Perceptual objects that have similar appearance are considered perceptually near each other, i.e. perceived objects that have perceived affinities or, at least, similar descriptions. In general, the term *affinity* means *close relationship based on a common origin or structure* [12]. In this paper, the term *affinity* means *close relationship between perceptual granules (particularly images) based on common description* [20].

In particular, this paper introduces a tolerance near set solution to the image correspondence problem [18], i.e. where one uses image matching strategies to establish a correspondence between one or more images. This is one of the central tasks in photogrammetry and computer vision. Recently, it has been shown that near sets can be used in a perception-based approach to discovering correspondences between images (see, e.g. [8, 9, 11, 18, 20, 22]). Sets of perceptual objects where two or more of the objects have matching descriptions are called near sets. Detecting image resemblance and formulating image description are part of the more general pattern recognition process enunciated by K. Cyran and A. Mrózek in 2001 [3]. Work on a basis for near sets began in 2002, motivated by image analysis and inspired by a study of the perception of the nearness of perceptual objects carried out in cooperation with Z. Pawlak in [14]. This initial work led to the introduction of near sets [17], elaborated in [15, 18, 19, 21, 22]. A perception-based approach to discovering resemblances between images leads to a tolerance space form of near sets that models human perception in a physical continuum.

The term *tolerance space* was coined by E.C. Zeeman in 1961 in modelling visual perception with tolerances [28]. A tolerance space is a set X supplied with a binary relation $\simeq$ (i.e., a subset $\simeq \subset X \times X$) that is reflexive (for all $x \in X$, $x \simeq x$) and symmetric (for all $x, y \in X$, $x \simeq y$ and $x \simeq y$) but transitivity of $\simeq$ is not required. For example, it is possible to define a tolerance space relative to subimages of an image. This is made possible by assuming that each image is a set of fixed points. Let O denote a set of perceptual objects (e.g. grey level subimages) and let $gr(x)$ = average grey level of subimage x. Then the tolerance relation $\simeq_{gr}$ is defined as $\simeq_{gr} = \{(x, y) \in O \times O \mid |gr(x) - gr(y)| \leq \varepsilon\}$ for some tolerance $\varepsilon \in \Re$ (reals). Then $(O, \simeq_{gr})$ is a sample tolerance space. A tolerance denoted by ε is directly related to the exact idea of closeness or resemblance (i.e. being within some tolerance) in comparing objects. The basic idea is to find objects such as images that resemble each other with a tolerable level of error. Sossinsky [27] observes that main idea underlying tolerance theory comes from **H**enri **P**oincaré [23]. Physical continua (e.g., measurable magnitudes in the physical world of medical imaging [5]) are contrasted with the mathematical continua (real numbers) where almost solutions are common and a given equation have no exact solutions. An *almost solution* of an equation (or a system of equations) is an object which, when substituted into the equation, transforms it into a numerical 'almost identity', i.e. a relation between numbers which is true only approximately (within a prescribed tolerance) [27]. Equality in the physical world is meaningless, since it can never be verified either in practice or in theory. Hence, the basic idea in a tolerance space view of images, for example, is to replace the indiscerniblity relation in rough sets [13] with a tolerance relation in partitioning images into homologous regions where there is a high likelihood of overlaps, i.e. non-empty intersections between image tolerance classes. The use of image tolerance spaces in this work is directly to recent work on tolerance spaces (see, e.g., [2, 4, 24–26, 29]). The contribution of this article is twofold, namely, a proposed tolerance near set-based approach to solving the image correspondence problem and a comparison the results obtained using several tolerance space-based image resemblance measures.

This paper has the following organization. A brief introduction to tolerance near sets is given in Sect. 2. This is followed in Sect. 3 by a presentation of four nearness measures. A report on the results of experiments with pairs of images using the Meghdadi image nearness toolset is given in Sect. 4.

2. TOLERANCE NEAR SETS

This section introduces the basic notions underlying a tolerance near set approach to detecting image resemblances.

Table 1: *Nomenclature*

Symbol	*Interpretation*
O, X	Set of perceptual objects, $X, Y \subseteq O$,
$\mathbb{F}, \mathcal{B}$	Sets of probe functions, $\mathcal{B} \subseteq \mathbb{F}$,
ε	$\varepsilon \in \Re$ (reals) such that $\varepsilon \geq 0$,
$\phi_i(x)$	ith probe function representing feature of x,
$\phi(x)$	$(\phi_1(x), \phi_2(x), \ldots, \phi_i(x), \ldots, \phi_l(x))$, $\phi_i \in \mathbb{F}$, $x \in O$, description of x,
$\cong_{\mathcal{B},\varepsilon}$	$\{(x, y) \in O \times O : \lVert\phi(x) - \phi(y)\rVert_2 \leq \varepsilon\}$ tolerance relation
$x/_{\cong_{\mathcal{B},\varepsilon}}$	$\{y \in X \mid x \cong_{\mathcal{B},\varepsilon} y\}$, tolerance class,
$X/_{\cong_{\mathcal{B},\varepsilon}}$	$\{x/_{\cong_{\mathcal{B},\varepsilon}} \mid x \in X\}$, covering of X,
$\mathbb{C}_i^X$	$x_i/_{\cong_{\mathcal{B},\varepsilon}}$, tolerance class for $x_i \in X$,
$\bigcap_{\mathcal{B},\varepsilon}$	$X \bigcap Y = \{(x, y) \in X \times Y : \lVert\phi(x) - \phi(y)\rVert_2 \leq \varepsilon\}$,
$\bigcap_{\cong_{\mathcal{B},\varepsilon}}$	$\mathbb{C}_i^X \bigcap_{\cong_{\mathcal{B}}} \mathbb{C}_j^Y = \{(x, y) \in \mathbb{C}_i^X \times \mathbb{C}_j^Y : \lVert\phi(x) - \phi(y)\rVert_2 \leq \varepsilon\}$,
$X \bowtie_{\cong_{\mathcal{B},\varepsilon}}$	X resembles Y.

2.1 Probe Function

Definition 1 Probe Function [16]: A probe function is a real-valued function representing a feature of a physical object.

Examples of probe functions are the colour, size, weight of an object. Probe functions are used to describe an object to determine the characteristics and perceptual similarity of perceivable physical objects. Perceptual information is always presented with respect to probe functions just as our senses define our perception of the world. For example, our ability to view light in the "visible spectrum" rather than infra red or microwaves spectra defines our perception of the world just as the selection of probe functions constrains the amount of perceptual information available for extraction from a set of objects.

2.2. Perceptual Systems

Definition 2 Perceptual System [21]: A perceptual system $\langle O, \mathbb{F} \rangle$ is a real valued total deterministic information system where O is a non-empty set of perceptual objects and $\mathbb{F}$ is a countable set of probe functions.

The notion of a perceptual system admits a wide variety of different interpretations that result from the selection of sample perceptual objects contained in a particular sample space O. Examples of perceptual objects include observable organism behaviour, growth rates, soil erosion, microfossils, events containing the outcomes of experiments such as energizing a network, microscope images, MRI scans, and the results of searches for relevant web pages.

2.3 L_2 Norm-based Object Description

The description of an object $x \in O$ is given by

$$\phi(x) = (\phi_1(x), \phi_2(x), \ldots, \phi_i(x), \ldots, \phi_l(x)),$$

where l is the length of the description vector ϕ and each $\phi_i(x)$ is a probe function that describes the object x. Let $d_i = \phi(x) - \phi(y)$ denote the ith vector difference relative to descriptions $\phi(x)$, $\phi(y)$ for objects $x, y \in O$. Then let $\mathbf{d}^T$, $\mathbf{d}$ denote row and column vectors of description differences, respectively, i.e.,

$$\mathbf{d}^T = (d_1 \ldots d_k),\ \mathbf{d} = \begin{bmatrix} d_1 \\ \cdots \\ d_k \end{bmatrix}$$

Finally, the overall distance computed using (1) is the L_2 norm $\|\mathbf{d}\|_2$ for a vector $\mathbf{d}$, i.e.,

$$\|\mathbf{d}\|_2 = \left(\mathbf{d}^T\mathbf{d}\right)^{\frac{1}{2}} = \sqrt{\sum_{i=1}^{k} d_i^2} \quad \text{... (1)}$$

In general, $\|\cdot\|_2$ denotes the length of a vector in L_2 space [10].

2.4 Perceptual Tolerance Relation

Definition 3 Perceptual Tolerance Relation [18]: Let $\langle O, \mathbb{F}\rangle$ be a perceptual system and let $\varepsilon \in \mathbb{R}$. For every $\mathcal{B} \subseteq \mathbb{F}$ the tolerance relation $\cong_{\mathcal{B}}$ is defined as follows:

$$\cong_{\mathcal{B},\varepsilon} = \{(x, y) \in O \times O: \|\phi(x) - \phi(y)\|_2 \leq \varepsilon\}.$$

For notational convenience, this relation can be written $\cong_{\mathcal{B}}$ instead of $\cong_{\mathcal{B},\varepsilon}$ with the understanding that ε is inherent to the definition of the tolerance relation.

As in the case with the indiscernibility relation, a tolerance class can be defined as

$$x/\cong_{\mathcal{B}} = \{y \in O \mid y \cong_{\mathcal{B}} x' \;\forall\, x' \in x/\cong_{\mathcal{B}}\}. \quad \text{... (2)}$$

Notice $(O, \cong_{\mathcal{B}})$ is an example of a tolerance space. The relation $\cong_{\mathcal{B}}$ in Defn. 3 defines a covering of O instead of partitioning O because an object can belong to more than one class. As a result, (2) is called a tolerance class instead of an elementary set. In addition, each pair of objects x, y in a tolerance class $x/\cong_{\mathcal{B}}$ must satisfy the condition $\|\phi(x)-\phi(y)\|_2 \leq \varepsilon$. Next, a quotient set for a given a tolerance relation is the set of all tolerance classes and is defined as

$$O/\cong_{\mathcal{B}} = \{x/\cong_{\mathcal{B}} \mid x \in O\}.$$

The following simple example highlights the need for a tolerance relation as well as demonstrates the construction of tolerance classes from sample data. Consider the 20 objects in Table 2 where $|\phi(x_i)| = 1$. Letting $\varepsilon = 0.1$ gives the following tolerance classes:

$$X / \cong_{\mathcal{B}} = \{\{x_1, x_8, x_{10}, x_{11}\}, \{x_1, x_9, x_{10}, x_{11}, x_{14}\},$$
$$\{x_2, x_7, x_{18}, x_{19}\},$$
$$\{x_3, x_{12}, x_{17}\},$$
$$\{x_4, x_{13}, x_{20}\}, \{x_4, x_{18}\},$$
$$\{x_5, x_6, x_{15}, x_{16}\}, \{x_5, x_6, x_{15}, x_{20}\},$$
$$\{x_6, x_{13}, x_{20}\}\}$$

Observe that each object in a tolerance class satisfies the condition $\|\phi(x) - \phi\|_2 \leq \varepsilon$, and that almost all of the objects appear in more than one class.

Definition 4 Tolerance Near Sets [18]: Let $\langle O, \mathbb{F}\rangle$ be a perceptual system, where O is a set of images (sets of points) and $\mathcal{B} \subseteq \mathbb{F}$ be a set of probe functions representing image features. Let $\mathcal{B}$ contains probe functions used to measure

Table 2: *Tolerance Class Example*

x_i	$\phi(x)$	x_i	$\phi(x)$	x_i	$\phi(x)$	x_i	$\phi(x)$
x_1	.4518	x_6	.6943	x_{11}	.4002	x_{16}	.6079
x_2	.9166	x_7	.9246	x_{12}	.1910	x_{17}	.1869
x_3	.1398	x_8	.3537	x_{13}	.7476	x_{18}	.8489
x_4	.7972	x_9	.4722	x_{14}	.4990	x_{19}	.9170
x_5	.6281	x_{10}	.4523	x_{15}	.6289	x_{20}	.7143

features of subimages in $X, Y \subset O$. A set X is perceptually near a Y *within* the perceptual system $\langle O, \mathbb{F}\rangle$ (i.e., $(X \underline{\bowtie}_{\mathbb{F},\varepsilon} Y)$) iff there are $x \in X$ and $y \in Y$ and there is $\mathcal{B} \subseteq \mathbb{F}$ such that $x \cong_{\mathcal{B},\varepsilon} y$.

3. RESEMBLANCE MEASURES

Tolerance classes can be viewed as structural elements in representing an image. The motivation for using tolerance classes in perceptual image analysis is the conjecture that visual perception in the human perception is performed in the class level rather than pixel level.

3.1 Description-based Set Intersection Nearness Measure

Because we want to consider the problem of comparing objects in pairs of disjoint sets such as separate images and, yet, we want to gather together from the disjoint sets those objects that have feature-values in common. This problem is of interest because its solution provides a formal basis for an image resemblance measure considered in the context of tolerance near sets. To solve this problem, we introduce a description-based intersection of sets.

Definition 5: Description-Based Tolerance Class Intersection [19].

Let $\langle O, \mathbb{F}\rangle$ be a perceptual system and let $X, Y \subseteq O$, $\mathcal{B} \subseteq \mathbb{F}$ and let $\mathbb{C}_i^X, \mathbb{C}_j^Y \subseteq X_{/\cong_\mathcal{B}}$, $Y_{/\cong_\mathcal{B}}$, respectively. The notation $\mathbb{C}_i^X$ denotes $x_i / \cong_\mathcal{B}$, a tolerance class for $x_i \in X$ and $\mathbb{C}_j^Y$ denotes $y_j / \cong_\mathcal{B}$, a tolerance class for $y_j \in Y$. Then

$$\mathbb{C}_i^X \underset{\cong_\mathcal{B}}{\cap} \mathbb{C}_j^Y = \left\{(x, y) \in C_i^X \times C_j^Y : \left\|\phi(x) - \phi(y)\right\|_2 \leq \varepsilon\right\}.$$

Table 3: *Feature Values for a Pair of Samples*

X	ϕ	Y	ϕ
x_1	0.7	y_1	0.8
x_2	0.75	y_2	0.85

Exam-'e 1: *Sample Tolerance Class Intersections*

$$\varepsilon = 0.2, \qquad \text{... (3)}$$

$$X_{/\cong_\phi, 0.2} = \{\{x_1, x_2\}\}, \qquad \text{... (4)}$$

$$Y_{/\cong_\phi, 0.2} = \{\{y_1, y_2\}\}, \qquad \text{... (5)}$$

$$\mathbb{C}_1^X \underset{\cong_{\mathcal{B}, 0.2}}{\cap} \mathbb{C}_1^Y = \{x_1, y_1\}, \qquad \text{... (6)}$$

$$\mathbb{C}_1^X \underset{\cong_{\mathcal{B}, 0.2}}{\cap} \mathbb{C}_2^Y = \{x_1, y_2\}, \qquad \text{... (7)}$$

$$\mathbb{C}_2^X \underset{\cong_{\mathcal{B}, 0.2}}{\cap} \mathbb{C}_1^Y = \{x_2, y_1\}, \qquad \text{... (8)}$$

$$\mathbb{C}_2^X \underset{\cong_{\mathcal{B},0.2}}{\cap} \mathbb{C}_2^Y = \{x_2, y_2\}. \qquad \text{... (9)}$$

Notice, then, that this example illustrates the fact that the magnitude of $\left|\mathbb{C}_i^X \underset{\cong_{\mathcal{B}}}{\cap} \mathbb{C}_j^Y\right|$ can equal the product of the sample sizes, i.e.,

$$\left|\mathbb{C}_i^X \underset{\cong_{\mathcal{B},0.2}}{\cap} \mathbb{C}_j^Y\right| = \left|\mathbb{C}_i^X\right| \cdot \left|\mathbb{C}_j^Y\right|.$$

Let $i, j \in I_X, I_Y$ be distinct indices used to identify the classes $\mathbb{C}_i^X$, $\mathbb{C}_j^Y$. Notice, also, that we can formulate a relationship between the description-based intersection $\mathbb{C}_i^X \underset{\cong_{\mathcal{B}}}{\cap} \mathbb{C}_j^Y$ and the product of the class sizes $\left|\mathbb{C}_i^X\right| \cdot \left|\mathbb{C}_j^Y\right|$.

Proposition 1

$$\sum_{i \in I_X,\, j \in I_Y} \left|\mathbb{C}_i^X \underset{\cong_{\mathcal{B},\varepsilon}}{\cap} \mathbb{C}_j^Y\right| \leq \sum_{i \in I_X,\, j \in I_Y} \left|\mathbb{C}_i^X\right| \cdot \left|\mathbb{C}_j^Y\right|.$$

A measure of the resemblance (i.e., image nearness measure $NM_{\cong_B}(X, Y)$) between a pair of images X, Y is then formulated in Def. 6.

Definition 6: *Tolerance Class Intersection-Based Nearness Measure [19]*

Let $\langle O, \mathbb{F}\rangle$ be a perceptual system and assume $X, Y \subseteq O$ with coverings defined using the tolerance relation $\cong_{\mathcal{B}}$. Let $\mathbb{C}_i^X$, $\mathbb{C}_j^Y$ denote tolerance classes in $X_{/\cong_{\mathcal{B}}}$, $Y_{/\cong_{\mathcal{B}}}$, respectively. Then a tolerance class intersection-based nearness measure ($tiNM_{\cong_{\mathcal{B}}}(X, Y)$) is given in (10) to measure the degree of nearness of X, Y is computed using

$$tiNM_{\cong_{\mathcal{B},\varepsilon}}(X, Y) = \frac{\displaystyle\sum_{i \in I_X,\, j \in I_Y} \left|\mathbb{C}_i^X \underset{\cong_{\mathcal{B},\varepsilon}}{\cap} \mathbb{C}_j^Y\right|}{\displaystyle\sum_{i \in I_X,\, j \in I_Y} \left|\mathbb{C}_i^X\right| \cdot \left|\mathbb{C}_j^Y\right|}.$$

3.2 Tolerance Class Overlap Distribution Nearness Measure

A tolerance class overlap distribution (TOD) nearness measure is proposed here based on statistical comparison of overlaps between tolerance classes at each subimage. The proposed method is as follows:

Suppose $X, Y \in O$ are two images (sets of perceptual objects). The sets of all tolerance classes for image X and Y are shown as follow and form a covering for each image.

$$X_{/\sim\mathcal{B},\varepsilon} = \{x_{/\sim\mathcal{B},\varepsilon} \mid x \in X\} \qquad \dots (11)$$

$$Y_{/\sim\mathcal{B},\varepsilon} = \{y_{/\sim\mathcal{B},\varepsilon} \mid y \in Y\} \qquad \dots (12)$$

Subsequently, the set of all overlapping tolerance classes corresponding to each object (subimage) x is named as $\Omega_{X_{\sim\mathcal{B},\varepsilon}}(x)$ and is defined as follows:

$$\Omega_{X_{\sim\mathcal{B},\varepsilon}}(x) = \{\mathcal{Z}_{/\sim\mathcal{B},\varepsilon} \in X_{/\sim\mathcal{B},\varepsilon} \mid x \in \mathcal{Z}_{/\sim\mathcal{B},\varepsilon}\} \qquad \dots (13)$$

Consequently, the normalized number of tolerance classes in $X_{/\sim\mathcal{B},\varepsilon}$ which are overlapping at x is named as ω and defined as follow:

$$\omega_{X_{\sim\mathcal{B},\varepsilon}}(x) = \frac{\left|\Omega_{X/\sim\mathcal{B},\varepsilon}(x)\right|}{\left|X_{/\sim\mathcal{B},\varepsilon}\right|} \qquad \dots (14)$$

Similarly, the set of all overlapping tolerance classes at every subimage $y \in Y$ is denoted by $\omega_{Y_{\sim\mathcal{B},\varepsilon}}(y)$. Assuming that the set of probe functions $\mathcal{B}$ and the value of ε are known, we use the more simplified notation of $\Omega_X(x)$ and $\omega_X(x)$ for the set $X_{/\sim\mathcal{B},\varepsilon}$ and the notations $\Omega_Y(y)$ and $\omega_Y(y)$ for the set $Y_{/\sim\mathcal{B},\varepsilon}$. Now, let $F_{\omega X}(\omega)$ and $F_{\omega Y}(\omega)$ be the empirical commutative distribution functions (CDF) of the functions $\omega_X(x)$ and $\omega_Y(y)$ respectively, when $x \in X$ and $y \in Y$. The *Tolerance Overlap Distribution* (*TOD*) nearness measure is defined as follow:

$$TOD = 1 - \left(\int_{\omega=0}^{\omega=1} \left(F_{\omega_X}(\omega) - F_{\omega_Y}(\omega) \right) d\omega \right)^{\gamma} \qquad \text{... (15)}$$

3.3 Tolerance Class Size-Based Nearness Measure

A tolerance class size-based nearness (tNM) measure is based on the idea that if one considers the union of two images as the set of perceptual objects, tolerance classes should contain almost equal number of subimages from each image. Therefore, the tolerance nearness measure between two image is calculated as follows:

Suppose X and Y are the sets of perceptual objects (subimages) in image 1 and image 2. $Z = X \cup Y$ is the set of all perceptual objects in the union of images and for each $z \in Z$ the tolerance class is show as:

$$z_{/\sim_{\mathcal{B},\varepsilon}} = \left\{ s \in Z \mid \left\| \phi_{\mathcal{B}}(z) - \phi_{\mathcal{B}}(s) \right\| \leq \varepsilon \right. \qquad \text{... (16)}$$

The part of the tolerance class that is a subset of X is named as $\left[z_{/\sim_{\mathcal{B},\varepsilon}} \right]_{\subseteq Y}$. and similarly, part of the tolerance class that is a subset of Y is named $\left[z_{/\sim_{\mathcal{B},\varepsilon}} \right]_{\subseteq Y}$. Therefore:

$$\left[z_{/\sim_{\mathcal{B},\varepsilon}} \right]_{\subseteq X} \triangleq \left\{ x \in z_{/\sim_{\mathcal{B},\varepsilon}} \mid x \in X \right\} \subseteq z_{/\sim_{\mathcal{B},\varepsilon}} \qquad \text{... (17)}$$

$$\left[z_{/\sim_{\mathcal{B},\varepsilon}} \right]_{\subseteq Y} \triangleq \left\{ y \in z_{/\sim_{\mathcal{B},\varepsilon}} \mid y \in Y \right\} \subseteq z_{/\sim_{\mathcal{B},\varepsilon}} \qquad \text{... (18)}$$

$$z_{/\sim_{\mathcal{B},\varepsilon}} = \left[z_{/\sim_{\mathcal{B},\varepsilon}} \right]_{\subseteq X} \cap \left[z_{/\sim_{\mathcal{B},\varepsilon}} \right]_{\subseteq Y} \qquad \text{... (19)}$$

Subsequently, the measure tNM is defined as the weighted average of the closeness between the cardinality (size) of sets $\left[z_{/\sim_{\mathcal{B},\varepsilon}} \right]_{\subseteq X}$ and the cardinality of $\left[z_{/\sim_{\mathcal{B},\varepsilon}} \right]_{\subseteq Y}$ where the cardinality of $z_{/\sim_{\mathcal{B},\varepsilon}}$ is used as the weighting factor.

$$tNM = \frac{1}{\sum_{z_{/\sim_{\mathcal{B},\varepsilon}}} \left| z_{/\sim_{\mathcal{B},\varepsilon}} \right|} \times$$

$$\sum_{z_{/\sim_{\mathcal{B},\varepsilon}}} \frac{\min\left(\left| \left[z_{/\sim_{\mathcal{B},\varepsilon}} \right]_{\subseteq X} \right|, \left| \left[z_{/\sim_{\mathcal{B},\varepsilon}} \right]_{\subseteq Y} \right| \right)}{\max\left(\left| \left[z_{/\sim_{\mathcal{B},\varepsilon}} \right]_{\subseteq X} \right|, \left| \left[z_{/\sim_{\mathcal{B},\varepsilon}} \right]_{\subseteq Y} \right| \right)} \times \left| z_{/\sim_{\mathcal{B},\varepsilon}} \right| \quad \text{... (20)}$$

3.4 Hausdorff Distance and Image Correspondence Measure

The Hausdorff distance is used to measure the distance between sets in a metric space [6] (see [7] for English translation), and is defined as

$$d_H(X, Y) = \max\left\{ \sup_{x \in X} \inf_{y \in Y} d(x, y), \sup_{y \in Y} \inf_{x \in X} d(x, y) \right\},$$

where sup and inf refer to the supremum and infimum, and $d(x, y)$ is the distance metric (in this case it is the l_2 norm). The Hausdorff distance is given graphically in Fig. 1.1. In this case, the distance from the $\mathbf{x} \in X$ to every element of set Y is determined, and the shortest distance is selected as the infimum. This process is repeated for every $x \in X$ and the largest distance (supremum) is selected as the Hausdorff distance of the set X to the set Y. Finally, this process is repeated for the set Y because the two distances will not necessarily be the same. For example, the Hausdorff distance of the set X to the set Y would be the distance from the lowercase letters in Fig. 1.1, but the result would be smaller for the distance of the set Y to the set X. The method of applying the Hausdorff distance to the image correspondence problem is the same as that described in Section 3.3. To reiterate, consider Fig. 1.2 where each rectangle represents a set of subimages (obtained by partitioning the original images X and Y) and the coloured areas represent some of the obtained tolerance classes[3]. Note, the

[3]The tolerance relation covers both images, but not all the classes are shown in the interest of clarity.

tolerance classes are created based on the feature values of the subimages, and consequently, do not need to be situated geographically near each other (as shown in Fig. 1.2). In the case of the tNM measure, the idea is that similar images should produce tolerance classes with similar cardinalities. Consequently, we are comparing the cardinalities of the portion of a tolerance class belonging to set X with the portion of the tolerance class belonging to set Y (represented in Fig. 1.2 as sets with the same colour). In contrast, the Hausdorff distance measures the distance between sets in some metric space. As a result, we measure distance in the feature space between the portion of a tolerance class belonging to set X with the portion of a tolerance class belonging to set Y (again, represented as a sets with the same colour). Here, the idea is that images that are similar should have tolerance classes are close to each other (in the Hausdorff sense) in the feature space. As a result, low Hausdorff distances are desirable. Finally, it is important to note, one may be inclined to intuitively think the Hausdorff distance should always be zero. However, this is not the case because we are dealing with tolerance where the feature values are similar, but not the same.

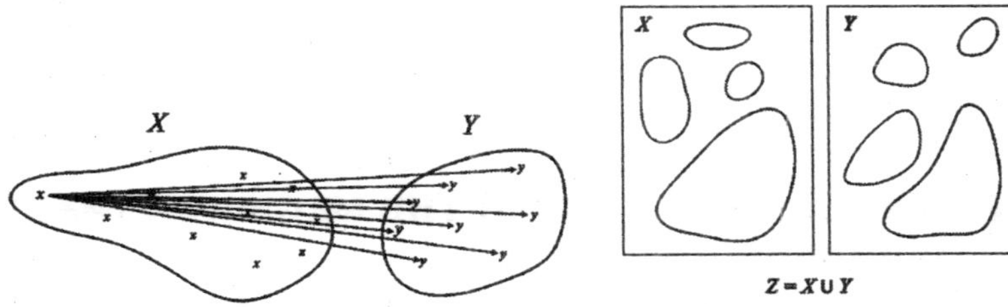

1.1: Hausdorff distance 1.2: Image Correspondence Problem

Fig. 1: *Graphical Interpretations*

2.1: Camera 2.2: covering$_C$ 2.3: Lena 2.4: covering$_L$

Fig. 2: *Nearness Measurement System*

3.1: Co1 3.2: Co2 3.3: Co3 3.4: Os1 3.5: Os2 3.6: Os3 3.7: Os4

Fig. 3: *Sample Microfossile Images*

Table 4: *Image Nearness Measurements*

Image 1	*Image (tol. classes)*	*Image 2*	*Image (tol. classes)*	*TOD*	*tNM*	*HdNM*	*wtAvg*
Fig. 2.1	Fig. 2.2 (282)	Fig. 2.3	Fig. 2.4 (236)	0.83	0.49	0.71	0.59
Fig. 3.1	Fig. 3.1 (544)	Fig 3.5	Fig 3.5 (576)	0.81	0.30	0.73	0.67
Fig. 3.3	Fig. 3.3 (738)	Fig. 3.2	Fig. 3.2 (2031)	0.93	0.27	0.58	0.59
Fig. 3.5	Fig. 3.5 (576)	Fig. 3.6	Fig. 3.6 (325)	0.83	0.59	0.85	0.80
Fig. 3.5	Fig 3.5 (576)	Fig. 3.7	Fig. 3.7 (348)	0.74	0.37	0.70	0.65
Fig. 3.7	Fig. 3.7 (348)	Fig. 3.4	Fig. 3.4 (290)	0.85	0.5	0.8	0.76

4. SAMPLE RESULTS WITH THE MEGHDADI IMAGE NEARNESS TOOLSET

Briefly, we illustrate the application of the image nearness measures for a collection of images. The coverings (with tolerance classes) for the camera man and Lena in Figs 2.1 and 2.3 are shown in Figs 2.2 and 2.4, respectively. For simplicity, we have chosen average greyscale level and entropy (information content) as the features used to define image coverings and measure similarities between images. The Meghdadi toolset for measuring nearness between images has been used to obtain the measurements reported in Table 4. This toolset was reported in [11] and not described again here. Nearness measurements using the tiNM measure are not included in the results reported in this section because the tiNM measure in (10) is mathematically easy to understand but, so far, the implementation of tiNMis computationally very slow. The images in Fig. 3 show two different types of microfossils [1], namely, Conodants in Figs 3.1, 3.2, 3.3 and Ostracods in Figs 3.4, 3.5, 3.6, 3.7, respectively. Among the comparisons of the pairs of microfossil images recorded in Table 4, the most remarkable is the comparison between Fig. 3.4 and Fig. 3.7 (these are microscope images of fossils from different rocks on micrometer scale). The image in fossil in Fig. 3.7 is broken up and quite indistinct compared with the image in Fig. 3.4, and, yet, each of the nearness measurements indicate a high degree of correspondence between the fossils in the two images. This suggests a very practical application of the nearness measures in comparing microfossils (a knowledge of the fossils at different levels in sedimentary deposits provide indicators of the presence or absence of natural gas and petroleum deposits).

5. CONCLUSION

This article focuses on an approach to solving the image correspondence problem by considering description based affinities between coverings of digital images. It introduces an approach to measuring the resemblance between pairs of images using several L_2 norm-based tolerance nearness measures. The proposed solution to the image correspondence problem is based on a comparison of the descriptions of elements of tolerance classes contained in coverings of pairs of digital images. It should be noted here that the proposed approach to measuring similarities between perceptual granules is not limited to digital images (this fact is implicit in the comparison given Table 4. The proposed approach also has promising implications for segmenting videos, especially in applications where grouping images in a video depends on very refined measurements overmany separate images contained in a video.

ACKNOWLEDGEMENTS

The insights and suggestions by Mihir Chakraborty, Amir H. Meghdadi, Christopher Henry, Homa Fashandi, Dan Lockery, Leszek Puzio, Andrzej Skowron, Piotr Wasilewski concerning topics in this article are gratefully acknowledged. This research has been supported by the Natural Sciences and Engineering Research Council of Canada

(NSERC) grants 185986, 194376, Canadian Arthritis Network grant SRI-BIO-05, Manitoba Hydro grant T277.

REFERENCES

[1] Armstrong H., Brasier M., "Microfossils". 2nd Ed. Blackwell Publishing, Oxford, UK (2005).

[2] Bartol W., Miró J., Pióro K., Rosselló, F.. "On the Coverings by Tolerance Classes". *Inf. Sci. Inf. Comput. Sci.* **166** (1–4), 193–211 (2004). *DOI http://dx.doi.org/10.1016/j.ins.2003.12.002.*

[3] Cyran K., Mrózek, A., "Rough Sets in Hybrid Methods for Pattern Recognition". *Int. J. of Intelligent Systems*, **16**, 149–168, (2001).

[4] Gerasin S.N., Shlyakhov V.V., Yakovlev S.V., "Set Coverings and Tolerance Relations". *Cybernetics and Sys. Anal.*, **44 (3)**, 333–340 (2008). *DOI http://dx.doi.org/10.1007/s10559-008-9007-y.*

[5] Hassanien A., Abraham A., Peters J., Schaefer G., Henry C., "Rough Sets and Near Sets in Medical Imaging: A Review". *IEEE Trans. Info. Tech. in Biomedicine (2009). Digital object identifier: 10.1109/TITB.2009.2017017, in press.*

[6] Hausdorff F., "Grundzüge der Mengenlehre". *Verlag Von Veit and Comp., Leipzig*, (1914).

[7] Hausdorff F., "Set Theory". Chelsea Publishing Company, New York (1962).

[8] Henry C., Peters J.. "Near Set Evaluation and Recognition (near) System". *Tech. rep., Computationa Intelligence Laboratory, University of Manitoba (2009). URL http://wren.ee.umanitoba.ca/images/ci reports/reportci-2009-015.pdf. UM CI Laboratory Technical Report No. TR-2009-015.*

[9] Henry C., Peters J., "Perception-based Image Analysis". *Int. J. of Bio-Inspired Computation*, **2 (2)**, (2009). in press.

[10] Jänich K., "Topology". Springer-Verlag, Berlin (1984).

[11] Meghdadi A., Peters J., Ramanna S., "Tolerance Classes in Measuring Image Resemblance". *Intelligent Analysis of Images and Videos (2009). in press.*

[12] Murray J., Bradley H., Craigie W., Onions C., "The Oxford English Dictionary". Oxford University Press, Oxford, UK (1933).

[13] Pawlak Z., "Rough Sets". *International Journal of Computer and Information Sciences*, **11**, 341–356 (1982).

[14] Pawlak Z., Peters J., "Jak Blisko (How Near)". *SystemyWspomagania Decyzji I*, **57**, **109**, (2002, 2007). ISBN 83-920730-4-5.

[15] Peters J., "Near Sets General Theory About Nearness of Objects. Applied Mathematical Sciences, **1 (53)**, 2609–2029. (2007).

[16] Peters J., "Near Sets General Theory About Nearness of Objects". Applied Mathematical Sciences, **1 (53)**, 2609–2029, (2007).

[17] Peters J., "Near Sets Special Theory About Nearness of Objects". *Fundamenta Informaticae*, **75**, (1–4), 407–433, (2007).

[18] Peters J., "Tolerance Near Sets and Image Correspondence". *Int. J. of Bio-Inspired Computation*, **4 (1)**, 239–445, (2009).

[19] Peters J., Henry C., "Feature-based Set Intersection in Measuring Image Resemblance". *Applied Mathematics and Computation*, (2009). *Submitted.*

[20] Peters J., Ramanna S., "Affinities Between Perceptual Granules: Foundations and Perspectives". In: A. Bargiela, W. Pedrycz (eds.) *Human-Centric Information Processing Through Granular Modelling SCI*, **182**, pp. 49–66. Springer-Verlag, Berlin (2009)

[21] Peters J., Wasilewski P., "Foundations of Near Sets. Information Sciences". *An International Journal*, **26**, 1–35 (2009). *Digital Object Identifier: doi:10.1016/j.ins.2009.04.018, in press.*

[22] Peters J.F., Puzio L., "Anisotropic Wavelet-based Image Nearness Measure". *Int. J. of Computational Intelligence Systems pp , in press (2009).*

[23] Poicaré H., "Mathematics and Science", Last Essays, trans. by J.W. Bolduc. Kessinger Pub., N.Y. (1913).

[24] Schroeder M., Wright M., "Tolerance and Weak Tolerance Relations. *Journal of Combinatorial Mathematics and Combinatorial Computing*, **11**, 123–160, (1992).

[25] Shreider Y.A., "Tolerance Spaces. Cybernetics and Systems Analysis", **6 (12)**, 153–758, (1970).

[26] Skowron A., Stepaniuk J., "Tolerance Approximation Spaces". *Fundamenta Informaticae*, **27 (2/3)**, 245–253, (1996).

[27] Sossinsky A., "Tolerance Space Theory and Some Applications". *Acta Applicandae Mathematicae: An International Survey Journal on Applying Mathematics and Mathematical Applications*, **5 (2)**, 137–167, (1986).

[28] Zeeman E., "The Topology of the Brain and the Visual Perception". Prentice Hall, New Jersey (1965). In K.M. Fort, Ed., Topology of 3-manifolds and Selected Topics, 240–256.

[29] Zheng Z., Hu H., Shi Z., "Tolerance Relation Based Granular Space". *Lecture Notes in Computer Science*, **3641**. 682, (2005).

9

ON THE THEORY OF VAGUE SOFT SETS

T. Som

Department of Applied Mathematics,
Institute of Technology,
Banaras Hindu University, Varanasi-221005
E-mail: som_tanmoy@yahoo.co.in

Abstract: *Many existing theories such as the probability theory, theory of fuzzy sets, theory of rough sets, theory of vague sets or the theory of approximate reasoning etc have been used to tackle with uncertainty or vagueness associated with a complex situation or problem in any discipline of Science, Social Science or Economics etc, which involve data that are not always crisp and precise. However the theory of fuzzy sets, rough sets or vague sets etc can deal with a kind of uncertainty which do not involve some parameters. Such situation can be dealt with the theory of soft sets given by Molodtsov in 1999. If these parameters also involve some variations in them, then such complex problems are dealt with the theory of fuzzy soft sets given by Maji et.al. If such parameters further involve some kind of vagueness, then it can be dealt with the present theory of vague soft sets given by Xu et. al. The present talk focuses on the concepts of vague soft sets and its properties.*

1. INTRODUCTION

Before going to the concept of rough soft set we give here the preliminary ideas on fuzzy set, rough set, soft set and fuzzy soft set in this section. Many real life problems may be from Engineering Science, Environmental Science, Economics or Social Science etc which involve data that are not very clear, can not be handled with usual mathematical techniques, but they can be addressed by either fuzzy set theoretic tool [19] or rough set tool [13] or vague set tool [6] or the theory of interval mathematics [3, 7] or intuitionistic fuzzy set theory [2] etc. depending on the type of the uncertainty or vagueness situation of the problem. However each of these non classical techniques have their own limitations. Proceeding in this direction Molodotsov [11] addressed such type of problems from a little different angle taking into account the parameters of the problem to model the problem's vagueness or uncertainty in some respect and gave soft set theory, which proved to have a potential application in the areas of social science, decision making, game theory etc. [12,15]. Recently Chen et.al. [5] gave a new definition of soft set parameterization reduction and compared their definitions with the related ones of attributes reduction in rough set theory and Aktas & Cagman [1] compared the concepts of a soft set with the concepts of fuzzy set and rough set. Taking a clue from these we go ahead in this direction and taking into account the concepts of vague sets given by Gau [6] we introduce the concepts of rough soft set to deal with the complexity of the parameters of the soft set, which is different from fuzzy soft set introduced by Maji et. al. [9, 10] in the same way the rough set approach is different from the fuzzy set approach. In case the parameterization is simple we take the help of soft set, but if the parameterization is complex we use rough soft set approach to tackle with such problems as given by Xu et. al. [17]. We further discuss some properties of the rough soft set in the respective section.

1.1 Fuzzy Set and Soft Set

Basically the theory of fuzzy sets is the theory of graded concepts, i.e., a theory in which every thing is a matter of degree or grade developed by Zadeh [19]. For every set $A \subset X$, its membership function $\mu_A(x): X \rightarrow [0, 1], x \in X$ gives the grade of belongingness of x to the set A. The membership value of x differs from person to person and other related factors. Due to this we may not get a good solution of the complicated problems using fuzzy set technique. In fact in the theory of fuzzy sets there is no consideration of different parameters associated with any complicated problem. To deal with such complicated problems Molodtsov (1999) gave the concept of Soft Set as a recently developed theory mentioned below.

Molodtsov [11] has introduced the Soft set theory in the following way.

Let U be an initial universe set and let E be a set of parameters. Let $P(U)$ denote the power set of U and $A \subset E$.

Definition 1.1: (see/7.) A pair (F, E) is called a soft set over U if and only if F is a mapping of E into the set of all subsets of the set U, i.e.

$$F : E \to P(U)$$

In other words, the soft set is a parameterized family of subsets of the set U. Every set $F(e)$, for $e \in E$, from the family may be considered as the set of e-elements of the soft set (F, E), or as the set of e-approximate elements of the soft set. Such a soft set we take as Level I soft set.

If we search for another parametric values out of e-approximate elements of the soft set, we land in the Level 2 soft set and so on.

Example 1.1: As an illustration, let us consider the following examples

1. Suppose a soft set (F, E) describes the attractiveness of the houses which Mr. X is going to buy.

   U = the set of houses under consideration.

   E = the set of parameters, each parameter is a word or a sentence.

   E = {expensive, beautiful, cheap, in the green surroundings, modern}.

In this case defining a soft set means to point out expensive houses, beautiful houses out of the set U and so on. It is worth noting that the sets $F(e)$ may be empty for some $e \in E$.

We consider below the same example in more detail for our next discussion

Suppose $U = \{h_1, h_2, h_3, h_4, h_5\}$ and $E = \{e_1, e_2, e_3, e_4, e_5\}$

where,

e_1 stands for the parameter 'expensive'
e_2 stands for the parameter 'beautiful'
e_3 stands for the parameter 'cheap'
e_4 stands for the parameter 'in the green surroundings'
e_5 stands for the parameter 'modern'.

Let

$$F(e_1) = \{h_2, h_4\}, \quad F(e_2) = \{h_1, h_3\}, F(e_3) = \{h_3, h_4, h_5\},$$
$$F(e_4) = \{h_1, h_3, h_5\}, F(e_5) = \{h_1\}.$$

So the soft set (F, E) is a parameterized family $\{F(e_i), i = 1, 2, ..., 5\}$ of subsets of the set U and gives us a collection of e-approximate descriptions of an object. These are Level I soft sets.

Now if we introduce a further parameter say 'interior decoration' of the houses in $F(e_1), F(e_2), F(e_3), F(e_4), F(e_5)$ we get the Level II soft set.

2. Zadeh's fuzzy set may be considered as a special case of the soft set. Let A be a fuzzy set of U with membership μ_A, where μ_A is a mapping of U into $[0, 1]$. Let us consider the family of α-level sets for the function μ_A given by

$$F(\alpha) = \{x \varepsilon U: \mu_A(x) \geq \alpha\}, \alpha \in [0, 1].$$

If we know the family F, we can find the functions $\mu_A(x)$ by means of the following formulae:

$$\mu_A(x) = \sup_{\alpha \in [0, 1], x \in F(\alpha)} (\alpha).$$

Thus the way of describing any object in the soft set theory principally differs from the way we use in classical mathematics. In classical mathematics, we construct a mathematical model of an object and define the notion of the exact solution of the model. Usually the mathematical model is too complicated and we cannot find the exact solution. So we introduce the notion of approximate solution and get it by numerical methods.

In the soft set theory, we have an opposite approach to this issue. The initial description of the object has an approximate nature and so we find an acceptable solution as we do not require to introduce the notion of exact solution.

The absence of any restrictions on the approximate description in soft set theory makes this theory very convenient and easily applicable in practice. We can use any parameterization we prefer with the help of words and sentences, real numbers, functions, mappings, and so on. It means that the problem of setting the membership function or any similar problem does not arise in the soft set theory.

Definition 1.2: For two soft sets (F, A) and (G, B) over a common universe U, we say that (F, A) is a soft subset of (G, B) if

(*i*) $A \subset B$, and

(*ii*) $\forall e \in A$, $F(e)$ and $G(e)$ are identical approximations.

We write this as $(F, A) \tilde{\subseteq} (G, B)$. (F, A) is said to be a soft super set of (G, B) if (G, B) is a soft subset of (F, A). We denote it by $(F, A) \tilde{\supseteq} (G, B)$.

Definition 1.3: EQUALITY OF TWO SOFT SETS. Two soft sets (F, A) and (G, B) over a common universe U are said to be soft equal if (F, A) is a soft subset of (G, B) and (G, B) is a soft subset of (F, A).

Definition 1.4: NOT SET OF A SET OF PARAMETERS. Let $E = \{e_1, e_2, e_3, ..., e_n\}$ be a set of a parameters. The NOT set of E denoted by $\rceil E$ is defined by $\rceil E = \{e_1, e_2, e_3, ..., e_n\}$ where $\rceil e_i =$ not $e_i \ \forall i$.

The following result is obvious.

Proposition 1.1:

1. $\rceil(\rceil A) = A$.
2. $\rceil(A \cap B) = (\rceil A \cap B)$
3. $\rceil (A \cup B) = (\rceil A \cup B)$.

Definition 1.5: COMPLEMENT SOFT SET OF A SOFT SET. The complement of a soft set $(F \cdot A)$ is denoted by $(F, A)^C$ and is defined by $(F, \rceil A)^C = (F, \rceil A)$, where F^C: $\rceil A \to P(U)$ is a mapping given by

$$F^C(\alpha) = U - F(\rceil \alpha), \forall \alpha \in \rceil A.$$

Let us call F^C to be the soft complement function of F. Clearly, $(FC)^C$ is the same as F and

$$((F, A)^C)^C = (F, A).$$

Example 1.2: Consider Example 1.1 (1).

Here $(F, E)^C = \{$not expensive houses $= \{h_1, h_3, h_5, h_6\}$, not beautiful houses $= \{h_2, h_4, h_5, h_6\}$, not cheap houses $= \{h_2, h_4, h_6\}$, not in the green surroundings houses $= \{h_2, h_3, h_4, h_5, h_6\}\}$.

Definition 1.6: NULL SOFT SET. A soft set (F, A) over U is said to be a NULL soft set denoted by Φ, if $\forall e \in A$, $F(e) = \phi$ (null set).

Definition 1.7: ABSOLUTE SOFT SET. A soft set (F, A) over U is said to be absolute soft set denoted by $\tilde{A}$ if $\forall\ e \in \tilde{A}$, $F(e) = U$.

Clearly, $\tilde{A}^C = \Phi$ and $\Phi^C = \tilde{A}$.

Definition 1.8: UNION OF SOFT SETS. The union of two soft sets (F, A) and (G, B) over the common universe U is the soft set (H, C), where $C = A \cup B$ and $\forall\ e \in C$

$$H(e) = \begin{cases} F(e) & \text{if } e \in A - B \\ G(e) & \text{if } e \in B - A \\ F(e) \cup G(e) & \text{if } e \in A \cap B \end{cases}$$

This we denote as $(F, A)\ \tilde{\cup}\ (G, B) = (H, C)$

Definition 1.9: INTERSECTION OF SOFT SETS. The intersection of two soft sets (F, A) and (G, B) over the common universe U is the soft set (H, C), where $C = A \cap B$ and $\forall\ e \in C$

$$H(e) = \begin{cases} F(e) & \text{if } e \in A - B \\ G(e) & \text{if } e \in B - A \\ F(e) \cap G(e) & \text{if } e \in A \cap B \end{cases}$$

This we denote as $(F, A)\ \tilde{\cap}\ (G, B) = (H, C)$.

We can easily show the following proposition.

Proposition 1.2: If EF, AF and EG, BF are two soft sets, then

(*i*) $\left((F, A)\,\tilde{\cup}\,(G, B)\right)^C = (F, A)^C\ \tilde{\cap}\,(G, B)^C$

(*ii*) $\left((F, A)\,\tilde{\cap}\,(G, B)\right)^C = (F, A)^C\ \tilde{\cup}\,(G, B)^C$

2. VAGUE SET

In this section we give the basic concepts on vague sets as given by Gau [4].

Definition 2.0: VAGUE SET. Let U be universe given by $U = \{u_1, u_2, \ldots, u_n\}$. Then a vague set V in U is characterized by a truth membership function t_v: $U \to [0, 1]$ and a false-membership function f_v: $U \to [0, 1]$, where $t_v(u_i)$ is a lower bound of the grade of membership of u_i derived from the evidence for u_i and $f_v(u_i)$ is a lower bound on the negation of u_i derived from the evidence against u_i and $t_v(u_i) + f_v(u_i) \leq 1$. The grade of membership of u_i in the vague set is bounded in a subinterval $[t_v(u_i), 1 - f_v(u_i)]$ of $[0, 1]$. The vague value $[t_v(u_i), 1 - f_v(u_i)]$ indicates that the exact grade of membership $\mu_v(u_i)$ of u_i may be unknown. But it is bounded by $t_v(u_i) \leq \mu_v(u_i) \leq 1 - f_v(u_i)$ with $t_v(u_i) + f_v(u_i) \leq 1$.

When the universe U is continuous, the vague set V can be written as

$$V = \int_U \left[t_A(u_i), 1 - f_A(u_i)\right]/u_i, \quad u_i \in U$$

And when the universe U is discrete, the vague set V' can be written as

$$V' = \sum_{i=1}^{n} \left[t_A(u_i), 1 - f_A(u_i)\right]/u_i, \quad u_i \in U$$

Definition 2.1: UNIT and ZERO VAGUE VALUE. Let x be a vague value in $[t_x, 1 - f_x]$, where $t_x \in [0, 1]$, $f_x \in [0, 1]$ and $0 \leq t_x \leq 1 - f_x \leq 1$. If $t_x = 1$ and $f_x = 0$ (i.e., $x = [1, 1]$, then x is called a unit vague value. If $t_x = 0$ and $f_x = 1$ (i.e., $x = [0, 0]$, then x is called a zero vague value.

Definition 2.2: EQUAL VAGUE VALUE. Let x and y be two vague values, where $x = [t_x, 1 - f_x]$, $y = [t_y, 1 - f_y]$. If $t_x = t_y$ and $f_x = f_y$, then the vague values x and y are called equal (i.e., $[t_x, 1 - f_x] = [t_y, 1 - fy]$).

Let A and B be two vague sets of the universe $U = \{u_1, u_2, ..., u_n\}$, where

$$A = [t_A(u_1), 1 - f_A(u_1)]/u_1 + [t_A(u_2), 1 - f_A(u_2)]/u_2 + ... + = [t_A(u_n), 1 - f_A(u_n)]/u_n \text{ and}$$

$$B = [t_B(u_1), 1 - f_B(u_1)]/u_1 + [t_B(u_2), 1 - f_B(u_2)]/u_2 + ... + = [t_B(u_n), 1 - f_B(u_n)]/u_n.$$

Definition 2.3: UNIT and ZERO VAGUE SETS. A vague set A of the universe U is called

(*a*) Unit vague set if $t_A(u_i) = 1$ and $f_A(u_i) = 0 \ \forall \ u_i \in U$, where $1 \leq i \leq n$ and

(*b*) Zero vague set if $t_A(u_i) = 0$ and $f_A(u_i) = 1 \ \forall \ u_i \in U$, where $1 \leq i \leq n$.

Definition 2.4: VAGUE SET COMPLEMENT. The complement of a vague set A, denoted by A', is defined by

$$t_{A'} = f_A \text{ and } 1 - f_{A'} = 1 - t_{A}.$$

Definition 2.5: EQUAL VAGUE SET. Two vague sets A and B of the universe U is said to be equal if

$$[t_A(u_i), 1 - f_A(u_i)] = [t_B(u_i), 1 - f_B(u_i)] \ \forall u_i \in U, \text{ where } 1 \leq i \leq n.$$

Definition 2.6: Let A and B be two vague sets of the universe U. If $\forall\ u_i \in \mathrm{U},\ t_A(u_i) \leq t_B(u_i)$ and $f_A(u_i) \leq 1 - f_B(u_i)$, where $1 \leq i \leq n$, then the vague set A is said to be included by the vague set B and this is denoted by $A \subseteq B$.

Definition 2.7: UNION OF VAGUE SETS. The union of two vague sets A and B of the universe U is the vague set C, written as $C = A \cup B$, whose truth-membership and false-membership functions are related to those of A and B by

$$t_C = \max(t_A, t_B) \text{ and } 1 - f_C = \max(1 - f_A, 1 - f_B)$$
$$= 1 - \min(f_A, f_B).$$

Definition 2.8: INTERSECTION OF VAGUE SETS. The intersection of two vague sets A and B of the universe U is the vague set C, written as $C = A \cap B$, whose truth-membership and false-membership functions are related to those of A and B by

$$t_C = \min(t_A, t_B) \text{ and } 1 - f_C = \min(1 - f_A, 1 - f_B)$$
$$= 1 - \max(f_A, f_B).$$

3. VAGUE SOFT SETS

This section deals with the recent concepts on vague soft set introduced by Xu et. al. [17], which is an extension of soft set with addition of the sense of vagueness in its parameter set.

As discussed earlier we know that a soft set is a mapping from a set of parameters to the power set of a universe set. However in whatever manner we take the parameters, the members of the soft set may differ from person to person, for every ones view point on the members with respect to the parameters may not be the same, but this can be tackled to a large extent by defining the fuzzy soft set as discussed by Maji et. al. [9, 10] and others. This situation can also be handled with the concept of vague soft set, which will be presented in this section with some basic properties and examples.

Let U be a universe set, E be the set of parameters, each parameter is a word or a sentence. $F(U)$ be the vague sets of U and $A \subseteq E$. Then we define the concept of vague soft set in the following way.

Definition 3.1: VAGUE SOFT SET. A pair $(\hat{F}, A)$ is called a vague soft set over U, where F is a mapping given by $\hat{F}: A \to F(U)$.

In other words, a vague soft set over U is a parameterized family of vague set of the universe U. For $e \in A$, $\mu_{\hat{F}(e)}: U \to [0, 1]$ is the set of e-approximate elements of the vague soft set $(\hat{F}, A)$.

Example 3.1: We consider a vague soft set $(\hat{F}, E)$, where U is the set of houses under consideration. E = {expensive, beautiful, cheap, in the green surroundings, modern}.

$$U = \{h_1, h_2, h_3, h_4, h_5\} \text{ and } E = \{e_1, e_2, e_3, e_4, e_5\}$$

where

e_1 stands for the parameter 'expensive'

e_2 stands for the parameter 'beautiful'

e_3 stands for the parameter 'cheap'

e_4 stands for the parameter 'in the green surroundings'

e_5 stands for the parameter 'modern'.

Suppose that

$$F(e_1) = ([0.1, 0.2]/h_1, [0.9, 0.2]/h_2, [0.4, 0.5]/h_3, [0.1, 0.2]/h_4, [0.5, 0.7]/h_5)$$

$$F(e_2) = ([0.8, 0.1]/h_1, [0.2, 0.7]/h_2, [0.5, 0.8]/h_3, [0.3, 0.5]/h_4, [0.6, 0.9]/h_5)$$

$$F(e_3) = ([0.6, 0.9]/h_1, [0.9, 1.0]/h_2, [0.3, 0.8]/h_3, [0.1,0.2]/h_4, [0.5, 0.7]/h_5)$$

$$F(e_4) = ([0.3, 0.6]/h_1, [0., 0.2]/h_2, [0.4, 0.5]/h_3, [0.2, 0.5]/h_4, [0.3, 0.5]/h_5),$$

$$F(e_5) = ([0.7, 1.0]/h_1, [0.6, 0.9]/h_2, [0.2, 0.5]/h_3, [0.3, 0.7]/h_4, [0.5, 0.7]/h_5)$$

Thus the vague soft set $(\hat{F}, E)$ is a parameterized family $\{\hat{F}(e_i), i = 1, 2, \ldots, 5\}$ of vague sets on U. So $\hat{F}(e_1)$ represents the expensive functional value members of the vague set given by ([0.1, 0.2]/h_1, [0.9, 0.2]/h_2, [0.4, 0.5]/h_3, [0.1, 0.2]/h_4, [0.5, 0.7]/h_5) and so on. We can also regard the vague soft set $(\hat{F}, E)$ as the vague approximations:

$(\hat{F}, E)$ = {Expensive houses = ([0.1, 0.2]/h_1, [0.9, 0.2]/h_2, [0.4, 0.5]/h_3, [0.1, 0.2]/h_4, [0.5, 0.7]/h_5),
Beautiful houses = ([0.8, 0.1]/h_1, [0.2, 0.7]/h_2, [0.5, 0.8]/h_3, [0.3, 0.5]/h_4, [0.6, 0.9]/h_5),
Cheap houses = ([0.6, 0.9]/h_1, [0.9, 1.0]/h_2, [0.3, 0.8]/h_3, [0.1, 0.2]/h_4, [0.5, 0.7]/h_5) ,
Houses in green surroundings = ([0.3, 0.6]/h_1, [0., 0.2]/h_2, [0.4, 0.5]/h_3, [0.2, 0.5]/h_4, [0.3, 0.5]/h_5),
Modern houses = ([0.7, 1.0]/h_1, [0.6, 0.9]/h_2, [0.2, 0.5]/h_3, [0.3,0.7]/h4, [0.5,0.7]/h5)}.

Definition 3.2: VAGUE SOFT SUBSET. Let $(\hat{F}, A)$ and $(\hat{F}, B)$ be two vague soft sets over a common universe U. Then $(\hat{F}, A)$ is said to be the vague soft subset of $(\hat{G}, B)$, denoted by $(\hat{F}, A)$ $(\hat{G}, B)$, if $A \subseteq B$ and $\forall\ e \in A$, $\hat{F}(e)$ and $\hat{G}(e)$ are identical vague approximations. Similarly $(\hat{F}, A)$ is said to be the vague soft superset of $(\hat{G}, B)$, denoted by $(\hat{F}, A)$ $(\hat{G}, B)$, if $A \supseteq B$ and $\forall e \in A$, $\hat{F}(e)$ and $\hat{G}(e)$ are identical vague approximations.

Definition 3.3: VAGUE EQUAL SOFT SET. Two vague soft sets $(\hat{F}, A)$ and $(\hat{G}, B)$ over a common universe U are said to be vague equal soft sets if $(\hat{F}, A)$ is vague soft subset of $(\hat{G}, B)$ and $(\hat{G}, B)$ is vague soft subset of $(\hat{F}, A)$.

Example 3.2: Let $A = \{e_1, e_2\}$ and $B = \{e_1, e_2, e_3\}$, then $A \subseteq B$. Let $(\hat{F}, A)$ and $(\hat{G}, B)$ be two vague soft sets over a common universe U with

$F(e_1)$ = ([0.1, 0.2]/h_1, [0.9, 0.2]/h_2, [0.4, 0.5]/h_3, [0.1, 0.2]/h_4, [0.5, 0.7]/h_5),

$F(e_2)$ = ([0.8, 0.1]/h_1, [0.2, 0.7]/h_2, [0.5, 0.8]/h_3, [0.3, 0.5]/h_4, [0.6, 0.9]/h_5),

$G(e_1) = ([0.1, 0.2]/h_1, [0.9, 0.2]/h_2, [0.4, 0.5]/h_3, [0.1, 0.2]/h_4, [0.5, 0.7]/h_5),$

$G(e_2) = ([0.8, 0.1]/h_1, [0.2, 0.7]/h_2, [0.5, 0.8]/h_3, [0.3, 0.5]/h_4, [0.6, 0.9]/h_5),$

$G(e_3) = ([0.6, 0.9]/h_1, [0.9, 1.0]/h_2, [0.3, 0.8]/h_3, [0.1, 0.2]/h_4, [0.5, 0.7]/h_5).$

Then clearly $(\hat{F}, A) \supseteq (\hat{G}, B)$.

Definition 3.4: VAGUE NULL SOFT SET. A vague soft set $(\hat{F}, A)$, over U is said to be vague null soft set if $(\hat{F}, A)$, denoted by $\hat{\Phi}$, if for every $e \in A$

$$t_{\hat{F}(e)}(x) = 0, \quad 1 - f_{\hat{F}(e)}(x) = 0, \cdot x \in U.$$

Definition 3.5: VAGUE ABSOLUTE SOFT SET. A vague soft set $(\hat{F}, A)$ over U is said to be vague absolute soft set, denoted by $\breve{A}$, if for every $e \in A$

$$t_{\hat{F}(e)}(x) = 1, \quad 1 - f_{\hat{F}(e)}(x) = 1, x \in U.$$

Definition 3.6: AND OPERATION. Let $(\hat{F}, A)$ and $(\hat{G}, B)$ be two vague soft sets over a common universe U. The "$(\hat{F}, A)$ and $(\hat{G}, B)$", denoted by $(\hat{F}, A) \wedge (\hat{G}, B)$, is defined as $(\hat{F}, A) \wedge (\hat{F}, B) = (\hat{H}, A \times B)$,

where $t_{\hat{H}(\alpha,\beta)}(x) = \max\left\{t_{\hat{F}(e)}(x), t_{\hat{G}(\beta)}(x)\right\}$, $1 - f_{\hat{H}(\alpha,\beta)}(x) = \max\left\{1 - f_{\hat{F}(e)}(x), 1 - f_{\hat{G}(\beta)}(x)\right\}$ $\forall (\alpha, \beta) \in (A \times B), x \in U$.

Definition 3.7: OR OPERATION. Let $(\hat{F}, A)$ and $(\hat{G}, B)$ be two vague soft sets over a common universe U. The "$(\hat{F}, A)$ or $(\hat{G}, B)$", denoted by $(\hat{F}, A) \vee (\hat{G}, B)$, is defined as $(\hat{F}, A) \vee (\hat{G}, B) = (\hat{H}, A \times B)$,

where $t_{\hat{H}(\alpha,\beta)}(x) = \max\left\{t_{\hat{F}(e)}(x), t_{\hat{G}(\beta)}(x)\right\}$, $1 - f_{\hat{H}(\alpha,\beta)}(x) = \max\left\{1 - f_{\hat{F}(e)}(x), 1 - f_{\hat{G}(\beta)}(x)\right\}$ $\forall (\alpha, \beta) \in (A \times B), x \in U$.

We can easily prove the following proposition:

Proposition 3.1: Let $(\hat{F}, A)$ and $(\hat{G}, B)$ be two vague soft sets on U, then

(*i*) $\left\{(\hat{F}, A)\vee(\hat{G}, B)\right\}^C = (\hat{F}, A)^C \wedge (\hat{G}, B)^C$

(*ii*) $\left\{(\hat{F}, A)\wedge(\hat{G}, B)\right\}^C = (\hat{F}, A)^C \vee (\hat{G}, B)^C$

Proof: (*i*) Let $(\hat{F}, A)\vee(\hat{G}, B) = (\hat{O}, A\times B)$, then we get

$\left\{(\hat{F}, A)\vee(\hat{G}, B)\right\}^C = (\hat{O}, A \times B)^C = (\hat{O}^C, \rceil(A \times B))$ and

$$(\hat{F}, A)^C \wedge (\hat{G}, B)^C = (\hat{F}^C, \rceil A) \wedge (\hat{G}^C, \rceil B) = (\hat{J}, \rceil A \times \rceil B)$$
$$= (\hat{J}, \rceil(wA \times B)),$$

where $t_{J(\rceil\alpha, \rceil\beta)}(x) = \min\left\{t_{\hat{F}^C(\rceil\alpha)}(x), t_{\hat{G}^c(\rceil\beta)}(x)\right\}$, $1 - f_{J(\rceil\alpha, \rceil\beta)}(x)$

$$= \min\left\{1 - f_{\hat{F}^C(\rceil\alpha)}(x), 1 - f_{\hat{G}^c(\rceil\beta)}(x)\right\}$$

$$\forall(\rceil\alpha, \rceil\beta) \in \rceil A \times \rceil B, x \in U.$$

Now taking $(\rceil\alpha, \rceil\beta) \in \rceil(A \times B)$ we get

$$t_{\hat{O}(\rceil\alpha, \rceil\beta)}(x) = f_{\hat{O}(\alpha, \beta)}(x)$$
$$= 1 - \max\left\{1 - f_{\hat{F}(\alpha)}(x), 1 - f_{\hat{G}(\beta)}(x)\right\}$$
$$= \min\left\{f_{\hat{F}(\alpha)}(x), f_{\hat{G}(\beta)}(x)\right\}$$
$$= \min\left\{t_{\hat{F}^C(\rceil\alpha)}(x), t_{\hat{G}^C(\rceil\beta)}(x)\right\} = t_{J(\rceil\alpha, \rceil\beta)}(x)$$

and

$$1 - f_{\hat{O}(\rceil\alpha, \rceil\beta)}(x) = 1 - t_{\hat{O}(\alpha, \beta)}(x)$$
$$= 1 - \max\left\{t_{\hat{F}(\alpha)}(x), t_{\hat{G}(\beta)}(x)\right\}$$

$$= \min\left\{1 - t_{\hat{F}(\alpha)}(x), 1 - t_{\hat{G}(\beta)}(x)\right\}$$

$$= \min\left\{1 - f_{\hat{F}^c(1\alpha)}(x), 1 - f_{\hat{G}^c(1\beta)}(x)\right\}$$

$$= 1 - f_{J(1\alpha, 1\beta)}(x).$$

Hence $\hat{O}$ and $\hat{J}$ are the same operator and this proves (*i*) Similarly we can prove (*ii*).

Following proposition can be proved easily from the definition.

Proposition 3.2: Let $(\hat{F}, A), (\hat{G}, B)$ and $(\hat{H}, C)$ be three vague soft sets on *U*, then

(*i*) $(\hat{F}, A) \vee ((\hat{G}, B) \vee (\hat{H}, C)) = ((\hat{F}, A) \vee (\hat{G}, B)) \vee (\hat{H}, C)$

(*ii*) $(\hat{F}, A) \wedge ((\hat{G}, B) \wedge (\hat{H}, C)) = ((\hat{F}, A) \wedge (\hat{G}, B)) \wedge (\hat{H}, C)$

(*iii*) $(\hat{F}, A) \vee ((\hat{G}, B) \wedge (\hat{H}, C)) = ((\hat{F}, A) \vee (\hat{G}, B)) \wedge ((\hat{F}, A) \vee (\hat{H}, C))$

(*iv*) $(\hat{F}, A) \wedge ((\hat{G}, B) \vee (\hat{H}, C)) = ((\hat{F}, A) \wedge (\hat{G}, B)) \vee ((\hat{F}, A) \wedge (\hat{H}, C))$

Definition 3.6: UNION OF VAGUE SETS. The union of two vague soft sets $(\hat{F}, A)$ and $(\hat{G}, B)$ over a universe *U* is a vague soft sets $(\hat{H}, C)$, where $C = A \cup B$ and $\forall e \in C$, given by

$$t_{\hat{H}(e)}(x) = \begin{cases} t_{\hat{F}(e)}(x) & \text{if } e \in A - B, x \in U \\ t_{\hat{G}(e)}(x) & \text{if } e \in B - A, x \in U \\ \max\left\{t_{\hat{F}(e)}(x), t_{\hat{G}(e)}(x)\right\} & \text{if } e \in A \cap B, x \in U \end{cases}$$

and

$$1 - f_{\hat{H}(e)}(x) = \begin{cases} 1 - f_{\hat{F}(e)}(x) & \text{if } e \in A - B, x \in U \\ 1 - f_{\hat{G}(e)}(x) & \text{if } e \in B - A, x \in U \\ \max\left\{1 - f_{\hat{F}(e)}(x), 1 - f_{\hat{G}(e)}(x)\right\} & \text{if } e \in A \cap B, x \in U \end{cases}$$

this we denote as $(\hat{F}, A) \tilde{\cup} (\hat{G}, B) = (\hat{H}, C)$.

Definition: INTERSECTION OF VAGUE SETS. The intersection of two vague soft sets $(\hat{F}, A)$ and $(\hat{G}, B)$ over a universe U is a vague soft sets $(\hat{H}, C)$, where $C = A \cup B$ and $\forall\, e \in C$, given by

$$t_{\hat{H}(e)}(x) = \begin{cases} t_{\hat{F}(e)}(x) & \text{if } e \in A - B, \quad x \in U \\ t_{\hat{G}(e)}(x) & \text{if } e \in B - A, \quad x \in U \\ \min\left\{t_{\hat{F}(e)}(x), t_{\hat{G}(e)}(x)\right\} & \text{if } e \in A \cap B, \quad x \in U \end{cases}$$

and

$$1 - f_{\hat{H}(e)}(x) = \begin{cases} 1 - f_{\hat{F}(e)}(x) & \text{if } e \in A - B, \quad x \in U \\ 1 - f_{\hat{G}(e)}(x) & \text{if } e \in B - A, \quad x \in U \\ \min\left\{1 - f_{\hat{F}(e)}(x), 1 - f_{\hat{G}(e)}(x)\right\} & \text{if } e \in A \cap B, \quad x \in U \end{cases}$$

this we denote as $(\hat{F}, A) \tilde{\cap} (\hat{G}, B) = (\hat{H}, C)$

The following propositions can be easily proved:

Proposition 3.3:

(*i*) $(\hat{F}, A) \tilde{\cup} (\hat{F}, A) = (\hat{F}, A)$ (*ii*) $(\hat{F}, A) \tilde{\cap} (\hat{F}, A) = (\hat{F}, A)$

(*iii*) $(\hat{F}, A) \tilde{\cup} \hat{\Phi} = (\hat{F}, A)$ (*iv*) $(\hat{F}, A) \tilde{\cap} \hat{\Phi} = \hat{\Phi}$

(*v*) $(\hat{F}, A) \tilde{\cup} \hat{\hat{A}} = \hat{\hat{A}}$ (*vi*) $(\hat{F}, A) \tilde{\cap} \hat{\hat{A}} = \hat{\Phi}$

Proposition 4.3: If $(\hat{F}, A)$ and $(\hat{G}, B)$ are two vague soft sets on U then

(*i*) $\left((\hat{F}, A) \tilde{\cup} (\hat{G}, B)\right)^C = (\hat{F}, A)^C \tilde{\cap} (\hat{G}, B)^C$

(ii) $\left(\left(\hat{F},A\right)\tilde{\cap}\left(\hat{G},B\right)\right)^{C}=\left(\hat{F},A\right)^{C}\tilde{\cup}\left(\hat{G},B\right)^{C}$

Proposition 4.4: If $\left(\hat{F},A\right)$, $\left(\hat{G},B\right)$ and $\left(\hat{H},C\right)$ are three vague soft sets on U then

(i) $\left(\hat{F},A\right)\tilde{\cup}\left(\left(\hat{G},B\right)\tilde{\cup}\left(\hat{H},C\right)\right)=\left(\left(\hat{F},A\right)\tilde{\cup}\left(\hat{G},B\right)\right)\tilde{\cup}\left(\hat{H},C\right)$

(ii) $\left(\hat{F},A\right)\tilde{\cap}\left(\left(\hat{G},B\right)\tilde{\cap}\left(\hat{H},C\right)\right)=\left(\left(\hat{F},A\right)\tilde{\cap}\left(\hat{G},B\right)\right)\tilde{\cap}\left(\hat{H},C\right)$

(iii) $\left(\hat{F},A\right)\tilde{\cup}\left(\left(\hat{G},B\right)\tilde{\cap}\left(\hat{H},C\right)\right)=\left(\left(\hat{F},A\right)\tilde{\cup}\left(\hat{G},B\right)\right)\tilde{\cap}\left(\left(\hat{F},A\right)\tilde{\cup}\left(\hat{H},C\right)\right)$

(iv) $\left(\hat{F},A\right)\tilde{\cap}\left(\left(\hat{G},B\right)\tilde{\cup}\left(\hat{H},C\right)\right)=\left(\left(\hat{F},A\right)\tilde{\cap}\left(\hat{G},B\right)\right)\tilde{\cup}\left(\left(\hat{F},A\right)\tilde{\cap}\left(\hat{H},C\right)\right)$

4. CONCLUSION

In this present talk the basic notions on soft set, vague set and vague soft set have been presented. The ideas of vague soft set are the recent extension of soft set and there is ample scope for further research about the relationship among soft set, vague set and vague soft set and the viable application of this new concept.

REFERENCES

[1] H. Aktas and N. Cagman, "Soft Sets and Soft Groups (to Appear in Information Sciences)".

[2] K. Atanassov, "Intuitionistic Fuzzy Sets, Fuzzy Sets and Systems", **20**, (1986), 87–96.

[3] K. Atanassov, "Operators Over Interval Valued Intuitionistic Fuzzy Sets, Fuzzy Sets and Systems, **64**, (1994), 159–174.

[4] H. Bustince and P. Burillo, "Vague Sets are Intuitionistic Fuzzy Sets, Fuzzy sets and Systems". **79**, (1996), 403–405.

[5] D. Chen, E.C.C. Tsang, D.S.Yeung and X. Wang, "The Parameterization Reduction of Soft Sets and its Applications", *Computers Math. with Appl.*, **49**, (2005), 757–763.

[6] W.L. Gau and D.J. Buchrer, "Vague Sets", *IEEE Tran. System Man Cybernet*, **23**, (1993), 610–614.

[7] M.B. Gorzalzany, "A Method of Inference in Approximate Reasoning Based on Interval Valued Fuzzy Sets, Fuzzy Sets and Systems", **21**, (1987), 1–17.

[8] D.H. Hong and C.H. Choi, "Multicriteria Fuzzy Decision Making Problems Based on Vague Set Theory. Fuzzy Sets and Systems", **114**, (2000), 103–113.

[9] P.K. Maji, R. Biswas and A.R. Roy, "On Soft Set Theory, Computers and Math with Appl.", **41**, (2003), 1–8.

[10] P.K. Maji, R. Biswas and A.R. Roy, "Fuzzy Soft Sets", *The Journal of Fuzzy Mathematics*, **9 (3)**, (2001), 589–602.

[11] D. Molodostov, "Soft Set Theory-first Results", *Computers Math. with Appl.*, **37**, (1999), 19–31.

[12] D. Molodostov, "The Theory of Soft Sets (in Russian)", URSS publishers, Moscow, 2004.

[13] Z. Pawlak, "Rough Sets", *Int. J. Infor. and Comp. Sc.*, **11**, (1982), 341–356.

[14] Z. Pawlak, "Hard Sets and Soft Sets", *ICS Research Report, Inst. of Comp. Sc.*, Poland, 1994.

[15] A.R. Roy and P.K. Maji, "A Fuzzy Soft Set Theoretic Approach to Decision Making Problems", *Jour. Computational and Appl. Math.*, **203**, (2007), 412–418.

[16] H. Thielle, "On the Concepts of Qualitative Fuzzy Sets", *1999 IEEE Inter. Sym. On Multivalued Logic*, Japan, 1999.

[17] W. Xu, J. Ma, S. Wang and G. Hao, "Vague Soft Sets and their Properties", (preprint).

[18] Y.Y. Yao, "Relational Interpretations of Neighbourhood Operators and Rough Set Approximation Operators", *Infor. Sc.*, **111**, (1998), 239–259.

[19] L.A. Zadeh, "Fuzzy Sets", *Infor and Control*, **8**, (1965), 338–353.

10

SOME EPIREFLECTIONS, COREFLECTIONS, AND INJECTIVES IN FUZZY TOPOLOGY*

Arun K. Srivastava

Department of Mathematics
and
Centre for Interdisciplinary Mathematical Sciences,
Banaras Hindu University, Varanasi-221005, India
E-mail: arunksrivastava@gmail.com

Abstract: *Some epireflective and coreflective categories, as well as some injective objects in fuzzy topology, having the common feature of being linked to the fuzzy Sierpinski space, are discussed.*

Keywords: *Fuzzy Sierpinski space; T_0-fuzzy-topological space; Fuzzy injective space; Specialization order; Completely distributive lattice.*

1. INTRODUCTION

Let **TOP** and $\mathbf{TOP}_0$ respectively denote the categories of topological and T_0-topological spaces. Some of the nice properties in topology of the usual two point Sierpinski space 2_S include (*i*) 2_S acts as 'open subset classifier' in **TOP**, (*ii*) for all $(X, T) \in |\mathbf{TOP}|$, T is the 'initial topology' on X with respect to $\mathbf{TOP}((X, T), 2_S)$, (*iii*) 2_S is an injective 'cogenerator' in $\mathbf{TOP}_0$ and the injective objects of $\mathbf{TOP}_0$ are precisely the retracts of products of copies of 2_S, (*iv*) the epireflective hull of 2_S in **TOP** is $\mathbf{TOP}_0$, (*v*) the epireflective hull of 2_S in $\mathbf{TOP}_0$ is the

* For the sake of completeness, the work reported here also includes a sketchy account of some published work with co-authors, besides some ongoing work with Veena Singh.

category of 'sober' topological spaces, (*vi*) the coreflective hull of 2_S in $\mathbf{TOP}_0$ is the category of 'Alexendro-discrete' spaces, and (*vii*) 'frames' are '*T*-algebras' induced (in some sense) by 2_S.

We give an account of our investigations on the counterparts of the properties in 'fuzzy topology', confining ourselves to fuzzy topology, fuzzy bitopology, and intuitionistic fuzzy topology.

A notion of a 'fuzzy Sierpinski space', leading subsequently to a notion of a T_0-fuzzy topological space (definitions recalled below), which turned out to satisfy the fuzzy topological counterparts of many of the above-mentioned properties, was found in [10].

Let $I = [0, 1]$ and let **FTS** denote the category of all fuzzy topological spaces and (fuzzy) continuous maps.

Definition 1.1: [10] I, together with the fuzzy topology on it generated by {id}, is called the fuzzy Sierpinski space.

By I_S, we shall denote the fuzzy Sierpinski space.

Theorem 1.1: [10] Given a fuzzy topological space (X, δ), (*i*) $\mu \in \delta$ iff $\mu: (X, \delta) \to I_S$ is continuous and (*ii*) δ is the 'initial fuzzy topology' on X with respect to **FTS** $((X, \delta), I_S)$.

Definition 1.2: [7] A fuzzy topological space (X, δ) is called T_0 if for all distinct $x; y \in X$, there exists $\mu \in \delta$ such that $\mu(x) \neq \mu(y)$.

The fuzzy Sierpinski space I_S is T_0.

We are assuming familiarity with some standard material on lattice theory (including the notions of *frame* and *completely distributive latticec*; [5]), fuzzy topology (cf. [3]) and category theory (including the notions of *epireflectivity, epireflective hulls, coreflectivity*, and *coreflective hulls*; cf. [1]).

2. SOME EPIREFLECTIVE HULLS IN FUZZY TOPOLOGY

Perhaps the first epireflective hull in fuzzy topology was computed in [7], with the following being its key result.

Theorem 2.1: [7] The epireflective hull in **FTS** of I_S is equal to the category $\mathbf{FTS}_0$ of T_0-fuzzy topological spaces.

Thus, using standard categorical arguments, T_0-ness is a productive and hereditary property in fuzzy topology.

Given a fuzzy topological space (X, δ), let $pt(\delta) = \{p : \delta \rightarrow I \mid p$ is a framemap$\}$. Define a mapping $\psi_X : X \rightarrow pt(\delta)$ as $\psi_X(x)(\mu) = \mu(x)$, $\forall x \in X, \forall \mu \in \delta$.

Definition 2.1: A fuzzy topological space (X, δ) is called sober if ψ_X is a bijection.

It turns out that ψ_X is injective iff (X, δ) is T_0. Thus every sober fuzzy space is T_0. Also, the fuzzy Sierpinski space I_S turns out to be sober. One of the key results on sober fuzzy spaces was proved in [13]

Theorem 2.2: [13] The epireflective hull in $\mathbf{FTS_0}$ of I_S is equal to the category **FSOB** of sober fuzzy topological spaces.

The above results were subsequently extended (cf. [6]) to the category **BFTS** of fuzzy bitopological spaces, i.e., spaces (X, δ_1, δ_2), where δ_1 and δ_2 are two fuzzy topologies on a set X, in the following manner.

Theorem 2.3: [6] There exists two fuzzy 'Sierpinski' bitopological spaces such that the epireflective hulls in **BFTS** of both of them are equal to be the category $\mathbf{BFTS_0}$ of $2T_0$-fuzzy bitopological spaces (i.e., fuzzy bispaces (X, δ_1, δ_2) such that the fuzzy space $(X, \delta_1 \vee 2)$ is T_0).

A fuzzy bispace (X, δ_1, δ_2) is called 'sober' if the map ψ_X defined earlier, canonically modified to and defined as a map $\psi_X : X \rightarrow pt(\delta_1 \vee \delta_2)$, is bijective.

Theorem 2.4: [6] The category **BSOB** of sober fuzzy bispaces is epireflective in $\mathbf{BFTS_0}$ (but it is not the epireflective hull in $\mathbf{BFTS_0}$ of any one of the two fuzzy Sierpinski bitopological spaces, mentioned above).

Remark 2.1: The epireflective hull in $\mathbf{BFTS_0}$ of one of the two fuzzy Sierpinski bitopological spaces turns out to be an interesting category, viz., that of–what have been called in [6] as–absolutely closed $2T_0$-fuzzy bitopological spaces. For further details, we refer to [6].

We close this section by giving an account of some analogous findings in intuitionistic fuzzy topology [11].

In order to explain the notations going to be used, we reacll only the most basic definitions pertaining to intuitionistic fuzzy sets (ifs, in short) and intuitionistic fuzzy topology (ifts, in short); for the remaining basic definitions and results on ifs's and ifts's, we refer to [2] and [4].

Definition 2.2: An intuitionistic fuzzy set u in a nonempty set X is a pair (u_1, u_2) of fuzzy sets in X, i.e., functions $u_1; u_2: X \to I$, such that $u_1(x) + u_2(x) \leq 1$, $\forall x \in X$.

We express u also as $u = (u_1, u_2)$ (in terms of its two 'parts') and it can frequently be also viewed as a function $u: X \to I \times I$, given by $u(x) = (u_1(x), u_2(x))$, $x \in X$. The ifs's $\tilde{0}$ and $\tilde{1}$ are given by $\tilde{0} = (\mathbf{0}, \mathbf{1})$ and $\tilde{1} = (\mathbf{1}, \mathbf{0})$, where $\mathbf{0}$ and $\mathbf{1}$ are respectively the 0-valued and the 1-valued constant fuzzy sets in X.

Definition 2.3: (1) The **supremum** and **infimum** of ifs's $u_j = (u_{j1}, u_{j2})$ in X, $j \in J$, are respectively defined as

(*i*) $\vee_j u_j = (\vee_j u_{j1}, \wedge_j u_{j2})$ and

(*ii*) $\wedge_j u_j = (\wedge_j u_{j1}, \vee_j u_{j2})$.

(2) The **complement** u^c of an ifs $u = (u_1, u_2)$ is defined as $u^c = (u_2, u_1)$.

Definition 2.4: A family τ of ifs's in a nonempty set X, containing $\tilde{0}$ and $\tilde{1}$ and closed under supremum and finite infimum, is called an intuitionistic fuzzy topology on the set X and the pair (X, τ), an intuitionistic fuzzy topological space. A function $f: (X, \tau) \to (Y, \phi)$ between two ift's is called (intuitionistic) continuous if $f^{-1}(\nu) \in \tau$ for all $\nu \in \phi$.

Let **IFTS** denote the category of ifts's.

Let $\mathcal{I} = \{(a, b) \in I \times I \mid a + b \leq 1\}$.

Consider the ifs π in $\mathcal{I}$, given by $\pi = (\pi_1, \pi_2)$, where $\pi_1, \pi_2: \mathcal{I} \to I$ are the two projection maps. It is easily verified that $\tau_S = \{\tilde{0}, \tilde{1}, \pi\}$ is an ift on $\mathcal{I}$.

Call the ifts $(\mathcal{I}, \tau_S)$ the Sierpinski intuionistic fuzzy topological space; the name being justified by the remaining results in this section.

Theorem 2.5: For any ifts (X, τ), $u \in \tau$ iff $u: (X, \tau) \to (\pi_1, \pi_2)$ is continuous. Moreover, if $\mathcal{F} = \{f: (X, \tau) \to (\mathcal{I}, \tau_S) \mid f \text{ is continuous}\}$, then τ is the smallest ift on X making each member of $\mathcal{I}$ continuous from (X, τ) to $(\mathcal{I}, \tau_S)$.

Definition 2.5: An ifts (X, τ) is called T_0 if for all distinct $x, y \in X$, there exists $u = (u_1, u_2) \in \tau$ such that $u_1(x) \neq u_1(y)$ or $u_2(x) \neq u_2(y)$.

Clearly, $(\mathcal{I}, \tau_S)$ is T_0. Also, it can be verified that T_0-ifts's are closed under subspaces and products.

Let $\mathbf{IFTS}_0$ denote the category of T_0 ifts's. Then:

Theorem 2.6: $\mathbf{IFTS}_0$ is the epireflective hull in **IFTS** of $(\mathcal{I}, \tau_S)$.

Define a partial order $\lhd$ on $\mathcal{I}$ as:

$$(a, b) \lhd (a', b') \text{ if } a \leq b \text{ and } a' \leq b'.$$

$(\mathcal{I}, \lhd)$ actually turns out be a frame.

Given any ifts (X, τ), let $pt(\tau) = \{p : \tau \to \mathcal{I} \mid p \text{ is a frame-map}\}$. Define a mapping $\psi_X : X \to pt(\tau)$ as $\psi_X(x)(u) = (u_1(x), u_2(x))$, $\forall x \in X$, $\forall u = (u_1, u_2) \in \tau$.

Definition 2.6: An ifts (X, τ) is called sober if ψ_X is a bijection.

It turns out that $(\mathcal{I}, \tau_S)$ is sober. Moreover:

Theorem 2.7: The category of all sober ifts's is the epireflective hull in $\mathbf{IFTS}_0$ of $(\mathcal{I}, \tau_S)$.

3. A COREFLECTIVE HULL

We recalled in the beginning that the category of Alexandrov-discrete topological spaces (i.e., spaces whose topologies are closed under arbitrary intersection) is the coreflective hull in **TOP** of the usual two-point Sierpinski space 2_S. Recently, Veena Singh in [8], has shown that the analogous result in **FTS** is not true, although the exact coreflective hull in **FTS** of I_S has been computed by her. We give a brief account of it.

Some standard category-theoretic reasonings show that a subcategory **C** of **FTS** is (*i*) coreflective iff it is closed under coproducts (≡ sum) and quotients and (*ii*) the coreflective hull of I_S iff it is homeomorphic to a quotient of a coproduct of copies of I_S.

Let *MC*-**FTS** denote the category of those fuzzy topological spaces, whose fuzzy topologies are closed under arbitrary meets. Then (*i*) has been used to prove:

Theorem 3.1: [8] *MC*-**FTS** is coreflective in **FTS**.

In view of (*ii*) above, the following example shows that *MC*-**FTS** is *not* the coreflecctive hull of I_S.

Example 3.1: Consider the fuzzy topology δ on I, generated by $\{id, 1 - id\}$.

Then (I, δ) is in MC-**TOP**. However, it can be shown that (I, δ) cannot be expressed as a quotient of a coproduct of copies of I_s's.

What then exactly is the coreflective hull of I_S, is shown in the following theorem:

Theorem 3.2: [8] A fuzzy topological space (X, δ) is in the coreflective hull of I_S in **FTS** iff the following two conditions are satisfied:

(*a*) $X = \bigcup_{i \in J} X_i$ for some index set J such that $|X_i| \leq |I|, \forall i \in J$,

(*b*) $\forall i \in J$, δ_i (the fuzzy subspace topology on X_i) is $\{\underline{0}, \underline{1}\}$ or $\{\underline{0}, \alpha_i, \underline{1}\}$, where α_i is a bijection.

4. INJECTIVE T_0-FUZZY SPACES

Given a class ε of morphisms of a category **C**, recall that an object X of **C** is called ε-injective if for any $e: Y \to Z$ in ε and for every **C**-morphism $f: Y \to X$, there is a **C**-morphism $f^*: Z \to X$ with $f = f^* \circ e$. For us here, **C** is always going to be the category of $\mathbf{FTS_0}$ or $\mathbf{BFTS_0}$, or $\mathbf{IFTS_0}$ and ε will always be the class of all (respective) **C**-embeddings. We shall use the term *injective* for ε-injective.

We recalled earlier that (*i*) 2_S is injective in $\mathbf{TOP_0}$ and that any injective object in $\mathbf{TOP_0}$ must be a retract of a product of copies of 2_S (cf. [14]) and (*ii*) every injective object in $\mathbf{TOP_0}$ is precisely a 'continuous lattice' endowed with the 'Scott topology' (cf. [14], [5]). The analogue of (*i*) in fuzzy topology was obtained as:

Theorem 4.1: [12] Injectives in $\mathbf{FTS_0}$ are precisely the retracts of I_S in $\mathbf{FTS_0}$.

Perfectly analogous results (and related ones) have been obtained both in fuzzy bitopology (unpublished) and in fuzzy intuitionistic topology ([11]), using the respective Sierpinski spaces. In what follows, we give an account of our ongoing work on the search for an analogue of (*ii*) in fuzzy topology.

The description of an injective $\mathbf{FTS_0}$-object, say (X, δ), in (*ii*) above has two 'components', a lattice structure (continuous lattice) and a topological structure (Scott topology) different from δ. So far, we have been able to show that the injectivity of every T_0-topological space X can be described in terms a specific property of a built in 'natural' partial order on X. We now explain the main points of our this observation.

Definition 4.1: Given a fuzzy topological space (X, δ), the specialization order on X is the relation $\leq_\delta$ on X given by:

$$x \leq_\delta y \Leftrightarrow \mu(x) \leq \mu(y), \forall \mu \in \delta.$$

Remark 4.1: $\leq_\delta$ is clearly reflexive and transitive. **Moreover,** $\leq_\delta$ **is a partial order iff** (X, δ) is T_0.

All T_0-fuzzy topological spaces (X, δ) will be assumed to be partially ordered with $\leq_\delta$.

Remark 4.2: If $f: (X, \delta) \to (Y, \delta')$ is a continuous map between two fuzzy topological spaces, then clearly, $x \leq_\delta y \Rightarrow f(x) \leq_{\delta'} f(y)$. Moreover, if f is a homeomorphism, then f is an order-isomorphism.

We can now state the main result.

Theorem 4.2: A T_0-fuzzy topological space (X, δ) is injective in $\mathbf{FTS}_0$ iff $(X, \leq_\delta)$ is a completely distributive lattice.

The proof is rather long and we give a very sketchy and brief outline. Owing to the T_0-ness of X, the 'evaluation map' $e: X \to I_S^\delta$ turns out to be an embedding. If we first assume X to be injective also, then Theorem 4.1 provides a retraction $r: I_S^\delta \to X$, with $r \circ e = id$. This helps us show that $e(X)$ is a completely distributive lattice in its own right, via which, we eventually show that X is also a completely distributive lattice. In the converse direction, i.e., when $(X, \leq_\delta)$ is assumed to be a completely distributive lattice, we somehow find a suitable subbase Δ of δ and then construct a continuous map $r: I_S^\Delta \to X$ such that for the evaluation map $e: X \to I_S^\Delta$ (note the change from the earlier δ to the present Δ), $r \circ e = id$. This enables us to express X as a retract of a product of copies of I_S, proving the injectivity of X (by using Theorem 4.1).

5. CONCLUSION

We have attempted to give an account in fuzzy topology of some epireflective and coreflective subcategories, and of injectivity, which are closely related to the respective fuzzy Sierpinski spaces. More and fuller details are given in the referred papers. Clearly, many results can be generalized to cases where I is replaced by more general lattices (e.g., as in [8]). Also, one can possibly (or should) carry out similar studies in fuzzy closure spaces, fuzzy limit spaces, fuzzy proximity spaces, ..., or in more general notions of fuzzy topological spaces, such as the one where every fuzzy set is assigned a 'grade' of openness. Indeed, some such studies have already been undertaken), but for want of space we have left that out. Two problems with which we are presently engaged are the following: (*i*) Can we associate something like the 'Scott topology' with an injective fuzzy T_0-space? (*ii*) In topology, injective T_0-spaces have been linked with some 'T-algebras' arising out of 2_S (cf. [9]). Is it possible that something similar happens in fuzzy topology also (with 2_S replaced by I_S)? Specifically, we have two functors $\boldsymbol{G}$: $\mathbf{SET}^{op} \to \mathbf{FTS}$ and F: $\mathbf{FTS} \to \mathbf{SET}^{op}$. G sends an object X to I_S^X and a morphism $f: X \to Y$ to $G(f): I_S^X \to I_S^Y$, given by $G(f)(g) = g \circ f$, while F sends an object $X = (X, \delta)$ to $C(X, I_S)$, where $C(X, I_S)$ denotes the family of all continuous functions from X to I_S and a morphism $f: X \to Y$, to $F(f): C(X, I_S) \to C(Y, I_S)$, given by $F(f)(\alpha) = \alpha \circ f$.

It can be easily verified that G and F are adjoint functors and hence induce a monad, say (T, η, μ), in **FTS**. The objective is to 'recognize' the 'T-algebras' given by this monad. Some initial findings by us seem encouraging regarding this second problem.

REFERENCES

[1] J. Adamek, H. Herrlich and G.E. Strecker, "Abstract and Concrete Categories", John Wiley, 1990.

[2] K. Atanassov, "Intuitionistic Fuzzy Sets, Fuzzy Sets and Systems", **20**, (1986), 87–96.

[3] C.L. Chang, "Fuzzy Topological Spaces", *J. Math. Anal. Appl.*, **24**, (1968), 182–189.

[4] D. Coker, "An Introduction to Intuitionistic Fuzzy Topological Spaces, Fuzzy Sets and Systems", **88**, (1997), 81–89.

[5] G. Gierz, K.H. Hofmann, K. Keimel, J.D. Lawson, M. Mislove and D.S. Scott, "A Compedium of Continuous Lattices", Springer-Verlag, 1980.

[6] A.S. Khastgir and A.K. Srivastava, "On two Epireflective Subcategories of the Category of $2T_0$-fuzzy Bitopological Spaces", *Questiones Math.*, **23**, (2000), 163–177.

[7] R. Lowen and A.K. Srivastava, **"FTS$_0$**: The Epireflective Hull of the Sierpinski Object in FTS, Fuzzy Sets and Systems", **29**, (1989), 171–176.

[8] V. Singh, "On a Coreflective Hull in *L*-**TOP**, Fuzzy Sets and Systems", **159**, (2008), 2611–2615.

[9] M. Sobral. "Contravaraint Hom-functors and Monadicity, in: Category Theory at Work", (H. Herrlich and H.-E. Porst, editors), (Heldermann-Verlag), (1991), 307–319.

[10] A.K. Srivastava, "Fuzzy Sierpinski Space", *J. Math. Anal. Appl.*, **103**, (1984), 103–105.

[11] A.K. Srivastava, "On the Category of IF Topological Spaces (Communicated)".

[12] A.K. Srivastava and S.P. Mishra, "Epis and Injectives in **FTS$_0$"**, *J. of Fuzzy Math.*, **1**, (1993), 389–393.

[13] A.K. Srivastava and A.S. Khastgir, "On Fuzzy Sobriety", *Information Sciences*, **110**, (1998), 195–205.

[14] D. Scott, "Continuous Lattices, in: Lecture Notes in Mathematics (Springer), **274**, (1972), 97–136.

11

NON-CLASSICAL LOGICS WITH REAL LIFE APPLICATIONS

Esko Turunen*

Tampere University of Technology,
P.O. Box 553, 33101 Tampere, Finland
E-mail: esko.turunen@tut.fi

Abstract: *A review of mathematical fuzzy logic, GUHA data mining logic and paraconsistent logic and their real life applications.*

Keywords: *Non-classical logics.*

1. INTRODUCTION

Boolean logic is the logic of mathematics; a natural number is either a prime number or is not (*tertium non datur*), a property either holds for all square matrices, or just one exception is enough to wreck a general law (*∀ and ∃ quantifiers*), from contradictory premises anything can be inferred (*ex contradictione quodlibet*). However, applying Boolean logic outside an ideal platonic realm causes problems:

- Symptoms of illness are mostly vague and obscure rather that crisp and clear: this leads us to study *many valued logics* and *fuzzy logic*.
- In economic science we accept the law of supply and demand, however, a few exceptions are tolerated: we are invited to learn more about non-classical quantifiers familiar in *GUHA data mining logic*.

* This paper is part of research project COST Action I0602 Algorithmic Decision Theory.

- In legal proceedings we usually meet paraconsistent testimonies; there are simultaneously testifies in favor of and against the prisoner at the bar, nevertheless, an unequivocal sentence can be passed. This makes *paraconsistent logic* worth studying.

In this paper we take a review of (*a*) mathematical fuzzy logic, in particular many-valued similarity, (*b*) logic based data mining related to mathematical fuzzy logic, (*c*) paraconsistent logic and its continuous valued extensions and (*d*) interaction and synergy of non-classical logics and their application to real world problems.

2. MATHEMATICAL FUZZY LOGIC AND MANY VALUED SIMILARITY

In the early papers of Zadeh on Fuzzy Set theory [27] and Fuzzy logic [28], the idea was to model vague phenomena such as *tall*, *green* or *young*. However, Zadeh's approach was rather technical, and a natural question of the mathematical foundations of fuzzy theories came up. One of the first milestones to this direction was Pavelka's study [16], where graded tautologies and graded probability in Lukasiewicz [0, 1]-valued logic [14] were introduced and many fundamental properties including consistency and completeness was proved. However, independent and important work was done also by Hájek [7, 8], Höhle [9, 10, 11], Mundici [3] Chakroborty [2, 19] and many others; in one way or another all these works are based on *residuated lattices*. At present there are hundreds of active researchers in the world studying various *mathematical fuzzy logics*. Owing to these studies mathematical fuzzy logic is today widely considered as a proper way of formalizing vague concepts and inference mechanisms, indeed, mathematical fuzzy logic enjoys both a large descriptive power and a strict mathematical foundation and methodology.

In [22] we took a more general approach to Pavelka's logic and gave a pure algebraic proof for many properties of Pavelka's fuzzy logic: these studies are based on the concept of *injective MV-algebra*, a special residuated lattice generalizing the standard Lukasiewicz structure on the real unit interval [0, 1]. In [13] we introduced an injective MV-algebra based *similarity algorithm* to construct fuzzy inference. The fundamental idea is that fuzzy IF.THEN inference system of type

Rule 1: IF x is A_1 and y is B_1 and z is C_1 THEN w is D_1,

Rule 2: IF x is A_2 and y is B_2 and z is C_2 THEN w is D_2,

...

...

Rule k IF x is A_k and y is B_k and z is C_k THEN w is D_k,

where $A_i = (X, \mu_{A_i})$, $B_i = (Y, \mu_{B_i})$, $C_i = (Z, \mu_{C_i})$ and $D_i = (W, \mu_{W_i})$, $i = 1, \ldots, k$, are the fuzzy subsets of non-void sets X, Y, Z and W, respectively can be modeled by several partial similarities which together generate a *total fuzzy similarity relation*. Indeed, as proved in [13], any fuzzy set generates a (partial) fuzzy similarity and, assuming that the algebraic framework is an injective MV-algebra, an aggregation of the partial similarities yield another fuzzy similarity: a reflective, symmetric and weakly transitive binary fuzzy relation.

The algorithm has been applied in several real world problem, e.g. in traffic signal control [15], to model real-time water reservoir operation [5] and in defining athlete's aerobic and anaerobic thresholds [12], moreover, several other applications are in the process of preparation.

3. GUHA-LOGIC FOR DATA MINING

The *GUHA method* for data mining was introduced in [6]. Assuming we have a data file composed of m columns and n rows, consider two Boolean attributes ϕ and ψ. *A four-fold table* $<a, b, c, d>$ related to these attributes is composed from numbers of objects in the data satisfying four different binary combinations of these attributes:

	ψ	$\neg\psi$
ϕ	a	b
$\neg\phi$	c	d

where

- a is the number of objects satisfying both ϕ and ψ,
- b is the number of objects satisfying ϕ but not ψ,
- c is the number of objects not satisfying ϕ but satisfying ψ,
- d is the number of objects not satisfying ϕ nor ψ.

Various relations between ϕ and ψ can be measured in the data by different *four-fold table quantifiers*, denoted by $\phi \sim \psi$, which are to be understood as functions with values in the real unit interval [0, 1]. The simplest and most well-known is the *basic implicational quantifier* characterized by an equation

$$\Rightarrow_{\phi}(a, b) = \frac{a}{a+b} \geq p.$$

An intuitive idea behind this *generalized quantifier* is that a proposition 'ϕ *implies* ψ' is accepted to be *true* (or, *supported by the data*) if, in the set of objects satisfying ϕ, the amount of objects satisfying ψ, too, is large enough or, equivalently, the number of objects *not* satisfying ψ is small enough, i.e. a fixed value $p \in (0, 1]$ is close to 1. In the opposite case the proposition 'ϕ *implies* ψ' is *false* (*not supported by the data*).

The GUHA method is based on a first-order logic that allows *generalized quantifier*: there are quantifiers that have a clear *statistical* justification, e.g. Fisher quantifier and χ^2 quantifier (cf. [6]) and many other such quantifiers, there are quantifiers that model situations like '*in a subset ϕ a certain property is considerably more common (rare) than in the whole data*' and '*in most cases, if ϕ occurs then also ψ occurs*'. Such quantifiers model situations where a few exceptions do not destroy a general rule.

The are several software applications based on GUHA logic; the most used GUHA procedure today is the 4ft-Miner procedure (cf. [18]), which mines for various types of association rules including conditional rules? Such software tools make it possible to very quickly compute various contingency tables. These procedures are included in the academic software system LISp-Miner (freely down loadable from http://lispminer.vse.cz.)

The LISp-Miner system has been applied to solve many real life practical tasks of data mining, as an example we mention to construct a traffic flow control system, cf. [4]; starting from several data matrices with tens of thousands of rows and tens of columns collected by automated traffic observation stations, we were able to construct a fuzzy IF-THEN inference system performing correctly in 90% of cases in predicting a proper conquest travel time.

At present several research activities are related to GUHA method and to the LISp-Miner system. They concern logical calculi for data mining, applications of semantics in data mining, analytical reports summarizing results of data mining and automatic converting generalized association rules into sentences of natural language; see http://lispminer.vse.cz/research/.

4. PARACONSISTENT LOGIC

Quoting from Stanford Encyclopedia of Philosophy [20] 'The contemporary logical orthodoxy has it that, from contradictory premises, anything can be inferred. To be more precise, let $\models$ be a relation of logical consequence, defined either semantically or proof.theoretically. Call $\models$ explosive if it validates $\{A, \neg A\} \models B$ for every A and B (ex contradictione quodlibet). The contemporary orthodoxy, i.e., classical logic, is explosive, but also some 'non-classical' logics such as intuitionist logic and most other standard logics are explosive.

The major motivation behind paraconsistent logic is to challenge this orthodoxy. A logical consequence relation, $\models$, is said to be paraconsistent if it is not explosive. Thus, if $\models$ is paraconsistent, then even if we are in certain circumstances where the available information is inconsistent, the inference relation does not explode into triviality. Thus, paraconsistent logic accommodates inconsistency in a sensible manner that treats inconsistent information as informative.

In Belnap's first order paraconsistent logic [1], four possible values associated with a formula α are true, false, contradictory and unknown: if there is evidence for α and no evidence against α, then α obtains the value true and if there is no evidence for α and evidence against α, then α obtains the value false.

A value contradictory corresponds to a situation where there is simultaneously evidence for α and against α and, finally, α is labeled by value unknown if there is no evidence for α nor evidence against α. More formally, the values are associated with ordered couples $T = <1, 0>$, $F = <0, 1>$, $K = <1, 1>$ and $U = <0, 0>$, respectively.

In [21] Tsoukias introduced an extension of Belnap's logic (named DDT) most importantly because the corresponding algebra of Belnap's original logic is not a Boolean algebra, while the extension is. Indeed, in that paper it was introduced and defined the missing connectives in order to obtain a Boolean algebra. Moreover, it was explained why we get such a structure. Among others it was shown that negation, which was reintroduced in [21] in order to recover some well known tautologies in reasoning, is not a complementation.

In [17] and [24], a continuous valued extension of DDT logic was studied.

The authors imposed reasonable conditions this continuous valued extension should obey and, after a careful analysis, they came to the conclusion that the graded values are to be computed via

$$t(\alpha) = \min\{a, 1 - b\}, \qquad \text{... (1)}$$

$$k(\alpha) = \max\{a + b - 1, 0\}, \qquad \text{... (2)}$$

$$u(\alpha) = \max\{1 - a - b, 0\}, \qquad \text{... (3)}$$

$$f(\alpha) = \min\{1 - a, b\}. \qquad \text{... (4)}$$

where an ordered couple $\langle a, b \rangle$, called *evidence couple*, is given. The intuitive meaning of a and b is the degree of evidence of a statement α and against α, respectively. Moreover, the set of 2×2 *evidence matrices* of a form

$$\begin{bmatrix} f(\alpha) & k(\alpha) \\ u(\alpha) & t(\alpha) \end{bmatrix}$$

is denoted by $\mathcal{M}$. The values $f(\alpha)$, $k(\alpha)$, $u(\alpha)$ and $t(\alpha)$ are values on the real unit interval [0, 1] such that $f(\alpha) + k(\alpha) + u(\alpha) + t(\alpha) = 1$. Their intuitive meaning is f = *falsehood*, k = *contradictory*, u = *unknown* and t = *truth* of the statement α. One of the most important features of paraconsistent logic is that truth and falsehood are not each others complements.

In [17] it is shown how such a fuzzy version of Belnap's logic can be applied in preference modeling. In [25] we show how paraconsistency is related to data mining.

In our resent paper [26] we show that, instead of Boolean structure, the valuation domain $\mathcal{M}$ should be equipped with a more general injective MV-algebra structure. This associates fuzzy extensions of four valued paraconsistent logics with Pavelka's fuzzy logic. As a consequence complete truth calculus is obtained and the corresponding paraconsistent sentential logic is Pavelka style fuzzy logic with new semantics. Thus, a rich semantics and syntax is available. For example, Lukasiewicz tautologies as well as Intuitionistic tautologies can be expressed in the framework of this logic. This follows by the fact that we have two sorts of logical connectives conjunction, disjunction, implication and negation interpreted either by the monoidal operations $\odot$, $\oplus$, $\rightarrow$,* or by the lattice operations $\wedge$, $\vee$, $\Rightarrow$, *, respectively (however, neither * nor * is a lattice complementation). Besides, there are many other logical connectives available.

REFERENCES

[1] Belnap N.D., "A Useful Four-valued Logic, in Epstein, G. and Dumme, J. (eds.) Modern uses of Multiple Valued Logics. D. Reidel, Dordrecht (1977), 8–37.

[2] Banerjee M., Mihir K. Chakraborty M.K., "Foundations of Vagueness : A Category-theoretic Approach. Electr. Notes Theor. Comput. Sci.", **82**, (2003), 142–149.

[3] R. Cignoli, R. D'Ottaviano V., Mundici D., "Algebraic Foundations of Many-valued Reasoning, Trends in Logic, 7, Kluwer Academic Publishers, Dordrecht, 2000.

[4] Coufal D. and Turunen E., "Short Term Prediction of Highway Travel Time Using Data Mining and Neuro-Fuzzy Methods", *Neural Network World*, 3-4 (2004), 221–231.

[5] Dubrovin T., Jolma. A. and Turunen E., "Fuzzy Model for Real-time Reservoir Operation", *Journal of Water Resources Planning and Management*, **128**, (2002), 66–73.

[6] Hájek P. and Havránek T., "Mechanizing Hypothesis Formation. Mathematical Foundations for a General Theory". Springer-Verlag, Berlin (1978), pp. 396.

[7] Hájek P., "Metamathematics of Fuzzy Logic, Trends in Logic", *Studia Logica Library*, Kluwer, Dordrecht, **4**, (1998).

[8] Hájek P., "Basic Fuzzy Logic and BL-algebras", *Soft Computing*, **2**, (1998), 124–128.

[9] Höhle U., "Monoidal Logic, in: Kruse R., Gebhardt J., and Palm R.(eds.): "Fuzzy Systems in Computer Science", Vieweg, Wiesbaden. (1994), 233–243.

[10] Höhle U., "Commutative, Residuated *l*-monoids", in: Höhle U. and Klement E. (Eds), "Non-Classical Logics and their Applications to Fuzzy Subsets". Kluwer, Dordrecht (1995), 53–106.

[11] Höhle, U., "On the Fundamentals of Fuzzy Set Theory", *J. Math. Anal. Applications*, (201), (1996), 786–826.

[12] Ketola J., Saastamoinen K., Turunen E., "Defining Athlete's Aerobic and Anaerobic Thresholds by Using Similarity Measures and Differential Evolution, 2004, *IEEE Int. Conference of Systems, Man and Cybernetics*, 1331–1335.

[13] Kukkurainen P., Turunen E., "Many-valued Similarity Reasoning. An Axiomatic Approach", *International Journal of Multiple Valued Logic*, **8**, (2002), 751–760.

[14] Lukasiewicz Jan, "Selected Works". Ludwik Borkowski. (ed.) (1970) North-Holland Pub. Co.

[15] Niittymöki, J. and Turunen, E., "Traffic Signal Control on Similarity Logic Reasoning, Fuzzy Sets and Systems", **133**, (2003), 109–131.

[16] Pavelka J., "On Fuzzy Logic I, II, III". *Zeitsch. f. Math. Logik*, **25**, (1979), 45–52, 119–134, 447–464.

[17] Perny P. and Tsoukias A., "On the Continuous Extensions of four Valued Logic for Preference Modeling", *Proceedings of the IPMU conference* (1998), 302–309.

[18] Rauch J., "Four-Fold Table Calculi for Discovery Science", in: Discovery Science. Eds. Arikawa, Setsuo–Motoda Hirishi. Berlin, Springer-Verlag (1998), 405–406.

[19] Sen J., Chakraborty M.K, "Linear Logic and Lukasiewicz-Valued Logic: A Logico-Algebraic Study", *Journal of Applied Non-Classical Logics*, **11**, (2001), 313–329.

[20] *http://plato.stanford.edu/entries/logic-paraconsistent/*

[21] Tsoukias A., "A First Order, four Valued, Weakly Paraconsistent Logic and its Relation to Rough Sets Semantics". *Foundations of Computing and Decision Sciences*, **12**, (2002), 85–108.

[22] Turunen E., "Well-defined Fuzzy Sentential Logic, Mathematical Logic Quarterly, **41**, (1995), 236–248.

[23] Turunen E., "Mathematics Behind Fuzzy Logic" Springer–Verlag (1999).

[24] Öztürk M., and Tsoukias A., "Modeling Uncertain Positive and Negative Reasons in Decision Aiding, Decision Support Systems", **43**, (2007), 1512–1526.

[25] Turunen E., "Interpreting GUHA Data Mining Logic in Paraconsistent Fuzzy Logic Framework (submitted)".

[26] Turunen E., "Paraconsistent Semantics for Pavelka Style Fuzzy Sentential Logic (submitted).

[27] Zadeh L., "Fuzzy Sets, Information and Control", **8**, (1965), 338–353.

[28] Zadeh L., "Fuzzy Logic, Neural Networks, and Soft Computing Comm. AMC". **37**, (1994), 77–84.

12

EMOTION RECOGNITION BASED ON ROUGH SET THEORY*

Guoyin Wang and Yong Yang

Institute of Computer Science and Technology, Chongqing University of Posts and Telecommunications, Chongqing, 400065, P.R. China

Abstract: *Emotion recognition is very important for human-computer intelligent interaction. It is often based on facial or audio information using artificial neural network, fuzzy set, support vector machine, hidden markov model, etc. Although some progress has already been made in emotion recognition, several unsolved issues still exist. For example, it is still an open problem which features are the most important for emotion recognition. It is a subject that was seldom studied in computer science. However, related research works have been conducted in cognitive psychology. In this paper, feature selection for emotion recognition is studied based on rough set theory, Firstly, a attribute reduction algorithm of classical rough set theory is taken as a feature selection method of emotion recognition, Secondly, a self-learning attribute reduction algorithm based on extended rough set theory named SARA is proposed based on extended rough set and domain oriented data-driven data-mining theory. Experimental results show that important and useful features for emotion recognition can be identified by rough set theory. It is found that the features concerning mouth are the most important ones in geometrical features for facial emotion recognition. Meanwhile, emotion recognition methods based on rough set and ensemble learning are also studied, a selective ensemble feature selection method named SEFSBRS and a dynamic ensemble feature selection method named DEFS are proposed. Experiment results show that emotion recognition methods based on rough set theory and ensemble learning are effective and efficient.*

Keywords: *Feature selection, Emotion recognition, Rough set, Domain-oriented data-driven data mining (3DM).*

* This paper is partially supported by National Natural Science Foundation of China under Grants No. 60773113, Natural Science Foundation of Chongqing under Grants No. 2007BB2445, No. 2008BA2017 and No. 2008BA2041.

1. INTRODUCTION

In recent years, there has been a growing interest in improving all aspects of the interactions between humans and computers. It is argued that to truly achieve effective human-computer intelligent interaction (HCII), there is a requirement for computers to be able to interact naturally with users, similarly to the way human-human interaction. HCII is becoming more and more important in such applications as smart home, smart office and virtual reality, and it will be popular in all aspects of daily life in the future. To achieve the purpose of HCII, it is essential for computers to recognize human emotion and to give a suitable feedback. Consequently, emotion recognition attracts significant attention in both industry and academia. There are several research works in this field in recent years, and some successful products such as AIBO, the popular robot dog produced by Sony. Usually, emotion recognition is studied by the methods of artificial neural network (ANN), fuzzy set, support vector machine (SVM), hidden markov model (HMM), and based on the facial or audio features, and the recognition rate often arrives at 64% to 98% [1 – 3]. Although some progress has been made in emotion recognition, several unsolved issues still exist. For example, it is still an open problem which features are the most important for emotion recognition. It is a subject that was seldom studied in computer science. However, related research works have been conducted in cognitive psychology [4 – 6].

There have been several research works related to the important features for emotion in cognitive psychology. Based on the results of psychological experiments, Sui et. al. argue that the information conveyed by different facial parts have diverse effects on the facial expression recognition and the eyes play the most important role [4]. Wang et al argue that the low spatial frequency information is important for emotion [5]. White argues that edge-based facial information is used for expression recognition [6].

In this paper, attribute reduction algorithms based on classical rough set theory are firstly used for feature selection of a facial emotion recognition system. Moreover, a novel attribute reduction algorithm based on extended rough set theory and domain oriented data driven data mining (3DM) is proposed as a feature selection method of emotion recognition. Experiment results show that important features for emotion recognition are found based on attribute reduction algorithms of rough set theory. Furthermore, efficient emotion recognition methods are also studied based on rough set theory and ensemble learning, a selective ensemble feature selection method and a dynamic ensemble feature selection method based on rough set theory are proposed.

The rest of this paper is organized as follows. In Section 2, feature selection for emotion recognition based on rough set theory is introduced. In Section 3, emotion recognition methods based in rough set theory and ensemble learning are proposed. Finally, conclusions and future works are discussed in Section 4.

2. FEATURE SELECTION FOR EMOTION RECOGNITION BASED ON ROUGH SET THEORY

Rough set (RS) is a valid mathematical theory for dealing with imprecise, uncertain, and vague information, it was developed by Professor Z. Pawlak in 1980s [7], [8]. RS has been successfully used in many domains such as machine learning, pattern recognition, intelligent data analyzing and control algorithm acquiring. The most advantage of RS is its great ability of attribute reduction (knowledge reduction, feature selection).

2.1 Feature Selection for Emotion Recognition Based on Classical Rough Set Theory

There are a lot of research works on the attribute reduction [9], [10]. In this paper, conditional entropy-based algorithm for reduction of knowledge without core (CEBARKNC) proposed by Wang Guo-Yin et. al. [11] is used for feature selection. Basic concepts of rough set theory are omitted due to the limited space. The detail of CEBARKNC algorithm can be found in [11]. Experiments are taken as follows for the purpose of proving the effectiveness of CEBARKNC as a method feature selection of emotion recognition.

Since there are few open facial emotional dataset, three facial emotional datasets are used in the experiments. The first dataset comes from the Cohn-Kanade AU-Coded Facial Expression (CKACFE) database [12] and the dataset is more representative of Caucasian to some extent. The second one is the Japanese female facial expression (JAFFE) database [13] and it is more representative of Asian women. The third one named CQUPTE [14] is collected from 8 graduate students in the Chongqing University of Posts and Communications in China, in which there are four female and four male.

Facial expression of human being is expressed by the shape and position of facial components such as eyebrows, eyes, mouth, nose, etc. The geometric features, appearance features, wavelet features and mixture features of facial are popular for emotion recognition in recent years. The geometric facial features represent the shape and locations of facial components, and it is used in the experiments since it is obvious and intuitionistic for the facial expression. 33 facial features are extracted for emotion recognition according to [15], [16].

There are three experiments in this section. In the first experiment, CEBARKNC [11] is taken as a method of feature selection for emotion recognition.. In this experiment, a greedy algorithm proposed by Nguyen [17] is taken as a discretization method and it is done on the platform RIDAS [18]. In the second experiment, a traditional feature selection method, Genetic Algorithm (GA) [19], is used as the feature selection method for emotion recognition. SVM are taken as classifiers for all the comparative experiments. SVM are given same parameters in all the experiments. 4-fold cross-validation is taken for all the experiments.

The results of correct recognition rate are shown in Table 1, and the results of feature selection is listed in Table 2. CRR is the percentage of the correct recognition rate, and RAN is the number of attributes after attribute reduction in Table 1.

Table 1: *The Results of Correct Recognition Rate of the Compared Experiments*

Database	*CEBARKNC + SVM*		*GA + SVM*	
	CRR	*RAN*	*CRR*	*RAN*
CKACFE	73.67	12.5	73.09	14.25
JAFFE	61.56	11	55.89	14.25
CQUPTE	89.10	13.5	88.95	14.75
average	74.78	12.33	72.64	14.42

Table 2: *The Results of Feature Selection of the Compared Experiments*

Database	*CEBARKNC*	*GA*
CKACFE	$x_1, x_8, x_{13}, x_{16}, x_{21}$	x_{14}, x_{25}
JAFFE	x_0, x_8, x_{13}	x_7, x_{25}, x_{32}
CQUPTE	$x_0, x_1, x_4, x_8, x_{13}, x_{26}, x_{27}, x_{32}$	$x_{11}, x_{14}, x_{22}, x_{23}, x_{24}, x_{32}$
Common Feature	x_8, x_{13}	–

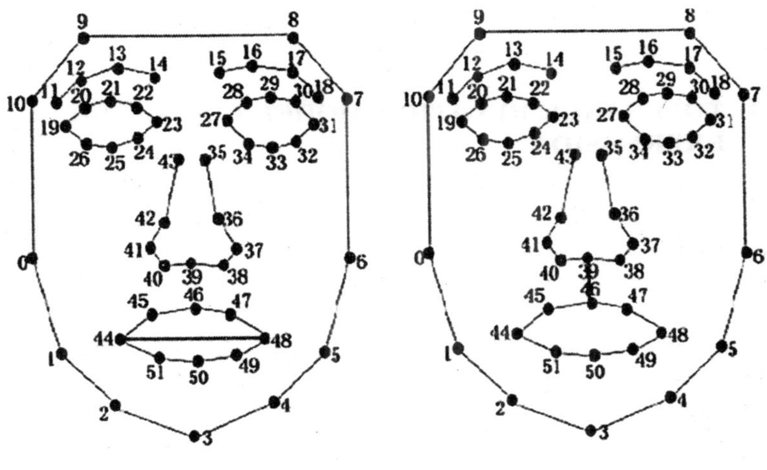

d_{13} d_8

Fig. 1: *The Important Features Selected by CEBARKN*

From the experiments listed above, we can find the attribute reduction algorithm can find the important features for emotion recognition and higher recognition rate can be gotten compared with traditional feature selection method GA.

2.2 Feature Selection for Emotion Recognition Based on Extended Rough Set Theory

In last section, attribute reduction algorithms based on classical rough set are used for the purpose of facial emotional feature selection. Some useful features are found. However, classical rough set theory is based on equivalence relation. There must be a process of discretization in equivalence relation since the measured facial features are continuous values. Information might be lost or changed in the discretization process, thereby affecting the result. To solve this problem, in this section, a novel feature selection method based on tolerance relation is proposed, which can avoid the process of discretization. Meantime, based on the idea of domain oriented data driven data mining (3DM) [20], [21], a method for finding suitable threshold of tolerance relation is introduced. Experimental results show that important and useful features for emotion recognition can be identified by the proposed method with a high recognition rate. It is found that the features concerning mouth are the most important ones in geometrical features for facial emotion recognition.

3. TOLERANCE RELATION MODEL FOR FACIAL EXPRESSION

Since all measured facial attributes are continuous value and imprecise to some extent, the process of discretization may affect the result of emotion recognition. We argue that it is suitable for the continuous value information systems that the attribute values are taken as equal if they are similar in some range. Based on this idea, a method based on tolerance relation that avoids the process of discretization is proposed in this paper.

Definition 1: Let an information system $S = (U, C \cup D, V, f)$ be a continuous value information system, a relation $R(x, y)$ is defined as:

$$R(x, y) = \{(x, y) | x \in U \wedge y \in U \vee \forall_{a \in C} (|a_x - a_y| \leq \varepsilon, 0 \leq \varepsilon \leq 1)\}$$

Apparently, R (x, y) is a tolerance relation since R (x, y) is symmetrical and reflextive. Since then, entropy and conditional entropy of tolerance relation R (x, y) can be defined, and detail properties of R (x, y) can be seen in [22].

The parameter of ε is a key issue if the tolerance relation is used for different information system. Usually, the parameter is set according to the prior knowledge of experts. But if there are no prior knowledge or the prior knowledge is hardly gotten, the tolerance relation can't used for the information system properly. In this section, a method for selected parameter of R (x, y) automatically according to the data only and without the prior knowledge based on 3DM is proposed as follows.

3DM is a data mining theory proposed by Guoyin Wang [20, 21]. According to the theory, knowledge could be expressed in different ways, i.e., some relationship exists between the different formats of the same knowledge. In order to keep the knowledge unchanged in a data mining process, the properties of the knowledge should remain unchanged during the knowledge transformation process [23]. Otherwise, mistake may occur in the process of knowledge transformation. Based on this understanding, knowledge reduction can be seen as a process of knowledge transformation, in which properties of the knowledge should be remained.

In the application of emotion recognition, no faces are entirely the same and nor are emotions. For any two different emotion samples, there must be some different features in the samples. Accordingly, an emotion sample belongs to an emotion state according to its features which are different to the others. From this standpoint, we argue that the discernability of the condition attribute set with respect to the decision attribute set can be taken as an important property of knowledge in the course of knowledge acquisition in emotion recognition. Based on the idea of 3DM, the discernability should be unchanged in the process of knowledge acquisition and attribute reduction. The attribute reduction algorithm based on 3DM is introduced as follows and detailed deduction is listed in [22].

The discernability of condition attribute set with respect to decision attribute set in tolerance relation is a fundamental feature of knowledge of a continuous value information system. The discernability should be unchanged according to 3DM. $E(D|C) = 0$ is a necessary and sufficient condition for keeping the discernability of condition attribute set with respect to decision attribute set in tolerance relation [22]. Therefore, a self-learning attribute reduction algorithm (SARA) is proposed for continuous value information systems, and detail of SARA can be seen in [22]. For the purpose of proving the effectiveness of SARA, compared experiments are taken as follows.

There are two experiments. In the first experiment, SARA is taken as the method of feature selection for emotion recognition. In the second one, CEBARKNC [11] is taken as a method of feature selection for emotion recognition. SVM are taken as classifiers for the comparative experiments. SVM are given same parameters in all the experiments. 4-fold cross-validation is taken for all the experiments.

The results of the comparative experiments are shown in Table 3

Table 3: *The Results of Comparative Experiments on the three Dataset*

Database	*SARA + SVM*		*CEBARKNC + SVM*	
	CRR	*RAN*	*CRR*	*RAN*
CKACFE	76.01	11.25	73.67	12.5
JAFFE	69.37	11.5	61.56	11
CQUPTE	92.45	14	89.10	13.5
average	79.28	12.25	74.78	12.33

When we compare the experimental results of SARA + SVM and CEBARKNC + SVM from the Table 3, we can find SARA selects as much features as CEBARKNC, but SARA gets a better correct recognition rate than CEBARKNC.

Common features reserved by the four feature selection methods are listed in Table 2.

Table 4: *The Common Features in the three Datasets*

Database	*SARA*	*CEBARKNC*
CKACFE	x_{13}, x_{14}, x_{15}	$x_1, x_8, x_{13}, x_{16}, x_{21}$
JAFFE	$x_5, x_{13}, x_{14}, x_{15}, x_{26}$	x_0, x_8, x_{13}
CQUPTE	$x_7, x_{13}, x_{14}, x_{15}, x_{24}, x_{30}, x_{31}$,	$x_0, x_1, x_4, x_8, x_{13}, x_{26}, x_{27}, x_{32}$
Common Feature	x_{13}, x_{14}, x_{15}	x_8, x_{13}

From Table 4, we can find the four feature selection algorithms can select different features for emotion recognition. Among all the experiment results, SARA selects three common features, x_{13}, x_{14} and x_{15} for all the three emotion datasets, meanwhile, CEBARKNC selects two common features, x_8 and x_{13}. Since better correct recognition rate can be achieved if SARA is used as a method of feature selection for emotion recognition. Therefore, x_{13}, x_{14}, x_{15} can be seen more important for emotion recognition. Although the features of x_{13}, x_{14}, x_{15} are normalized features, the importance of original features of d_{13}, d_{14}, d_{15} are also evident. Since the feature of d_{13}, d_{14}, d_{15} are all concerning mouth, therefore, we can draw a conclusion that the geometrical features concerning mouth are the most important features for emotion recognition. The original selected features of SARA and CEBARKNC are shown in Fig. 2.

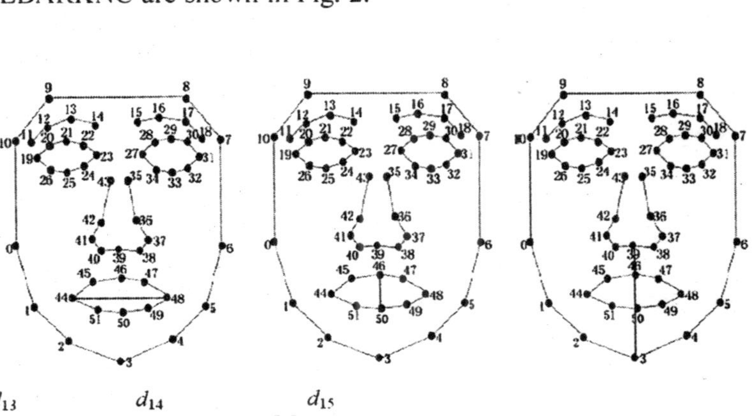

d_{13} d_{14} d_{15}

Fig. 2: *Common Features Selected by SARA*

3.1 Experiments for the Features Concerning Mouth for Emotion Recognition

From the last section, we draw a conclusion that the geometrical features concerning mouth are important for emotion recognition. In this section, there are four experiments for the purpose of testing the importance of the geometrical feature concerning mouth for emotion recognition. In the first experiment, all the 33 facial features are used for emotion recognition. In the second experiment, only the features selected by SARA are used for emotion recognition. In the third experiment, all the features concerning mouth are deleted and there are 19 features are used for emotion recogñition, in which there are 17 features concerning eyes and two features concerning cheek but not mouth. In the fourth experiment, all the features concerning eyes are deleted and there are 12 features are used for emotion recognition, in which there are 10 features concerning mouth and two features concerning cheek but not eyes. SVM is taken as classifier in the four experiments and is given the same parameters. Experiment results are listed in Table 3.

Table 5: *The Results of Comparative Experiments on the three Dataset*

	SARA Reserved		*ALL Features*		*No Mouth*		*No Eyes*	
	CRR	*RAN*	*CRR*	*RAN*	*CRR*	*RAN*	*CRR*	*RAN*
CKACFE	76.01	11.25	79.80	33	45.18	19	75.15	12
JAFFE	69.37	11.5	74.46	33	53.32	19	58.29	12
CQUPTE	92.45	14	93.86	33	69.02	19	89.72	12
Average	79.28	12.25	82.71	33	55.84	19	74.39	12

From the Table 5, we can find that the correct recognition rate is decreased greatly if there is no feature concerning mouth. Therefore, it is concluded that the features concerning mouth are the most important geometrical features for emotion recognition. On the other hand, we can find the correct recognition rate does not affected so much if there are no features concerning eyes. Therefore, the geometrical features concerning eyes do not play an important role in emotion recognition. But from the psychological experiments of

[4], Sui et. al. found the eyes play an important role in emotion, therefore, we may draw a conclusion that the geometrical features concerning mouth are the most important in the geometrical features for emotion recognition, and the geometrical features concerning eyes are not so important. Furthermore, the important features concerning eyes for emotion recognition should be discovered and used in emotion recognition in the further work. Meanwhile, we can find that the correct recognition rate is decreased in CKACFE more than in JAFFE and CQUPTE. Therefore, we can draw a conclusion that the geometrical features concerning mouth are more important for emotion expression for the Caucasian than the eastern people.

4. EMOTION RECOGNITION METHODS BASED ON ROUGH SET THEORY AND ENSEMBLE LEARNING

4.1 A Selective Ensemble Feature Selection Method for Emotion Recognition

Ensemble learning has been a hot research topic in machine learning since 1990s' [24]. Ensemble methods are learning algorithms that construct a set of base classifiers and then classify new objects by integrating the prediction of the base classifiers. An ensemble system is often much more accurate than each base classier. As a popular machine learning method, ensemble methods are often used in pattern recognition, network security, medical diagnosis, etc.

A necessary and sufficient condition for an ensemble system to be more accurate than any of its base classifier is that the base classifiers are accurate and diverse. An accurate classier is one that has an error rate less than random guessing on new instances. Two classifiers are diverse if they make different prediction errors on unseen objects. Besides accuracy and diversity, another important issue for building an effective ensemble system is the choice of the function for combining the predictions of the base classifiers. There are many techniques for the integration of an ensemble system, such as majority voting, weighted voting, reliability-based weighted voting, etc.

There are many methods proposed for ensemble. The most popular way for ensemble is to get different subset of the original dataset by resampling the training data set many times. Bagging, boosting and cross-validation are all such ensemble methods. These methods work well especially for unstable learning algorithms, such as decision trees, neural network. Some other methods are also studied, such as manipulating the output targets, injecting randomness into classifiers. Besides these methods, there is another effective approach for ensemble, which uses different feature subsets, and is usually called ensemble feature selection. Ensemble feature selection (EFS) is also a classical ensemble method. It takes different feature subset as the input features for a base classifier construction.

There are two methods for generating base classifiers and integrating the predictions of base classifiers. One is called direct strategy, the other is called over producing and choosing strategy. The direct strategy aims to generate an ensemble of base classifiers directly in the training period. The over producing and choosing strategy is also called selective ensemble, which creates a lot of base classifiers at first, and then chooses a subset of the most suitable base classifiers and generates the final prediction. In this paper, selective ensemble is discussed.

In this section, a selective ensemble feature selection method is proposed based on rough set theory. Firstly, all possible reducts of the original information system can be generated with the attribute reduction algorithm based on dicernibility matrix and logical operation. Using these reducts, classifiers could be built which have the same classification ability as the original whole decision table. Therefore, these classifiers can be taken as candidates of an ensemble system. In the second, a diversity measure named the double fault (DF) measure is used [25], and all base classifiers are clustering on validation set. Therefore, a pair of base classifiers, which are the most diverse among two clusters, are chosen from each two clusters. In the third, the classifiers which are more accurate and more diverse among all the classifiers are chosen for the ensemble system. At last, majority voting is taken as the integration method for ensemble, and the final prediction can be taken on the testing set. The detailed algorithm is can be seen in [26].

Compared experiments are taken for the purposed of proving the effective of SEFSBRS as a method of emotion recognition. There are three experiments. In the first experiment, the proposed method

SEFSBRS are taken a method of emotion recognition. In the second experiment, a method named integrate all are used as a method of emotion recognition, which take all the classifiers trained from the algorithm of dicernibility matrix and logical operation, and combined with the method of majority voting. In the third experiment, a single classifier is taken as the method of emotion recognition and the CEBARKNC is taken a feature selection method. SVM is taken as classifier for the three experiments. The experiment results are shown in Table 6, and NOC means the number of classifers.

Table 6: *The Results of the Compared Experiments*

Database	*SEFSBRS*		*Ensemble All*		*CEBARKNC*	
	NOC	*CRR*	*NOC*	*CRR*	*NOC*	*CRR*
CKACFE	8.75	83.674	17.75	81.35	1	73.07
JAFFE	5.88	70.16	11.50	68.88	1	63.17
COUPTE	5.38	89.566	10.75	90.63	1	78.83
Average	6.67	81.13	13.33	80.29	1	71.69

From Table 6, we can see that the proposed method SEFSBRS can get higher correct recognition rate than integrate all and single classifier. Therefore, it is a suitable emotion recognition method.

4.2 A Dynamic Ensemble Feature Selection Method for Emotion Recognition

Since it is difficult for defining and choosing a suitable diversity measure for different information systems, a dynamic ensemble feature selection method named DEFS is proposed in this section. It use the local ability of an unseen sample, that is to say, if the neighborhood of an unseen sample are corrected classified by a certain classifier, the unseen sample is more likely correctly classified by the classifier. Meanwhile, an algorithm for finding multiple reducts is proposed based on tolerance relation model of the continuous value information system and 3DM theory. Detail of the algorithm can be seen in [27], therefore, multiple reducts can be gotten and candidate classifiers for ensemble system are gotten accordingly. After that, the most suitable classifier for a unseen sample is choosing according to the local ability of the unseen sample on the validation set. Finally, the dynamic choosing classifier is used for the unseen samples. The detail algorithm of DEFS can be seen in [27].

Compared experiments are taken for the purposed of proving the effective of DEFS as a method of emotion recognition. There are three experiments. In the first experiment, the proposed method DEFS is taken a method of emotion recognition. In the second experiment, a method named integrate all are used as a method of emotion recognition, which take all the classifiers trained from the algorithm of finding multiple reducts based on extended rough set theory is [27], and combined with the method of majority voting. In the third experiment, a single classifier is taken as the method of emotion recognition and the SARA is taken a feature selection method. SVM is taken as classifier for the three experiments. The experiment results are shown in Table 7.

Table 7: *The Results of the Compared Experiments*

Database	*DEFS*		*Ensemble all*		*SARA*	
	NOC	*CRR*	*NOC*	*CRR*	*NOC*	*CRR*
CKACFE	1	81.57	3	78.28	1	76.01
JAFFE	1	69.48	2.75	66.48	1	69.37
CQUPTE	1	89.40	5	87.53	1	92.45
Average	1	80.15	3.58	77.43	1	79.28

From Table 7, we can see that the proposed method DEFS can get higher correct recognition rate than integrate all and single classifier. Therefore, it is a suitable emotion recognition method.

5. CONCLUSION

In this paper, feature selection based on rough set theory for emotion recognition is studied. Rough set theories and methods are proved effective for feature selection for emotion recognition. Important features for emotion recognition can be found. Moreover, the geometrical features concerning mouth are found to be the most important geometrical features for emotion recognition. Meanwhile, a selective ensemble feature selection method and a dynamic ensemble feature selection method are proposed. Experiment results show that emotion recognition methods based on rough set theory and ensemble learning are effective and efficient.

In the next, Cognition theories should be taken more attention in emotion recognition, especially for the feature selection for emotion recognition.

REFERENCES

[1] R.W. Picard, "Affective Computing". Cambridge: MIT Press, 1997.

[2] R.W. Picard, "Affective Computing: Challenges". *International Journal of Human-Computer Studies*, 2003, **59 (1)**, 55–64.

[3] R.W. Picard, E. Vyzas, and J. Healey, "Toward Machine Emotional Intelligence: Analysis of Affective Physiological State". *IEEE Transactions on Pattern Analysis and Machine Intelligence*, 2001, **23 (10)**, 1175–1191.

[4] X. Sui, Y.T. Ren, "Online Processing of Facial Expression Recognition". *Acta Psychologica Sinica*, 2007, **39 (1)**, 64–70 (In Chinese).

[5] Y.M. Wang, X.L. Fu, "Recognizing Facial Expression and Facial Identity: Parallel Processing or Interactive Processing", *Advances in Psychological Science*, 2005, **13 (4)**, 497–516 (In Chinese).

[6] M. White, "Effect of Photographic Negation on Matching the Expressions and Identities of Faces", *Perception*, 2001, 30: 969–981.

[7] Z. Pawlak, "Rough Sets". *International J. Comp. Inform. Science*, **11**, (1982), 341–356.

[8] Z. Pawlak, "Rough Classification". *International Journal of Man-Machine Studies*, **5**, (1984) 469–483.

[9] R.W. Swiniarski, A. Skowron, "Rough Set Methods in Feature Selection and Recognition", *Pattern Recognition Letters*, **24**, (2003), 833–849.

[10] Z. Ning, A. Skowron, "A Rough Set-Based Knowledge Discovery Process", *Int. J. Appl. Math. Comput. Sci.*, 2001, **11 (3)**, 603–619.

[11] G.Y. Wang, H. Yu, D.C. Yang, "Decision Table Reduction Based on Conditional Information Entropy". *Journal of Computers*, 2002, **25 (7)**, 759–766 (In Chinese).

[12] "The Cohn-Kanade AU-Coded Facial Expression Database". *http://vasc.ri.cmu.edu/idb/html/face/facial_expression/index.html*.

[13] "The Japanese Female Facial Expression (JAFFE) Database". *http://www.kasrl.org/jaffe.html*.

[14] "Chongqing University of Posts and Telecommunications Emotional Database (CQUPTE)". *http://cs.cqupt.edu.cn/users/904/docs/9317-1.rar*.

[15] Y. Yang, G.Y. Wang, P.J. Chen, et. al., "Feature Selection in Audiovisual Emotion Recognition Based on Rough Set Theory". *Transaction on Rough Set*, **7**, 2007, 283–294.

[16] P.J. Chen, G.Y. Wang, Y. Yang, et. al. "Facial Expression Recognition Based on Rough Set Theory and SVM". *Proceedings of RSKT 2006*, 2006, 772–777.

[17] H.S. Nugyen, A. Skowron, "Quantization of Real Value Attributes: Rough Set and Boolean Reasoning Approach", *Proceedings of the Second Joint Annual Conference on Information Science*, Wrightsville Beach, NC, USA, 34–37.

[18] G.Y. Wang, Z. Zheng, Y. Zhang, "RIDAS-A Rough Set Based Intelligent Data Analysis System", *The First Int. Conf. on Machine Learning and Cybernetics*, (ICMLC2002), 2002, 646–649.

[19] D.E. Goldberg, "Genetic Algorithms in Search, Optimization and Machine Learning". Boston, MA, USA: Addison-Wesley Longman Publishing Co., Inc., 1989.

[20] G.Y. Wang, Y. Wang, "3DM: Domain-oriented Data-driven Data Mining", *Fundamenta Informaticae*, 2009, **90 (4)**, 395–426.

[21] G.Y. Wang, "Introduction to 3DM: Domain-Oriented Data-Driven Data Mining", *Proceedings of RSKT 2008*, 2008, 25-26.

[22] Yong Yang, Guoyin Wang, Hao Kong. "Data Driven Facial Emotional Feature Selection based on Rough Set Theory, Mathematical Problems in Engineering". *Accepted and to be appeared.*

[23] S. Ohsuga, "Knowledge Discovery as Translation". T.Y. Lin, et. al. (Eds.): *Foundations of Data Mining and Knowledge Discovery*, Springer, 2005, 1–19.

[24] T.G. Ditterrich. "Ensemble Methods in Machine Learning". In: Kittler, J., Roli, F., (Eds.), *Multiple Classi¯er Systems. LNCS 1857*, Springer, Berlin (2001), 1–15.

[25] L.I. Kuncheva, C.J. Whitaker. "Measures of Diversity in Classifier Ensembles and their Relationship with the Ensemble Accuracy". *Machine Learning*, **2**, (2003), 181–207.

[26] Y. Yang, G.Y. Wang, K. He. "An Approach for Selective Ensemble Feature Selection Based on Rough Set Theory". *Proceedings of Second International Conference on Rough Sets and Knowledge Technology (RSKT2007)*, 2007, 518–525.

[27] Y. Yang, G.Y. Wang, H. Kong. "Emotion Recognition Based on Dynamic Selective Ensemble Learning". *Proceedings of 2009 International Conference on Man-Machine Interactions, Accepted and to be appeared.*

13

A PROFOUND THEORY OF FUZZY INTERPOLATION

Irina Perfilieva

University of Ostrava, Institute for Research and Applications of Fuzzy Modeling, 30. dubna 22, 70103 Ostrava 1, Czech Republic
E-mail: Irina.Perfilieva@osu.cz

Abstract: *We propose a general framework for the interpolation problem stemming from its classical elaboration. We introduce the notion of a fuzzy function and show, how it can be constructed. We also propose a solution of interpolation at a point being disjoint with any of fuzzy sets in antecedents of the given sparse fuzzy rule base.*

1. INTRODUCTION

It is well known that a fuzzy rule base is a characterization of a partially given mapping (fuzzy function) between fuzzy universes. This mapping is known at a set of fuzzy nodes where arguments in antecedents and values in consequences of the rule base are expressed by fuzzy predicates. For practical applications, it is desirable to interpolate the partially given function in order to compute its values at an arbitrary point (fuzzy or crisp). Moreover, interpolation requires that in the case of coincidence between a point and an argument in one of given fuzzy nodes, a computation method should produce a result which coincides with the corresponding value in the same node.

To summarize, the *interpolation problem* considered in this paper consists in the extension of a partially given fuzzy function to a total fuzzy function defined on a fuzzy space. This formulation requires a precise definition of a fuzzy space and a fuzzy function which will be given below.

In our contribution, we propose a general framework for the interpolation problem stemming from its classical elaboration. We restrict ourselves to solutions of the interpolation problem which can be expressed by fuzzy functions represented by fuzzy relations [8, 9]. In this particular case, the interpolation problem is equivalent to the solvability problem of a system of fuzzy relation equations. We introduce a notion of a functional fuzzy relation and show, how it can be constructed. We also propose a solution of interpolation at a point being disjoint with any of fuzzy sets in antecedents of the given sparse fuzzy rule base.

2. FUZZY FUNCTIONS

The notion of fuzzy function closely relates and generalizes the notion of function. With this respect, let us first remind how the latter is defined. Two equal definitions are known and both assume that two sets: a domain X and a range Y are given. According to the first definition, a *correspondence* $f: x \mapsto y$ is a *function* if for any $x \in X$, the value $y = f(x)$ is uniquely determined. The latter means that for all $x_1, x_2 \in X$ such that $x_1 = x_2$, respective values $f(x_1), f(x_2)$ are equal as well, i.e. $f(x_1) = f(x_2)$. The second definition introduces a function as a *special binary relation* $\rho \subseteq X \times Y$ such that for all $x \in X$, and all $y_1, y_2 \in Y$, $(x, y_1) \in \rho$ and $(x, y_2) \in \rho \Rightarrow y_1 = y_2$.

An immediate observation shows that the first definition (declarative) is focused on a computational aspect of a function, while the second definition (descriptive) is focused on specification of a function as a binary relation. Another observation shows that both definitions are based on the notion of set which is equipped with equality. Therefore, a generalized notion of function should be based on a generalized notion of *L-valued set* (Ω-valued set in [6]) where L is a lattice of truth values.

Since the notion of fuzziness requires a scale of degrees (a lattice of truth values) and the notion of fuzzy equality/equivalence requires an algebra on such a scale, we will step aside and introduce those notions.

We choose a complete residuated lattice as the basic algebra of operations.

Definition 1: A residuated lattice is an algebra

$$\mathcal{L} = \langle L, \vee, \wedge, *, \rightarrow, \mathbf{0}, \mathbf{1} \rangle.$$

with four binary operations and two constants such that

- $\langle L, \vee, \wedge, \mathbf{0}, \mathbf{1} \rangle$ is a lattice where the ordering $\leq$ defined using operations $\vee$, $\wedge$ as usual, and $\mathbf{0}$, $\mathbf{1}$ are the least and the greatest elements, respectively;
- $\langle \boldsymbol{L}, *, \mathbf{1} \rangle$ is a commutative monoid, that is, * is a commutative and associative operation with the identity $a * \mathbf{1} = a$;
- the operation $\rightarrow$ is a residuation operation with respect to *, i.e.

  $a*b \leq c$ iff $a \leq b \rightarrow c$.

A residuated lattice is complete if it is complete as a lattice.

The following binary operation of biresiduation can be additionally defined in $\mathcal{L}$:

$$x \leftrightarrow y = (x \rightarrow y) \wedge (y \rightarrow x).$$

Let us remark that in [5]) the above notion of residuated lattice has been introduced as integral commutative, residuated *l*-monoid (or *cl*-monoid, if the underlying lattice is complete). The term "residuated lattice" has been introduced by Dilworth and Ward in [3].

The well known examples of residuated lattices are boolean algebras, Gödel, Łukasiewicz, and product algebras. In the particular case $L = [0, 1]$, multiplication * is known as a *t*-norm. In the sequel, we will fix a complete residuated lattice $\mathcal{L}$.

Definition 2: An L-valued fuzzy set A of X (fuzzy set, shortly) is a map $A: X \rightarrow L$ which is known as a membership function of A as well. (Thus a fuzzy set is identified with its membership function.) A fuzzy set A is normal if there exists $x_A \in X$ such that $A(x_A) = 1$.

Two fuzzy sets A and B are equal ($A = B$) or one is less than the other one (e.g., $A \leq B$) if for all $x \in X$, $A(x) = B(x)$ or $A(x) \leq B(x)$, respectively.

We denote L^X a class of L-valued fuzzy sets of X. If $A, B \in L^X$ the a *fuzzy equality* $A \equiv B$ is defined as follows:

$$(A \equiv B) = \bigwedge_{x \in X} \left(A(x) \leftrightarrow B(x)\right)$$

Fuzzy equality $(A \equiv B) \in L$ is a degree of coincidence between two membership functions $A(x)$ and $B(x)$. It is known that

$(A \equiv B) = 1$ iff $A = B$.

Definition 3: The couple $(L^X, \equiv)$ is a fuzzy space of X.

Now we can define a fuzzy function (see also [8]) and an extensional fuzzy function, both as a special correspondence between two fuzzy spaces.

Definition 4: Let $\mathcal{L}$ be a complete residuated lattice and $(L^X, \equiv)$, $(L^Y, \equiv)$ fuzzy spaces of X and Y respectively.

- A map $f\colon L^X \to L^Y$ is a fuzzy function if for every $A, B \in L^X$, $A = B$ implies $f(A) = f(B)$.
- A fuzzy function $f\colon L^X \to L^Y$ is extensional if for every $A, B \in L^X$,

  $$(A \equiv B) * f(A) \leq f(B).$$

By the definition of extensionality, $f\colon L^X \to L^Y$ is extensional if and only if $(A \equiv B) \leq (f(A) \equiv f(B))$. Below, we give an example of an extensional fuzzy function.

Example 1: Let $(L^X, \equiv)$, $(L^Y, \equiv)$ be as in the definition above, $R \in L^{X \times Y}$ a fuzzy relation.

For every $A \in L^X$, we denote

$$(A \circ R)(y) = \bigvee_{x \in X} \left(A(x) * R(x, y)\right) \qquad \dots (1)$$

Then $f_R\colon A \mapsto A \circ R$ is a fuzzy function. We will say that f_R is a fuzzy function represented by fuzzy relation R. Moreover, it can be proved that f_R is an extensional fuzzy function. The class of fuzzy functions represented by fuzzy relations will be denoted by $\mathcal{F}_{\text{rel}}$.

3. CLASSICAL AND FUZZY APPROACHES TO INTERPOLATION

3.1 Interpolation in a Class of Functions

We will consider the easiest formulation of interpolation problem when all functions are real functions of real arguments and, moreover, they depend on one argument. Let X and Y be subsets of the set of reals and $f\colon X \to Y$ a function known at finite set of nodes $A = \{x_1, \dots, x_n\} \subseteq X$. Let P be a subset of functions with predetermined properties (continuity, being represented by polynomials, etc.) which we choose for interpolating functions. We say that a function $g \in P$ interpolates g if $|A = f|_A$ or

$$g(x_i) = f(x_i),\ i = 1, \ldots n. \qquad \ldots (2)$$

Hence, the interpolation problem can be formulated as follows:

*For a given set of pairs $\{x_i, f(x_i)) \mid i = 1, \ldots, n\}$ and a set **P** of functions, find $g \in P$ such that (2) holds true. In this case, g is called an interpolating function for f.*

It is easy to see that a solution (if it exists) could not be unique. This fact pushes forward those sets of functions where there is only one solution of a given interpolation problem, e.g. sets of polynomials.

3.2 Interpolation in a Class of Binary Relations

Let X, Y and $f: X \rightarrow Y$ be as above and $\mathcal{R}$ be a class of all binary relations on $X \times Y$. The following requirement should be put on $R \in \mathcal{R}$ in order to say that R interpolates f at nodes $x_1, \ldots, x_n$:

$$(y = f(x_i)) \Rightarrow (R(x_i, y) = 1),\ i = 1, \ldots, n. \qquad \ldots (3)$$

Therefore, the interpolation problem in a class of all binary relations, can be formulated as follows:

For given set of pairs $\{(x_i, f(x_i)) \mid i = 1, \ldots, n\}$ and class $\mathcal{R}$ of binary relations on $X \times Y$, find $R \in \mathcal{R}$ such that (3) holds true. In this case, R is called an interpolating binary relation for f.

It is easy to see that if we do not put other requirements on interpolating relations then each relation $Q: X \times Y \rightarrow \{0, 1\}$ such that $\check{R} \leq Q \leq \hat{R}$ where

$$\check{R}(x, y) = \begin{cases} 1, & \text{if } x = x_i,\ y = f(x_i),\quad i = 1, \ldots, n, \\ 0, & \text{otherwise} \end{cases} \qquad \ldots (4)$$

and

$$\hat{R}(x, y) = \begin{cases} 0, & \text{if } x = x_i,\ y \neq f(x_i),\quad i = 1, \ldots, n, \\ 1, & \text{otherwise} \end{cases} \qquad \ldots (5)$$

fulfils requirement (3) and can be taken as an interpolating relation for f. A natural requirement on an interpolating relation is functionality. If $\mathcal{R}^{\text{fun}}$ denotes a class of functional binary relations on $X \times Y$ then the interpolation problem can be considered in $\mathcal{R}^{\text{fun}}$ as well.

If $\chi_i: X \to \{0, 1\}$ and $\psi_i: Y \to \{0, 1\}$ are characteristic functions of $\{x_i\}$ and $\{f(x_i)\}$ respectively, then

$$\check{R}(x, y) = \bigvee_{i=1}^{n} (\chi_i(x) \wedge \psi_i(y)), \qquad \text{... (6)}$$

$$\hat{R}(x, y) = \bigwedge_{i=1}^{n} (\chi_i(x) \to \psi_i(y)) \qquad \text{... (7)}$$

where $\wedge$, $\vee$, $\to$ are Boolean operations. Moreover, for each $i = 1, ..., n$,

$$\psi_i(y) = \sup_{x \in X} (\chi_i(x) \wedge \hat{R}(x, y)), \text{ and}$$

$$\psi_i(y) = \sup_{x \in X} (\chi_i(x) \wedge \hat{R}(x, y)) \qquad \text{... (8)}$$

If the expressions (8) are extended for arbitrary characteristic functions of points $t \in X$, i.e.

$$\sup_{x \in X} (\chi_t(x) \wedge \check{R}(x, y)) = \check{R}(t, y)$$

or

$$\sup_{x \in X} (\chi_t(x) \wedge \hat{R}(x, y)) = \hat{R}(t, y)$$

then the way of computation via sup–min composition (widely accepted in fuzzy relational calculus) can be easily recognized. In general, neither $\check{R}(t, y)$, nor $\hat{R}(t, y)$ is a functional binary relation and so, they do not determine an interpolating function for f.

3.3 Interpolation in a Class of Fuzzy Binary Relations

Let X, Y be as above, $\mathcal{L}$ a complete residuated lattice on L, $(L^X, \equiv)$, $(L^Y, \equiv)$, $(L^{X \times Y}, \equiv)$ the respective fuzzy spaces. Let a fuzzy function $F: L^X \to L^Y$ be known at finite set $\mathcal{A} = \{A_1, ..., A_n\}$ of fuzzy nodes from L^X so that $F(A_i) = B_i$, $i = 1, ..., n$. The interpolation problem in the class $L^{X \times Y}$ of fuzzy binary relations is formulated as follows:

For a given set of pairs $\{(A_i, B_i) | i = 1, ..., n\}$ and a set $\mathcal{R}_{fuz}$ of binary fuzzy relations, find $R \in L^{X \times Y}$ such that

$$B_i(y) = (A \circ R)(y) = \sup_{x \in X} \left(A_i(x) * R(x, y)\right) \qquad \text{... (9)}$$

holds true. In this case, R is called an interpolating fuzzy relation for the set of pairs $\{(A_i; B_i) | i = 1, ..., n\}$.

Unlike the interpolation problem in a class of binary relations above, (9) is not always solvable with respect to R. Therefore, much effort has been paid to discovering criteria and/or sufficient conditions of solvability of (9) with respect to an unknown R (see e.g., [1, 2, 4, 7, 9–12]).

3.4 Interpolation in a Class of Fuzzy Functions Represented by Fuzzy Relations

Let X, Y be as above, $\mathcal{L}$ a complete residuated lattice on L, $(L^X, \equiv)$, $(L^Y, \equiv)$, $(L^{X \times Y}, \equiv)$ respective fuzzy spaces. Let fuzzy function $F: L^X \to L^Y$ be known at finite set $\mathcal{A} = \{A_1, ..., A_n\}$ of fuzzy nodes from L^X so that $F(A_i) = B_i$, $i = 1, ..., n$. The interpolation problem in the class $\mathcal{F}_{rel}$ of fuzzy functions represented by fuzzy relations is formulated as follows:

For a given set of pairs $\{(A_i, B_i) | i = 1, ..., n\}$ and the class $\mathcal{F}_{rel}$ of fuzzy functions represented by fuzzy relations, find $f_R \in \mathcal{F}_{rel}$ such that

$$f_R(A_i) = B_i,\ i = 1, ..., n, \qquad \text{... (10)}$$

holds true. The fuzzy function f_R satisfying (10) is called interpolating fuzzy function.

It is easy to see that an interpolating fuzzy function f_R, represented by a fuzzy relation R exists if and only if R is an interpolating fuzzy relation for the same set of pairs $\{(A_i, B_i) | i = 1, ..., n\}$. For certainty, we will concentrate on finding of an interpolating fuzzy function.

4. INTERPOLATION BETWEEN NODES

The interpolation problem is very important in applications because its solution (an interpolating function or interpolating binary relation) enables a computation of a partially given function (ordinary or fuzzy) at points different from the nodes in $\mathcal{A}$. Evidently, a value of an interpolating function at any point approximates a respective (unknown) value of an original function. Below, we will discuss how the problem of estimation of a quality of approximation (by an interpolating function) can be formulated in the case of fuzzy interpolation.

4.1 Interpolation by Fuzzy Functions Represented by Fuzzy Relations

In this subsection, we assume that the interpolation problem is considered in a space of fuzzy functions represented by fuzzy relations. First, we will discuss how a closeness measure between two elements from $\mathcal{F}_{rel}$ can be chosen.

Let $R, S \in L^{X \times Y}$ be two binary fuzzy relations and f_R, f_Q respective fuzzy functions from $\mathcal{F}_{rel}$. For each $A \in L^X$, let us define a measure of closeness between $f_R(A)$ and $f_Q(A)$ by $f_R(A) \equiv f_Q(A)$. Then the problem of estimation of a *quality of approximation* (by an interpolating function) can be transformed into estimation of $f_R(A) \equiv f_Q(A)$ in the space $\mathcal{F}_{rel}$. The following theorem gives an estimation of $f_R(A) \equiv f_Q(A)$.

Theorem 1: Let fuzzy function $F \in \mathcal{F}_{rel}$, known at finite set $\mathcal{A} = \{A_1, ..., A_n\}$ of fuzzy nodes from L^X, be represented by fuzzy relation $Q \in L^{X \times Y}$ so that for each $A \in L^X$, $F(A) = f_Q(A)$. Let $R \in L^{X \times Y}$ be another fuzzy relation such that

$$f_Q(A_i) = f_R(A_i),\ i = 1, ..., n,$$

or in other words, fuzzy function f_R interpolates f_Q. Then for each $A \in L^X$,

$$(f_R(A) \equiv f_Q(A)) \geq (R \equiv Q). \qquad ...(11)$$

4.2 Interpolation in the Vicinity of a Node

Although Theorem 1 gives an answer to the problem of estimation of a quality of approximation by an interpolating function, it does not answer the question about interpolation between nodes where the stress is put on the word "between". Theorem 1 estimates closeness between original and interpolating functions at arbitrary (fuzzy point) irrespective of its position with respect to the given nodes. This limitation is due to the chosen universe of fuzzy sets which is not linearly ordered so that it is not always possible to verify that a given fuzzy set (point) lies between two fuzzy nodes. However, it is possible to verify that a given fuzzy set is close to some fuzzy node. A measure of closeness (e.g., similarity) can be used for that. Therefore, we can consider a problem of estimation of a measure of closeness between $f_R(A)$ and $f_Q(A)$ with additional assumption about closeness between A and some of A_i, $i = 1, ..., n$.

Theorem 2: Let a fuzzy function $F \in \mathcal{F}_{rel}$ known at finite set $\mathcal{A} = \{A_1, ..., A_n\}$ of fuzzy nodes from L^X be represented by a fuzzy relation $Q \in L^{X \times Y}$ so that for each $A \in L^X$, $F(A) = f_Q(A)$. Let $R \in L^{X \times Y}$ be another fuzzy relation such that

$$f_Q(A_i) = f_R(A_i) = B_i,\ i = 1, ..., n.$$

Then for each $A \in L^X$,

$$f_Q(A_i) \equiv F_R(A) \geq (A_i \equiv A). \qquad ...\ (12)$$

The following formula gives a two-side restriction on the value of $F_R(A)$ at arbitrary $A \in L^X$.

$$\bigvee_{i=1}^{n} \left((A_i \equiv A) * B_i \right) \leq f_R(A) \leq \bigwedge_{i=1}^{n} \left((A_i \equiv A) \to B_i \right) \qquad ...\ (13)$$

4.3 Interpolation with Better Quality of Approximation

Theorem 2 does not restrict the value of $F_R(A)$ in the case when for every $i = 1, ..., n$, $(A_i \equiv A) = 0$ (a sparse rule base). In this subsection, we will show how an interpolating fuzzy function with better quality of approximation can be constructed.

Assume that a set of pairs $\{(A_i, B_i) \mid i = 1, ..., n\}$ is given and that nodes $A_1, ..., A_n$ are normal fuzzy sets which establish a semi-partition of L^X, i.e. for all $i, j = 1, ..., n$,

$$\bigvee_{x \in X} \left(A_i(x) * A_j(x) \right) \leq \left(\bigwedge_{x \in X} \left(A_i \leftrightarrow A_j(x) \right) \right)$$

In this case, fuzzy relations

$$\check{R}(x,y) = \bigvee_{i=1}^{n}(A_i * B_i), \quad \hat{R}(x,y) = \bigwedge_{i=1}^{n}(A_i \to B_i).$$

are interpolating for the set of pairs $\{(A_i, B_i) \mid i = 1, ..., n\}$ if and only if

$$\check{R} \leq \hat{R}.$$

By (11),

$$\left(f_{\check{R}}(A) \equiv f_{\hat{R}}(A)\right) \geq \left(\check{R} \equiv \hat{R}\right).$$

Theorem 3: Let a set of pairs $\{(A_i, B_i) \mid i = 1, ..., n\}$ be as above and $A_1, ..., A_n$ and $A'_1, ..., A'_n$ be two sets of normal fuzzy sets, both establishing different semi-partitions of L_X. Let moreover, $A_i \leq A'_i$, $i = 1, ..., n$, and

$$\check{R} \leq \check{R}' \leq \hat{R}' \leq \hat{R}$$

where

$$\check{R}'(x,y) = \bigvee_{i=1}^{n}\left(A'_i * B_i\right), \quad \hat{R}'(x,y) = \bigwedge_{i=1}^{n}\left(A'_i \to B_i\right).$$

If a normal fuzzy set $A \in L^X$ fulfils the inequality

$$\sup_{x \in X} \bigvee_{i=1}^{n}\left(A(x) * A'_i(x)\right) > 0$$

then each fuzzy relation R such that $\check{R}' \leq R \leq \hat{R}'$ is interpolating for the set of pairs $\{(A_i, B_i) \mid i = 1, ..., n\}$. Moreover,

$$\bigvee_{i=1}^{n}\left(\left(A'_i \equiv A\right) * B_i\right) \leq F_R(A) \leq \bigwedge_{i=1}^{n}\left(\left(A'_i \equiv A\right) \to B_i\right).$$

ACKNOWLEDGMENT

The paper has been supported partially by the grant IAA108270902 of GA AV ČR and partially by the project 1M0572 of the MŠMTČR.

REFERENCES

[1] B. De Baets, "Analytical Solution Methods for Fuzzy Relation Equations", in: D. Dubois, H. Prade (eds.), *The Handbooks of Fuzzy Sets Series*, **1**, Kluwer, Dordrecht, 2000, pp. 291 – 340.

[2] A. Di Nola, S. Sessa, W. Pedrycz, E. Sanchez, "Fuzzy Relation Equations and Their Applications to Knowledge Engineering", Kluwer, Boston, 1989.

[3] R.P. Dilworth, M. Ward, "Residuated Lattices, Trans. Amer. Math. Soc., **45**, (1939), 335–354.

[4] S. Gottwald, "On the Existence of Solutions of Systems of Fuzzy Equations", *Fuzzy Sets and Systems*, **12**, (1984), 301–302.

[5] U. Höhle, "Commutative Residuated l-monoids", in: U. Höhle, E.P. Klement (eds.), "Non-Classical Logics and Their Applications to Fuzzy Subsets". *A Handbook of the Mathematical Foundations of Fuzzy Set Theory*, Kluwer, Dordrecht, 1995, pp. 53 – 106.

[6] U. Höhle, L.N. Stout, "Foundations of Fuzzy Sets", *Fuzzy Sets and Systems*, **40**, (1991), 257–296.

[7] F. Klawonn, "Fuzzy Points, Fuzzy Relations and Fuzzy Functions", in: V. Novák, I. Perfilieva (eds.), "Discovering the World with Fuzzy Logic", Springer, Berlin, 2000, pp. 431–453.

[8] I. Perfilieva, "Fuzzy Function as an Approximate Solution to a System of Fuzzy Relation Equations", *Fuzzy Sets and Systems*, **147**, (2004), 363–383.

[9] I. Perfilieva, S. Gottwald, "Solvability and Approximate Solvability of Fuzzy Relation Equations", *Int. J. of General Systems*, **32**, (2003), 361–372.

[10] I. Perfilieva, A. Tonis, "Compatibility of Systems of Fuzzy Relation Equations", *Int. J. of General Systems*, **29**, (2000), 511–528.

[11] E. Sanchez, "Resolution of Composite Fuzzy Relation Equations", *Information and Control*, **30**, (1976), 38–48.

[12] X. Wang, "Infinite Fuzzy Relational Equations on a Complete Brouwerian Lattice", *Fuzzy Sets and Systems*, **138**, (2003), 657–666.

14

SOME MORE RESULTS ON FUZZY-ROUGH OSCILLATORY REGION AND IT'S APPLICATION IN DECISION MAKING

Sharmistha Bhattacharya (Halder) and Susmita Roy

Department of Mathematics, Tripura University
Suryamaninagar, Tripura-799130, India.
E-mail: halder_731@ rediffmail.com and rsusmita@ yahoo.com

Abstract: *In this paper we proposed a new rough set model namely fuzzy rough oscillatory region. Some of its properties are shown. It is a data mining technique and also helpful in decision making process. In this paper we consider those data sets which contain linguistic attributes. The linguistic attributes are first converted to a numerical value lying between [0, 1] by "Direct method with one expert" [2]. The decision attribute of the objects are crisp i.e {0, 1}. Using this concept of fuzzy rough oscillatory region we are finding the pattern of the attributes of an object. With the help of this pattern we can make a decision of an unknown object.*

Keywords: *Fuzzy-rough oscillatory range, height of oscillation, etc.*

1. INTRODUCTION

The concept of fuzzy set theory has been introduced by Lotfi A Zadeh in 1965 [8] and the concept of Rough set theory has been introduced by Z. Pawlak in 1980 [4]. Both of the topic deals with uncertainty or vague concept.

A data set consists of three parts.

(*i*) The object about which the data is to be collected.

(*ii*) The attribute set, the quality of the objects.

(*iii*) The decision set.

To arrive at a decision point, data may be collected by various techniques from various sources and so data becomes huge and also sometimes some data are missing. Sometimes it may so happen that data set contain linguistic variable. And hence we get lots of problems. To deal with such type of problem a new model namely fuzzy rough oscillatory region has been introduced in this paper. With this concept, decision about an unknown object can be drawn. Since the data set may be huge in number a *C*-algorithm is also developed.

2. ON FUZZY-ROUGH OSCILLATORY REGION

In this section at first we construct a *C*-algorithm to find the set of lower approximation, upper approximation, boundary, etc. This lower approximation set forms the topology along with Î and ϕ, where Î is the membership of the Universal set and ϕ is the membership of null set. Next we introduce the concept of fuzzy-rough oscillation. *AC*-algorithm is developed in this connection.

The algorithm of the *C*-program is as follows:

Step 1: Input dimension (m, n).

Step 2: Input Objects and attributes $(x_{ij})_{mxn}$

Step 3: Input decision attributes d_i

Step 4: Find indiscernible elements:

For $i = 1$ to m

If attribute values of x_i and x_k $(k = i + 1)$ are same then x_i and x_k are indiscernible.

Step 5: If (x_i and x_k are indiscernible and their decisions are positive)

Then x_i and x_k are in lower approximation.

Else if (decisions of x_i and x_k are negative)

Then x_i and x_k are in outside region.

Else x_i and x_k are in boundary region.

End

Definition 2.1: An operator $\Lambda: I^x \to I^x$ is defined as

$$\Lambda_{a_j}(x) = \inf\{\mu_{a_j}(x_i): \mu_{a_j}(x_i) \geq \mu_{a_j}(x), x_i \in G, G \text{ is an open set}, j = 1, 2, \ldots, n\},$$
$$= \hat{I}, \text{ otherwise}$$

where $\mu_{a_j}(x_i)$ is the membership value of any particular attribute a_j of any object x_i and $\mu_{a_j}(x)$ is the membership value of unknown object for a particular attribute a_j.

Example 2.2: Let us consider a data set as follows:

Object	a_1	a_2	a_3	a_4	*D*
x_1	0.3	0.1	0.2	0.4	1
x_2	0.5	0.1	0.6	0.8	1
x_3	0.4	0.1	0.3	0.4	1
x_4	0.4	0.1	0.3	0.4	0

Here $X = \{x_1, x_2, x_3, x_4\}$, the set of objects, d is decision set and $A = \{a_1, a_2, a_3, a_4\}$ is the set of attributes. Therefore $LA = \{x_1, x_2\}$ and $UA = \{x_1, x_2, x_3, x_4\}$ of the set $\{x_1, x_2, x_3\}$.

Let P be any object with attribute values $a_1(0.4)$, $a_2(0.0)$, $a_3(0.3)$, $a_4(0.2)$.

Then

$$\Lambda_{a_1}(p) = \inf\{0.5\} = 0.5,\ \Lambda_{a_2}(p) = \inf\{0.1, 0.1\} = 0.1.$$

$$\Lambda_{a_3}(p) = \inf\{0.6\} = 0.6,\ \Lambda_{a_4}(p) = \inf\{0.4, 0.8\} = 0.4.$$

Definition 2.3: An operator Int: $I^x \to I^x$ is defined as

$$Int_{a_j}(x) = \text{Sup}\ \{\mu_{a_j}(x_i): \mu_{a_j}(xi) \le \mu_{a_j}(x),\ x_i \in G,\ G \text{ is an open set}, j = 1, 2, \ldots, n\},$$

$$= \phi, \text{ otherwise}$$

Where $\mu_{a_i}(x_i)$ is the membership value of any particular attribute a_j of any object x_i and $\mu_{a_j}(x)$ is the membership value of unknown object for a particular attribute a_j.

Example 2.4: From example 2.2, $Int_{a_1}(p) = 0.3$, $Int_{a_2}(p) = \phi$, $Int_{a_3}(p) = 0.2$, $Int_{a4}(p) = \phi$.

Definition 2.5: Let us consider an operator O^o: $I^x \to I^x$ such that $O^o_{a_j}(x_i) = \Lambda_{a_j}(x_i)\text{-}Int_{a_j}(x_i)$.

This operator is said to be fuzzy rough open oscillatory operator.

Example 2.6: From example 2.2, $O^o_{a_1}(p) = 0.2$, $O^o_{a_2}(p) = 0.1$, $O^o_{a_3}(p) = 0.4$, $O^o_{a_4}(p) = 0.4$

Remark 2.7: The above open oscillating operator helps us in some decision-making problems. The open oscillating operator implies the length of oscillation of each attribute for any particular object in lower approximation. But by this oscillating operator we cannot get the decision full correctly. In some cases it may so happen that the membership value of each attribute of the object whose decision we are to determine is greater than the lower approximation. Then this method fails and in this case $O^o_{a_j}(x_i) = \hat{1}\text{-}Int_{a_j}(x_i)$. Also sometimes the attribute of the object may have so small membership value that if $Int_{a_j}(x_i) = \phi$, then $O^o_{a_j}(x_i)\text{-}\phi = \Lambda_{a_j}(x_i)$. Also sometimes the cases may arise that $\Lambda_{a_j}(x_i) = \hat{1}$ and $Int_{a_j}(x_i) = \phi$. Then $O^o_{a_j}(x_i) = \hat{1}\text{-}\phi = \hat{1}$. In this case the oscillations are very high and it is a critical situation. Also for some object $O^o_{a_j}(x_i) = \phi$, i.e. the object is in lower approximation.

Definition 2.8: An operator Cl: $I^x \to I^x$ is defined as

$$Cl_{a_j}(x) = \inf\{\mu_{a_j}(x_i): \mu_{a_j}(x_i) \geq \mu_{a_j}(x), x_i \in G, G \text{ is an closed set}, j = 1, 2, \ldots, n\},$$
$$= \hat{1}, \text{ otherwise}$$

where $\mu_{a_j}(x_i)$ is the membership value of any particular attribute a_j of any object x_i and $\mu_{a_j}(x)$ is the membership value of unknown object for a particular attribute a_j.

Example 2.9: From example 2.2, closed sets are $\{x_1, x_2, x_3, x_4\}$. Therefore $Cl_{a_1}(p) = 0.4$, $Cl_{a_2}(p) = 0.1$, $Cl_{a_3}(p) = 0.3$, $Cl_{a_4}(p) = 0.4$.

Definition 2.10: An operator V: $I^x \to I^x$ is defined as

$$V_{a_j}(x) = \sup\{\mu_{a_j}(x_i): \mu_{a_j}(x_i) \leq \mu_{a_j}(x), x_i \in G, G \text{ is an closed set}, j = 1, 2, \ldots, n\},$$
$$= \phi, \text{ otherwise}$$

where $\mu_{a_j}(x_i)$ is the membership value of any particular attribute a_j of any object x_i and $\mu_{a_j}(x)$ is the membership value of unknown object for a particular attribute a_j.

Example 2.11: From example 2.2,

$$V_{a_1}(p) = 0.4,\ V_{a_2}(p) = 0,\ V_{a_3}(p) = 0.3,\ V_{a_4}(p) = 0$$

Definition 2.12: Let us consider an operator O^c: $I^x \to I^x$ such that $O^c_{a_j}(x_i) = Cl_{a_j}(x_i)\text{-}V_{a_j}(x_i)$. This operator is said to be fuzzy rough closed oscillatory operator.

Example 2.13: $O^c_{a_1}(p) = 0.1$, $O^c_{a_2}(p) = 0.1$, $O^c_{a_3}(p) = 0.1$, $O^c_{a_4}(p) = 0.4$

Remark 2.14: Similar to Remark 2.7, we can agree that, $\mathcal{O}^c_{a_j}(x_i) = \hat{I}-V_{a_j}(x_i)$, $\mathcal{O}^c_{ai}(x_i) = Cl_{a_j}(x_i)-\phi = Cl_{a_j}(x_i)$, $\mathcal{O}^c_{a_j}(x_i) = \hat{I}$, $\mathcal{O}^c_{a_j}(x_i) = \phi$. It is not mandatory that if $\mathcal{O}^c_{a_j}(x_i) = \hat{I}$ then $\mathcal{O}^c_{a_j}(x_i) = \hat{I}$.

Theorem 2.15: For any x_i, if $\mathcal{O}^o_{aj}(x_i) = 0$ then $\mathcal{O}^c_{a_j}(x_i) = 0$, the object is stable, i.e full conclusions may be drawn about the object.

Proof: Let us consider an object x_i having attributes a_j with membership values $\mu_{a_j}(x_i)$.

Let if possible $\mathcal{O}^o_{a_j}(x_i) = 0$, i.e $\Lambda_{a_j}(x_i)-Int_{a_j}(x_i) = 0$, i.e. $\Lambda_{a_j}(x_i) = Int_{aj}(x_i)$, i.e the infimum of all open sets containing x_i is same as the supremum of all open sets contained in x_i.

This is possible iff the object itself is lower approximation and in this case

$$\mathcal{O}^c_{a_j}(x_i) = Cl_{a_j}(x_i)-V_{a_j}(x_i) = 0.$$

Theorem 2.16: For any object x_i, $\mathcal{O}^c_{a_j}(x_i) = 0$ iff the object is either in the boundary or in the lower approximation.

Proof: Let $\mathcal{O}^c_{a_j}(x_i) = 0$, i.e $Cl_{a_j}(x_i)-V_{a_j}(x_i) = 0$, i.e $Cl_{a_j}(x_i) = V_{a_j}(xi)$, i.e infimum of all closed sets containing x_i is same as the supremum of all closed sets contained in x_i. This is possible iff the object is in the boundary or lower approximation. Converse part is obvious.

Note: $\Lambda a_j(x_i) \geq \mu_{a_j}(x_i) \geq Int_{a_j}(x_i)$, $V_{a_j}(x_i)$ and $Cl_{a_j}(x_i) \geq \mu_{a_j}(x_i)$ $Int_{a_j}(x_i)$, $V_{a_j}(x_i)$.

Now if we put $U_{a_j}(x_i) = \inf\{\Lambda a_j(x_i), cl_{a_j}(x_i)\}$ and $L_{a_j}(x_i) = \sup\{Int_{a_j}(x_i), V_{a_j}(x_i)\}$, then the height of oscillation may be denoted as $h_{a_j}(x_i) = U_{a_j}(x_i)-L_{a_j}(x_i)$.

From Example 2.2, $h_{a_1}(p) = 0.1$, $h_{a_2}(p) = 0.1$, $h_{a_3}(p) = 0.1$, $h_{a_4}(p) = 0.4$.

Remark2.17: We have, $h_{a_j}(x_i) = \inf\{\Lambda a_j(x_i), Cl_{a_j}(x_i)\}-\sup\{Int_{a_j}(x_i), V_{a_j}(x_i)\}$

From the above relation following cases may arise:

Case I: $h_{a_j}(x_i) = \Lambda a_j(x_i)-Int_{a_j}(x_i)$, the attribute may lie in boundary region.

So we need to check the numerical value of $h_{a_j}(x_i)$.

Here three cases may arise:

(*i*) If $h_{a_j}(x_i)$ 0.5, the attribute tends to go towards the boundary region.

(*ii*) If $h_{a_j}(x_i) < 0.5$, the attribute tends towards lower approximation.

(*iii*) If $h_{a_j}(x_i) = 0$, the attribute is definitely in the lower approximation

Case II: $h_{a_j}(x_i) = \Lambda_{a_j}(x_i) - V_{a_j}(x_i)$, the attribute may lie in boundary region. So we need to check the numerical value of $h_{a_j}(x_i)$. Here also two cases may arise:

(*i*) If $h_{a_j}(x_i) \geq 0.5$, the attribute tends to go outside region.

(*ii*) If $h_{a_j}(x_i) < 0.5$, the attribute tends to go towards the lower approximation.

Remark: In this case $h_{a_j}(x_i) \neq 0$, since if $h_{a_j}(x_i) = 0$ then $\Lambda_{a_j}(x_i) = V_{a_j}(x_i)$, which is not possible.

Case III: $h_{a_j}(x_i) = Cl_{a_j}(x_i) - Int_{a_j}(x_i)$,

In this case we need to check the numerical value of $h_{a_j}(x_i)$. The attribute lie on boundary.

Here two cases may arise:

(*i*) If $h_{a_j}(x_i) \geq 0.5$, then the attribute is in the boundary region but tends to go towards the outside region.

(*ii*) If $h_{a_j}(x_i) < 0.5$, then the attribute is tending to go towards the lower approximation.

Remark: $h_{a_j}(x_i) \neq 0$ when $h_{a_j}(x_i) = Cl_{a_j}(x_i) - Int_{a_j}(x_i)$, since if $h_{a_j}(x_i) = 0$ then $Cl_{a_j}(x_i) = Int_{a_j}(x_i)$, which is not possible.

Case IV: $h_{a_j}(x_i) = Cl_{a_j}(x_i) - V_{a_j}(x_i)$. In this case, the attribute is in boundary region and the decision is same as Case III.

Remark: $h_{a_j}(x_i) = 0$ is possible here. And if it happens then the object lies completely in the boundary.

Case V: $h_{a_j}(x_i) = \hat{1} - V_{a_j}(x_i)$. In this case the attributes lie on outside region.

To draw decision we need to check the numerical value of the difference between the membership value of the attribute a_j and $V_{a_j}(x_i)$ of the object x_i.

Let this difference $d = V_{a_j}(x_i) - \mu_{a_j}(x_i)$. From this three cases may arise:

(*i*) If $d \geq 0.5$, the attribute lie in outside region.

(*ii*) If $d < 0.5$, the attribute lie in boundary region .

(*iii*) If $d = 0$, the attribute lies completely in the boundary region.

Case VI: $h_{a_j}(x_i) = \hat{1} - Int_{a_j}(x_i)$, the attribute may lie in outside or boundary.

In this case we need to check the numerical value of d, the difference between the membership value of the attribute a_j and $Int_{a_j}(x_i)$ of the object. Let this difference be $d = Int_{a_j}(x_i) - \mu_{a_j}(x_i)$.

Here also three cases may arise:

(*i*) If $d \geq 0.5$, the attribute lie in outside region.

(*ii*) If $d < 0.5$, the attribute lie in boundary region.

(*iii*) If $d = 0$, the attribute is in the lower approximation.

Case VII: $h_{a_j}(x_i) = \Lambda_{a_j}(x_i) - \phi$. In this case, we need to find the difference between $\Lambda_{a_j}(x_i)$ and the membership value of the attribute a_j of the unknown object. Let $d = \Lambda_{a_j}(x_i) - \mu_{a_j}(xi)$.

Hence three cases may arise:

(*i*) If $d \geq 0.5$, the attribute tends to go outside region.

(*ii*) If $0 < d < 0.5$, the attribute tends to go lower approximation.

(*iii*) If $d = 0$, the attribute must lie in L.A.

Case VIII: $h_{a_j}(x_i) = cl_{a_j}(x_i) - \phi$. In this case the attribute lie on Upper approximation.

Let $d = Cl_{a_j}(x_i) - \mu_{a_j}(x_i)$.

Hence three cases may arise:

(*i*) If $d \geq 0.5$, the attribute tends to go outside region.

(*ii*) If $0 < d < 0.5$, the attribute tends to go lower approximation.

(*iii*) If $d = 0$, the attribute must lie completely in the boundary region.

Case IX: $h_{a_j}(x_i) = \hat{I} - \phi$. This is an unstable case. In this case we cannot make any decision because our dataset does not permit to deal with such type of object.

So while drawing any decision by the help of height of oscillation at first we have to check the membership values of the height of oscillation and then the structure. Finally we can draw a conclusion about an unknown object or about the pattern of that unknown object with the help of value of height of oscillation as follows:

Let us distribute these cases in three parts:

1. Stable: In this case we can make perfect decision about an unknown object.

Case (a): If the height of oscillation of each attribute lies on the lower approximation then the object lie on lower approximation and full decision can be drawn in favour of lower approximation.

Case (b): If the height of oscillation of each attribute lies outside, then the object is in outside region.

2. Unstable: In this case we cannot make any decision because our dataset does not permits to deal with such type of object, e.g the Case IX is unstable case.

3. Oscillating: In this case we can make decision with some error.

Case (a): If the height of oscillation of 50% attributes lies in LA and 50% in outside region, then we cannot make perfect decision. So this is an oscillating object.

Case (b): If less than 50% attribute lie in LA and greater than 50% attribute outside region, then the decision of the object can draw in favour of outside region with some error.

Case (c): If more than 50% attribute is in LA and less than 50% in outer region then the decision of the object can draw in favour of LA with some error.

Case (d): If all the attributes lies in the boundary region then the object also lies in the boundary region and no definite conclusions can be given.

Case (e): If more then 50% of the attribute are in the boundary region and others in the lower approximation and conversely then the decision may be drawn in favour of LA with error terms more or less according to the cases.

Case (f): If more then 50% of the attribute are in the boundary and other in the outer region and conversely then the decision may be drawn in favour of outer region with errors terms more or less according to the cases.

The above cases are oscillating.

Now we can formulate C-programming for the above concept.

The algorithm is as follows:

Step I: Input the no. of object and attribute in lower approx (mxn).

Input the number of object and attribute in upper approx (rxs).

Step II: Input lower approximation-$(\text{low}_{ij})_{mxn}$.

Input upper approximation-$(\text{upper}_{ij})_{rxs}$.

Input the membership value of unknown person-$(b_i)_n$.

Step III: For $i = 1$ to loop.

Find lembda $[i]$, $Int\ [i]$, $Cl\ [i]$, $V\ [i]$.

End loop.

Step IV: For $i = 1$ to loop.

Find height of oscillation of each attributes a_i.

Find oscillating pattern of each attributes a_i.

End loop.

Step V: Making decision about an object x_i.

End.

3. APPLICATION OF FUZZY-ROUGH OSCILLATION

In this section an example is taken to show the application of fuzzy-rough oscillation in the field of data set consist of attributes whose values are fuzzy membership value but decision has only values {0, 1}. The example is taken from the paper [6] to apply the concept introduced in this paper. Suppose we have data about 6 patients as shown in table below. The attribute values are linguistic variables. The values are converted to fuzzy values by "Direct methods with one expert" and we get the following table:

Table 1

Object	*Headache*	*Muscle-pain*	*Temperature*	*Flu*
x_1	0.4	0.7	0.7	0.7
x_2	0.6	0.3	0.8	0.8
x_3	0.9	0.8	1	0.9
x_4	0.2	0.6	0.5	0.5
x_5	0.6	0.3	0.8	0.4
x_6	0.5	0.8	0.9	1

In the above Table 1, for normal we take 0.5, for high and very high we take > 0.5, for low < 0.5.

The Lower and Upper approximation is:

From the above table $X = \{x_1, x_2, x_3, x_4, x_5, x_6\}$

By C-programming the lower approximation and upper approximation are LA = $\{x_1, x_3, x_6\}$.

$UA = \{x_1, x_2, x_3, x_5, x_6\}$ resp., whereas the boundary line cases are the patients = $\{x_2, x_5\}$.

Let us find the decision of an unknown person whose attributes are as follows;

Let P be any object whose attribute values are as follows:

Headache = 0.65, muscle-pain = 0.55, temperature = 0.85.

$h_H(P)$ = inf {0.9, 0.9}-sup {0.5, 0.6} = 0.3 < 0.5 [$\Lambda - V$], by Case II, the attribute tends to go toward lower approximation.

$h_M(P)$ = inf {0.7, 0.6}-sup {ϕ, 0.3} = 0.6 – 0.3 = 0.3 [$cl - V$], by Case IV, the attribute tends to go toward lower approximation.

$h_T(P)$ = inf {0.9, 0.9}-sup {0.7, 0.8} = 0.9 – 0.8 [$\phi - V$] = 0.1 < 0.5, by Case II, the attribute tends to go toward lower approximation, i.e. by Stable case (a) the object has possibility to have the disease flu.

4. CONCLUSION

The concept studied in this paper helps us to draw conclusion about the pattern of an unknown object So a computer programming is required for this purpose. This method helps us to deal with all type of data set. But stochastic process deals with t only time dependent data. This program will help to draw conclusion about an unknown object easily. Though there are lots of software as Rosetta, KDD, LERS, LEM1, LEM2, MLEM2, etc. but they are costly, not available and not so easy to deal. But this program can be done by any mathematical programming language like C, Mathematica, Matlab software which is available everywhere and also decision can be drawn easily which is not possible by the other software available in market.

REFERENCES

[1] Halder S., Jan 27th to 29th 2007; "Oscillation and Stability of a Fuzzy Set", *International Conference on Fuzzy Mathematics and its application*, Ahmednagar, New Arts Commerce and Science college

[2] Klir George J., Yuan Bo, "Fuzzy Sets and Fuzzy Logic", pp. 280–285.

[3] Mukherjee A., and Halder S, "Some More Results on Oscillatory Region of a Fuzzy Set", *International conference on Modeling and Simulation CIT*, Coimbatore, 27–29 August 2007, pp. 665–670.

[4] Pawlak Z., (1982), "Rough Sets". *International Journal of Computer and Information Sciences*, **11**, 341–356.

[5] Pawlak Z., (1991), "Rough Sets—Theoretical Aspects of Reasoning About Data", Kluwer Academic publishers, Boston, London, Dordrecht. 1991.

[6] Pawlak Z., (1991), "Rough Sets", *http://grammars.grlmc.com/GRLMC/reports/rep29.doc.*

[7] Wasilewski Anita, "Topological Rough Algebras", *Department of Computer Science*, State University of NewYork, Stony brook NY, USA 11794.

[8] Zadeh L., "Fuzzy Sets", *Information and Control*, 8, 333–353, 1965.

15

KNOWLEDGE ACQUISITION—A ROUGH SET APPROACH

Debadutta Mohanty[1] and Nilaratna Kalia[2]

[1]Lecturer in Mathematics
Department of Mathematics
Seemanta Mohavidyalaya, Jharpokhoria
Mayurbhanj, Orissa-757086, India
E-mail: debadutta.mohanty@rediffmail.com

[2]Lecturer in Mathematics
Department of Mathematics and CS
Upendra Nath College, Nalagaja
Mayurbhanj, Orissa-757073, India
E-mail: nilaratna_math@rediffmail.com

Abstract: *In this paper we study some properties on complement of Rough Set Theory through algebraic method and extend the results of Grzymala-Busse ([3]). Further rough set is defined on the instuitionistic fuzzy set and a theorem on inclusion is established.*

Keywords: *Rough Set, Axiomatic approach, Bottom, Top-Inclusion, RIFS.*

1. INTRODUCTION

Rough set theory was first introduced by Z. Pawlak in 1982 [5]. The theory of rough set is an extension of set theory for the study of intelligent system characterized by insufficient and incomplete information. The idea of rough set consists of approximation of a set by a pair of sets, called lower and upper approximation. Using the concepts of lower and upper approximation, knowledge hidden in the information system may be unraveled and expressed in the form of decision rules. The basic assumption of rough set theory is that human knowledge about an universe depends upon their capability to classify the objects of the universe under consideration. Equivalence relations are considered to induce classifications on universes.

Let $U \neq \varphi$ be a finite set of objects, called the universe of discourse and R be a knowledge (be an equivalence relation) on U. Then U/R denotes (the family of all equivalence classes of U with respect to the relation R) a classification or a partition of U called as concepts or categories of knowledge R. For any $x \in U$, $[x]_R$ be an equivalence class (a concept or a category) of R in U. Any family of concepts in U will be referred to as abstract knowledge about U. A relational system $\mathcal{K} = (U, \mathcal{R})$ is called the knowledge base where $\mathcal{R}$ be a family of equivalence relations over U.

For any subset $\mathcal{P} \subseteq \mathcal{R} (\mathcal{P} \neq \varphi)$, IND ($\mathcal{P}$) is the intersection of all equivalence relation in $\mathcal{P}$, called as indiscernibility relation over $\mathcal{P}$. By IND ($\mathcal{K}$) we denote the family of all equivalence relations defined in $\mathcal{K}$.

For any set $X \subseteq U$ and an equivalence relation $R \in$ IND ($\mathcal{K}$), we define

$$\underline{R}X = \cup \{Y \in U/R : Y \subseteq X\} \text{ and } \bar{R}X = \cup \{Y \in U/R : Y \cap X \neq \varphi\}$$

are called R-lower approximation and R-upper approximation of X, respectively.

The R-boundary region of X be denoted by $BN_R(X)$ defined by $BN_R(X) = \bar{R}X - \underline{R}X$. We say that X is rough with respect to R if and only if $\bar{R}X \neq \underline{R}X$ and X is said to be R-definable if and only if $\bar{R}X = \underline{R}X$, that is $BN_R(X) = \varphi$, at that time X becomes a crisp set. This is known as the constructive definition of rough set.

In 1987 a new approach, that is algebraic or axiomatic definition to rough set was due to by Iwinski [4].Let P, Q be two such sets that $P \subset Q \subset U$. Then the pair (P, Q) form a rough set for which P be the below (Lower) and Q be the above (Upper) approximation concept. Applying some operational criteria to P, Q it can be converted to Pawlak's Rough set $(\bar{R}X, \underline{R}X)$. In this article we use the algebraic definition of rough set.

2. DEFINITION AND PROPERTIES

In an algebraic or axiomatic approach, the system $\left(2^U, \cap, \cup, \sim, L, H\right)$ be called a rough set algebra where L, H: $2^U \to 2^U$ are dual unary operators on the power set 2^U and $\cap$, $\cup$, $\sim$ be the set intersection, union and complement respectively.

Definition 2.1 [7]: Let L, H: $2^U \to 2^U$ are unary operators. They are dual if

1. $L \sim A = \sim HA$
2. $H \sim A = \sim LA$

The dual unary operators L and H satisfy

3. $LU = U$
4. $H\varphi = \varphi$
5. $L (A \cap B) = LA \cap LB$
6. $H (A \cup B) = HA \cup HB$, where A, B are the subsets of U

Also L and H satisfy the weaker conditions

7. $L (A \cup B) \supseteq LA \cup LB$
8. $H (A \cap B) \subseteq HA \cap HB$ and
9. $A \subseteq B \Rightarrow LA \subseteq LB$
10. $A \subseteq B \Rightarrow HA \subseteq HB$

Definition 2.2: Let U be any non empty finite set of objects. Let LA and HA be two elements of 2^U such that $LA \subset A \subset HA$. Then the pair (LA, HA) forms a rough set of A on U, where L, H: $2^U \to 2^U$ are dual unary operator which satisfy (1) to (4) of the Definition 2.1.

Definition 2.3: Let $\mathbb{A} = (LA, HA)$, $\mathbb{B} = (LB, HB)$ be two rough sets of A and B respectively defined on U then the union, $\mathbb{A} \cup \mathbb{B}$ and the intersection, $\mathbb{A} \cap \mathbb{B}$ be defined by $\mathbb{A} \cup \mathbb{B} = (L (A \cup B), H (A \cup B))$ and $\mathbb{A} \cap \mathbb{B} = (L (A \cap B), H (A \cap B))$ respectively.

Definition 2.4: Let $\mathbb{A} = (LA, HA)$, $\mathbb{B} = (LB, HB)$ be two rough sets defined on U then the subset $\mathbb{A} \subset \mathbb{B}$ is defined by $LA \subset LB$ and $HA \subset HB$. Two rough sets $\mathbb{A}$, $\mathbb{B}$, are equal if and only if $\mathbb{A} \subset \mathbb{B}$ and $\mathbb{B} \subset \mathbb{A}$.

Definition 2.5: If $\mathbb{A} = (LA, HA)$ be a rough set then the complement of $\mathbb{A}$ denotes $\sim\mathbb{A}$, be defined by $\sim\mathbb{A} = (\sim HA, \sim LA)$ where $\sim LA$, $\sim HA$ are respective complement of LA and HA in U.

The following properties come immediate.

Proposition 2.1: For any three rough sets $\mathbb{A}$, $\mathbb{B}$ and $\mathbb{D}$

(*i*) $\mathbb{A} \cup \mathbb{B} = \mathbb{B} \cup \mathbb{A}$

(*ii*) $\mathbb{A} \cap \mathbb{B} = \mathbb{B} \cap \mathbb{A}$

(*iii*) $\mathbb{A} \cup (\mathbb{B} \cup \mathbb{D}) = (\mathbb{A} \cup \mathbb{B}) \cup \mathbb{D}$

(*iv*) $(\mathbb{A} \cap \mathbb{B}) \cap \mathbb{D} = \mathbb{A} \cap (\mathbb{B} \cap \mathbb{D})$

(*v*) $\mathbb{A} \cap (\mathbb{B} \cup \mathbb{D}) = (\mathbb{A} \cap \mathbb{B}) \cup (\mathbb{B} \cap \mathbb{D})$

Now we produce some properties on rough definability of a classification or partition.

Theorem 2.1: Let $\{A_1, A_2, A_3, ... A_n\}$ be a partition of U. Then $LA_i \neq \varphi$ for each i, if and only if $H\ (U_{j \neq i}\ A_j) \neq U$, where $i, j \in \{1, 2, 3 ... n\}$.

Theorem 2.2: Let $\{A_1, A_2, A_3, ..., A_n\}$ be a partition of U. Then for each i, $1 \leq i \leq n$, $HA_i \neq, U$ if and only if $L\ (U_{f \neq l} A_l) + \varphi$, for $1 \leq j \leq n$,

Note: Now we turn back to Pawlak's definition (constrictive method) of rough set. Let R be an equivalence relation on U and $R \in$ IND ($\mathcal{K}$). A few properties on the concept of rough definability of classifications of the set of objects are given below.

Proposition 2.2: Let $F = \{X_1, X_2 ..... X_n\}$ be a classification of U and $R \in$ IND ($\mathcal{K}$). Then for each fixed $i \in \{1, 2, 3, ... n\}$, $\underline{R}X_i \neq \varphi$ if and only if $U_{j \neq i}\bar{R}X_j \neq U$ for each $i \in \{1, 2, 3, ... n\}$.

Equivalently $\underline{R}X_i \neq \varphi$ if and only if $\bar{R}(\sim X_i) \neq U$ for each $i \in \{1, 2, 3, ... n\}$.

Proof: Suppose that, $RX_i \neq \varphi$, then there exits $x \in U$ such that $[x]_R \subseteq X_i$ for fixed i, $1 < i < n$. Thus $[x]_R \cap X_j = \varphi$ for each $j \neq i$. That is, $\bar{R}X_j \cap [x]_R = \varphi$ for each $j \neq i$. Hence $U_{j \neq i}\ \bar{R}X_j \cap [x]_R = \varphi$. This gives $U_{j \neq i}\ \bar{R}X_j \neq U$. Conversely, suppose that $U_{j \neq i}\ \bar{R}X_j \neq U$ for i, $j \in \{1, 2, 3 ... n\}$. That is, $\bar{R}(U_{j \neq i}\ X_j) \neq U$. This implies, there exits $x \in U$ such that $[x]_R \cap U_{j \neq i} X_j \neq \varphi$ which yields $[x]_R \subseteq X_i$. Hence $\underline{R}X_i \neq \varphi$. Hence the proposition.

This is the necessary and sufficient condition to the PROPOSITION 1 of Jerzy W. Grzymda-Busse [3]. In similar manner we can extend the PROPOSITION 2, PROPOSITION 2' and PROPOSITION 1'. of Grzymala-Busse [3] whose proves are straight forward.

Proposition 2.3: Let $\{X_1, X_2 \,.....\, X_n\}$ be the classification of U and $R \in \text{IND}(\mathcal{K})$. Then for each $i \in. \{1, 2, 3, \,.....\, n\}$, $\bar{R}X_i = U$ iff $\underline{R}(U_{j \neq i} X_j) = \varphi$. Equivalently $\bar{R}X_i = U \Leftrightarrow \underline{R}(\sim X_i) = \varphi$.

Proposition 2.4: Let $F = \{X_1, X_2 \,.....\, X_n\}$ be a classification of U and $R \in \text{IND}(\mathcal{K})$. Then $\underline{R}(U_{j \neq i} X_j) \neq \varphi$ if and only if $\underline{R}X_i \neq U$.

Proposition 2.5: Let $F = \{X_1, X_2 \,.....\, X_n\}$ be a classification of U and R be an equivalence relation on U. Then for all $i \in \{1, 2, 3 \,...\, n\}$ $\underline{R}X_i = \varphi$ if and only if $U_{j \neq i} \bar{R}X_i = U$.

Applying the operational criteria to L and H, that is, if we denote $LX = \underline{R}X$ and $HX = \bar{R}X$, then the Theorem 2.1 of this article becomes Proposition 2.2 which is the necessary and sufficient condition to Proposition 1 of Grzymala-Busse [3] hence on other theorem also.

We conclude this section after this theorem.

Theorem 2.5: Let $\{A_1, A_2, A_3, \,.....,\, A_n\}$ be a partition of U. Then for each $i_1, i_2, i_3, \,.....\, i_k \in \{1, 2, 3 \,.....\, n\}$, $H(U_m A_{i_m}) = U$ if and only if $L(U_{j \neq i_m} A_j) = \varphi$.

If we have positive examples of every category in the approximation of classification, then we must have also negative examples for each categorie. For example if categories are colours say, $c_1, c_2, c_3 \,.....\, c_n$ and if we have examples of, $c_1, c_2, c_3 \,.....\, c_n$ objects in our knowledge base, then we must have, also, the examples of not $c_1, c_2, c_3 \,.....\, c_n$ or c_1, not $c_2, c_3 \,.....\, c_n$ or c_1, c_2 not $c_3 \,.....\, c_n$, etc. Moreover if we have the knowledge about the objects of colour $c_{i_2}, c_{i_2}, c_{i_2}, \,.....\, c_{i_k}$ from the knowledge base then we must have the knowledge about the objects of remaining colours, that is $U^n_{j=1} c_j - U^k_{i=1} c_{i_k}$.

3. EQUALITIES AND INCLUSIONS

We first define the rough equalities of set through algebraic method. Let $\mathbb{A} = (LA, HA)$ and $\mathbb{B} = (LB, HB)$ be two rough sets of A and B respectively, on U.

(*a*) Sets A and B are bottom equal, that is, $A \eqsim B$ if and only if $LA = LB$.

(*b*) Sets A and B are top equal, that is, $A \simeq B$ if and only if $HA = HB$.

(*c*) Sets A and B are equal if and only if $A \eqsim B$ and $A \simeq B$ that is if and only if $LA = LB$ and $HA = HB$.

Also the rough inclusion of sets through algebraic approach be defined as follows:

(*a'*) Set A is bottom included by B, that is $A \subset_* B$ if $LA \subseteq LB$.

(*b'*) Set A is top included by the set B that is $A \subset^* B$ if $HA \subseteq HB$.

(*c'*) Set A is included by B, that is $A \subset_*^* B$ if and only if $A \subset_* B$ and $A \subset^* B$.

Definition 3.1: Let $\mathbb{A} = (LA, HA)$ and $\mathbb{B} = (LB, HB)$ be two rough sets on U. Then the difference denote $\mathbb{A} - \mathbb{B}$ be defined by $\mathbb{A} - \mathbb{B} = (L(A - B), H(A - B))$

Notice that $\mathbb{A} - \mathbb{B} = \mathbb{A} \cap (\sim\mathbb{B}) = (LA, HA) \cap (L\sim B, H\sim B) = (L(A \cap \sim B), H(A \cap \sim B)) = (L(A - B), H(A - B))$

Proposition 3.1: Let $L, H: 2^U \rightarrow 2^U$ be dual unary operators which satisfy condition (1) to (10) and let A, B are subsets of U, then

(*i*) $L(A - B) = LA - HB \subseteq LA - LB$

(*ii*) $H(A - B) \subseteq HA - LB$

(*iii*) $A \subset_* B$ if and only if $\sim B \subset^* \sim A$

(*iv*) $A \in^* B$ if and only if $\sim B \subset_* \sim A$

(*v*) $A \subset_* B \Rightarrow A - B \eqsim \varphi$

Proof: We have $L(A - B) = L(A \cap \sim B) = LA \cap L(\sim B) = LA \cap \sim HB = LA - HB \subseteq LA - LB$.

Hence (*i*) is proved. Proof of (*ii*), (*iii*), and (*iv*) are similar.

Next suppose that $A \subset_* B \Leftrightarrow LA \subseteq LB \Rightarrow LA - LB = \varphi$

But $L(A - B) \subseteq LA - LB = \varphi$ which implies $L(A - B) = \varphi = L\varphi$,

Hence $A - B \eqsim \varphi$. This completes the proof of (*v*). We claim that $A \approx B \Leftrightarrow \sim A \eqsim \sim B$.

We have $A \simeq B \Leftrightarrow HA = HB \Leftrightarrow \sim L\sim A = \sim L\sim B \Leftrightarrow L\sim A = L\sim B \Leftrightarrow \sim A \eqsim \sim B$.

Theorem 3.1: Let $A, B, A_1, B_1 \subset U$, if $A \approx A_1$ and $B \simeq B_1$ then $A - B \approx A_1 - B_1$.

Proposition 3.2: Let U be the universe, R be an equivalence relation on U, $R \in \text{IND}(\mathcal{K})$. Then for any set $X, Y \subseteq U$, we have

(*i*) $\underline{R}(X - Y) - \underline{R}X - \bar{R}Y \subset \underline{R}X - \underline{R}Y$

(*ii*) $\bar{R}(X - Y) \subseteq \bar{R}X - \underline{R}Y$

(*iii*) $X \subset_* Y \Leftrightarrow \sim Y \subset^* \sim X$

(*iv*) $X \subset^* Y \Leftrightarrow \sim Y \subset_* \sim X$

(*v*) $X \subset_* Y \Leftrightarrow X - Y \approx \varphi$.

4. ROUGH INTUITIONISTIC FUZZY SETS (RIFS)

Definition 4.1 ([1], [2]): An intuitionistic fuzzy set A is characterized by two functions μ_A and ν_A, called the membership and non-membership function of A such that $\mu_A: U \to [0, 1]$, $\nu_A: U \to [0, 1]$ where $0 \le \mu_A + \nu_A \le 1$ for all $x \in U$.

The set $A = \{< x, \mu_A(x), \nu_A(x) >: x \in U\}$ is called intuitionistic fuzzy set (IFS).

The function π_A, called the hesitation function of A be defined by $\pi_A(x) = 1 - \mu_A(x) - \nu_A(x)$ for all $x \in U$. When $\pi_A(x) = 0$ for all x, the IFS A reduces to a fuzzy set.

Definition 4.2: Let A be an IFS defined on U. The complement of A denoted by $\sim A$ and defined by the function $\mu_{\sim A} = 1 - \mu_A(x)$ and $\nu_{\sim A}(x) = 1 - \nu_A(x)$ for all $x \in U$.

We find the Iwinski definition for RIFS as

Definition 4.3: Let U be any non empty set and $\mathcal{A}$ be the set of all IFS in U. The pair $(U, \mathcal{A})$ is the rough intuitionistic fuzzy universe.

Let LA and HA be two elements of $\mathcal{A}$ such that $LA \subset A \subset HA$. Then the pair $\mathbb{A} = (LA, HA)$ is called a rough intuitionistic fuzzy set (RIFS) on U, where the dual unary operators satisfy the conditions of Definition 2.1.

Definition 4.4: Let $\mathbb{A} = (LA, HA)$ and $\mathbb{B} = (LB, HB)$ be two RIFS on U. Then the subset, $\mathbb{A} \subseteq \mathbb{B}$ be defined by $LA \subseteq LB$ and $HA \subseteq HB$ which is equivalent to $\mu_{LA} \le \mu_{LB}$, $\nu_{LA} \ge \nu_{LB}$, and $\mu_{HA} \le \mu_{HB}$, $\nu_{HA} \ge \nu_{HB}$.

Proposition 4.1: Let $\mathbb{A} = (LA, HA)$, $\mathbb{B} = (LB, HB)$ be two RIFS on U. Then

(*i*) $\mathbb{A} \subset_* \mathbb{B}$ if and only if $\sim\mathbb{B} \subset^* \sim\mathbb{A}$

(*ii*) $\mathbb{A} \subset^* \mathbb{B}$ if and only if $\sim\mathbb{B} \subset_* \sim\mathbb{A}$

Proof: Suppose that $\mathbb{A} \subset^* \mathbb{B}$. This is equivalent to $LA \subseteq LB$ so that $\mu_{LA}(x) \leq \mu_{LB}(x)$ and $\nu_{LA}(x) \geq \nu_{LB}(x)$ for all $x \in U$.

This implies and implied by $1 - \mu_{LA}(x) \geq 1 - \mu_{LB}(x)$ and $1 - \nu_{LA}(x) \leq 1 - \nu_{LB}(x)$ for all $x \in U \Leftrightarrow \sim LA \supseteq \sim LB \Leftrightarrow H \sim A \supseteq H \sim B \Leftrightarrow \sim \mathbb{A} \subset^* \sim \mathbb{B}$.

5. CONCLUSION

The constructive approach is more suitable for practical applications of rough sets, while the algebraic approach to rough set is appropriate for studying the structure of rough set algebra. The algebraic approach deals with axioms that must be situated by approximation operators without explicitly referring to a binary relation. Some rules are generated separately through algebraic method and this extends the results of knowledge acquisition of G. Busse ([3]) to obtain properties of approximation of classification.

REFERENCES

[1] Atanassov K, "Intuitionistic Fuzzy Sets in" V. Sagureved VII, ITKR'S SECTION Sofio, June (1983).

[2] Atanossov K and Stoeva S., "Intuitionistic Fuzzy Sets in", *Polish Symp. On Interval and Fuzzy mathematics*, Poznan (Aug, 1983), pp. 23–28.

[3] Grezymala-Busse, Jerry W., "Knowledge Acquisition Under Uncertanty–Rough Set Approach". *Journal of Intelligent and Robotic System*, **1**, (1988), pp. 3–16.

[4] Iwinski T, "Algebraic Approach to Rough Sets", *Bull. Polish Acad. Sci., Math*, **35**, (1987), pp. 673–783.

[5] Pawlak Z., "Rough Sets", *Int. Jour. Inf. Comp. Sc.*, **2**, (1982), pp. 341–356.

[6] Tripathy B.K., "Rough Sets on Intuitionistic Fuzzy Approximation Spaces", *Proceeding of 3rd International IEEE Conference*, "On Intelligent Systems" (ISO6), Landon, Sept. 4–6, (2006), pp. 776–779.

[7] Yao, Y.Y., "Constructive and Algebraic Methods of Theory of Rough Sets", *Jour. of Internation Science*, **109**, (1998), pp. 21–47.

16

AN INTELLIGENT WEB MINING MODEL—A ROUGH SET BASED APPROACH

Biswajit Saha

Department of Computer Science and Engineering,
Tripura Institute of Technology, Narsingarh, Tripura (W)-799009
E-mail: sahabiswa@yahoo.com

Abstract: *Intelligent web mining model is a kind of mining to server logs. And its aim is getting useful users' access information in logs, the paper advances an architecture of web usage mining, expanded some key matters in system design. In addition pattern discovery using Rough set based approach are introduced for serving user to the best of its abilities.*

1. INTRODUCTION

The World Wide Web constitutes the largest source of information on a wide range of topics and is able to serve clients with various interests. This large information system is constantly expanding, as new sites are added to the existing ones at an extremely high rate. The addition of these sites continuously strengthens the ability of the Web to cover clients' information needs. Taking into account the vastness of the Web nowadays, two problems of utmost importance to the users are the finding of contents strictly according to personal interests, and the perception of the least latency possible when downloading these contents.

At present, there are about three kinds of search engines in Internet. The one is search engine based on artificial intelligence, such as Yahoo, Google. The other is search engine based on Robot/agent, for example, AltaVista, Lycos and Excite. And the third is meta-search engine, i.e, Byte-search, Meta Crawler, etc. Although the present search engines have taken great conveniences for people's searching information, it has little good effect. On the other hand, data mining technology is becoming more and more mature after many years' development. And it may use computer forwardly to distill valuable data modes from enormous data to meet people' different demands. So currently it is an important task to import data mining into Web information retrieval/web mining.

2. WEB MINING

Web mining is mining of data related to WWW. It is a technology on finding out implicit pattern *p*, from vast Web document structures and master, *C*. If we see *C* as export and ρ as import, then, the process of Web mining can be thought a mapping from import to export: $\xi: C \rightarrow p$, the objects of Web mining are vast, heterogeneous and distributing documents. Secondly, because the logistic structure of Web is a graph structured by documents and hyperlinks, the mining results may be is on Web contents or Web structures. Lastly, Web document is semi-structure do run structured itself and is short of semantic understood by machines. So the objects of Web mining are made up of all kinds of documents and users access information in Web. According to different mining objects, Web mining can be divided into three kinds, these are Web content mining, Web structure mining and Web usage mining.

In this paper, we are mainly concerned with Web Usage Mining that undertake the building of a user profile and represents individual interest. We study it by using Rough-*C* means clustering algorithm to generate clusters from preprocessed web logs, then, these clusters are used as training examples for a rough set based inductive learning to generate rules, which represent usage patterns. The clustering of web data tend to have imprecise boundaries. Therefore. The role of soft computing technique such as fuzzy theory and rough theory is highlighted.

2.1 Web Usage Mining (WUM)

Web usage mining is about finding patterns of page views by web users or to find the usage of a particular website.

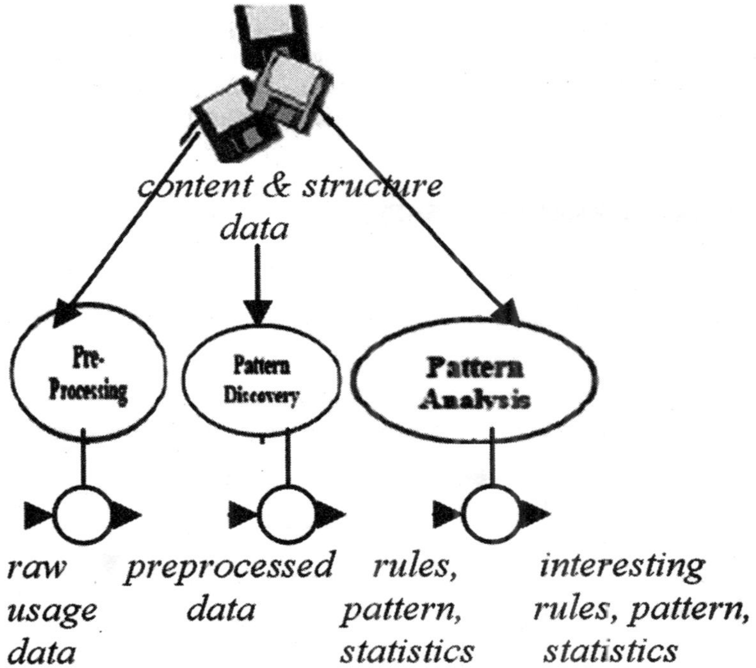

Fig. 1: *Web Usage Mining System*

The best way to find out how a user browses web pages is to put a cookie into his or her computer. The cookie will keep a log of page visits and send information to the webmaster. Other way page view information, access pattern etc. can be collected from the vast server web log files. These patterns can be used to predict users' intention, and to provide timely help for using features available on a web site.

As shown in Fig. 1, WUM is divided into three core phases [5]: data pre-processing, pattern discovery and pattern analysis. Using the data set yielded by the preprocessing phase, the pattern discovery phase applies mining methods and algorithms in search of useful, new and significant patterns.

The resulting patterns have to be interpreted and validated during the pattern analysis phase. Effectiveness of the proposed mining system is evaluated by using log files collected from the NIC web server. In the following section, we present the Data preprocessing.

Section 4 describes the workflow of the proposed mining system. Section 5 describes the pattern discovery with rough set approach. Experimental analysis are presented in Section 6. Conclusions and Future work are given in Section 7, followed by references in Section 8.

3. DATA PREPROCESSING

These are general rules, and they may change according to the purpose of the mining system. As shown in Fig. 2, there are four steps in data preprocessing: data cleaning, user identification, session identification, and data filtering.

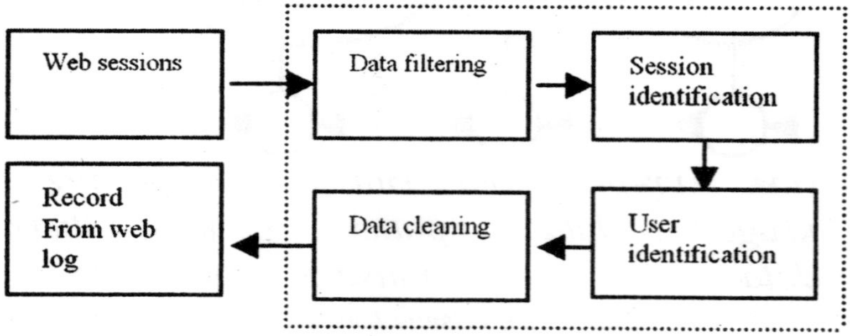

Fig. 2: *Data Preprocessing Component*

3.1 Data Cleaning

Web logs are designed for debugging purposes in a way the web accesses are recorded in the order they come [1] and due to the connectionless nature of the HTTP (i.e. each request is handled in a separate connection), for a single user to surf a--website that will result in huge number of records in the web page that are not necessary in sequence. So for each page component a separate record is recorded in the web log file. Usually each record in the web log file has the following standard formats [2]:

3.1.1 Common Web Log Format

Table 1: *Common Web Log Format*

Remote Host	Log Name	User Name	Data	Required	Stat	Byte

- Remote host is the remote hostname or its IP address,
- Log name is the remote log name of the user,
- Date is the date and time of the request,
- Request is the exact request line as it came from the user,
- Status is the HTTP status code returned to the client, and
- Byte is the content-length of the document transferred.

3.1.2 Extended Log File Format

An extended common log format file simply adds two additional fields to the end of the line, the referrer Universal Resource Locator (URL) and the user agent fields. Referrer URL is the page the visitor was on when they clicked to come to this page. User Agent is whatever software the visitor used to access this site. It's usually a browser, but it could equally be a web robot, a link checker, an FTP client or an offline browser. The format of the extended common log file is shown in Table 2.

Table 2: *Format of Extended Log File*

Remote Host	Log Name	User Name	Date	Request	Status	Byte	Referrer	User Agent

Usually for web mining purposes the only interesting element are the HTML pages and the scripting pages. In the cleaning phase the pages types that are related to the navigation structure are kept and the other pages are eliminated. The status filed in the web log can be used to keep the successfully fulfilled request and delete the others. Finally the mining process can be limited to certain time or date so only transaction during such time and date will be considered.

3.2 User Identification

Because web logs are recorded in a sequential manner as they arrive, therefore, records for a specific user are not necessary recorded in consecutive order rather they could be separated by records from other users. User is defined as a unique client to the server during a specific period of time. The relationship between users and records is one to many. Each user has one or more records to identify it. The users are found based on two assumptions:

(*i*) Each user has a unique IP address during his browsing time span that means other users can have the same IP address during different time span.

(*ii*) Maximum of 30 minutes for the user to be in ideal state after that the user will be considered that he left the website.

3.3 Session Identification

Using the website ontology we can identify different sessions in a single user's visit. We assume that the website ontology is already available. Work on retrieving web site ontology can be found in [6, 7, 8]. A web site is defined as a triple *W* = (*P*, *L*, *F*), where *P* = (*P* ..... *Pk*) web site pages, which is a list of *k* pages, and *k* is an arbitrary integer.

L: web site links, which is the group of links for the web site. Each link 1 = (*P*, ... *Pd*) is defined by two pages the source page (*ps*) where the link starts from and the destination page (*Pd*) where the links ends to. *F* = <*fo*, *f*1, ..., *fn*, 1>, web site functionalities, where *Vf F f* = <*Pss* ...*Pe*>.

The web functionality *f* consist of at least two pages the start page and the end page, and zero or more pages in between. The session identification algorithm will take the users identified and divide them into different sessions using the website functionalities.

From the website functionality we can identify some pages that are considered as breaking point for the session, such as sign in or sign out page.

3.4 Data Filtering

Filtering is the preprocessing of data file before mining starts. After the data is been separated into different sessions filtering is done based on removing the house keeping pages, which are necessary for the web application to run properly. These pages shall be identified by the website designer and can be found using the website ontology. Removing the pages can result in redundant pages which can be sometimes misleading to the distance function.

4. CLUSTERING AND CLASSIFICATION

Searching on the web is tedious and time consuming. Search engines cannot index the huge and highly dynamic web contents. The user's "intention behind the search" is not clearly expressed which results in too general, short queries. Clustering is a useful technique for grouping web sessions such that sessions within a single group/cluster have similar characteristics, while sessions in different groups are dissimilar. For most data mining tasks, we start out with a pre classified training set and attempt to develop a model capable of predicting how a new record will be classified. In clustering, assembling of vast data items into classes acquire least similarity degree between classes and maximal similarity degree of a class. The ultimate branch point between classification and clustering is that the categorical attribute values of a data-base's data items are already known in classification, but in clustering we had to find out the categorical attribute values by using some algorithms. The workflow is shown in Fig. 3.

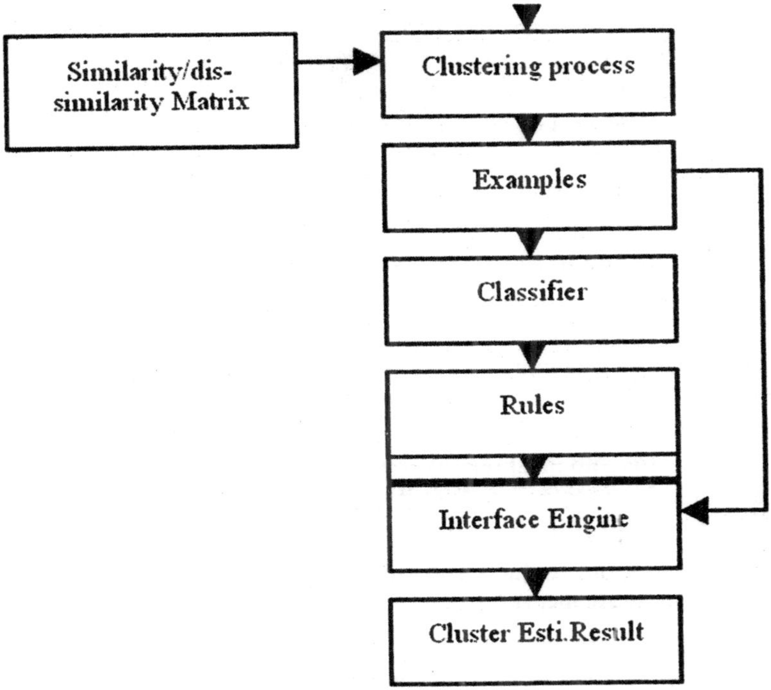

Fig. 3: *Web Mining System Workflow*

Usually there is no pre-classified data and no distinction between independent and dependent variables. Instead, we are searching for groups of records that are similar to one another, in the expectation that similar records represent similar thing that will behave in similar ways.

The rules are used for both predicting users' behavior based on past navigation history and describing the cluster of users' usage sequences in terms of what functionality the user is trying to do.

5. PATTERN DISCOVERY USING ROUGH SET BASED APPROACH

The knowledge discovery from real-life web databases is a multi-phase process. As more and more information is being stored digitally the ability to collect data far outweighs the ability for a person to analyze it. Prior to the emergence of the data-mining field, it has been common practice to either design a database application on online data or use a statistical (or an analytical) package along with a domain expert to interpret the results.

A piece of knowledge is a relationship or pattern or feature among data elements that is potentially interesting and useful. In general, discovery means finding something that is hidden or previously unknown.

Z. Pawlak developed rough set theory in the early 1980's. Rough set deals with classification of discreet data table in a supervised learning environment. Although in theory rough set deals with discreet data, rough set is commonly used in conjunction with other technique to do discretization on the data set. The main feature of rough set data analysis is non-invasive, and the ability to handle qualitative data. This fits into most real life application nicely. Rough sets have seen light in many researches. Web Mining with rough set is also consist of multi-phase process of mainly:

Discretization, Reducts and rules generation on training set, & Classification on test set

Because rough set theory is a symbolical method rather than a numerical method, rough set theory cannot process continuous data. Discretization is a process that converts continuous data into discreet intervals to be used in rough set. There a couple of popular techniques that is used to discretize data. The reducts and rules generation is the core of the rough set. The algorithm to generate reducts and rules are discussed at 5.1. In Rough Set terminology, a data table is also called an information system.

If some of the attributes are interpreted as outcomes of classification, it is also called a decision system. More formally, a data table is a tuple (U, A, Va) where, U is a set of objects, A is a set of attributes a: U ... Va, and Va is a set of values for the attribute a.

We can split the set of attributes in two subsets $C A$ and $D = A - C$, respectively the conditional set of attributes and the decision (or class) attribute (s). Condition attributes represent measured features of the objects, while the decision attribute is an a posteriori outcome of classification. Data cleaning and preprocessing as mentioned earlier is one of the application fields of rough sets. It is known that prior knowledge is very important in this process. On the other side, one of the main features of rough set data analysis is that it does not use information outside the target data set. These two things seem to be incompatible. At first, rough sets are used to reduce and clean data with minimal model assumptions. Then, the result is used as a basis for further analysis performed with other methods. Missing values can be handled before RS data analysis. We can use some data preparation methods, to obtain a complete system [3].

5.1 Reduct Algorithm

In this paper, Quick reduct and VPRS algorithms are proposed for feature selection from the web log file. The third stage applies the proposed reduct algorithms for finding the smallest set of attributes after reconstructing the decision table.

5.1.1 Quick Reduct Algorithm

The reduction of attributes is achieved by comparing equivalence relations generated by sets of attributes. Attributes are removed so that the reduced set provides the same predictive capability of the decision feature as the original.

A reduct is defined as a subset of minimal cardinality Rmin of the conditional attribute set C such that $\gamma R\ (D) = \gamma C\ (D)$, $R = \{X\ C;\ \gamma X\ (D) = \gamma C\ (D)\}$

$$R\min = \{X{:}X\ R;\ YR;\ |X| \leq |Y|\}$$

The intersection of all the sets in Rmin is called the core, the elements of which are those attributes that cannot be eliminated without introducing more contradictions to the dataset. In this method a subset with minimum cardinality is searched for. The pseudo code of the Quick reduct is given in [10, 11, and 12].

5.1.2 Variable Precision rough Set (VPRS)

Variable precision rough sets (VPRS) extend rough set theory by the relaxation of the subset operator. It was proposed to analyze and identify data patterns which represent statistical trends rather than functional. The VPRS approach may also be found in [4, 9]. The main idea of VPRS is to allow objects to be classified with an error smaller than a certain predefined level. This introduced threshold relaxes the rough set notion of requiring no information outside the dataset itself. Let X, Y U, the relative classification error is defined by

$$c\,(X, Y) = 1 - \{|X \cap Y|/|X|\} \qquad \text{... (1)}$$

Observe that $c\,(X, Y) = 0$ if and only if $X = Y$. A degree of inclusion can be achieved by allowing a certain level of error, β in classification:

$$X\,\beta Y \text{ if } fc\,(X, Y) \leq \beta,\ 0 \leq \beta < 0.5 \qquad \text{... (2)}$$

Using β instead of, the β-upper and β-lower approximations of a set X can be defined as:

$$\underline{R}\beta X = \cup\,\{[x]_R^{\in}\ \ U/R \mid [x]\ \ \beta X\} \qquad \text{... (3)}$$

$$\overline{R}\beta X = \cup\,\{[x]_R^{\in}\ \ U/R \mid c\,([x]\,R, X) < 1 - \beta\} \qquad \text{... (4)}$$

Note that $\underline{R}\beta X = \underline{R}X$ for $\beta = 0$. The positive, negative and boundary regions in the original rough set theory can now be extended to:

$$POSR, \beta\ (X) = R\beta X \qquad \text{... (5)}$$

$$NEGR, \beta\ (X) = U - R\beta X \qquad \text{... (6)}$$

$$BNDR, \beta\ (X) = R\beta X - R\beta X \qquad \text{... (7)}$$

Consider a decision Table $A = (U, C \cup D)$, where C is the set of conditional attributes and D the set of decision attributes. The β-positive region of an equivalence relation Q on U may be determined by

$$POSR, \beta\ (Q) = \cup\, X\ \ U/QR\beta X \qquad \text{... (8)}$$

where R is also an equivalence relation on U. This can then be used to calculate dependencies and thus determine β-reducts. The dependency function becomes:

$$\gamma R, \beta\ (Q) = |\ \mathrm{POSR}, \beta f\,(Q)\ |\ /\ |U| \qquad \ldots (9)$$

6. EXPERIMENTAL ANALYSIS

The Result analysis is carried out using an evolutionary rough c-means clustering method. This will group user usage patterns $\{S = (s_1, s_2, \ldots s_n)$. Where $s_i\ (1 < i < n)\}$ from web log file. The user usage pattern behavior of Ith user can be represented as $s_i = \{(Urli_1, ti_1), (Urli_2, ti_2), \ldots, (Url_{il}, t_{il})\}\ (1 < i < n)$, where Url_{ik} denotes kth visited web page, t_{ik} denotes the time duration on Url_{ik}, l is the total number of visited web pages. Distance between any two users' patterns, is calculated by converting each pattern into a fuzzy vector $\{V_i = \langle v^l_{i1}, v^l_{i2} \ldots\ldots v^l_{im}\rangle$ with the same length, in which each element represents the visited webpage and time duration on it. Then VPRS reduct algorithm is applied to select the reduced feature from the resulting table as effective knowledge.

Downloaded web log file used for this experiment and total 2164 access pattern are extracted, which are grouped into 4 cluster a portion of the result are shown in Table 3.

Table 3: *Clustering Result*

Cluster No	*Web Access Patterns*	*Rules generation using VPRS*	
		No. Pattern	*Name of Pattern*
1	526	5	Remote host , user, date request user agent
2	735	5	Remote host , user, date request user agent
3	678	3	Remote host Request user
4	234	3	Remote host Request user

7. CONCLUSION AND FUTURE WORK

This paper studies the dynamic mining of users' interest navigation pattern to guide the user to select appropriate web pages. Here rough c-means and reduct algorithm are applied to cluster user access patterns from server logs and describe a session. It will also group the similar surfing behavior together into one class. To evaluate the effectiveness of our model, we utilize these recommendation to built a user navigation model. Secondly we apply rough set for mining users interest patterns dynamically.

Our experiment have verified the methodology's effectiveness. It can be suitable for applications in *E*-business for designing personalized service. However many problems remain such as Time and Space complexity, etc. We also plan to perform extensible testing and investigate possible improvements for solving these issues in the future.

REFERENCES

[1] Kohavi R, "Mining E-commerce Data: The Good, the Bad, and the Ugly", *7th International Conference on Knowledge Discovery*, San Francisco, California, 2001.

[2] WWW Consortium, "The Common Log File Format", *http://www.w3.org*.

[3] Cooley R., B. Mobasher, and J. Srivastava, "Data Preparation for Mining World Wide Web Browsing Patterns", *Knowledge and Info. Syss*, **1**, pp. 5–32, 1999.

[4] M.J. Beynon. "An Investigation of β-reduct Selection within the Variable Precision Rough Set Model". *Proceedings of Second International Conference on Rough Sets and Current Trends in Computing*, 114–122, 2000.

[5] Chan. C.C. and S. Santhosh, "BLEM2: Learning Bayes Rules from Examples Using Rough Sets", *Proc. NAFIPS2003, 22nd Int. Conference of the North American Fuzzy Information Processing Society*, July 24–26, 2003, Chicago, Illinois, pp. 187–190.

[6] P. Clerkin, P. Cunningham, and C. Hayes, "Ontology Discovery for the Semantic Web Using Hierarchical Clustering", *Semantic Web Mining WorkshopatECML/PKDD-2001*, Freiburg, Germany, 2001.

[7] Crave M., D. DiPasquo, D. Freitag, A. McCallum, T. Mitchell, K. Nigam, and S. Slattery, "Learning to Construct Knowledge Bases from the WorldWide Web", *Artificial Intelligence*, **118**, pp. 69–113, 2000.

[8] Maedche A. and S. Staab, "Discovering Conceptual Relation from Text", *Euorpean Conference on Artificial Intelligence (ECAIOO)*, Berlin, 2000.

[9] M.J. Beynon, "Reducts within Variable Precision Rough Sets Model: A Further Investigation". *European Journal of Operational Research*, **134 (3)**, 592–605, 2001.

[10] R. Jensen, "Combining Rough and Fuzzy Sets for Feature Selection". *Ph.D Thesis, School of Informatics*, University of Edinburgh, 2005.

[11] R. Jensen and Q. Shen, "Fuzzy-rough Attributes Reduction with Application to Web Categorization". *Fuzzy Sets and Systems*, **141 (3)**, 469–485, 2004.

[12] R. Jensen and Q Shen, "Semantics Preserving Dimensionality Reduction: Rough and Fuzzy-Rough-Based Approaches". *IEEE Transactions on Knowledge and Data Engineering*. **16 (12)**, 2004.

[13] "Data Mining—a Tutorial Based Approach", by Roiger Geatz Pearson Education 2006.

[14] "Data Mining-Methods and Techniques". By Ali and Wasimi, Cenage Learning 2007.

[15] YiMinxin, W. Sheng, "Data Preparation in Web Usage Mining", *Computer Engg. and Appl. Beijing*, 2003.

[16] Gondy Leroy, A.M Lally, "The use of Dynamic Contexts to Improve Casual Internet Searching", *ACM Transaction on Information System*, 2003 (21) pp. 229–253.

[17] Wang Yuzen, "Research on Web Usage Mining", *Computer Application*, China 2003, pp. 86–88.

[18] Xu Jianchao, "Methods for Information Extraction and Data Mining from Web", *Journal of Changchun Univ.*, China 2002, pp. 49–53.

[19] T.W. Yan, M. Jacobsen, H. Garcia-Molina, and U. Dayal. "From User Access Patterns to Dynamic Hypertext Linking. *In Proc. of 5th WWW*, pp. 1007–1014, 1996.

[20] Zhang Huiying, Liang Wei, "An Intelligent Algorithm of Data Pre-processing in Web Usage Mining", *Proceedings of the 5th World Congress an Intelligent Control and Automation*, Hangzhou, P.R. China. June 15–19, 2004.

17

EXTRACTION METHOD OF DECISION RULES FOR TURBINE-GENERATOR UNIT FAULT DIAGNOSIS BASED ON ROUGH SET THEORY

Hossam A. Nabwey[3], E.A. Rady[1], A.M. Kozae[2], A.N. Abady[3]

[1]*I.S.S.R, Cairo University, Cairo, Egypt.*

[2]*Department of Mathematics, Faculty of Science, Tanta University, Tanta, Egypt.*

[3]*Department of Engineering Basic science, Faculty of Engineering, Menofia University, Menofia, Egypt.*

Abstract: *Extraction of simple, effective and efficient decision rules for fault diagnosis is one of the most important problems in mechanical fault diagnosis because available information is often inconsistent and redundant. The rough set theory is a new mathematical tool to deal with uncertainty and vagueness of decision systems and it has been applied successfully in many fields. This paper presents a method to extract decision rules for Turbine-Generator Unit Fault Diagnosis. First, a decision table for fault diagnosis is obtained by discretization of continuous symptom attributes from original data; second, the discretized fault symptom attributes are reduced using rough set methodology; finally, a set of maximally generalized decision rules is generated by using a rule induction algorithm based on the symbolic value partition technique and the Generalized Distribution Table (GDT). The proposed method effectively reduces both the number of attributes and the size of attributes domains. Furthermore, it help computing smaller rule sets with better coverage and better classification accuracy rates compared with those of the attribute reduction approaches which only reduce the number of attributes.*

Keywords: *Knowledge discovery–Rule Induction–Rough sets–fault diagnosis.*

1. INTRODUCTION

The efficient use of a machine system requires reliable information about its current operating conditions. This information is often used as a basis for machine condition monitoring and fault diagnosis [11]. In reality, the raw knowledge gathered from a machine system may be inconsistent or redundant. In addition, many extracted fault symptom attributes are needed in order to completely describe the fault condition of the machine. But the importance of different attributes is often varied and some of the symptom attributes are even redundant, which is a waste of diagnostic resources. Using these redundant attributes will directly affect the generation of simple, effective and efficient decision rules and may reduce the validity of fault diagnosis. It is obvious that inconsistency and redundancy of symptom attributes are common problems in fault diagnosis. Establishing a method of extracting simple, effective and efficient decision rules from the symptom attributes is very valuable in the practical application of fault diagnosis.

Many approaches have been proposed for extracting hidden relationships holding among pieces of information stored in a given database [12]. Rough set theory is a relatively new mathematical and AI technique introduced by Pawlak and Skowron [17], [15]. The technique is particularly suited to reasoning about imprecise or incomplete data and discovering relationships in this data. The main advantage of rough set theory is that it does not require any preliminary or additional information about data-like probability in statistics, basic probability assignment in DS theory or the value of possibility in fuzzy set theory.

2. PRELIMINARIES

2.1 Decision Tables

Data are often presented as a table, columns of which are labeled by attributes, rows by objects of interest and entries of the table are attribute values. This paper only concerns decision tables with only one decision attribute. Formally, a decision table is a triple $S = (U, C, \{d\})$ where $d \notin C$ is the decision attribute and elements of C are called conditional attributes or conditions for briefness. Table 1 lists a decision table where $U = \{x_1, x_2 \dots x_9\}$, $C = \{C_1, C_2, C_3\}$ and the decision attribute is d. It will be employed throughout the paper.

2.2 M-Relative Reduct

The number of relative reducts of a decision table may be large [1], [2], and there may exist many minimal reducts of a decision table. Bazan presented the concept of dynamic reduct [3] to obtain stable reducts.

Definition 1: Given a decision table $S = (U, C, \{d\})$ and a set of specified attributes $M \subseteq C$, any $B \subseteq C$ is called an M-relative reduct of S iff:

1. $M \subseteq B$;
2. $POS_B(\{d\}) = POS_C(\{d\})$;
3. $\forall\ a \in (B - M)$, $POS_{B - \{a\}}(\{d\}) \subset POS_C(\{d\})$.

The set of all M-relative reducts of S is denoted by Red (S, M).

Similar to the definition of a reduct [6], we propose two concepts as follows. A partition scheme P is consistent iff:

$POS_C^P(\{d\}) = POS_C(\{d\})$. A decision table S is unpartitionable iff there does not exist a partition scheme P such that P is consistent and rank $(S^P) <$ rank (S). A partition reduct P is optimal iff rank (S^P) is minimal.

2.3 Discretization

Data sets frequently have attributes with numerical domains that make them unsuitable for certain data mining algorithms that deal mainly with nominal attributes such as decision trees and rough set theory. To use such algorithms we need to replace numerical attributes with nominal attributes that represent intervals of numerical domains with discrete values. This process, known as "discretization" has received a great deal of attention in data mining literature and includes a variety of ideas ranging from fixed k-interval discretization [10], fuzzy discretization [8, 9], Shannon-entropy discretization due to Fayyad and Irani presented in [13, 14]. Many discretization methods have been put forward in the past few years. Existent methods can be classified by the following three different axes: global versus local, supervised versus unsupervised and static versus dynamic [14].

2.4 Scaling

In some theories, especially Formal Context Analysis [2], there is a need to transform a many-valued attribute into a number of binary valued attributes. This process is called scaling. The scaling process has no influence on the indiscernibility relations or positive regions of attribute sets. Here we require that the decision attribute not to be changed in the scaling process.

Definition 3: Given a decision table $S = (U, C, \{d\})$, the set of scaled attributes of $Q \subseteq C$ is:

$$Q_B = \{(a, v) | a \in Q, v \in V_a\} \quad \text{... (1)}$$

Where $(a, v): U \rightarrow \{0, 1\}$ and

$$(a, v)(u) = \begin{Bmatrix} 1 & \text{if } a(u) = v \\ 0 & \text{otherwise} \end{Bmatrix} \quad \text{... (2)}$$

2.5 Generalized Distribution Table

Generalized Distribution Table (GDT) is a table in which the probabilistic relationships between concepts and instances over discrete domains are represented [18]. Any GDT consists of three components: possible instances, possible generalizations of instances, and probabilistic relationships between possible instances and possible generalizations.

The possible instances, which are represented at the top row of GDT, are defined by all possible combinations of attribute values from a database, and the number of the possible instances is

$$\prod_{i=1}^{m} n_i \quad \text{... (3)}$$

Where m is the number of attributes, n is the number of different data values in each attribute.

The possible generalizations for instances, which are represented by the left column of a GDT, are all possible cases of generalization for all possible instances, and the number of the possible generalizations is

$$\left(\prod_{i=1}^{m}(n_i+1)\right)-\left(\prod_{i=1}^{m}n_i\right)-1 \qquad \text{... (4)}$$

A wild card `*' denotes the generalization for instances, for simplicity, the wild card will sometimes be omitted in the paper. For example, the generalization $a_0 * c_0$ means that the attribute b is superfluous (irrelevant) for the concept description. In other words, if an attribute b takes values from $\{b_0, b_1\}$ and both $a_0b_0c_0$ and $a_0b_1c_0$ describe the same concept, the attribute b is superfluous, i.e. the concept can be described by a_0c_0. Therefore, the generalization $a_0{*}c_0$ used to describe the set $\{a_0b_0c_0, a_0b_1c_0\}$.

The probabilistic relationships between possible instances and possible generalizations, represented by entries G_{ij} of a given GDT, are defined by means of a probabilistic distribution describing the strength of the relationship between every possible instance and every possible generalization. The prior distribution is assumed to be uniform if background knowledge is not available. Thus, it is defined by

$$G_{ij}=\begin{Bmatrix}\dfrac{1}{N_{PG_i}} & \text{if } PG_i \text{ is a generalization of } PI_j \\ 0 & \text{otherwise}\end{Bmatrix} \qquad \text{... (5)}$$

Where: PI_j is the jth possible instance, PG_i is the Ith possible generalization, and N_{PG_i} is the number of the possible instances satisfying the ith possible generalization, that is,

$$N_{PG_i}=\prod_{j}^{m}n_j \qquad \text{... (6)}$$

Where $j = 1, \dots, m$, and j # the attribute that is contained by the ith possible generalization (i.e., j just contains the attributes expressed by the wild card).

Rule Strength: In this paper, the rules are expressed in the following form: $X \rightarrow Y$ with S. That is, "if X then Y with strength S". Where: X: denotes the conjunction of the conditions that a concept must satisfy, Y: denotes a concept that the rule describes, and S: is a "measure of strength" of which the rule holds. The strength of a given rule reflects the incompleteness and uncertainty in the process of rule inducing influenced by both unseen instances and noise. It is defined by

$$S(X \rightarrow Y) = s(X) \cdot [1 - r(X \rightarrow Y)] \quad \text{... (7)}$$

where $s(X)$: The strength of the generalization X and r: noise rate function.

3. THE ALGORITHM

3.1 Algorithm Structure

The main structure of the algorithm is repeating the three steps, namely, scaling, reduction and converting back. Until the new decision table is unpartitionable. In other words, a single group partition reduct (SGPR) of the new decision table should be computed recursively and new values such as 2, 3... should be assigned to k in Eq. (8). By doing so an optimal partition reduct can be obtained [6], [4], where the optimal metric is defined by the cardinality sum of new attribute domains.

Definition 4: A partition scheme $P = [P_1 \ldots\ldots P_{|C|}]$ of S is called a single group partition scheme (SGPS) if for any i $\{1, \ldots, |C|\}$, $|W_{ai}| = 1$. Given an SGPS $P = [P_1, \ldots, P\ |C|]$, for any i $\{1, \ldots, |C|\}$, P_i essentially divides V_{ai} into two disjoint subsets $V_{a_i}^F$ and $V_{a_i}^G$, and

$$P_i(v) = \begin{Bmatrix} v & \text{if } v \in V_{a_i}^F \\ k & v \in V_{a_i}^G \end{Bmatrix} \text{ where } k \in w_{ai} \quad \text{... (8)}$$

In order to make the algorithm easier to comprehend, an example will be analyzed in the next section.

3.2 An Example

Before using rough set theory to diagnose mechanical fault, the attributes field must be specified to establish the decision table.

According to a typical vibration signal obtained from a turbine-generator unit, the frequency range of the vibration signal can be broadly divided into three hands based on the running frequency of the machine, namely $(0.3-0.44)f$, $(0.45-0.6)f$, $1f$, within each of the bands, the peak value of power spectrum is indicative of a fault symptom at a particular frequency. The fault symptom attribute set is denoted $C = \{c_1, c_2, c_3\}$, The decision attribute set are the three major types of mechanical faults denoted $D = \{D_1, D_2, D_3\}$ where D_1 is oil film oscillation, D_2 is imbalance and D_2 is asymmetry. The decision table is shown in Table 1

Step 1: *Scaling:* Table 2 lists the scaled decision table $S_B = (U, C_B, \{d\})$ of the decision table S listed in Table 1, where (c_1, s), (c_1, d), (C_3, n) stand for (c_1, low), (c_1, very high) ...., (C_3, no), respectively.

Step 2: *Reduction of the scaled decision table:* One can apply attribute reduction to the scaled decision table. For example, $R_1 = \{(c_1, vh), (c_1, m), (c_1, h), (c_2, l)\}$ is a minimal reduct of S_B listed in Table 2 and a reduced decision table $(U, R^1, \{d\})$ can be obtained.

Step 3: *Converting back to a "normal" decision table:* However, it is more interesting to convert $(U, R^1, \{d\})$ back to a "normal" decision table as listed in Table 3, where duplicated objects are removed. Because $(c1, l) \notin R^1$ and $(c1, e) \notin R^1$, we do not distinguish low from exceeded. From semantic point of view, low and exceeded can be replace by others, for the extending purpose 1 is used instead.

Step 4: Construct the partition scheme P^1: In fact, a partition scheme $P^1 = [\{(l, 1), (vh, vh), (m, m), (h, h), (e, e)\}, \{(l, l), (h, 1), (n, 1)\}, \{(y, 1), (n, 1)\}]$ could be constructed such that the decision table listed in Table 3 is just S^{P1}.

Step 5: *Repeating steps from Step 1 to Step 4:* The scaling, reduction and converting back process (where new values such as 2, 3 ... instead of 1 should be used) could be repeated until the cardinality of any attribute cannot be reduced further $S_B^{P_1}$ is obtained as listed in Table 4. $R^2 = \{(c1^{P1}, e), (c1^{P1}, h), (c2, 1)\}$ is a reduct of $S_B^{P_1}$ and S^{P2} is obtained as listed in Table 5 where $P^2 = [\{(l, 1), (vh, 2), (m, 2), (h, t), (e, 1)\}, \{(l, 2), (h, 1), (n, 1), \{(y, 1), (n, 1)\}]$. $S_B^{P_2}$ is listed in Table 6. $R^3 = \{(c1^{P2}, 1), (c1^{P2}, 2), (c2^{P2}, 1)\}$ is a reduct of $S_B^{P_2}$ and S^{P3} is obtained as listed in Table 7 where $P^3 = [\{(l, 1), (vh, 2), (m, 2),$

$(h, 3), (e, 1)\}, \{(l, 2), (h, 1), (n, 1)\}, \{(y, 1), (n, 1)\}]$. Since S^{P3} is unpartitionable, the whole process terminates and $P = P^3$ is a partition reduct.

Step 6: *Obtain the Optimal Single Group Partition (OSGP):* Since the whole process terminates and $P = P^3$ is an optimal partition reduct, R^3 is an Optimal Single Group Partition. A more comprehensive version of S^{P3} is listed in Table 8.

Step 7: *Create the GDT:* For simplicity, the attribute "$c3$" can be eliminated as shown from Table 8. Table 9 lists GDT for the decision table listed in Table 8. where *l*, *vh*, *h*, *e*, *m*, *h*, *n*, *L* stand for low, very high, high, exceeded, medium, high, normal, low, respectively.

Step 8: *Simplify the GDT:* By deleting all of the instances and generalizations un-appeared in Table 8 So the simplified GDT is shown in Table 10

Step 9: *Group the generalizations:* Generalizations can be divided into four groups contradictory, belonging to class *D*1, belonging to class *D*2, and belonging to class *D*3. The contradictory generalizations, containing the instances belonging to different decision classes, cannot be used as the rules. Hence they are ignored. In other words, we are just interested in the generalizations belonging to class *D*1 or *D*2 or *D*3, which will be selected as the rules.

Table 11 lists the generalizations belonging to class *D*1, Table 12 lists the generalizations belonging to class *D*3 and Table 13 lists the generalizations belonging to class *D*2.

Step 10: *Rule Selection:* The algorithm has an advantage that it allows to the user to select the decision rules based on the previous experience or any required demands. There are several possible ways for rule selection. For example:

- Selecting the rules that contain as many instances as possible.
- Selecting the rules in the levels of generalization as high as possible according to the number of " * " in a generalization.
- Selecting the rules with larger strengths.

Hence All the decision rules that can be induced are:

The induced decision Rules for class D1 are:

Rule 1: **If** *C*1 is very high **Then** *d* is ***D1***, with *S* = 1

Rule 2: **If** *C*1 is medium **Then** *d* is ***D1***, with *S* = 1

Rule 3: **If** *C*2 is is high **Then** *d* is ***D1***, with *S* = 2/5

Rule 4: **If** *C*2 is is normal **Then** *d* is ***D1***, with *S* = 2/5

The induced decision Rules for class D2 are:

Rule 1: If *C*1 is low Then *d* is **D2**, with *S* = 1/3

Rule 2: If *C*1 is high Then *d* is **D2**, *S* = 1

Rule 3: If *C*1 is exceeded Then *d* is **D2**, with *S* = 1/3

Rule 4: If *C*2 is high Then *d* is **D2**, with *S* = 1/5

Rule 5: If *C*2 is normal Then *d* is **D2**, with *S* = 1/5

Rule 6: If *C*2 is normal Then *d* is **D2**, with *S* = 1

The induced decision Rules for class D3 are:

Rule 1: If *C*1 is low Then *d* is **D3**, with *S* = 2/3

Rule 2: If *C*1 is exceeded Then *d* is **D3**, with *S* = 2/3

Rule 3: If *C*2 is high Then *d* is **D3**, with *S* = 3/5

Rule 4: If *C*2 is normal Then *d* is **D3**, with *S* = 1/5

Table 1: *A Decision Table S*

U	*C1*	*C2*	*C3*	*d*
***x*1**	low	low	yes	*D*2
***x*2**	Very high	high	no	*D*1
***x*3**	medium	high	yes	*D*1
***x*4**	medium	normal	yes	*D*1
***x*5**	high	normal	no	*D*2
***x*6**	high	normal	yes	*D*2
***x*7**	exceeded	normal	yes	*D*3
***x*8**	low	normal	no	*D*3
***x*9**	low	high	no	*D*3

Table 2: S_B, *the Scaled Decision Table of S*

u	(C1, l)	(C1, vh)	(C1, m)	(C1, h)	(C1, e)	(C2, l)	(C2, h)	(C2, n)	(C3, y)	(C3, n)	d
x1	1	0	0	0	0	1	0	0	0	1	D2
x2	0	1	0	0	0	0	1	0	0	1	D1
x3	0	0	1	0	0	0	1	0	1	0	D1
x4	0	0	1	0	0	0	0	1	1	0	D1
x5	0	0	0	1	0	0	0	1	0	1	D2
x6	0	0	0	1	0	0	0	1	1	0	D2
x7	0	0	0	0	1	0	0	1	1	0	D3
x8	1	0	0	0	0	0	0	1	0	1	D3
x9	1	0	0	0	0	0	1	0	0	1	D3

Table 3: S^{P^1}

U	$C1^{P1}$	$C2^{P1}$	$C3^{P1}$	d
x1	1	low	1	D2
x2	Very high	1	1	D1
x3	medium	1	1	D1
x5	high	1	1	D2
x7	1	1	1	D3

Table 4: $S_B^{P^1}$, *the Scaled Decision Table of* S^{P^1}

u	($C1^{P1}$, l)	($C1^{P1}$, d)	($C1^{P1}$, n)	($C1^{P1}$, t)	($C2^{P1}$, l)	($C2^{P1}$, l)	(C_3^{P1}, l)	d
x1	1	0	0	0	1	0	1	D2
x2	0	1	0	0	0	1	1	D1
x3	0	0	1	0	0	1	1	D1
x5	0	0	0	1	0	1	1	D2
x7	1	0	0	0	0	1	1	D3

Table 5: S^{P2}

U	$C1^{P1}$	$C2^{P1}$	$C3^{P1}$	d
x1	1	2	1	*D*2
x2	2	1	1	*D*1
x5	teacher	1	1	*D*2
x7	1	1	1	*D*3

Table 6: S_B^{P2}, *the Scaled Decision Table of* S^{P2}

u	$(C1^{P2}, 1)$	$(C1^{P2}, 2)$	$(C1^{P2}, t)$	$(C2^{P2}, 1)$	$(C2^{P2}, 2)$	$(C3^{P2}, 1)$	d
x1	1	0	0	0	1	1	*D*2
x2	0	1	0	1	0	1	*D*1
x5	0	0	1	1	0	1	*D*2
x7	1	0	0	1	0	1	*D*3

Table 7: $S^B = S^{P3}$

U	$C1^{P3}$	$C2^{P3}$	$C3^{P3}$	d
x1	1	2	1	*D*2
x2	2	1	1	*D*1
x5	3	1	1	*D*2
x7	1	1	1	*D*3

Table 8: *A More Comprehensive Version of* S^{P3}

U	$C1$	$C2$	$C3$	d
x1	{low, exceeded}	Low	{yes, no}	*D*2
x2	{very high, medium}	{high, normal}	{yes, no}	*D*1
x5	high	{high, normal}	{yes, no}	*D*2
x7	{low, exceeded}	{high, normal}	{yes, no}	*D*3

Table 9: *The GDT for the Decision Table Shown in Table 8*
(***Note:*** The elements that are not displayed are all zero)

	lh	*lL*	*ln*	*vhh*	*vhL*	*vhn*	*hh*	*hL*	*hn*	*eh*	*eL*	*en*	*mh*	*mL*	*mn*
l*	1/3	1/3	1/3												
vh*				1/2		1/2									
h*							1/2		1/2						
e*										1/3	1/3	1/3			
m*													1/2		1/2
***h**	1/5			1/5			1/5			1/5			1/5		
***n**			1/5			1/5			1/5			1/5			1/5
***L**		1/2									1/2				

Table 10: *The Simplified GDT*
*(**Note:** The elements that are not displayed are all zero)*

	lh	*lL*	*ln*	*vhh*	*vhn*	*vhh*	*hn*	*eh*	*eL*	*en*	*mh*	*mn*
l*	1/3	1/3 s	1/3									
vh*				1/2	1/2							
h*						1/2	1/2					
e*								1/3	1/3	1/3		
m*											1/2	1/2
***h**	1/5			1/5		1/5		1/5			1/5	
***n**			1/5		1/5		1/5			1/5		1/5
***L**		1/2							½			

Table 11: *The Generalizations Belonging to Class D1*

	vhh	*vhn*	*mh*	*mn*
vh*	1/2	1/2		
m*			1/2	1/2
*** h**	1/5		1/5	
***n**		1/5		1/5

Table 12: *The Generalizations Belonging to Class D3*

	lh	*ln*	*eh*	*en*
l*	1/3	1/3		
e*			1/3	1/3
***h**	1/5		1/5	
***n**		1/5		1/5

Table 13: *The Generalizations Belonging to Class D2*

	lL	*hh*	*hn*	*eL*
l*	1/3			
h*		1/2	1/2	
e*				1/3
***h**		1/5		
***n**			1/5	
***L**	1/2			1/2

4. CONCLUSION

This paper introduces a fault diagnosis method for turbine generator unit based on rough set theory, in which minimal generalized decision rules can be extracted from original data, which includes inconsistent and redundant information. A rule extracted from a data set and the corresponding features can be considered as one of many models describing a data set.

A Turbine-generator unit vibration fault is used to illustrate the establishment and application of the method. The new algorithm is both efficient for the space and time complexities and tends to give suboptimal results. It helps obtaining small rule sets with good performance for most datasets.

REFERENCES

[1] A. Skowron, C. Rauszer, "The Discernibility Matrices and Functions in Information Systems", in: Slowinski [4], pp. 331–362.

[2] B. Ganter, R. Wille (Eds.), "Formal Concept Analysis: Mathematical Foundations", Springer-Verlag, New York, 1999.

[3] C. Xu, F. Min, Weighted reduction for decision tables, in: L. Wang, L. Jiao, G. Shi, X. Li, J. Liu (Eds.), Proceedings of The Third International Conference on Fuzzy Systems and Knowledge Discovery (FSKD 2006), LNCS 4223, Springer, 2006, pp. 246–255.

[4] Du Jim, Sun Caixin, Bi Weimin, et. al. "A Steam Turbine Generator Vibration Fault Diagnosis Method Based on Rough Set", *Proceeding of the 2002 International Conference on Power System Technology (Power Con)*, **3**, Kuming, China, October 2002, pp. 1532–1534.

[5] F. Min, Q. Liu, C. Fang, J. Zhang, "Reduction Based Symbolic Value Partition", *in: ICHIT 2006 Electronic Proceeding, Soft Computing and Rough Sets*, 2006, pp. 1–10.

[6] F. Min, Q. Liu, C. Fang, "Rough Sets Approach to Symbolic Value Partition", *Internat. J. Approx. Reason*, 2008.

[7] F. Min, Q. Liu, H. Tan, L. Chen, "The M-relative Reduct Problem", in: Wang et. al. [27], pp. 170–175.

[8] I. Kononenko, "Inductive and Bayesian Learning in Medical Diagnosis". *Appl Artif Intell*, **7**, (1993), 317–337.

[9] I. Kononenko, "Naive Bayes Classifier and Continuous Attributes Informatica, **16**, (1992), 1–8.

[10] J. Dougherty, R. Kohavi, M. Sahami., "Supervised and Unsupervised Discretization of Continuous Features". *Proc. of the 12th International Conference on Machine Learning*, (1995), 194–202.

[11] L.P. Khao, L.Y. Zhai, "A Rough Set Approach to the Treatment of Continuous Valued Attributes in Multi-concept Classification for Mechanical Diagnosis", Artificial Intelligence for Engineering Design, Analysis and Manufacturing, **15**, No. 3, pp. 211–221, 2001.

[12] Pawlak Z., Busse J.G., R. Slowinski and Ziarko W., "Rough Sets, Communication of the ACM", **38**: 11, pp. 89–95, 1995.

[13] Pawlak Z., and Skowron A., (2007a). "Rough Sets and Boolean Reasoning". *Information Sciences*, **177**, 41–73.

[14] Pawlak Z., and Skowron A., (2007b). "Rough Sets: Some Extensions". *Information Sciences*, **177**, 28–40.

[15] U. Fayyad, G. Piatetsky-Shapiro, P. Smyth, and R. Uthurusamy. "Advances in Knowledge Discovery and Data Mining". AAAI Press/MIT Press, 1996.

[16] U. Fayyad, K. Irani, "Multiinterval Discretization of Continuous-valued Attributes for Classification Learning", *Proc. of the 12th International Joint Conference on Artificial Intelligence*, (1993), 1022–27.

[17] U. Fayyad, "On the Induction of Decision Trees for Multiple Concept Learning". *Ph.D. thesis, University of Michigan*, (1991).

[18] Z. Weihua, Wei Zhang, Yunqing Fu, "An Incomplete Data Analysis Approach using Rough Set Theory", *Proceedings of the 2004 international Conference on Intelligent Mechatronics and Automation*, Chengdu , China August 2004.

18

MODELING DNA MUTATION USING FUZZY RELATIONS

Tazid Ali[1] **and Chandra Kanta Phukan**[2]

Department of Mathematics, Dibrugarh University, Assam-786004, India.

E-mail: tazidali@yahoo.com[1]

E-mail: chandra_kanta25@yahoo.com[2]

Abstract: *DNA mutation processes includes various types of mutations namely substitution, deletion, insertion and inversion of a section of the sequence. The processes have been modeled using the language of probability. In this paper we have attempted to generalize the existing probability model using fuzzy relations.*

Keywords: *DNA mutation,* ***S****ubstitution, deletion, insertion, fuzzy transition matrix, and membership matrix.*

1. INTRODUCTION

Molecular evolution is a wide range of biological processes. Depending on the Nature, species accumulate and lose external as well as internal characters, behaviors, etc. from generation to generation in the processes of evolution. In molecular level also species carry some underlying variability in their genetic set up, especially in DNA molecules of cells. Emphasis on DNA level mutation processes, a probability model has already been developed to deduce the effect genetic variability of ancestral generation on future generation.

In our study, we have developed a model to generalize the existing model, by removing some kind of restrictions from it, and have created fuzzy relations among the sites of the DNA sequences of two successive generations. In addition to this, the model has been extended to study three different types of mutation namely, deletion, insertion and both together respectively.

2. DNA BACKGROUND

DNA (Deoxyribonucleic acid) is found in the cells of all living organisms except plant virus. The main role of DNA in cell is long-term storage of genetic information, which is copied during cell division. DNA is composed of three types of compounds such as Sugar molecules (Pentose Sugar), Phosphoric acid and Four nitrogenous bases namely Adenine, Guanine, Cytosine, and Thymine represented by letters A, G, C and T respectively. Because of chemical similarity, Adenine and Guanine are contained in purines and that of Cytosine and Thymine in pyrimidines.

A DNA molecule forms a double helix, i.e. a twisted ladder-like structure. Each pole of the ladder is composed of sugar molecules and each sugar molecule attaches one phosphate group at c-3 and c-5 carbon atom of it alternatively. While c-1 carbon atom of sugar is attached by one of the four bases A, G, C, and T. The attached bases of these two poles can also be joined through hydrogen bond and these form the rungs of the ladder. The unique feature of base paring is $A \rightarrow T$, $T \rightarrow A$, $G \rightarrow C$ and $G \rightarrow C$. Therefore the sequence of bases of one side is enough to deduce the other. For, e.g. if bases along one pole is AGTCGCTA then the other have the complementary sequence TCAGCGAT. Some segments of DNA sequence that carry genetic information are called genes. Genes encodes instruction for the manufacturing of protein through messenger RNA. Three consecutive bases of genes forms codons, each codon specify a particular amino acid say GCT-Alamine.

3. FUZZY SETS AND FUZZY RELATION

A fuzzy set of a universal set X is a function $\mu: X \rightarrow (0, 1)$. For $x \in X$, $\mu(x)$ is called grade of membership of x. The height $h(\mu)$ of a fuzzy set A is the largest membership grade obtained by an element of that set. Formally, $h(\mu) = \sup_{x \in X} \mu(x)$. The fuzzy μ is said to be normal if $h(\mu) = 1$. If we divide each grade of membership of $x \in X$ by $h(\mu)$, then we get a normal fuzzy set. This process of obtaining a normal fuzzy set from a given fuzzy set is called normalization and the new fuzzy set obtained is called normalized fuzzy set. If we are concerned with only the relative grade of membership of different elements of the universal set, then we can replace a given fuzzy set by its normalized form.

For two universal sets X, Y a fuzzy relation from X to Y is a fuzzy subset of $X \times Y$, i.e. a function $\mu: X \times Y \rightarrow (0, 1)$. $\mu(x, y)$ will represent the degree of association of x to y. A convenient representation of fuzzy relation is by membership matrices $M = (r_{xy})$, where $r_{xy} = \mu(x, y)$. Another useful representation of fuzzy relation is a sagittal diagram. For a fuzzy relation μ from $X \rightarrow Y$, the inverse fuzzy relation $\mu^{-1}: Y \times X \rightarrow (0, 1)$ is defined as $\mu^{-1}(y, x) = \mu(x, y)$. Consequently the membership matrix of μ^{-1} is the transpose of the membership matrix of μ. Let be a fuzzy relation from $X \rightarrow Y$ and ν be a fuzzy relation from $Y \rightarrow Z$ then the composite relation from $X \rightarrow Z$ is given by $(\gamma\, o\, \nu)(x, z) = \max \{[\min \gamma(x, y), \nu(y, z)]: y\, e\, Y\}$. Other types of composition can also be used in fuzzy relation such as max-max, min-min, max-product, and max-average, etc. But the max-min composition has become the best known and the most frequently used one. The physical interpretation of the composition can be given as the strength of a set of chains linking x to z. Each chain has the form x-y-z. The strength of such a chain is that of the weakest link. The strength of the relation between x and z is that of the strongest chain between x and z.

4. OVERVIEW OF EXISTING PROBABILITY MODEL (1)

The most common mutation that occurs in the coping of DNA sequences is base substitutions. This is nothing but the replacement of a base by another at a certain site in the sequence. For example if AGATC in an ancestral sequence becomes ACATC in descendent, the only substitution occurs is $G \rightarrow C$. Focusing on base substitution (no deletion, insertion and inversion occurs) and ignoring the effect of codons in protein, a probability model has been carried out. Given an ancestral sequence, the probability of each of the four bases A, G, C and T to be present in the sequences are P_A, P_G, P_C, and P_T respectively and denoted by a vector P_o that is $P_o = (P_A, P_G, P_C, P_T)$, where $P_A + P_G + P_C + P_T = 1$ (sum of the total probability). This vector describes ancestral probability distribution of bases. It is observed that 16 conditional probabilities have been specified through out the processes. These are $P\,(S_1 = i/S_0 = j)$ for $i, j = A, G, C$ and T. Now these can be arranged in a 4×4 matrix of vectors to make it mathematically tractable, i.e.

$$\begin{array}{cc} & S_o \rightarrow \quad A \qquad G \qquad C \qquad T \\ \begin{array}{r} A \\ \uparrow G \\ S_1\, C \\ T \end{array} & \begin{pmatrix} P_{A/A} & P_{A/G} & P_{A/C} & P_{A/T} \\ P_{G/A} & P_{G/G} & P_{G/C} & P_{G/T} \\ P_{C/A} & P_{C/G} & P_{C/C} & P_{C/T} \\ P_{T/A} & P_{T/G} & P_{T/C} & P_{T/T} \end{pmatrix} = M\,(\text{say}) \end{array}$$

where, $P_{x/y} = P\,(S_1 = x/S_o = y) = \dfrac{P\left(S_1 = x \text{ and } S_o = y\right)}{P\left(S_o = y\right)}$

The matrix $M_{4 \times 4}$ is considered as transitional matrix. On ordinary multiplication of matrix $M_{4 \times 4}$ with $P_{o\,4 \times 1}$ (the column matrix of initial probability), one 4×1 matrix can be obtained and denoted it by P_1, which gives the probabilities of various bases occurring in the sequences S_1.

5. GENERALIZED MODEL FOR DNA MUTATION

Suppose we are given a DNA sequence having m sites in S_o generation as

$$S_o = AGTCGATCCGTAGAGG.......(m \text{ site}),$$

where mutation will occur during molecular evolution. Now the effect of mutations of S_o generation in the next S_1 generation is noticeable. We assume the sequence to be a series of m distinct holes, where each hole can be filled up by one of the four bases A, G, C and T. And each site is independent from any other site in the sequence, i.e. biologically omitting the effect of codons in protein in the sequences. The previous model has the restriction of sum of the probabilities equal to 1. In our model we will relax this restriction by defining some degree of membership, which takes values in [0, 1] but there is no restriction on total sum.

5.1 Model for Base Substitution

As already mentioned a DNA sequence can be considered a sequence of m distinct holes. The nth site of S_o generation will be denoted by N_o. Concentrating on an arbitrary nth hole that is nth site of the sequence, we can fill up this by any one of four bases A, G, C, and T with some grade of membership say, $\mu_o(A)$, $\mu_o(G)$, $\mu_o(C)$, $\mu_o(T)$, i.e. we can consider the nth site as fuzzy subset of $\{A, G, C, T\}$. This can also be looked upon as fuzzy relation $\mu_o: N_o \to \{A, G, C, T\}$. Similarly the nth site of S_1 generation will be denoted by N_1. There the membership of A, G, C, T will be denoted by $\mu_1(A)$, $\mu_1(G)$, $\mu_1(C)$, $\mu_1(T)$ which also gives a fuzzy relation say $\mu_1: N_1 \to \{A, G, C, T\}$. Let

$$\begin{bmatrix} \mu_o(A) \\ \mu_o(G) \\ \mu_o(C) \\ \mu_o(T) \end{bmatrix} = M_o \qquad \begin{bmatrix} \mu_1(A) \\ \mu_1(G) \\ \mu_1(C) \\ \mu_1(T) \end{bmatrix} = M_1$$

M_o and M_1 will be called membership matrix of S_o generation and S_1 generation respectively. When the mutation takes place, any of the bases A, G, C, T can mutate to any one of A, G, C, T. There will be 16 possible mutations. This can be represented by a fuzzy relation μ_{ij}: $\{A, G, C, T\} \times \{A, G, C, T\} \to [0, 1]$ $i = A, G, C, T$ $j = A, G, C, T$. Now the fuzzy relation μ_1 can be obtained as a composite of fuzzy relations μ_o and μ_{ij}, which can be express by a sagittal diagram as below

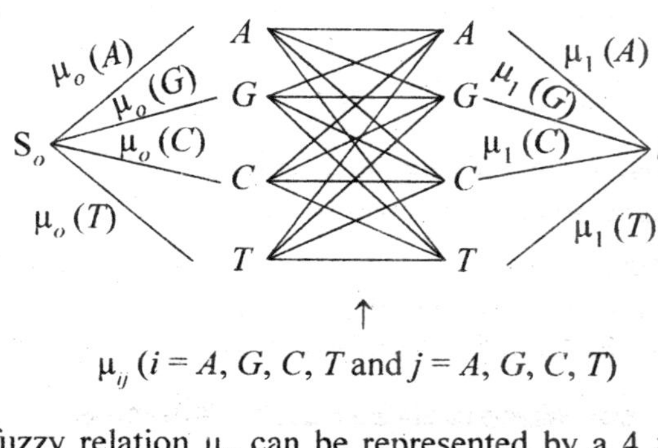

The fuzzy relation μ_{ij} can be represented by a 4×4 matrix as shown below. We call this matrix a fuzzy transition matrix (FTM).

$$\begin{array}{c} \\ A \\ G \\ C \\ T \end{array}\begin{array}{c} \begin{array}{cccc} A & G & C & T \end{array} \\ \begin{pmatrix} \mu_{AA} & \mu_{GA} & \mu_{CA} & \mu_{TA} \\ \mu_{AG} & \mu_{GG} & \mu_{CG} & \mu_{TG} \\ \mu_{AC} & \mu_{GC} & \mu_{CC} & \mu_{TC} \\ \mu_{AT} & \mu_{GT} & \mu_{CT} & \mu_{TT} \end{pmatrix} \end{array} = F$$

On operating the FTM F and the membership matrix M_o, applying max.min composition, we have obtained one 4×1 matrix M_1. And each entry of M_1 represents the membership grade for nth site (N_1) of S_1 generation. This is given by

$$\mu_1(j) = \max [\min \{(\mu_{Aj}, \mu_o(A)), (\mu_{Gj}, \mu_o(G)), (\mu_{Cj}, \mu_o(C)), (\mu_{Tj}, \mu_o(T))\}], j = A, G, C, T.$$

5.2 Model for Substitution and Deletion

This model is an extension of 5.1. In this case we assume the nth site of S_1 generation may disappear in the process of mutation although it is already being filled by one of A, G, C and T in S_o. We introduce a letter B as one more site in S_1 to represent the disappearances. In a similar manner of 5.1 the FTM is defined as relation from the (A, G, C, T) to (A, G, C, T, B). The corresponding matrix is shown below

$$\begin{array}{c} \\ A \\ G \\ C \\ T \\ B \end{array}\begin{array}{c} \begin{array}{cccc} A & G & C & T \end{array} \\ \begin{pmatrix} \mu_{AA} & \mu_{GA} & \mu_{CA} & \mu_{TA} \\ \mu_{AG} & \mu_{GG} & \mu_{CG} & \mu_{TG} \\ \mu_{AC} & \mu_{GC} & \mu_{CC} & \mu_{TC} \\ \mu_{AT} & \mu_{GT} & \mu_{CA} & \mu_{TT} \\ \mu_{AB} & \mu_{GB} & \mu_{CB} & \mu_{TB} \end{pmatrix} \end{array} = F$$

Now applying max.min composition between F and M_o we will get a 5×1 membership matrix whose entries will be

$\mu_1(j) = \max [\min \{(\mu_{Aj}, \mu_o(A)), (\mu_{Gj}, \mu_o(G)), (\mu_{Cj}, \mu_o(C)), (\mu_{Tj}, \mu_o(T))\}], j = A, G, C, T, B.$ And they represent the membership grades of the bases A, G, C, and T in the nth site of S_1.

5.3 Model for Substitution and Insertion

To model this situation we have introduced a new membership grade in an arbitrary site of S_o to be a blank site, where the insertion may take place. Here membership matrix M_o of S_o will have five components $[\mu_o(A), \mu_o(G), \mu_o(C), \mu_o(T), \mu_o(B)$ and the FTM is of 4×5 order, i.e.

$$\begin{matrix} & A & G & C & T & B \\ \begin{matrix} A \\ G \\ C \\ T \end{matrix} & \left(\begin{matrix} \mu_{AA} \\ \mu_{AG} \\ \mu_{AC} \\ \mu_{AT} \end{matrix}\right. & \begin{matrix} \mu_{GA} \\ \mu_{GG} \\ \mu_{GC} \\ \mu_{GT} \end{matrix} & \begin{matrix} \mu_{CA} \\ \mu_{CG} \\ \mu_{CC} \\ \mu_{CT} \end{matrix} & \begin{matrix} \mu_{TA} \\ \mu_{TG} \\ \mu_{TC} \\ \mu_{TT} \end{matrix} & \left.\begin{matrix} \mu_{BA} \\ \mu_{BG} \\ \mu_{BC} \\ \mu_{BT} \end{matrix}\right) \end{matrix} = F \quad \text{Here} \quad M_o = \begin{pmatrix} \mu_o(A) \\ \mu_o(G) \\ \mu_o(C) \\ \mu_o(T) \\ \mu_o(B) \end{pmatrix}$$

Similarly applying max.min composition we will get a 4×1 membership matrix of S_1 with entries $\mu_1(j) = \max[\min\{(\mu_{Aj}, \mu_o(A)), (\mu_{Gj}, \mu_o(G)), (\mu_{Cj}, \mu_o(C)), (\mu_{Tj}, \mu_o(T)), (\mu_{Bj}, \mu_o(B))\}], j = A, G, C, T.$

Combination of all the above three cases gives us an extended model, where membership matrix of S_o has five components $\mu_o(A)$, $\mu_o(G)$, $\mu_o(C)$, $\mu_o(T)$, $\mu_o(B)$ with a 5×5 order FTM as

$$\begin{matrix} & A & G & C & T & B \\ \begin{matrix} A \\ G \\ C \\ T \\ B \end{matrix} & \left(\begin{matrix} \mu_{AA} \\ \mu_{AG} \\ \mu_{AC} \\ \mu_{AT} \\ \mu_{AB} \end{matrix}\right. & \begin{matrix} \mu_{GA} \\ \mu_{GG} \\ \mu_{GC} \\ \mu_{GT} \\ \mu_{GB} \end{matrix} & \begin{matrix} \mu_{CA} \\ \mu_{CG} \\ \mu_{CC} \\ \mu_{CT} \\ \mu_{CB} \end{matrix} & \begin{matrix} \mu_{TA} \\ \mu_{TG} \\ \mu_{TC} \\ \mu_{TT} \\ \mu_{TB} \end{matrix} & \left.\begin{matrix} \mu_{BA} \\ \mu_{BG} \\ \mu_{BC} \\ \mu_{BT} \\ \mu_{BB} \end{matrix}\right) \end{matrix} = F \quad \text{and}$$

$\mu_1(j) = \max[\min\{(\mu_{Aj}, \mu_o(A)), (\mu_{Gj}, \mu_o(G)), (\mu_{Cj}, \mu_o(C)), (\mu_{Tj}, \mu_o(T)), (\mu_{Bj}, \mu_o(B))\}], j = A, G, C, T, B.$

The entry μ_{BB} of FTM represents the grade of association of a blank site of S_o to a blank site in the sequence of S_1. Finally this model becomes the most generalized model in our study.

6. CONCLUSION

In this paper we have attempted to construct a model by which we can predict the membership matrix of the next generation, if we know the membership matrix of the previous generation. The model involves many membership values, e.g. μ_{AA}, μ_{GA}, μ_{CA}, etc. These values will vary from species to species. Even for the same species these values will be influenced by various biological factors. So these values will have to be supplied by experts in the field. However we can make certain assumption to make the model a bit simpler. For example it is usually observed that there is a natural tendency for a base to remain unchanged. So we can assume that $\mu_{AA} \geq \{\mu_{AG}, \mu_{AC}, \mu_{AT}, \mu_{AB}\}$. Another assumption that we can make is that $\mu_{AA} = \mu_{GG} = \mu_{CC} = \mu_{TT} = \mu_{BB}$. Again we are more concerned with relative membership values in the membership matrix so we can divide each entry of the transition matrix by μ_{AA}. The third assumption we can make is that $\mu_{AG} = \mu_{GA}$ and etc. This assumption is in fact essential since it is inherent in the very definition of inverse of a fuzzy relation. With the above assumptions our transition matrix will be a symmetric matrix with 1 in each entry of its diagonal. Our model will in fact predict membership matrix of the previous generation if we are provided with information on the membership matrix of the new generation. Further we note that if we replace entries of the membership matrix by probability and conditional probabilities as entries of the fuzzy transition matrix, i.e. the grade of associations as conditional probability, then we get the probability model as a particular case of our generalized model.

REFERENCES

[1] Elizabeth S. Allman, Jhon A. Rhodes, (2004), "Mathematical Modeling in Biology, an Introduction", Cambridge University Press.

[2] George J. Klir, Tina A. Floger (1993), "Fuzzy Set, Uncertainty and Information", Prentice-Hall of India Privet Limited, New Delhi.

[3] Richard Durrett, "Probability and its Application, Probability Model for DNA Sequence Evolution", Second Edition, Springer Science + Business Media, LLC233 Springer Street, New York, NY 10013, USA.

[4] Veer Bala Rastogi, "Fundementals of Molecular Biology", Ane Book India, 4821, Parwana Bhawan, 1st Floor, 24 Ansari Road, Darya Ganj, New Delhi, India.

19

COMPUTATION OF BOUNDS OF RISK IN CONTAMINANT TRANSPORT THROUGH GEOLOGICAL REPOSITORY USING TYPE 2 FUZZY SET

D. Datta and H.S. Kushwaha

Health Physics Division, Bhabha Atomic Research Centre, Mumbai-400085, India,

E-mail: ddatta@barc.gov.in

Abstract: *This study developed the risk assessment approach based on fuzzy modeling for analyzing the time dependent environmental risk arising from water-borne nuclide migration from a geological repository. Model supporting the case study includes the most important aspects of radionuclide migration, contains components representing the source term, geosphere transport and biosphere transport for a single member radionuclide chain such as Technetium (Tc^{99}). The fuzzy modeling is used to examine the impacts of system uncertainties on model outputs as the model contains imprecise physical parameters. The risk assessment based on fuzzy modeling is useful for comprehensively evaluating the bounds of risk associated with the transport of such contaminants through geological media. In this paper, fuzzy parameters of the model are addressed by Type 2 fuzzy sets. The lower and upper membership functions of the input fuzzy variables and the vertex method are used to calculate the Type 2 membership values of the model output. Paper describes the application of Type 2 fuzzy set in evaluating the risk bounds.*

Keyword: *Risk Assessment, Type 2 fuzzy sets, Geological repository.*

1. INTRODUCTION

Subsurface contamination by leakage and spill of radioactive liquid effluent from any nuclear industry has resulted in many environmental concerns for a period of many years. Environmental risks of such contamination need to be analyzed to provide support for decisions related to remediation of the contamination problems. However, many models and relevant field studies have recognized that the contaminant fate in subsurface is substantially influenced by uncertainties inherent in natural porous media which makes the related risk assessment complicated and difficult to quantify accurately. Therefore, handling uncertainties in this domain is essential for generating reliable and realistic outcomes. Mathematical models used for carrying out risk assessment in this domain (risk analysis of a waste repository) include imprecise parameters having values varying within some ranges. Quantification of related uncertainties in risk assessment so far has been carried out using probability theory, in which risks are computed as the sum total of the product of magnitude of possible consequences or contaminations and the corresponding probabilities. Examples of probabilistic approach of risk assessment are elsewhere found [1]. In this context, the assessment and presentation of the effects of uncertainty are widely recognized as important parts of analyses for such complex systems [1]. Generally, most of the earlier studies used the stochastic approaches to accommodate the uncertainties associated with contaminant transport models [2, 3]. However, probability theory often suffers from lack of data or imperfect knowledge of the model parameters, and hence is not applicable for risk assessment using imprecise measurement of physical parameters of a representative model [3]. Risk assessment of the model containing imprecise parameters is carried out using fuzzy set theory [4]. A number of risk assessment studies for petroleum waste management through fuzzy set theory have been reported. For example, an interval parameter fuzzy relation analysis model for environmental risk assessment of petroleum-contaminated aquifers has been developed by Huang et. al. [5]. Therefore, fuzzy-set techniques are not only powerful in handling uncertainties associated with the transport modeling process, but also capable of quantifying environmental risks through fuzzy membership grades. However, in all the previous studies simulation of contaminant transport under uncertainty has been reported using Type 1 fuzzy sets. In the Type 1 fuzzy modeling, the values used in developing the membership functions of Type 1 fuzzy numbers are often overly precise, because they require that each element of the universe should be assigned a specific value of membership [5, 6]. But, when the level of information is not adequate to specify membership functions with this precision, it is necessary to use Type 2 fuzzy numbers to represent the uncertainties of the model variables [5, 6]. The computation of functions of Type 1 fuzzy numbers is a well-known field of fuzzy set theory and has been documented in several publications [6]. However, usage of Type 2 fuzzy number for risk assessment in the present field of interest is not available.

The paper aims to estimate the bounds of the risk associated with the migration of radionuclide transport through geological waste repository. The structure of the paper is as follows. In Section 2, the general definition and characteristics of Type 2 fuzzy sets are described. Section 3 presents the computation of functions of Type 2 fuzzy sets. Section 4 describes a case study to demonstrate the utility of Type 2 fuzzy sets in environmental risk assessment. Finally, conclusions are presented in Section 5.

2. FUZZY SETS (TYPE 1 AND TYPE 2)

2.1 Type 1 Fuzzy Sets

A conventional Type 1 fuzzy set A, defined on the universe of discourse X is represented by an ordered set of pairs $(x, \mu(x))$, where, the element $x \varepsilon X$ and $\mu(x)$ denotes the degree of membership [6]; $\mu(x)\ \varepsilon\ [0, 1]$. Membership function, $\mu(x)$ in Type 1 fuzzy set is characterized as two dimensional membership function which is totally crisp. Alternate mathematical representation of Type 1 fuzzy set, A can be written as

$$A = \{(x, \mu_A(x)) \mid \forall x \in X\}, \quad A = \int_{x \in X} \mu_A(X)/x \quad \ldots (1)$$

2.2 Type 2 and Interval Type 2 Fuzzy Sets

A Type 2 fuzzy set $\tilde{A}$ is represented by a three-dimensional membership function called as Type 2 membership function which itself is fuzzy [6]. Mathematical representation of Type 2 fuzzy set $\tilde{A}$ is written as

$$\tilde{A} = \{((x, u), \mu_{\tilde{A}}(x, u)) \mid \forall x \in X, \forall u \in J_x \subseteq [0,1]\},$$

$$\widetilde{A} = \int_{x\in X}\int_{u\in J_x} \mu_{\widetilde{A}}(x,u)/(x,u) \quad \text{... (2)}$$

where, $\mu_{\widetilde{A}}(x)$ is the secondary membership function (Type 1 fuzzy set) for a generic element $x \in X$ and can be represented by $\mu_{\widetilde{A}}(x) = \int_{u\in J_x} f_x(u)/u$, where, $J_x \subseteq [0,1]$. The union ($\cup$) of all primary membership of $\widetilde{A}$ is a bounded region called fingerprint of uncertainty (FOU) which is defined as

$$FOU(\widetilde{A}) = \bigcup_{x\in X} J_x,\quad \overline{\mu}_{\tilde{A}}(x) = \overline{FOU(\tilde{A})} = \bigcup_{x\in X} \overline{J_x},$$

$$\underline{\mu}_{\tilde{A}}(x) = \underline{FOU(\tilde{A})} = \bigcup_{x\in X} \underline{J_x} \quad \text{... (3)}$$

Type 2 fuzzy set, $\widetilde{A}$ in discrete space of crisp values can be expressed by

$$\widetilde{A} = \sum_{i=1}^{n}\left[\sum_{k=1}^{m} f_{x_i}(u_{ik})/u_{ik}\right]/x_i \quad \text{... (4)}$$

Details of Type 2 fuzzy set, interval Type 2 fuzzy numbers can be elsewhere found in [7]. If $f_x(u) = 1,\ \forall u \in J_x \subseteq [0,\ 1],\ \forall x \in X$, [Eq. (4)], the secondary membership functions are interval sets and $\widetilde{A}$ is called an interval Type 2 fuzzy set. Interval secondary membership functions reflect a uniform uncertainty at the primary membership of x. This feature popularizes interval Type 2 fuzzy sets as the most widely used Type 2 fuzzy sets to date [8].

2.3 Interval and Functions of Type 2 Fuzzy Numbers

A Type 2 fuzzy number is broadly defined as a Type 2 fuzzy set that has a numerical domain. An interval Type 2 fuzzy set can be defined as $A_\alpha = \{[a^\alpha, b^\alpha], [c^\alpha, d^\alpha]\},\ \forall\alpha \in [0,\ 1],\ \forall a^\alpha, b^\alpha, c^\alpha, d^\alpha \in R$. In this work, we have considered interval Type 2 lognormal fuzzy numbers with geometric mean, (GM) and uncertain geometric standard deviation, σ (GSD). Lower and upper membership functions satisfy their convexity and normality property with this definition of

Type 2 fuzzy numbers. Lower and upper membership functions of an interval Type 2 lognormal fuzzy number with uncertain GSD can be expressed as

$$\bar{\mu}_{\tilde{A}}(x) = \exp\left[-\frac{1}{2}\left(\frac{\ln x - \log(GM)}{(\sigma_2)}\right)^2\right] \text{ and}$$

$$\underline{\mu}_{\tilde{A}}(x) = \exp\left[-\frac{1}{2}\left(\frac{\ln x - \log(GM)}{(\sigma_1)}\right)^2\right] \quad \text{... (5)}$$

Fuzziness of any physical variable occurred in real practice is described by a fuzzy function. Three kinds of fuzzy functions occur: (*a*) Ordinary functions having fuzzy properties or satisfying fuzzy constraints, (*b*) functions that just "carry" the fuzziness in their argument(s) without generating extra fuzziness themselves and (*c*) ill-known functions of nonfuzzy arguments [9].

3. COMPUTATION OF TYPE 2 FUZZY SETS USING VERTEX METHOD

Calculation of Type 2 membership functions described in Section 2 can be carried out using the vertex method [10], the fuzzy weighted average (FWA) method [10] or transformation method [10]. For the present study, we have used the vertex method. The vertex method is based on the combination of the α-cut concept and standard interval analysis. The endpoint pairs of each interval intersect in the *n*-dimensional space and form the vertices of the *n*-dimensional Cartesian space. The coordinates of these vertices are the values used in the vertex method to calculate the output interval for each α-cut.

4. RISK ASSESSMENT MODEL-CASE STUDY

An environmental risk assessment model is considered as a case study for computing the bounds of risk using Type 2 fuzzy sets. Model demonstrates the time dependent risk evolving from migration of water-borne radionuclide (Technetium 99) through a repository [11, 12]. Model includes the radionuclide migration, advection, dispersion, equilibrium sorption and radioactive decay [13]. Analytical expression of the time dependent risk [13] is given by

$$C(t) = 0.5 * B * k * M_0 * \exp(-\lambda t) * \exp(L/2d) * \exp(-RL^2/4d\nu t) * \exp(-\nu t/4dR) * \Omega \quad \text{... (6)}$$

where,

$$\Omega = [\phi(z_1) + \phi(z_2)], \ \phi(z) = \exp(z^2) * erfc(z),$$

$$z_1 = \left(\frac{RL^2}{4d\nu t}\right)^{1/2} + \left(\frac{\nu t}{4dR} - kt\right)^{1/2}, \text{ and}$$

$$z_2 = \left(\frac{RL^2}{4d\nu t}\right)^{1/2} - \left(\frac{\nu t}{4dR} - kt\right)^{1/2}$$

B = biosphere risk conversion factor (deaths/Bq), k = fractional release rate (/y), M_0 = Initial inventory (Bq), λ = decay constant of radionuclide (/y), L = geosphere path length (m), R = retardation factor, d = dispersivity (m), ν = groundwater velocity (m/y) and T_0 = containment period (y). The time dependent consequence, $C(t)$ (Risk) in Eq. (6) means the probability of deaths per year-beyond the initial containment period ($t > T_0$). Parameters, k, ν and B are expressed as Type 2 lognormal fuzzy number. Lower and upper membership values of these parameters are computed with median as the most likely and 5% and 95% of the standard deviation as two different spreads, σ_1 and σ_2. Table 1 presents the lower and upper interval of all these parameters for various alpha cuts (0.2, 0.3, 0.5, 0.7 and 0.9). Tables 2*A* and 2*B* show the values of the fixed and fuzzy parameters used in calculation. Vertices are constructed (2^3 = 8 for 3 parameters) from the combination of [k-low, k-high, ν-low, ν-high, B-low and B-high] at the specified alpha cuts. We have computed the lower and upper values of time dependent risk, $C(t)$ for a time period of t = 20000 years at each alpha cuts using all these vertices and fuzzy vertex method. Table 3 presents the lower and upper values of risk at various alpha cut levels. Figures 1 and 2 present the cumulative probability plot of computed minimum and maximum values of risk normalized to 1.0^{-6} deaths/year. Plot of the membership function of the fuzzy risk is shown in Fig. 3. Shape of the membership function tells that the model output (risk) also follows a log-normal pattern.

Table 1: *Lower and Upper Intervals of Type 2 Fuzzy Parameters*

Alpha cut	*k-low*	*k-high*	*v-low*	*v- high*	*B-low*	*B-high*
0.2	3.38e-4	1.67e-4	0.1035	0.1926	1.109e-18	7.112e-18
0.3	3.35e-4	9.78e-5	0.1030	0.1762	1.093e-18	5.456e-18
0.5	3.31e-5	7.45e-5	0.1023	0.1537	1.070e-18	3.623e-18
0.7	3.26e-5	5.85e-5	0.1016	0.1361	1.050e-18	2.518e-18
0.9	3.22e-5	4.41e-5	0.1009	0.1183	1.027e-18	1.652e-18

Table 2A: *Values of the Parameters Used in the Model*

Parameters	*Description*	*Fixed values*
M_0	Initial inventory (Bq)	5.0e16
T_0	Containment period (y)	316
λ	Decay constant ($/y$)	3.25e-6
R	Retardation factor	10.0
d	Dispersivity (m)	20.0
L	Geosphere path length (m)	316

Table 2B: *Values of Fuzzy Parameters*

Fuzzy parameters	*Description*	*Median*	*Spread*	*Sigma 1*	*Sigma 2*
k	Fractional release rate ($/y$)	3.16e-5	0.766	0.038	0.728
v	Groundwater velocity (m/y)	0.1	0.384	0.019	0.365
B	Biosphere conv. Factor (deaths/y)	1.0e-18	1.151	0.058	1.093

Table 3: *Minimum and Maximum Risk*

Alphacut	*Risk (deaths/y) (Minimum)*	*Risk (deaths/y) (Maximum)*
0.2	8.98e-8	3.69e-6
0.3	7.62e-8	3.02e-6
0.5	2.92e-8	6.75e-7
0.7	2.73e-8	3.03e-7
0.9	2.56e-8	1.04e-7
1.0	2.35e-8	2.35e-8

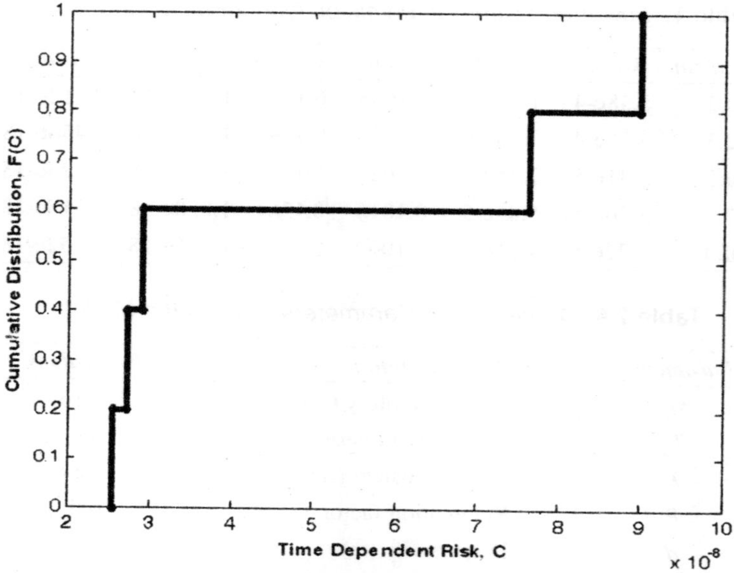

Fig. 1: *Cumulative Probability Plot of Minimum of Risk*

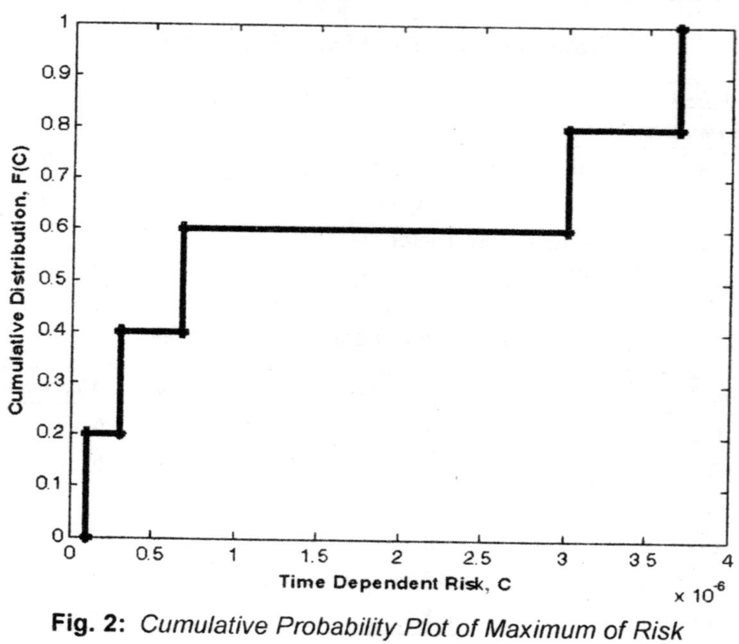

Fig. 2: *Cumulative Probability Plot of Maximum of Risk*

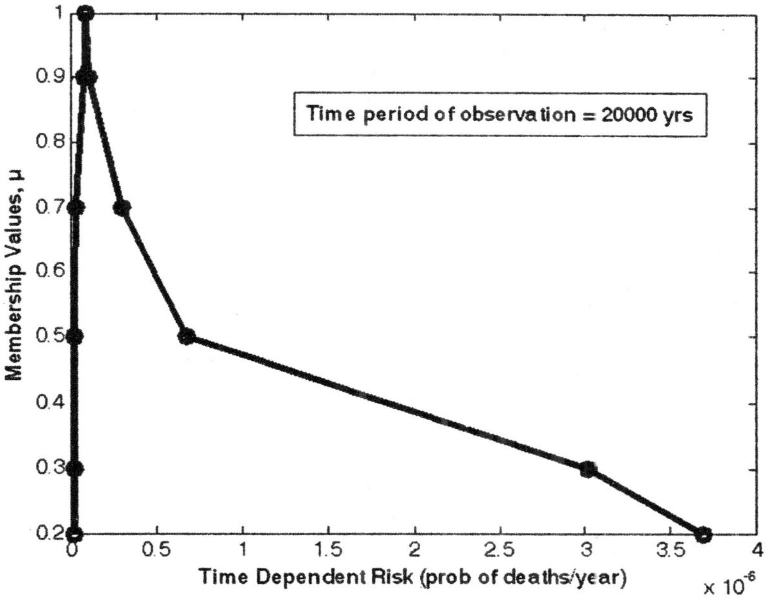

Fig. 3: *Membership Function of Risk*

5. CONCLUSIONS

In this work, usage of Type 2 fuzzy set in the form of functions of interval Type 2 fuzzy number for quantifying the uncertainty of the model parameters in assessing the bounds of risk has been presented. Interval Type 2 fuzzy sets can be interpreted as a collection of an uncountable number of embedded Type 2 fuzzy sets. Interval Type 2 lognormal variables for uncertain parameters of the representative model are used to represent the upper and lower membership functions of Type 2 fuzzy number. Bounds of risk may facilitate the user to calculate belief and plausibility in the risk analysis, thus providing a robust estimation methodology of uncertainty analysis [14]. Moreover, interval Type 2 fuzzy numbers have the advantage to be general enough to facilitate the modeling of the available approximate knowledge. Results of risk bounds are evaluated for only one time period of observations, i.e. $t = 20000$ yrs. Results for the case study indicated that the fuzzy parameters will have large influence on modeling outputs, where the degrees of influences varied under different conditions. Computations may repeat for different time period of observations to generate a spectrum of risk bounds. The results obtained by using the proposed approach may provide a tool to assess the performance criteria of waste disposal facility or geological repository.

REFERENCES

[1] NCRP (National Council on Radiation Protection and Measurements), "A Guide for Uncertainty Analysis in Dose and Risk Assessments Related to Environmental Contamination". *NCRP Commentary No. 14, Bethesda, MD: National Council on Radiation Protection and Measurements*; 1996.

[2] Chen, Z., Huang, G.H., Chakma, A., "Hybrid Fuzzy-Stochastic Modeling Approach for Assessing Environmental Risks at Contaminated Groundwater Systems", *Journal of Environmental Engineering*, **129 (1)**, pp. 79–88, 2003.

[3] Vose, D., "Quantitative Risk Analysis-A Guide to Monte Carlo Simulation Modeling", Wiley, New York, 1996.

[4] Chen, Z., Huang, G.H., Chakma, A., "Integrated Environmental Risk Assessment for Petroleum-Contaminated Sites–A North American Case Study", *Water Science and Technology*, **38**, pp. 131–138, 1998.

[5] Huang, G.H., Chen, Z., Tontiwachwuthikul, P. and Chakma, A., "Environmental Risk Assessment for Underground Storage Tanks Through an Interval Parameter Fuzzy Relation Analysis Approach", *Energy Sources*, **21**, pp. 75–96, 1999.

[6] T.J. Ross, "Fuzzy Logic with Engineering Applications", 2nd Ed. New York: Wiley, 2004.

[7] Mendel, J. M. and R.I. John, "Type-2 Fuzzy Sets Made Simple", *IEEE Trans. Fuzzy Systems,* **10**, No. 2, pp. 117–127, April 2002.

[8] J. Romero Aguero, and A. Vargas, "Calculating Functions of Interval Type-2 Fuzzy Numbers for Fault Current Analysis", *IEEE Trans. Fuzzy Sysems,* **15**, No. 1, pp. 31–40, Feb., 2007.

[9] D. Dubois and H. Prade, "*Fuzzy Sets Syst.*", New York: Academic, 1980.

[10] Dong, W. and H. Shah, "Vertex Method for Computing Functions of Fuzzy Variables", *Fuzzy Sets Syst.*, **24**, No. 1, pp. 65–78, Oct. 1987.

[11] Robinson, P.C. and D.R. Hodgkinson, "Exact Solutions for Radionuclide Transport in the Presence of Uncertainties", *UK Atomic Energy Agency Report No. AERE R 12125*, 1986.

[12] Mishra S., "Sensitivity Analysis with Correlated Inputs–An Environmental Risk Assessment Example", *Paper Presented in the Proceedings of the Crystal Ball User Conference*, 2004.

[13] Helton, J.C., "Uncertainty and Sensitivity Analysis Techniques for use in Performance Assessment for Radioactive Waste Disposal", *Rel. Eng. System Safety*, **42**, pp. 327–367, 1993.

[14] Iman, R.L., Hora, S.C., "A Robust Measure of Uncertainty Importance for use in Fault Tree System Analysis", *Risk Anal.*, **10 (3)**, pp. 401–406, 1990.

20

RULE MODUS PONENS VIS-A-VIS EXPLOSIVENESS IN GRADED PERSPECTIVE

Soma Dutta and Mihir K. Chakraborty
Department of Pure Mathematics
University of Calcutta

Abstract: *In this paper the connection between the notion of explosiveness and the rule Modus Ponens (MP) from the perspective of graded consequence has been investigated.*

Keywords: *Graded consequence, Fuzzy rule MP, Graded rule MP, Notion of explosiveness.*

1. INTRODUCTION

The primitive rule of inference of classical logic is Modus Ponens (MP) i.e. 'From α and $\alpha \supset \beta$ derive β'. Not only in classical logic, MP is prevalent in most of the logics. MP determines the interrelation of the connective $\supset$ (implication) with the notion of derivation or consequence. On the other hand, a notion of consequence is said to be explosive in strong sense if for any α, $\{\alpha, \sim\alpha\}$ yields any formula β. That is, the notion of explosiveness determines the interrelation of $\sim$ (negation) with the notion of consequence.

Classically, $\{\alpha, \alpha \supset \beta\} \vdash \beta$, i.e. $\{\alpha, \sim\alpha \vee \beta\} \vdash \beta$ implies $\{\alpha, \sim\alpha\} \vdash \beta$, that means MP due to classical logic axioms implies explosiveness. One of the aims of this paper is to look into the nature of this connection in many-valued scenario.

Though, MP seems to have a very natural demand, while dealing with vague predicates, it generates a paradox which is known as

Sorites paradox. Consider a set of premises consisting of the sentences 'A 15 year old person is young' and 'Whoever is one day older than a young person is young'. Then application of Modus Ponens several number of times leads at some stage to the conclusion 'A 90 year old person is young'.

There has been a lot of research on this paradox and various solutions are suggested. Applying many-valued logics, i.e. admitting truth values other than true (1) and false (0) for the object level formulae, many researchers [19] came up with several solutions for this situation. On the other hand, fuzzy logics introduce the idea of many-valued rule/fuzzy rule Modus Ponens which also takes care of this paradox. Section 3 is devoted to some discussions in this regard.

The theory of graded consequence [2, 3, 4, 5, 6], has been different from these existing many-valued logics/fuzzy logics in its basic philosophy. This theory proposes a notion of manyvalued consequence between a set of formulae and a single formula. It generally may have values other than 0 and 1. In section 1 we will introduce the notion of graded consequence briefly, to make the paper a self-contained one.

According to graded consequence, the rule Modus Ponens is a fuzzy relation, i.e. for any α, β, from the premise $\{\alpha, \alpha \supset \beta\}$ the conclusion β follows to a specified degree, known as the degree/grade of the rule Modus Ponens. How does this graded Modus Ponens rule help in resolving Sorites paradox will be discussed in Section 3.

It should be emphasized that while doing a logic activity, in general three levels of statements are needed. The Level 0 (or object level) consists of all formulae, all sets of formulae, etc. Level 1 (or meta level) deals with sentences like 'A formula is a consequence of a set of formulae', 'A set of formulae is inconsistent', etc. And the Level 2 (or meta-meta level) contains the relational statements about Level 1 entities for instance 'If X is a subset of Y and α is a consequence of X then α is a consequence of Y'. Different levels of activity may employ different kinds of reasoning/logic and hence should in general be associated with different algebraic structures. That, this explicit differentiation between levels is significant and that the theory of graded consequence has been developed with full awareness in this regard, are discussed in detail in [6, 7]. In this paper we only mention as much of the ideas regarding level distinction as would be required to explain the notion of graded Modus Ponens rule.

Furthermore, in Section 4 we shall investigate the connection between the graded Modus Ponens rule and the graded counterpart of the notion of explosiveness viz. GC4. We shall also explore a case for finding out the grade of Modus Ponens.

2. INTRODUCTION TO GRADED CONSEQUENCE RELATION

A language with a notion of consequence constitutes a logic. One can say that in logic, the latter one is the basic concept. Tarski in 1956 introduced axiomatically the notion of consequence as an operator which maps a set of formulae to its set of consequences [18]. On the other hand, Gentzen in 1969 presented the notion of consequence as a binary relation between a set of formulae and a single formula through some postulates [9]. These two notions are equivalent in the sense that taking any one as primitive the other can be obtained.

Chakraborty in 1986 offered a generalization to Gentzen's notion of consequence in a technical report subsequently published in 1988 [3]. In many-valued context a consequence relation, known as graded consequence relation, is an L-fuzzy relation from the power set of formulae P(F) to the set of formulae F over any alphabet satisfying conditions

GC1: if $\alpha \in X$ then gr $(X \mid\sim \alpha) = 1$ (Reflexivity)

GC2: if $X \subseteq Y$ then gr $(X \mid\sim \alpha) \leq$ gr $(Y \mid\sim \alpha)$ (Monotonicity)

GC3: $\inf_{\beta \in Z}$ gr $(X \mid\sim \beta)*$ gr $(X \cup Z \mid\sim \alpha) \leq$ gr $(X \mid\sim \alpha)$ (Cut)

where in general $(L, *, \rightarrow, 0, 1)$, a complete residuated lattice, is taken as an algebraic structure for the meta language containing the binary predicate $\mid\sim$. The interpretation is that for any set of formulae X and a single formula α, gr $(X \mid\sim \alpha)$ that is an element of L, denotes the degree to which α is a consequence of X and $*$ and 'inf' are taken to compute the meta-linguistic 'and' and 'for all' respectively. As indicated by the names, GC1-GC3 are respectively the generalized versions of classical reflexivity, monotonicity and cut conditions of classical consequence relation.

It has already been mentioned in the introduction, that in the context of graded consequence object level formulae as well as the meta-linguistic sentence 'α is a consequence of X', denoted by $X|\sim\alpha$ may have values other than 0 and 1. Let us now concentrate on the graded counterpart of the notion of semantic consequence, as proposed in the theory of graded consequence.

The notion of semantic consequence $|=$ in classical two-valued logic is given by: "$X|=\alpha$ if and only if for all valuations T, if every member of X is true then α is also true" where X is a set of wffs and α is a single wff.

Shoesmith and Smiley [16] generalized this notion by relativizing $|=$ in terms of an arbitrary collection $\{T_i\}_{i\in I}$ of valuations as follows:

$X|=\{T_i\}_{i\in I}\,\alpha$ if and only if for all valuations $T\in(T_i)_{i\in I}$, if every member of X is true then α is also true.

A valuation T_i may be identified with the set of wffs which are true under T_i. So T_i may be considered to be a subset of the set of wffs. Thus the definition of $|=\{T_i\}_{i\in I}$ can be given by $(\Sigma)\ X|=\{T_i\}_{i\in I}\,\alpha$ if and only if for all valuations $T\in\{T_i\}_{i\in I}$, if $X\subseteq T$ then $\alpha\in T$. This definition can be considered as a logical consequence because it satisfies the axioms proposed by Gentzen.

In order to have the generalized version of (Σ) in many-valued context, we consider the following.

1. The valuation function T_i's are now many-valued. T_i's are functions (homomorphism) from the algebra of wffs F to the same set L but with an appropriate algebraic structure, generally different from the algebraic structure for the meta language of $|\sim$. For instance, we shall observe in section 3 that another implication operator $\rightarrow_o$ has been used for computing object level implication $\supset$.
2. The value set for the meta language is taken to be a complete residuated lattice as mentioned before. In many-valued context the defining sentence on the right hand side of 'if and only if' in (Σ) is evaluated by

$$inf_i\,[gr\,(X \subseteq T_i) \rightarrow$$
$$gr\,(\alpha \in T_i) = inf_i\,\{inf_{x \in F}[gr\,(x \in X) \rightarrow gr\,(x \in T_i)] \rightarrow gr\,(\alpha \in T_i)\}$$
$$= inf_i\,\{inf_{x \in F}[X(x) \rightarrow T_i(x)] \rightarrow T_i(\alpha)\}$$

where 'inf' and '$\rightarrow$' are used to compute the meta linguistic 'for all' and 'if-then' of the defining sentence (Σ).

So, the above value is to be considered as the grade to which α follows from the premise set X, and since a complete residuated lattice is taken as the value set for meta-level (level-1) the above reduces to

(Σ') $\mathrm{gr}(X| \approx \{T_i\}_{i \in I}\ \alpha) = \inf_i\,[\inf_{x \in X} T_i(x) \rightarrow T_i(\alpha)]$.

The link between the notion $|\sim$, the axiomatic graded consequence relation and $|\approx \{T_i\}_{i \in I}$, the semantic graded consequence relation is given by the following results [3].

(*i*) For any collection $(T_i\}_{i \in I}$ of fuzzy subsets over F, the fuzzy relation $|\approx \{T_i\}_{i \in I}$, generated by the above definition (Σ') is a graded consequence relation.

(*ii*) Given any graded consequence relation $|\sim$, there exists $\{T_i\}_{i \in I}$, a collection of fuzzy subsets over F such that $|\approx \{T_i\}_{i \in I}$ coincides with $|\sim$.

3. MODUS PONENS IN FUZZY LOGICS AND IN GRADED CONSEQUENCE

Pavelka in [14] introduced the notion of fuzzy rule of inference. In designing a fuzzy rule of inference, he took motivation from Goguen [11], where the Modus Ponens rule intended to capture the following idea.

"If you know P is true at least to the degree a and $P \supset Q$ at least to the degree b then conclude that Q is true at least to the degree $a.b$".

['.' is the multiplication operator in the unit interval [0, 1]].

An immediate formalized structure of the above idea of many-valued Modus Ponens, is

$$\frac{\begin{array}{c}(P, a)\\(P \supset Q, b)\end{array}}{(Q, a.b)}$$

That is, a many-valued rule of inference can be viewed as a subset of $P(F \times [0, 1]) \times (F \times [0, 1])$, which is a crisp relation and hence a crisp rule of inference.

Pavelka's own interpretation of a many-valued rule of inference is as follows: a many-valued rule of inference r consists of two components $<r', r''>$, where the first (grammatical) component r' operates on formulas and the second (evaluation) component r'' operates on truth values and says how the truth value of the conclusion is to be computed from the truth values of the premises. Pavelka, himself puts a many-valued rule of Modus Ponens as,

$$\frac{\begin{array}{c} P(a) \\ P \supset Q(b) \end{array}}{Q(a * b)}$$

This is a replica of the form proposed by Goguen.

('*' is the multiplication operator in a complete residuated lattice)

According to Peter Hájek [12], the rules of fuzzy logic intend that 'from partially true premises partially true conclusion can be deduced'. Identifying the pair (P, a) with the formula $\boldsymbol{a} \supset P$, he has proposed the fuzzy rule Modus Ponens as follows:

$$\frac{\begin{array}{c} \boldsymbol{a} \supset P \\ \boldsymbol{b} \supset (P \supset Q) \end{array}}{\boldsymbol{a} * \boldsymbol{b} \supset Q}$$

where $\boldsymbol{a}$ is the wff denoting the truth value a. The novelty of Hajek lies in adopting names for each object language truth value within the language and identification of the pair (P, a) with the formula $\boldsymbol{a} \supset P$. But basically this corroborates the idea, initiated by Goguen. One can not claim that a grade has been attached with the rule MP.

On the other hand, in the context of graded consequence, the fuzzy rule Modus Ponens is a fuzzy relation which assigns a fixed value to $\{\alpha, \alpha \supset \beta\} | \sim \beta$, for any α, β. This fixed value, read as the degree of the rule Modus Ponens, depends upon the collection $\{T_i\}_{i \in I}$ which generates $|\sim$. To explain more clearly, let us consider $\rightarrow_o$ to be the operator for computing $\supset$. Then with respect to the collection $\{T_i\}_{i \in I}$ the grade of Modus Ponens, i.e. $\inf_{\alpha, \beta} (gr \{\alpha, \alpha \supset \beta\} | \sim \beta)$, denoted by $|MP|$ would be:

$inf_{\alpha,\beta}\,[inf_i\,[(T_i(\alpha) \wedge T_i(\alpha \supset \beta)) \to T_i(\beta)]]$ i.e.

$inf_{\alpha,\beta}\,[inf_i\,[(T_i(\alpha) \wedge T_i(\alpha) \to_o T_i(\beta)) \to T_i(\beta)]]$. [see ($\Sigma'$)]

Without entering into the technical details, let us present the solution to the Sorites paradox, from the perspective of fuzzy logics as well as graded consequence. According to the fuzzy logics (following Goguen) the chain of arguments to resolve Sorites paradox would be:

1. (A 15 year old person is young, a_1).
2. (A 15 year old person is young $\supset$ A 15 year and one day old person is young, a_2).
3. (A 15 year and one day old person is young, $a_1 * a_2$).

:

The truth value of the sentences of the form 'An m year old person is young' is determined by the degree of belongingness of the number m in the fuzzy subset, interpreting the predicate 'young'. So, as the truth value of the antecedent sentence must be greater or equal to that of the consequent sentence, the truth value of the conditional statement is not necessarily 1. So, as the number of steps increases the truth value of the sentence of the form 'An m year old person is young' decreases and hence the truth value of the conclusion viz. 'A 90 year old person is young' tends to zero.

According to the theory of graded consequence [5, 6], derivation of a formula say α_n from a set of formulae say X is an ordered pair of sequences viz. ($<\alpha_1, \alpha_2, \ldots. \alpha_n>$, $<|\alpha_1|, |\alpha_2|, \ldots..|\alpha_n>$) where the first sequence consists of formulae used in the derivation and the second sequence indicates the values associated with each step. The value associated with ith step will be 1 if α_i comes from the premise or the value will be tautologihood degree of α_i viz. $\inf_i T_i(\alpha)$ if α_i is an axiom or the value will be the degree $|R|$of a rule R if α_i is obtained by that rule. So, solution to the Sorites paradox can be presented as below:

1. A 15 year old person is young. ...1
2. A 15 year old person is young $\supset$ A 15 year and one day old person is young. ...1
3. A 15 year and one day old person is young. ...|MP|

(n) A 90 year old person is young. ...|MP|

Hence from the given premises 'A 90 year old person is young' can be derived to the degree |MP|* |MP|*.....* |MP|, taking product (*) of values associated with each step under consideration. In this context, the value of 'An *m* year old person is young' (say, α) is greater or equal to the value of 'An *m* year and one day old person is young' (say, β). Besides, to compute the value of the conditional statement 'An *m* year old person is young ⊃ An *m* year and one day old person is young', i.e. $\alpha \supset \beta$, we need to employ an fuzzy implication say $\rightarrow_o$. And as, all most all fuzzy implications satisfy $a \rightarrow_o b \geq b$ the value of $\alpha \supset \beta$ should be greater or equal to that of β. Hence $T_i(\alpha) \wedge (T_i(\alpha) \rightarrow_o T_i(\beta))$ is greater or equal to $T_i(\beta)$. It should be noted, that due to the gradually decreasing nature of the 'young' predicate with the increase of 'age' there should be some instant, where the value of $T_i(\alpha) \wedge (T_i(\alpha) \rightarrow_o T_i(\beta))$ is not 1. So, with reference to the calculation for |MP|, the value of |MP| is not 1 and hence the value of derivation of the conclusion approaches to zero as the number of steps increases.

The basic difference between these two approaches is as follows. For the first one, to know the value associated with a step, obtained by applying MP, it is necessary to know the value associated with the steps on which MP has been applied. Whereas for the latter approach value associated with the step, obtained by MP is independent of the values associated with the steps on which MP has been applied. In that sense value of the derivation in the second approach depends upon the form of the derivation, not on the exact formulae used in the derivation.

4. INTERRELATION OF MODUS PONENS AND THE NOTION OF EXPLOSIVENESS

In the previous section, while resolving Sorites paradox from the perspective of graded consequence we mentioned that |MP| is not necessarily 1. The following theorem gives a necessary condition for |MP| = 1.

4.1. Theorem: Let $inf_{\alpha,\beta}\, gr(\{\alpha, \alpha \supset \beta\} \mid\!\sim \beta) = 1$ and $\mid\!\sim$ be generated by $\{T_i\}_{i \in I}$. Then $a \rightarrow_o b \leq a \rightarrow b$ for all $a, b \in T_i(F)$, for each i.

Proof: $gr(\{\alpha, \alpha \supset \beta\} \mid\sim \beta) = 1$ for any α, β.

If and only if $inf_i\, [(T_i(\alpha) \wedge T_i(\alpha \supset \beta)) \to T_i(\beta)] = 1$

If and only if $[(T_i(\alpha) \wedge T_i(\alpha \supset \beta)) \to T_i(\beta)] = 1$ for each i

If and only if $(T_i(\alpha) \wedge T_i(\alpha \supset \beta)) \leq T_i(\beta)$ for each i

If and only if $T_i(\alpha) \wedge (T_i(\alpha) \to_o T_i(\beta)) \leq T_i(\beta)$ for each i ...(1)

(1) implies $T_i(\alpha) * (T_i(\alpha) \to_o T_i(\beta)) \leq T_i(\beta)$ for each i

Hence $T_i(\alpha) \to_o T_i(\beta) \leq T_i(\alpha) \to T_i(\beta)$

Now given $a, b \in T_i(F)$ for some i, there exists α, β such that $T_i(\alpha) = a$ and $T_i(\beta) = b$.

Hence for each i, for all $a, b \in T_i(F)$, $a \to_o b \leq a \to b$.

4.2. Corollary: If $inf_{\alpha,\beta}\, gr(\{\alpha, \alpha \supset \beta\} \mid\sim \beta) = 1$ for all $\mid\sim$ defined over the complete residuated lattice $(L, *, \to, 0, 1)$ then $|MP| = 1$ implies $\to_o \leq \to$ over L.

4.3. Theorem: If complete Heyting algebra (i.e. a residuated lattice L with the monoidal composition as lattice meet) is taken as the algebra of the meta language and $\to_o \leq \to$ then for any $\mid\sim$ over L, $inf_{\alpha,\beta}$ $gr(\{\alpha, \alpha \supset \beta\} \mid\sim \beta) = 1$.

Proof: Let $(L, \wedge, \to, 0, 1)$ be the algebra for the meta language and $\to_o \leq \to$. Hence for the collection $\{T_i\}_{i \in I}$, generating $\mid\sim$,

$$\begin{aligned} gr(\{\alpha, \alpha \supset \beta\} \mid\sim \beta) &= inf_i\, [(T_i(\alpha) \wedge T_i \alpha \supset \beta)) \to T_i(\beta)] \\ &= inf_i\, [(T_i(\alpha) \wedge (T_i(\alpha) \to_o T_i(\beta))) \to T_i(\beta)] \\ &\geq inf_i\, [(T_i(\alpha) \wedge (T_i(\alpha) \to T_i(\beta))) \to T_i(\beta)] \\ &\quad \text{(Since } \to \text{ is antitone in the 1st variable)} \\ &\geq inf_i\, [T_i(\beta) \to T_i(\beta)] \quad [a \wedge (a \to b) \leq b \\ &\quad \text{and } \to \text{ is antitone in the 1st variable]} \\ &= 1. \end{aligned}$$

As mentioned in the introduction, classically, Modus Ponens and the notion of explosiveness have a close connection. Let us explore the situation in the context of graded consequence.

Assuming the presence of a negation (~) in the object language, graded consequence has been extended in [4, 5] and the following is one of those additional axioms.

GC4: There is some $k > 0$ such that, $inf_{\alpha,\beta}\, gr(\{\alpha, \sim\alpha\} \mid\sim \beta) = k$.

GC4 is the graded counterpart of the explosiveness condition and GC4 proposes that for any α, $\{\alpha, \sim\alpha\}$ yields any β at least to a non-zero degree k. The implication of GC4 may be understood by the following theorem.

4.4. Theorem: Let $C_i(X) = \{\beta / gr(X \mid\sim \beta) \geq i\}$. Then for any $i\ (\in L) \leq k$, $C_i(\{\alpha, \sim\alpha\}) = F$ for any α, and $C_i(\{\alpha, \sim\alpha\}) \neq F$ for some α, when $i > k$ or i is non-comparable to k.

Proof is fairly long but straightforward by considering all possible cases.

Let $\mid\sim$ be characterized by GC1 to GC4 and let there be another connective $\veebar$ in the object language. Let $(+)$ be the operator in L, corresponding to $\veebar$ in the respective structure for object language and the property imposed on $(+)$ be $a, b \leq a\ (+)\ b$. It is to be noted again that the value set of both the meta-level language and the object level language are the same but the structures are in general different.

4.5. Proposition: For any collection $\{T_i\}_{i \in I}$, generating a graded consequence relation $\mid\sim$ $gr(X \cup \{\alpha \veebar \beta\} \mid\sim \gamma) \leq gr(X \cup \{\alpha\} \mid\sim \gamma)$ and $gr(X \cup \{\alpha \veebar \beta\} \mid\sim \gamma) \leq gr(X \cup \{\beta\} \mid\sim \gamma)$

Proof: $gr(X \cup \{\alpha \veebar \beta\} \mid\sim \gamma)$

$$= inf_i\, [(inf_{\delta \in X}\, T_i(\delta) \wedge T_i(\alpha \veebar \beta)) \rightarrow T_i(\gamma)]$$

$$= inf_i\, [(inf_{\delta \in X}\, T_i(\delta) \wedge (T_i(\alpha)\ (+)\ T_i(\beta))) \rightarrow T_i(\gamma)]$$

$$\leq inf_i\, [(inf_{\delta \in X}\, T_i(\delta) \wedge T_i(\alpha)) \rightarrow T_i(\gamma)]$$

[Since $a, b \leq a\ (+)\ b$ and $\rightarrow$ is antitone in 1st variable]

$$= gr(X \cup \{\alpha\} \mid\sim \gamma)$$

Note: Let us denote either of the above properties in the statement of Proposition 4.5 by Pr [$\veebar$]. Using *Pr* [$\veebar$] the following theorems can be proved.

4.6. Proposition: If $\alpha \supset \beta \equiv \sim\alpha \veebar \beta$ then GC4 and Pr [$\veebar$] imply $inf_{\alpha, \beta}\, gr(\{\alpha, \alpha \supset \beta\} \mid\sim \beta) \leq k$.

Proof: $gr(\{\alpha, \alpha \supset \beta\} \mid\sim \beta) = gr(\{\alpha, \sim\alpha \veebar \beta\} \mid\sim \beta) \leq gr(\{\alpha, \sim\alpha\} \mid\sim \beta)$ {By *Pr* [$\veebar$]}

$inf_{\alpha, \beta}\, gr(\{\alpha, \alpha \supset \beta\} \mid\sim \beta) \leq inf_{\alpha, \beta}\, gr(\{\alpha, \sim\alpha\} \mid\sim \beta) = k$

Let us call the following condition as *Pr* [$\supset$].

There is some $k' > 0$ such that $inf_{\alpha, \beta}\, gr(\{\alpha, \alpha \supset \beta\} \mid\sim \beta) = k'$.

4.7. Proposition: If $\alpha \supset \beta \equiv \sim\alpha \veebar \beta$ then $Pr\,[\supset]$ implies $inf_{\alpha,\beta}\, gr\,(\{\alpha, \sim\alpha\} \mid\!\sim \beta) \geq k'$.

Note: That is, according to (4.6) if we assume GC4 then the grade of MP will be at most k, as proposed by GC4. Whereas, in (4.7) it is ascertained that if we assume a grade of MP say k' (> 0)in advance, then $inf_{\alpha,\beta}\, gr\,(\{\alpha, \sim\alpha\} \mid\!\sim \beta) \geq k'$, i.e. $inf_{\alpha,\beta}\, gr\,(\{\alpha, \sim\alpha\} \mid\!\sim \beta)$ will assume some non-zero value–this is exactly the condition GC4.

Note: Usually, fuzzy implications can be generated from standard t-norms, t-conorms, and complementation operators [13]. The above definition of $\supset$ gives one such if $\veebar$ and $\sim$ are taken as some standard t-conorm and complementation operator. Let us consider another definition of $\supset$ which is given by $\alpha \supset \beta \equiv \sim\alpha \veebar (\alpha \text{ and } \beta)$. So, we assume existence of a negation ($\sim$), disjunction ($\veebar$) and a conjunction (&) in the object language.

4.8. Proposition: If $\alpha \supset \beta \equiv \sim\alpha \veebar (\alpha \,\&\, \beta)$ then independent of the interpretation of & and $\sim$, (4.6) and (4.7) hold.

Note: In case of many-valued logics, the above definitions generate two different kinds of implication [13], but classically both represent the same connective $\supset$.

These interdependencies between GC4 and graded rule MP, mentioned in (4.6) and (4.7) may help in determining grade of the rule MP. As stated, the grade of the rule MP will be determined by taking 'inf' over all formulae, which is basically an infinite set. So, for practical necessity, finding out actual grade of MP is problematic. The following theorem gives the solution in a special case.

4.9. Theorem: Let $\{p_1, p_2, \ldots p_n\}$ be n propositional variables and $\sim$ and $\veebar$ be two connectives in the object language. Let $\supset$ be defined by $p_i \supset p_j \equiv \sim p_i \veebar p_j$ $(i, j = 1, 2, \ldots n)$. Let us consider $\{T_i\}_{i \in I}$, a collection of fuzzy subsets such that for each p_i $(i = 1, 2, \ldots n)$ $T_i(p_i) \in [0, 1]$ and the operators for $\sim$ and $\veebar$ be $\neg$ i.e. $\neg a = 1 - a$ and 'max' respectively.

Then $inf_{\alpha,\beta}\, gr\,(\{\alpha, \alpha \supset \beta\} \mid\!\sim \beta) = inf\, p_i, p_j\, gr\,(\{p_i', p_i' \supset p_j'\} \mid\!\sim p_j')$, where p_i' denotes any one of p_i and $\sim p_i$ for each $i = 1, 2, \ldots, n$.

Proof is by applying induction on the complexity of formulae.

Note: One can observe that $\supset$, the implication in the theorem 4.9, is the Kleene-Dienes implication [13].

Example: Let $\{p_1, p_2\}$ be the propositional variables of a language and ~ and $\vee$ be two connectives which are computed by $\neg a = 1 - a$ and 'max' respectively. Let $\supset$ be defined by $p_i \supset p_j \equiv \sim p_i \vee p_j$ ($i = 1, 2$) Let $\{T_1, T_2\}$ be a collection of fuzzy subsets over the set of formulae, generated from the above mentioned alphabet and defined by:

	p_1	p_2	$p_1 \supset p_2$	$p_2 \supset p_1$	$\sim p_1 \supset p_2$	$\sim p_2 \supset p_1$	$p_1 \supset p_1$	$p_2 \supset p_2$	$p_1 \supset \sim p_2$	$p_2 \supset \sim p_1$
T_1	.7	.8	.8	.7	.8	.8	.7	.8	.3	.3
T_2	.8	.9	.9	.8	.9	.9	.8	.9	.2	.2

We have not considered the formulas $\sim p_1 \supset \sim p_1$, $\sim p_2 \supset p_2$, $\sim p_2 \supset \sim p_1$, $\sim p_1 \supset \sim p_2$ in the above table as they have the same truth values as $p_1 \supset p_1$, $p_2 \supset p_2$, $p_1 \supset p_2$ and $p_2 \supset p_1$ respectively.

We now calculate the grade of MP taking ([0, 1], *, →, 0, 1), an MV-algebra as the meta-level algebraic structure where → is defined by

$$a \to b = 1, \quad \text{if } a \le b$$
$$= (1 - a + b), \text{ otherwise.}$$

As $inf_{\alpha,\beta}\, gr\,(\{\alpha, \alpha \supset \beta\} \mid\!\sim \beta)$

$$= inf_{pi,\, pj}\, gr\,(\{p_i', p_i' \supset p_j'\} \mid\!\sim p_i')$$
$$= inf_{pi,\, pj}\, [inf_i\, \{(T_i(p_i') \wedge T_i(p_i' \supset p_j')) \to T_i(p_j')\}]$$
$$= .9$$

5. CONCLUSION

To Deal with vagueness of natural language and human reasoning, many-valued logics as well as fuzzy logics come up with several solutions. All these logics are generalizations of classical logic in the sense that classical set up can be obtained by restricting the value set into [0, 1]. Graded consequence is also one such generalization. But the basic differences of graded consequence from these logics are as follows.

1. Both object level as well as meta-level languages are in general many-valued.

2. As classically, the notions of consequence and inconsistency are equivalent, [17] in graded situation also a similar result has been established [5, 6].
3. Classically, rules are relations between a set of formulae and a single formula, identified by a specified form. According to fuzzy logics many-valued rule MP is a relation between a set of formulae, tagged with truth values and a single formula, tagged with some truth value. That is, MP, so introduced does not depend on the syntactic form only; rather it depends on the exact formulae, involved in each instance. Whereas, graded rule MP depends on the syntactic form of the formulae involved in it and each instance of the rule gets the same value, known as the grade of the rule derived from the context $\{T_i\}_{i \in I}$.
4. Paradoxes seem to generate a situation where from true premises, following valid arguments conclusion turns out to be false. Fuzzy logic treatment to Sorites paradox is so designed that one of the premises viz. the conditional statement of the form 'An m year old person is young $\supset$ An m year and one day old person is young' is supposed to assume values not necessarily 1. That means, before proposing the solution to the problem the paradoxical situation is being removed, assuming one of the premises to be not true (i.e. not necessarily 1). In this respect, the solution, proposed by graded consequence is more tenable.

REFERENCES

1. Basu S., "A Study in Logical and Philosophical Implications of Graded Consequence in Many-valued Systems". *A Ph.D thesis*. Jadavpur University. 2003.
2. Basu S., Chakraborty M.K., Vagueness. "Sorites and Graded Consequence, in Fuzzy Logic and its Application in Technology and Management". Debjani Chakraborty, S. Nanda, D. Dutta Majumdar (eds.). pp. 15-20, 2007.
3. Chakraborty M.K., "Use of Fuzzy Set Theory in Introducing Graded Consequence in Multiple Valued Logic", in M.M. Gupta and T. Yamakawa (eds.). "Fuzzy Logic in Knowledge-Based Systems, Decision and Control", Elsevier Science Publishers, B.V. (North Holland) 247-257. 1988.
4. Chakraborty M.K., "Graded Consequence: Further Studies". *Journal of Applied Non-Classical Logics*, **5 (2)**, 127-137. 1995.

[5] Chakraborty M.K., and Basu S., "Introducing Grades to Some Metalogical Notions", *Proc. Fuzzy Logic Laboratorium*, Linz, 1996.

[6] Chakraborty M.K., and Basu S., "Graded Consequence and Some Metalogical Notions Generalized", *Fundamenta Informaticae*, **32**, 299–311, 1997.

[7] Chakraborty M.K., Dutta S., "Grade in Meta Logical Notions: A Comparative Study of Fuzzy Logics", *in the proceedings of Studia Logica International Conference*, Trends in Logic V: Manyvalued Logics and Cognition, sponsored by Studia Logica, Institute of Philosophy and Sociology of the Polish Academy of Sciences and Institute of Logic and Cognition, Sun Yat-sen University, Guangzhou, China, July 6–9, 2007.

[8] Chakraborty M.K., Dutta S., "Theory of Graded Consequence and Fuzzy Logics, Logic Algebra and Truth Degrees", *First Conference of working Group on Mathematical Fuzzy Logic*, Siena, September, 2008.

[9] Gentzen G., "Investigations into Logical Deductions, in the Collected Papers of G. Gentzen", M.E. Szabo, ed., 68–131, North Holland Publications, Amsterdam, 1969.

[10] Gerla G., "Fuzzy Logic: Mathematical Tools for Approximate Reasoning", Kluwer Academic Publishers, 2001.

[11] Goguen J.A., "The Logic of Inexact Concepts, Synthese", **19**, 325–373, 1968.

[12] Hájek P., "Metamathematics of Fuzzy Logic", Kluwer Academic Publishers, 1998.

[13] Klir J.G., Yuan B., "Fuzzy Sets and Fuzzy Logic, Theory and Applications".

[14] Pavelka J., "On Fuzzy Logic I, II, III, Zeitschrift für Math. Logic and Grundlagen d. Math", **25**, 45–52, 119–134, 447–464, 1979.

[15] Pelta C., "Wide Sets, Deep Many-valuedness and Sorites Arguments", *Mathware and Soft Computing*, **11**, 5–11, 2004.

[16] Shoesmith D.J., Smiley T.J., "Multiple Conclusion Logic", Cambridge University Press, 1978.

[17] Surma, S.J., "The Growth of Logic out of the Foundational Research in Mathematics, in Modern Logic-a Survey", ed E. Agazzi, D. Reidel Publishing Co., Dordrecht, 1981, pp. 15–33.

[18] Tarski A., "Methodology of Deductive Sciences, in Logic, Semantics", Metamathematics, 60–109, Clavendon Press, 1956.

[19] Tye Michael, "Vague Objects", Mind 99, 1990.

21

CLAIMS VALIDATION SYSTEM— FUZZY APPROACH

Sanjeev Kumar[1] **and Pooja Pathak**[2]

[1]*Department of Mathematics, Dr. B.R. Ambedkar University, IBS, Khandhari, Agra-282002, India*
E-mail: sanjeevibs@yahoo.co.in

[2]*Department of Applied Sciences, GLA Institute of Technology and Management, Mathura, India*
E-mail: pooja.pathak@glaitm.org

Abstract: *The markets are dynamically changing with changing consumer behavior. End users in all sectors, be it product or service find themselves empowered with wide range of options available to them. And therefore in today's world nothing exists like monopoly. Insurance today is not what it used to be decade ago. It has evolved now, where consumers are more concerned about the quality, customer handling, transparent and easy process and proactive value added customer service. Insurance claim settlement process tops the priority list of all policy holders. From insurance company perspective, it is one of the most tedious business processes which directly impacts profit margins, policy holder satisfaction and business growth in turn.*

In order to reduce turnaround time of claim settlements process claims need to be verified and validated besides settling the claim in time. Till now it is ensured by human adjusters and this process makes claim settlement vulnerable to subjective judgment and the use of discretion while finalizing a claim. Trends show adjusters get biased to claimants and settle claims in their favor and impact monetary gains of insurers, to counter this growing trend and limiting cost on human expertise, companies are in need of reliable expert systems that can help them in validating or authenticating processed claims. This model of Claims Validation System will provide indicative result on the authenticity of claim process and the result from this system will also help auditors during settlement process. The methodology is also supported with an example.

Keywords: *Fuzzy logic, insurance, inference system, claim validation, MATLAB, soft computing, index of vagueness.*

1. INTRODUCTION TO PROBLEM AND SIGNIFICANCE

The insurance company has two Key Process Areas (KPA) while controlling their costs,

(*a*) Cost of claims and

(*b*) Reducing rate of return on investment [7].

With prevailing low rate of interest in markets it is difficult for insurers to raise their return on investment [5]. In this unfavorable condition, it becomes imperative for insurers to make sure that any amount spent in claim settlement, is indisputable in all respects. All claims raised for settlement must be verified and validated to ensure that a claimant is not attempting to obtain false claims through some external illegitimate means or by entering into collusion with the claim adjustors of insurance firms [4].

Some recent applications of fuzzy logic based techniques include pattern recognition [1] automobile body injury claims fraud [2], risk assessment of management fraud [3]. Soft computing techniques have also found applications in detecting fraud in various settings. The proposed system works in conjunction with internal database of an insurer's some external reference data repository. The use of soft computing in the insurance industry is not new as many major insurers in developed countries are practicing for long time [6].

This system will enable insurer to oversee their claim settlement business process and evaluate role played by their customer care support executive at call centre and a claim adjustors employed by insurers while assessing and deciding the claims. This system can evaluate claim of any size, already settled by the insurers. For design purpose claim size is restricted to a specific amount, which is expected to take care of small claims arising out of insured causes. This work with an example presents an approach which may help internal auditors to identify claims that carry an element of fraud from amongst all settled claims.

2. METHODOLOGY

The fuzzy logic expert system for detecting fraud in insurance claims is designed around the "index of vagueness" concept. This index reflects the degree of vagueness or elusiveness in the information furnished by the claimant and the information collected by the claim adjustors from various other sources. The index is used as a measure of genuineness in the insurance claims adjustment. Because of the subjective nature of the parameters used in the estimation of the claims, fuzzy parameters have been used in order to quantify the vagueness associated with information. An index known as "vagueness index" has been formulated in order to recognize genuineness while settling such claims. Fuzzy logic can quantify the contribution of a set of information to various parameters in terms of fuzzy membership.

3. SYSTEM ARCHITECTURE

Here we have developed a fuzzy logic based expert system for validation in insurance claims. Figure 1 shows the control mechanism of such system. The fuzzy logic based expert system consists of four components: fuzzifier, inference engine, defuzzifier, and the rule base. The role of fuzzifier is to convert a crisp input variable into linguistic variables that is ready to be processed by the inference engine.

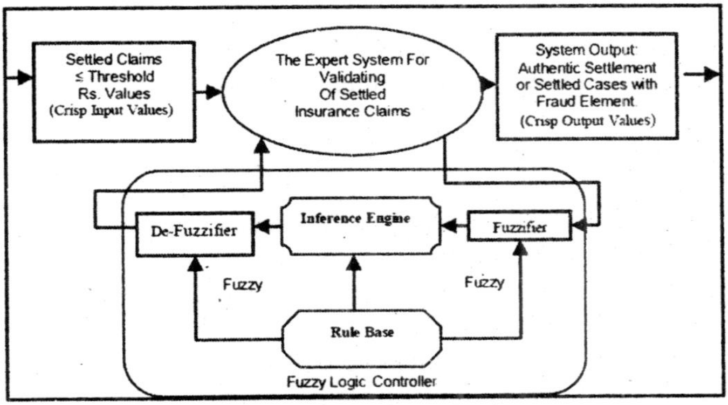

Fig. 1: *Fuzzy Claims Validation System*

The validation process starts by entering two sets of data; one furnished by the claimant and the other obtained by the auditor reviewing the claim. This information is obtained from a standard claim form that is used by the insurance company.

This form contains different sections, each section containing a set of information such as personal information of the claimant, financial information, and nature of the claim etc. All the settled claims that are below a certain value are settled immediately, whereas claims involving high amount are considered for verification by the expert system. The expert system verifies the genuineness of the settled claim by using various measures such as: vagueness index, degree of incomplete information of the claim, level of judgment used by the claim settlers in settling the claim. These qualitative measures are quantified and converted into linguistic variables with corresponding membership functions. For example, the vagueness index for the information section *i*th is given by [6]:

$$X_1 = \frac{\left(\sum_{i=1}^{I}\sum_{j=1}^{J} W_{ij}\,\Delta_{ij}\right)}{I} \quad \text{... (1)}$$

where W_{ij} is the weightage or impact factor given to the *j*th information of the *i*th section, and Δ_{ij} is a 0 – 1 variable (Δ_{ij} = 1 if there is any deviation/difference in the information furnished by the claimant and the one obtained by the auditor, 0 otherwise). It is worthwhile noting that the information that is crucial in deciding the settlement of the claim is given higher weightage/impact factor. Also all the weights for a set of *i*th information, $\sum_{j=1}^{J} W_{ij}$ added to unity. Similarly, the values of the other inputs can be determined. The normalized values of these measures are used as inputs to the expert system. The degree of membership corresponding to a value of input is determined by the use of triangular membership functions because of their simplicity and good result obtained by simulation. These membership functions are designed on the basis of available information.

Figure 2, shows the definition of the fuzzy sets of the input and the output functions. A rule base is then constructed which is based on all the applicable input parameters and for each decision, several rules are to be fired. Table 1 shows a sample rule base for the system under consideration which emphasize on the fact that in real life situations, the expertise of the human auditors will be used in the construction of the rule base.

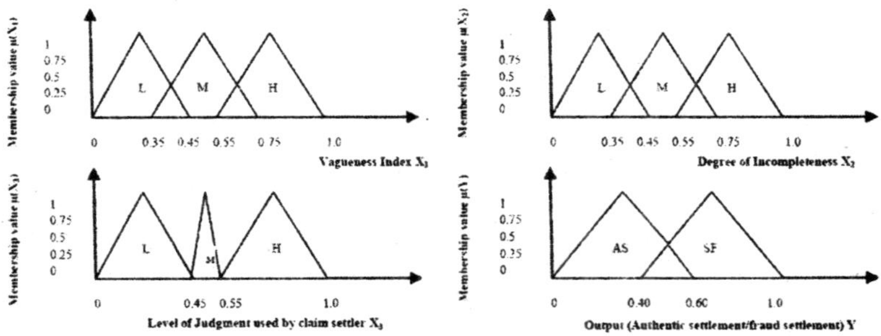

Fig. 2: *Membership Functions of Inputs and Outputs Functions*

Table 1: *Sample Rule Base for the Fuzzy Logic Based Expert System*

Rule No.	*INPUTS*			*OUTPUT*
	X_1	X_2	X_3	Y
1	Low	Low	Low	AS
2	Low	Low	Medium	AS
3	Low	Low	High	SF
4	Low	Medium	Low	AS
5	Low	Medium	Medium	AS
6	Low	Medium	High	SF
7	Low	High	Low	AS
8	Low	High	Medium	SF
9	Low	High	High	SF
10	Medium	Low	Low	AS
11	Medium	Low	Medium	AS
12	Medium	Low	High	SF
13	Medium	Medium	Low	AS
14	Medium	Medium	Medium	AS
15	Medium	Medium	High	SF
16	Medium	High	Low	AS
17	Medium	High	Medium	SF
18	Medium	High	High	SF
19	High	Low	Low	AS
20	High	Low	Medium	AS

Table Contd.

Table 1 Contd.

21	High	Low	High	SF
22	High	Medium	Low	AS
23	High	Medium	Medium	SF
24	High	Medium	High	SF
25	High	High	Low	AS
26	High	High	Medium	SF
27	High	High	High	SF

4. ALGORITHM–USING FUZZY APPROACH

The steps of the expert system are summarized below:

1. ***Input:*** The crisp value of the insurance claim settlement and other information obtained.
2. ***Compare against threshold:*** If the value of the claim settlement is less than certain pre-determined value as determined by the auditors (say α = Rs. 1,000), then the settled claim is to be considered authentic and go to Step 8.
3. ***Evaluate the authenticity of claim settlement:*** Determine the vagueness index X_1, degree of incomplete information of the claims X_2, level of judgment used by the claim settlers X_3.
4. ***Fuzzify the crisp values of inputs:*** Through the use of membership functions defined for each fuzzy set for each linguistic variable (Fig. 2), determine the degree of membership of a crisp value in each fuzzy set. Each of these three ambiguity indices have been divided into three fuzzy sets (LOW–L, MEDIUM–M and HIGH–H). The equations for computing memberships are:

$$\mu\,(X_i)_L = \begin{cases} \max\left\{0, \dfrac{X_i - a_i^L}{c_i^L - a_i^L}\right\} & \text{if } X_i < c_i^L \\ \max\left\{0, \dfrac{b_i^L - X_i}{b_i^L - c_i^L}\right\} & \text{if } c_i^L \le X_i \end{cases} \quad \dots (2)$$

$$\mu\,(X_i)_M = \begin{cases} \max\left\{0, \dfrac{X_i - a_i^M}{c_i^M - a_i^M}\right\} & if\ X_i < c_i^M \\ \max\left\{0, \dfrac{b_i^M - X_i}{b_i^M - c_i^M}\right\} & if\ c_i^M \le X_i \end{cases} \qquad \text{... (3)}$$

$$\mu\,(X_i)_H = \begin{cases} \max\left\{0, \dfrac{X_i - a_i^H}{c_i^H - a_i^H}\right\} & if\ X_i < c_i^H \\ \max\left\{0, \dfrac{b_i^H - X_i}{b_i^H - c_i^H}\right\} & if\ c_i^H \le X_i \end{cases} \qquad \text{... (4)}$$

where (a, c, b) are the vertices of the triangular membership function while L, M and H represents the fuzzy set LOW, MEDIUM, and HIGH, respectively.

5. ***Fire the rule bases that correspond to these inputs:*** All expert systems which is based on fuzzy logic uses IF-THEN rules. The "IF" part is known as antecedent or premise, whereas the "THEN" part is termed as a consequence or conclusion. Since all the three inputs have three fuzzy sets, therefore 27 (3 × 3 × 3) fuzzy decisions are to be fired. There are two outputs: AUTHENTIC SETTLEMENT–AS, CASES SETTLED WITH FRAUD ELEMENT OR ELEMENT OF SUSPECT–SF.

6. ***Execute the inference engine:*** Once all crisp input values have been fuzzified into their respective linguistic values, the inference engine will access the fuzzy rule base of the fuzzy expert system to derive linguistic values for the intermediate as well as the output linguistic variables. The two main steps in the inference process are aggregation and composition.

During aggregation, each condition in the IF part of a rule is assigned a degree of truth based on the degree of membership of the corresponding linguistic term. From here, product (PROD) of the degrees of truth of the conditions are computed to clip the degree of truth from the IF part. This is assigned as the degree of truth of the THEN part. The next step in the inference process is to determine the degrees of truth for each linguistic term of the output linguistic variable.

7. ***Defuzzification:*** The last phase in the fuzzy expert system is the defuzzification of the linguistic values of the output linguistic variables into crisp values. The most common techniques for defuzzification are center-of-maximum (CoM) and center-of-area (CoA).
8. ***Output of the decisions of the expert system:*** In our case, the types of the outputs are: genuine settlement of the claim and claims settled with element of suspect or fraud. The auditor further investigates into the claims settled with element of suspect or fraud. However, in principle, in all the fuzzy logic based expert system, we explore the implicit and explicit relationships within the system by mimicking human thinking and subsequently develop the optimal fuzzy control rules as well as knowledge base.

5. EXAMPLE

For the purpose of illustration, we consider that the insurance company uses three inputs – vagueness index X_1, degree of incomplete information of the claims X_2, level of judgment used by the claim settlers X_3.

1. Input. Value of the claim = 5,000 Rs.
2. Compare against threshold. Since the value is greater than a pre-determined value (say α = Rs. 1,200). The authenticity of claim settled needs to be verified by the expert system.
3. Evaluate the authenticity of claim settlement. The values of the inputs of the claims have to be evaluated, $X_1 = 0.40$; $X_2 = 0.70$ and $X_3 = 0.50$ (say).
4. Fuzzification of the crisp values of inputs. Through the use of membership functions defined for each fuzzy set for each linguistic variable (Fig. 2), the degree of membership of a crisp value in each fuzzy set is determined as follows:

$$\mu\,(X_1)_L = \max\left\{0, \frac{b_1^L - X_1}{b_1^L - c_1^L}\right\} = 0.22$$

$$\mu\,(X_1)_M = \max\left\{0, \frac{X_1 - a_1^M}{c_1^M - a_1^M}\right\} = 0.25$$

$$\mu\,(X_1)_H = \max\left\{0,\ \frac{X_1 - a_1^H}{c_1^H - a_1^H}\right\} = 0$$

$$\mu\,(X_2)_L = \max\left\{0,\ \frac{b_2^L - X_2}{b_2^L - c_2^L}\right\} = 0$$

$$\mu\,(X_2)_M = \max\left\{0,\ \frac{b_2^M - X_2}{b_2^M - c_2^M}\right\} = 0.25$$

$$\mu\,(X_2)_H = \max\left\{0,\ \frac{X_2 - a_2^H}{c_2^H - a_2^H}\right\} = 0.667$$

$$\mu\,(X_3)_L = \max\left\{0,\ \frac{b_3^L - X_3}{b_3^L - c_3^L}\right\} = 0$$

$$\mu\,(X_3)_M = \max\left\{0,\ \frac{b_3^M - X_3}{b_3^M - c_3^M}\right\} = 1$$

$$\mu\,(X_3)_M = \max\left\{0,\ \frac{X_3 - a_3^H}{c_3^H - a_3^H}\right\} = 0$$

Where

$$(a_1^L, c_1^L, b_1^L) = (0, 0.225, 0.45);$$

$$(a_1^M, c_1^M, b_1^M) = (0.35, 0.55, 0.75);$$

$$(a_1^H, c_1^H, b_1^H) = (0.55, 0.775, 1.0)$$

$$(a_2^L, c_2^L, b_2^L) = (0, 0.225, 0.45);$$

$$(a_2^M, c_2^M, b_2^M) = (0.35, 0.55, 0.75);$$

$$(a_2^H, c_2^H, b_2^H) = (0.55, 0.775, 1.0)$$

$$(a_3^L, c_3^L, b_3^L) = (0, 0.225, 0.45);$$

$$(a_3^M, c_3^M, b_3^M) = (0.45, 0.50, 0.55);$$

$$(a_3^H, c_3^H, b_3^H) = (0.55, 0.775, 1.0)$$

5. Fire the rule bases that correspond to these inputs. Based on the value of the fuzzy membership function values for the example under consideration, the following rules apply:

   Rule 5: If X_1 is LOW, X_2 is MEDIUM, X_3 is MEDIUM then Y is an AUTHENTIC SETTLEMENT (AS).

   Rule 8: If X_1 is LOW, X_2 is HIGH, X_3 is MEDIUM then Y is a SETTLEMENT WITH ELEMENT OF FRAUD (SF).

   Rule 14: If X_1 is MEDIUM, X_2 is MEDIUM, X_3 is MEDIUM then Y is an AUTHENTIC SETTLEMENT (AS).

   Rule 17: If X_1 is MEDIUM, X_2 is HIGH, X_3 is MEDIUM then Y is a SETTLEMENT WITH ELEMENT OF FRAUD (SF).

6. Execute the Inference Engine. We use the "root sum squares" (RSS) method to combine the effects of all applicable rules, scale the functions at their respective magnitudes. The respective output membership function strengths (range: 0 – 1) from the possible rules (R1 – R27) are:

$$\text{"Authentic Settlement"} = \sqrt{\sum_{i \in AS} (\mu_{R_i})^2}$$

$$AS = \{1, 2, 4, 7, 8, 10, 11, 13, 16, 17, 19, 20, 22, 25\}$$

$$= \sqrt{(0.055)^2 + (0.0625)^2} = 0.083$$

"Settlement with element of fraud suspect"

$$= \sqrt{\sum_{i \in SF} (\mu_{R_i})^2}$$

$$SF = \{3, 5, 6, 9, 12, 14, 15, 18, 21, 23, 24, 26, 27\}$$

$$= \sqrt{(0.147)^2 + (0.167)^2} = 0.222$$

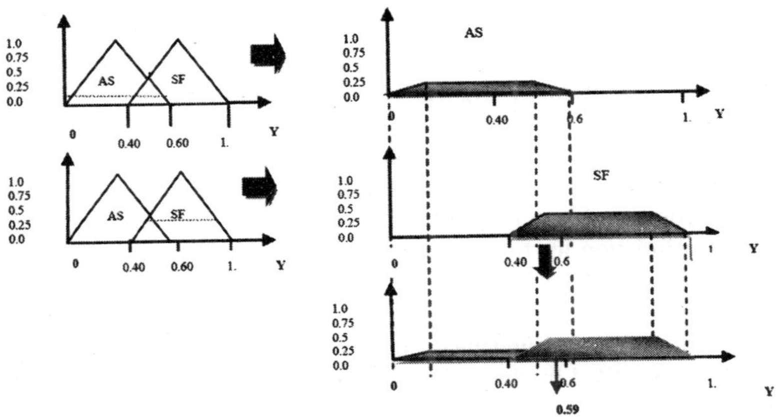

Fig. 3: *Output of the Decision of the Expert System*

7. Defuzzification. We use "Weighted average method" for defuzzification. The defuzzification of the data into crisp output is accomplished by combining the results of the inference process. The weighted average method is formed by weighting each membership function in the output by its respective *max* membership value and the result is taken as the crisp output. Figure 3 shows the crisp output of 0.59. The crisp output belongs to the set of settlement with fraud (SF) more than the set of authentic settlement (AS) (as evident from its membership function). Hence, the decision in this case is SF.
8. Output the decisions of the expert system. For the cases under consideration, the system indicates the case has been fraud settled.

In the above example we are towards settlement with fraud element (SF), i.e. the element of fraud are involved in the insurance claim settlement. So the insurance company will give only the threshold amount, i.e. the amount which is predetermined by the insurance company for a particular case. So the insurance company will give only Rs. 1,200 to the claimant.

6. CONCLUSION

The development of a fuzzy based expert system for fraud detection in settled insurance claims is reported in this work. Our future efforts will be on the improvement of the performance of the system by adjusting the membership function of the inputs. It would be interesting to tune the rule base using data from real life problems so that the performance of the system is optimized. We propose to use neural networks that can produce an optimum surface representing all the combination points from a few of the tested combinations. Another direction for future extension would be to take into account input factors such as: time taken by the claim adjusters to settle the claim, whether the approval of a higher authority or peers was taken into account while settling the claim, number of similar types of claims that has been filed by the claimant, etc. It is worthwhile noting that inclusion of these factors would increase the size of the rule base to the point that the tuning of the rule base using data from real life scenarios will be deemed necessary to optimize the performance of the system. The system proposed through this work is evaluated on hypothetical data.

REFERENCES

[1] Bezdek J.C. (1981), "Pattern Recognition with Fuzzy Objective Function Algorithms," Plenum Press, New York, NY.

[2] Brockett P.L., Xia X. and Derrig R. (1998), "Using Kohonen's Self Organizing Feature Map to Uncover Automobile Bodily Injury Claims Fraud", *The Journal of Risk and Insurance*, **65**, No. 2, pp. 245–274.

[3] Deshmukh A. and Lakshminarayana T. (1998), "A Rule-Based Fuzzy Reasoning System for Assessing the Risk of Management Fraud", *International Journal of Intelligent Systems in Accounting*, Finance and Management, **4**, pp. 231–41.

[4] Derrig R. and Ostaszewski K. (1995), "Fuzzy Techniques of Pattern Recognition in Risk and Claim Classification", *The Journal of Risk and Insurance*, **62**, No. 3, pp. 447–482.

[5] Greenwald, J. (2003), "P/C Insurers See Strong Quarter", *Business Insurance*, **37**, pp. 13–22.

[6] Pathak J., Vidhyarthi N. and Summers S.L. (2005), "A Fuzzy-Based Algorithm for Auditors to Detect Elements of Fraud in Settled Insurance Claims", *Managerial Auditing Journal*, **20**, No. 6, pp. 632–644.

[7] Zyl S.V. (2003), "Canadian P/C Insurers in Intensive Care?", *National Underwriter*, **107**, No. 22, pp. 1–30.

22

ON INTUITIONISTIC FUZZY REGULAR CLOSED m_x SET

Sharmistha Bhattacharya (Halder) and Jayanti Tripathi

Department of Mathematics, Tripura University,
Suryamaninagar, Tripura-799130
E-mail: halder_731@rediffmail.com

Abstract: *The purpose of this paper is to construct the concept of intuitionistic fuzzy regular closed $m_x$set in intuitionistic fuzzy topological space. Some characterization theorems and properties of this newly constructed set are studied. Also the concept of intuitionistic fuzzy regular closed m_x supra continuous function is introduced and some related theorems are studied through out this paper.*

Keywords: *Intuitionistic fuzzy $m_x$structure, Intuitionistic fuzzy topological space, Intuitionistic fuzzy supracontinuous function, etc.*

AMS Subject Classification: *54C08, 54C60.*

1. INTRODUCTION

Zadeh introduced the concept of fuzzy sets and Atanassov generalized it to intuitionistic fuzzy sets. Coker introduced intuitionistic fuzzy topological space and other related concepts. T. Noiri and V. Popa have given new class of m-continuous function. The concept of minimal open sets is introduced by I.M. James. F. Nakaoka and N. Oda have shown some applications of it. In this paper we introduce the concept of intuitionistic fuzzy regular closed m_x set and some properties of it are studied. In Section 2, some preliminaries related to the topic is given. In section 3,the concept of intuitionistic fuzzy regular closed m_x set in intuitionistic fuzzy topological space is introduced. In Section 4, the concept of intuitionistic fuzzy regular closed m_x-supra continuity is introduced and some related theorems are studied.

2. PRELIMINARIES

In this section, we study some ready results and definitions for references.

Definition 2.1: Let X is a nonempty set and I the unit interval (0, 1). An intuitionistic fuzzy set (IFS for short) U is an object having the form $U = \{< x, \mu_u(x), \gamma_u(x) >: x \in X\}$ where the functions $\mu_u: X \to I$ and $\gamma_u: X \to I$ denote respectively the degree of membership and degree of non-membership of each element $x \in X$ to the set U, and $0 \leq \mu_u(x) + \gamma_u(x) \leq 1$ for each $x \in X$.

Definition 2.2: An intuitionistic fuzzy topology (IFT for short) on a nonempty set X is a family τ of IFS's in X containing $0_\sim$, $1_\sim$ and closed under arbitrary infimum and finite supremum.

In this case the pair (X, τ) is called an intuitionistic fuzzy topological space (IFTS for short) and each IFS in τ is known as an intuitionistic fuzzy open set (IFOS for short).

The compliment U^C of an IFOS U in an IFTS (X, τ) is called an intuitionistic fuzzy closed set (IFCS, for short) in X.

Definition 2.3: Let X be a nonempty set and let IFS's U and V be in the form $U = \{< x, \mu_u(x), \gamma_u(x) >: x \in X\}$ and $V = \{< x, \mu_v(x), \gamma_v(x) >: x \in X\}$. Then

1. $U^C = \{< x, \gamma_u(x), \mu_u(x) >: x \in X\}$
2. $U \cap V = \{< x, \Lambda [\mu_u(x), \mu_v(x)], V [\gamma_u(x), \gamma_v(x)] >, x \in X\}$
3. $U \cup V = \{< x, V [\mu_u(x), \mu_v(x)], \Lambda [\gamma_u(x), \gamma_v(x)] >: x \in X\}$
4. $1_\sim = \{< x, 1, 0 >: x \in X\}, 0_\sim = \{< x, 0, 1 > x \in X\}$
5. $(U^C)^C = U, 0_\sim^C = 1_\sim, 1_\sim^C = 0_\sim$

Definition 2.4: Let (X, τ) be a topological space. A subfamily m_x of power set $P(X)$ is said to be a m_x structure if ϕ and X are members of m_x. Each member of m_x is said to be m_x open set and compliment of it is said to be a m_x closed set.

Definition 2.5: A family τ of intuitionistic fuzzy sets on X is called an intuitionistic fuzzy supra topological space (IFST for short) on X if $0_\sim \in \tau$, $1_\sim \in \tau$ and τ is closed under arbitrary supremum. Each member of τ is called an intuitionistic fuzzy supra open set and compliment of an intuitionistic fuzzy supra open set is an intuitionistic fuzzy supra closed set. And a family τ of intuitionistic fuzzy sets is called an intuitionistic fuzzy m_x structure if $0_\sim$ and $1_\sim$ are the members of τ.

Definition 2.6: Let (X, τ) be an IFTS and m_x an intuitionistic fuzzy m_x structure on X. For a subset A of (X, τ).

1. $m_x - cl\,(A) = \cap\,\{F: A \subseteq F, F^C \in m_x\}$ and (2) $m_x -$ int $(A) = \cup\,\{G: G \subseteq A, G \in m_x\}$.

Lemma 2.7: Let (X, τ) be an IFTS and m_x an intuitionistic fuzzy m_x structure on X. For a subset A of (X, τ).

1. $m_x - cl\,(A^C) = \{m_x - int\,(A)\}^C$ and $m_x - int\,(A^C) = [m_x - cl\,(A)]^C$.
2. If $A^C \in m_x$, then $m_x - cl\,(A) = A$ and if $A \in m_x$ then $m_x - int\,(A) = A$.
3. $m_x -$ int $(0_\sim) = 0_\sim$ and $m_x - int\,(1_\sim) = 1_\sim$.
4. $m_x - cl\,(0_\sim) = 0_\sim$ and $m_x - cl\,(1_\sim) = 1_\sim$.
5. If $A \subseteq B$ then $m_x - cl\,(A) \subseteq m_x - cl\,(B)$ and $m_x - int\,(A) \subseteq m_x - int\,(B)$.
6. $A \subseteq m_x - cl\,(A)$ and $m_x - int\,(A) \subseteq A$.
7. $m_x - cl\,[m_x - cl\,(A)] = m_x - cl\,(A)$ and $m_x - int\,[m_x - int\,(A)] = m_x - int\,(A)$.

Definition 2.8: Let A be an IF Subset of $X \cdot A$ is said to be an IF Open m_x set if $m_x - int\,(A) = A$ and IF closed m_x set if $m_x - cl\,(A) = A$.

Remark 2.9: The m_x open sets are open m_x sets but the converse need not be true. The following structural relationship holds.

IF m_x open set $\Rightarrow$ IF open m_x set $\Rightarrow$ IF open set.

Definition 2.10: A subset A of an IFTS (X, τ) is said to be IF generalized closed if $Cl\,(A) \subseteq B$ whenever $A \subseteq B$, where B is an IFOS of τ.

Definition 2.11: Let A is an IFS of X. Then A is said to be IF α open if $A \subseteq int\,[cl\,(int\,A)]$.

Throughout this paper intuitionistic fuzzy topological space is shortly written as IFTS and is denoted by (X, τ), (Y, σ) and (Z, α) and complement of a set U is denoted by U^C.

3. ON INTUITIONISTIC FUZZY REGULAR CLOSED m_x SET

In this section we are studying the structure of the operators formed by combining interior and closure with the corresponding m_x interior and m_x closure. It can be shown that m_x interior, if operated upon interior of an IFS A of X gives the operator m_x interior. If we combine m_x interior with closure operator then we get a new type of operator. In this section we are dealing with this type of operator and corresponding topology.

Theorem 3.1: For an intuitionistic fuzzy m_x structure (X, m_x) defined on an IFTS (X, τ).

1. $m_x - int\,(int\,A) = m_x - int\,A$.
2. $m_x - cl\,(cl\,A) = m_x$ - $cl A$.

Proof:

1. Let A is an IFS of X, then we know that $int\,(A) \subseteq A$ and $m_x - int\,A \subseteq int\,A$ (as $m_x \subseteq t$), i.e. $m_x - int\,A \subseteq$ int $A \subseteq A \Rightarrow m_x - int\,(m_x - int\,A) \subseteq m_x - int\,(int\,A) \subseteq m_x - int\,A \Rightarrow m_x - int\,A \subseteq m_x - int\,(int\,A) \subseteq m_x - int\,A \Rightarrow m_x - int\,(int\,A) = m_x - int\,A$.
2. $m_x - clA \supseteq cl\,A$ (as $m_x \in \tau$) and $cl\,A \supseteq A$, so $m_x - cl\,A \supseteq cl\,A \supseteq A \Rightarrow m_x - cl\,(m_x cl\,A)\ m_x - cl\,(cl\,A)\ m_x - cl\,A \Rightarrow m_x - cl\,A\ m_x - cl\,(cl\,A)\ m_x - cl\,A \Rightarrow m_x - cl\,A = m_x - cl\,(cl\,A)$.

Remark 3.2: $int\,(m_x - int\,A) = m_x - int\,A$.

Lemma 3.3: Let (X, τ) be an IFTS and m_x an intuitionistic fuzzy m_x structure on X formed from the IFTS (X, τ).

1. $cl\,[m_x - int\,(0_\sim)] = 0_\sim$ and $cl\,[m_x - int\,(1_\sim)] = 1_\sim$.
2. $cl\,[m_x - int\,(A^C)] = \{int\,[m_x - cl\,(A)]\}^C$.
3. $cl\,\{m_x - int\,[cl\,(mx - int\,(A))]\} = cl\,[m_x - int\,(A)]$.
4. $cl\,\{m_x - int\,[\cup\,(A_i : i \in I)]\} \cup \{cl\,[m_x - int\,(A_i) : i \in I]$.
5. $cl\,[m_x - int\,(\cap\,A)] \subseteq \cap\,\{cl\,[m_x - int\,(A_i), i \in I\}$.

Proof: Proof is obvious.

With the help of above results we are now in position to define intuitionistic fuzzy regular closed m_x set.

Definition 3.4: Let (X, τ) be an IFTS and (X, m_x) an intuitionistic fuzzy m_x structure on X. An IFS A is said to be an IF regular closed m_x

set iff $cl\ [m_x - int\ (A)] = A$ and an IF regular open m_x set iff $int\ [m_x - cl\ (A)] = A$.

Example 3.5: Let $A = (< x, 0.3, 0.5 >, x \in X)$ and $B = (< x, 0.2, 0.4 >, x \in X)$, be two IF subset of X. $m_x = (0_\sim, 1_\sim, A, B)$ and $\tau = (0_\sim, 1_\sim, A, B, A \cup B, A \cap B)$ be the IF m_x structure and the IF topology defined on X. Consider $C = (< x, 0.4, 0.3 >, x \in X)$, $m_x - int\ C = (< x, 0.3, 0.4 > x \in X)$, $cl\ (m_x - \text{int}\ C) = (< x, 0.4, 0.3 > x \in X) = C$.

Hence C is IF regular closed m_x set.

Remark 3.6: $0_\sim$ and $1_\sim$ are IF regular closed m_x set as well as IF regular open m_x set.

Theorem 3.7: Collection of all IF regular closed m_x set defined on an IF m_x structure forms an IF supra topology.

Proof: Let ψ denote the collection of all IF regular closed m_x sets.

(*i*) $0_\sim$ and $1_\sim$ are the members of ψ (by remark 3.6).

(*ii*) Let $(A_i, i \in I) \in \psi. \Rightarrow cl\ (m_x\ int\ A_i) = A_i$, for every $i \in I$.

Now $\cup A_i \supseteq A_i$.

$\Rightarrow cl\ [m_x\ int\ (\cup A_i)]\ cl\ [m_x\ int\ (A_i)] \Rightarrow cl\ [m_x\ int\ (\cup A_i)] \cup cl\ [m_x\ int\ (A_i)]\ \ cl\ [m_x\ int\ (\cup A_i)] \cup (A_i)$.

But $\cup A_i \subseteq cl\ [m_x\ int\ (\cup A_i)]$. Therefore $\cup A_i \in \psi$, i.e. $\cup A_i$ is also an IF regular closed m_x set. Hence ψ is forming an IF supra topology.

Remark 3.8: The intersection of two IF regular closed m_x sets need not be an IF regular closed m_x set. This can be shown with the help of an example.

Let $X = (0, 1)$, $A = (< x, 0.3, 0.5 >, x \in X)$ and $B = (< x, 0.2, 0.4 >, x \in X)$ be two IF subsets of X, the corresponding m_x structure be $m_x = (0_\sim, 1_\sim, A \cdot B)$ and the corresponding topology be $\tau = (0_\sim, 1_\sim, A, B, A \cup B, A \cap B)$. Let $C = (< x, 0.3, 0.4 >, x \in X)$. Here $cl\ [m_x - int\ (C)] = C$, i.e. C is an IF regular closed m_x set. Consider $D = < x, 0.2, 0.4 >$, clearly D is also an IF open m_x set. Now $C \cap D = < x, 0.2, 0.5 >$.$m_x - int\ (D) = 0_\sim$, $int\ [m_x - int\ (D)] = 0_\sim \neq C \cap D$.

Theorem 3.9: Every IF regular closed m_x set is IFCS but the converse need not be true.

Proof: Let A is an IF regular closed m_x set. So $cl\ (m_x - int\ A) = A \Rightarrow cl\ cl\ (m_x - int\ A) = cl A \Rightarrow cl\ (m_x int\ A) = cl A \Rightarrow A = cl A \Rightarrow A$ is IFCS.

The converse of the above is not true. Consider Example 3.5.Take $C = (< x, 0.5, 0.3 > x \in X)$, here C is IFCS, then $cl\,(m_x - int\,C) = < x,$ $0.4, 0.3 > \neq C$.

Theorem 3.10: Every IF regular closed $m_x$set is IF α closed.

Proof: Let A is an IF regular closed m_x set. So $cl\,(m_x - int\,A) = A$. Now $m_x\,int\,A \subseteq cl\,A \Rightarrow int\,(m_x - int\,A) \subseteq (\text{int}\,cl\,A) \Rightarrow cl\,(m_x - int\,A) \subseteq$ $cl\,(\text{int}\,cl\,A) \Rightarrow A \subseteq cl\,(\text{int}\,cl\,A) \Rightarrow A$ is IFαCS.

Remark 3.11: The converse of the above theorem need not be true. This can be shown with the help of an example. Consider set C in the previous example. Here $C = (< x, 0.5, 0.3 > x \in X)$. $ClC = (<$ $x, 0.5, 0.3 > x \in X)$, $int\,(cl\,C) = cl\,[\text{int}\,(clC)] = (< x, 0.4, 0.3 > x \in X)$. So $C \subseteq cl\,[int\,(clC)]$. Hence C is IF α closed. But C is not an IF regular closed m_x set.

Theorem 3.12: Let A be an IF regular closed m_x set of X then A is an IFGCS.

Proof: Let A is an IF regular closed m_x set and $A \subseteq B$ (B is some IFOS). Then $clA = cl\,(cl\,m_x\,int\,A) = cl\,(m_x\,int\,A) = A$, so $cl\,A \subseteq B$. Hence A is IFGCS.

The converse of above need not be true.

This can be shown with the help of an example. Let $X = (0, 1)$, $A = (< x, 0.2, 0.5 >, x \in X)$ and $B = (< x, 0.4, 0.6 >, x \in X)$ $m_x = (0_\sim,$ $1_\sim, A, B)$ and $\tau = (0_\sim, 1_\sim, A, B, A \cap B, A \cup B)$. Consider a set $C = (< x,$ $0.2, 0.4 >, x \in X)$, $C \subseteq 1_\sim$ and $cl\,(C) \subseteq 1_\sim$. Hence C is an IFGCS. Now $m_x\,int\,C = (< x, 0.2, 0.5 > x \in X)$ and $cl\,m_x\,int\,C = (< x, 0.5, 0.4$ $> x \in X) \neq C$. Hence C is not an IF regular closed m_x set.

Theorem 3.13: Let A be an IF regular closed m_x set of X then A is an IF Cm_xS.

Proof: Let A is an IF regular closed m_x set. Then $m_x\,cl\,A = m_x\,cl$ $(cl\,m_x\,int\,A) = A$ [as $m_x cl\,(cl\,A) = m_x\,cl\,A$), so $m_x\,cl\,A = A$. Hence A is IF Cm_xS.

The converse of the above need not be true. We can take an example to prove this. $X = (0, 1)$, $A = (< x, 0.3, 0.5 > x \in X)$ and $B = (< x, 0.2,$ $0.4 >, x \in X)$, $m_x = (0_\sim, 1_\sim, A, B)$ and $\tau = (0_\sim, 1_\sim, A, B, A \cap B, A \cup B)$. Let $C = (< x, 0.5, 0.3 > x \in X)$, then C is IF $m_x$closed set. $cl\,m_x\,int\,C$ $= (< x, 0.4, 0.3 > x \in X) \neq C$. Hence C is not an IF regular closed m_x

So we are getting following structural relation

$$\begin{array}{c} \text{IFGCS} \\ \Uparrow \\ \text{IF Regular closed } m_x \text{ set} \Rightarrow \text{IF closed } m_x \text{ set} \Rightarrow \text{IF closed set} \\ \Downarrow \\ \text{IF } \alpha \text{ closed set} \end{array}$$

Theorem 3.14: If A is an IF regular closed m_x set and m_x int $A \subseteq B \subseteq A$, then B is IF semi open set.

Proof: m_x *int* $A \subseteq m_x$ *int* $B \subseteq m_x$ *int* $A. \Rightarrow m_x$ *int* $A = m_x$ *int* B. As $B \subseteq A = cl\ (m_x\ int\ A) = cl\ (m_x\ int\ B) \Rightarrow B \subseteq cl\ (m_x\ int\ B) \subseteq cl\ int\ B. \Rightarrow$ B is IF semi open set.

4. ON INTUITIONISTIC FUZZY REGULAR CLOSED m_x SUPRA CONTINUITY

In this section the concept of Intuitionistic Fuzzy Regular Closed m_x Supra Continuity is introduced and some related theorems are studied.

Definition 4.1: A function $f : (X, \tau) \to (Y, \sigma)$ is called IF Regular closed m_x supra continuous function if inverse image of every closed set V in Y is IFRegular closed m_x set in X.

Theorem 4.2: Let a function $f: (X, \tau) \to (Y, \tau)$ be IF Regular closed m_x continuous function. Then f is IF continuous function.

Proof: Let V be an IF m_x closed set in Y. Since f is an IF Regular closed m_x supra continuous function, $f^{-1}(V)$ is an IF Regular closed m_x set of X. Since IF Regular closed m_x set is IF closed set, so $f^{-1}(V)$ is an IF closed set, i.e. inverse image of every IF Regular closed m_x set is IF closed set. Hence f is an IF continuous function.

Theorem 4.3: Let $f: (X, \tau) \to (Y, \sigma)$ be a mapping, then the following are equivalent

1. f is an IF Regular closed m_x supra continuous function.
2. Inverse image of any IFCS in Y is IF Regular closed m_x set in X.
3. $f\,[cl\ (m_x\, int\ A)] \subseteq cl\, f\,(A)$.
4. $cl\ [m_x\ int\, f^{-1}\,(B)] \subseteq cl\, f^{-1}\,(B)$.

Proof: (1) $\Rightarrow$ (2) from the definition of IF Regular closed m_x continuity.

$(2) \Rightarrow (3)$

Let A is an IFS of X. Since $A \subseteq f^{-1}[f(A)]$, we have $A \subseteq f^{-1}\{cl[f(A)]$, $cl[f(A)]$ is an IFCS of $Y \Rightarrow f^{-1}[cl\, f(A)]$ is an IF regular closed m_x set in X and it contains A.

$\Rightarrow [cl\,(m_x\ int\ A)] \subseteq f^{-1}[cl\, f(A)] \Rightarrow f[cl\,(m_x\ int\ A)] \subseteq f\{f^{-1}[cl\, f(A)]\} \Rightarrow f[cl\,(m_x\ int\ A)] \subseteq cl\, f(A)$

$(3) \Rightarrow (4)$

$\Rightarrow f[cl\,(m_x\ \mathrm{int}\ A)] \subseteq cl\, f(A) \Rightarrow f^{-1}\{f[cl\,(m_x\ int\ A)]\} \subseteq f^{-1}[cl\, f(A)] \Rightarrow cl\,(m_x\ int A) \subseteq f^{-1}[cl\, f(A)]$.

Let us take $A = f^{-1}(B)$, $\Rightarrow cl\,[m_x\ int f^{-1}(B)] \subseteq f^{-1}(cl\, B)$.

$(4) \Rightarrow (2)$

$cl\, B$ is an IFCS of Y. From (4) $cl\,[m_x\, int f^{-1}(B)] \subseteq f^{-1}(cl\, B)$, but $cl\,[m_x\ int f^{-1}(B)] \supseteq f^{-1}(cl\, B)$.

$\Rightarrow cl\,[m_x\, int f^{-1}(B)] = f^{-1}(cl\, B) \Rightarrow f^{-1}(clB)$ is an IF regular closed m_x set.

Theorem 4.4: Let f: $(X, \tau) \to (Y, \sigma)$ be an IF regular closed m_x Supra continuous function. Then

1. f is IF Generalized continuous.
2. f is IF α continuous.

Proof: Since f is IF regular closed m_x supra continuous function, $f^{-1}(B)$ is an IF regular closed m_x set for every IFCS B of Y. From the structural relation in Section (3) each IF regular closed m_x set is IFGCS, therefore $f^{-1}(B)$ is IFGCS and hence f is an IFG continuity. Again since every IF regular closed m_x set is IFclosed set, hence $f^{-1}(B)$ is IF α closed set. Therefore the function f is IF α continuous. Hence the result.

Theorem 4.5: Let f: $(X, \tau) \to (Y, \sigma)$ be an IF regular closed m_x supra continuous function and g: $(Y, \sigma) \to (Z, \alpha)$ be an IF continuous function then gof: $(X, \tau) \to (Z, \alpha)$ is IF regular closed m_x supra continuous function.

Proof: Since g is IF continuous function $\Rightarrow g^{-1}(W)$ is an IF closed set in X for every IFCS W in $Z \Rightarrow (gof)^{-1}(W)$ is IF regular closed m_xS of $X \Rightarrow (gof)^{-1}$ is IF regular closed m_x supra continuous function.

Theorem 4.6: Let f: $(X, \tau) \rightarrow (Y, \sigma)$ be an IF continuous function and g: $(Y, \sigma) \rightarrow (Z, \alpha)$ be an IF regular closed m_x supra continuous function then gof: $(X, \tau) \rightarrow (Z, \alpha)$ is IF continuous function.

Proof: Since g is an IF regular closed m_x supra continuous function $\Rightarrow g^{-1}(W)$ is If regular closed m_x set in Y, for every W in Z, and as every IF regular closed m_x set is IFCS as well $\Rightarrow g^{-1}(W)$ is IF closed set in Y. $\Rightarrow (gof)^{-1}(W)$ is IFC set in X (given that f is IF continuous function) $\Rightarrow (gof)$ is continuous function.

5. CONCLUSION

Throughout this paper we have introduced the concept of IF Regular closed m_x supra topological space and IF Regular closed m_x supra continuity. In ordinary topology this type of spaces have helped a lot in digital topology. So IF Regular closed m_x supra topological space being a generalized form of Regular closed space of ordinary topology may give more application. In this paper some continuities are also shown.

REFERENCES

[1] D. Coker, "An Introduction to Intuitionistic Fuzzy Topological Spaces". *Fuzzy Sets and Systems*, **88**, 1997, No. 1, 81–89.

[2] D. Coker and A.H.Es, "Fuzzy Compactness in Intuitionistic Fuzzy Topological Space". *J. Fuzzy Math*, **3**, 1995, No.4, 899–909.

[3] F. Nakaoka and Nobuyuki Oda; "Some Applications of Minimal Open Sets". *Int. J. Math Sci.* **27 (8)**, 2001, 471–476.

[4] H. Maki, "On Generalizing Semi Open and Preopen Sets; Meeting on Topological Spaces, Theory and it's Applications" 24-25 August, 1996, Yatsushiro Coll. Tech, 13–18.

[5] K. Balachandran, P. Sundaram and H. Maki, "Generalized Continuous Maps in Topological Spaces", *Mem. Fac. Sci. Kochi Univ. Ser. A, Math*, **12 (1991)**, 5–13.

[6] K.K. Azad, "Fuzzy Semi Continuity, Fuzzy Almost Continuity". *JMAA*, **82** 14–32 (1981).

[7] K.T. Atanassov, "Intuitionistic Fuzzy Sets". *Fuzzy Sets and Systems*, **20**, 87–96, 1986.

[8] L.A. Zadeh, "Fuzzy Sets Information and Control", **8**, 338–353, 1965.

[9] N. Levine, "Generalized Closed Sets in Topology, Rend". *Circ. Mat. Palermo*, **19 (2)** (1970), 89–96.

[10] N. Turanli; "Intuitionistic; Fuzzy Supra Topological Space". *International Conference on Modeling and Simulation*, Spain, **2**, 69–77, 1999.

[11] Sharmistha Bhattacharya (Halder) and Jayanti Tripathi, "On Intuitionistic Fuzzy m_x Structure", *Communicated to Punjab University*, Pakistan.

[12] Sharmistha Bhattacharya (Halder) and Jayanti Tripathi, "On Intuitionistic Fuzzy Minimal Open Set", *presented in International Conference on frontier of pure and applied Mathematics*, University of Burdwan, India.

[13] S.J. Lee and E.P. Lee, "Intuitionistic Fuzzy Proximity Space". *IJMMS* 2004: **49**, 2617–2628.

23

ON INTUITIONISTIC FUZZY α-IRRESOLUTE MAPPINGS

R.N. Bhaumik[1] **and A. Bhattacharjee**[2]

[1]*Department of Mathematics, Tripura University, Suryamaninagar, Tripura West-799130*
E-mail: bhaumik_r_n @ yahoo.co.in.

[2]*Department of Mathematics, D.D.M. College, Khowai, Tripura West-799201*

Abstract: *In 1994, the Fuzzy a-irresolute mappings were introduced in a fuzzy topological space by R. Prasad, S.S. Thakur and R.K. Saraf. The objective of this paper is to introduce the concept of α-irresolute mappings in the intuitionistic fuzzy topological spaces and to study the properties of these mappings.*

1. INTRODUCTION

The concept of fuzzy sets was introduced by Zadeh [9] and later Atanassov [1] introduced the fundamental concept of intuitionistic fuzzy sets. In 1997 Coker [4] introduced the notion of intuitionistic fuzzy topological space, some types of intuitionistic fuzzy continuity and some related concepts. Very recently R.N. Bhaumik and A. Bhattacharjee [3] introduced the concept strongly α-continuous function in intuitionistic fuzzy topological space. In 1994, Fuzzy α-irresolute mappings were introduced by R. Prasad, S.S. Thakur and R.K. Saraf [8]. In Section 3, α-irresolute mappings in intuitionistic fuzzy topological spaces are defined and their properties are studied.

2. PRELIMINARIES

Definition 2.1 [2]: Let X be a non empty fixed set. An intuitionistic fuzzy set (IFS in short) A is an object having the form $A = \{< x, \mu_A(x)$,

$\nu_A(x) >: x \in X\}$, where the function $\mu_A(x): X \to I$ and $\nu_A(x): X \to I$ denote the degree of membership and degree of non membership of each element $x \in X$ and $0 \le \mu_A(x) + \nu_A(x) \le 1$.

Definition 2.2 [4]: Let X and Y be two non empty sets and $f: X \to Y$ be a function. If $B = \{<y, \mu_B(y), \nu_B(y)>: y \in Y\}$ is an IFS in Y, then the pre image of B under f is $f^{-1}(B) = \{<x, f^{-1}(\mu_B)(x), f^{-1}(\nu_B)(x)>: x \in X\}$.

Definition 2.3 [5]: Let X be a non empty set and $c \in X$ a fixed element in X. If $\gamma \in (0, 1]$ and $\beta \in [0, 1)$ are two fixed real numbers such that $\gamma + \beta \le 1$, then the IFS $c(\gamma, \beta) = <x, c_\gamma, 1 - c_{1-\beta}>$ is called an intuitionistic fuzzy point (IFP in short) in X, where γ denotes the degree of membership of $c(\gamma, \beta)$, β the degree of non membership of $c(\gamma, \beta)$ and $c \in X$ the support of $c(\gamma, \beta)$.

Definition 2.4 [5]: Let $c(\gamma, \beta)$ be an IFP in X such that $\gamma, \beta \in (0, 1)$ and $A = <x, \mu_A, \nu_A>$ be an IFS in X, $c(\gamma, \beta)$ is said to be properly contained in A [$c(\gamma, \beta) \in A$ for short] if and only if $\gamma < \mu_A(c)$ and $b > \nu_A(c)$.

Definition 2.5 [6]: An IFS A in an IFTS (X, τ) is called

(*a*) an intuitionistic fuzzy semi open set (IFSOS) if $A \subseteq cl(int(A))$.

(*b*) an intuitionistic fuzzy α-open set (IFαOS) if $A \subseteq int(cl(int(A)))$.

The set of all IFαO subsets of X will be denoted by IFα (X). The complement of IFSO (IFαO) set is called intuitionistic fuzzy semi closed (α-closed).

Definition 2.6: A function $f: (X, \tau_1) \to (Y, \tau_2)$ is called an

(*a*) Intuitionistic fuzzy continuous [6] if inverse image of each IFOS of Y is an IFOS in X.

(*b*) Intuitionistic fuzzy α-continuous [6] if inverse image of each IFOS of Y is an IFαOS in X.

(*c*) Intuitionistic fuzzy strongly α continuous [3] if inverse image of each IFSOS of Y is an IFαOS in X.

Definition 2.7 [7]: Let X, Y be two non empty sets and $A = <x, \mu_A(x), \nu_A(x)>$, $B = <y, \mu_B(y), \nu_B(y)>$ IFS's of X and Y respectively.

Then $A \times B$ is an IFS of $X \times Y$ defined by $(A \times B)_{(x, y)} = \langle (x, y), \min [\mu_A(x), \mu_B(y)], \max [\nu_A(x), \nu_B(y)] \rangle$.

3. INTUITIONISTIC FUZZY α-IRRESOLITE MAPPINGS

Definition 3.1: A mapping $f: (X, \tau_1) \to (Y, \tau_2)$ from an IFT space (X, τ_1) to an IFT space (Y, τ_2) is said to be intuitionistic fuzzy -irresolute if the inverse image of every IFα-open set of Y is an IF α-open set in X.

The following examples show that the concept of IF continuous mappings and IF α-irresolute mappings are independent.

Example 3. 2: Let $X = \{a, b\}$, $Y = \{p, q\}$ be two non empty sets and

$$G_1 = \langle x, (a/.3, b/.4), (a/.5, b/.3) \rangle$$

$$G_2 = \langle x, (a/.2, b/.6), (a/.1, b/.3) \rangle$$

$$U_1 = \langle y, (p/.4, q/.4), (p/.3, q/.5) \rangle$$

$$U_2 = \langle y, (p/.6, q/.2), (p/.3, q/.1) \rangle$$

Then the family $\tau_1 = \{0_\sim, 1_\sim, G_1\}$ and $\tau_2 = \{0_\sim, 1_\sim, U_1\}$ are IFTS's on X and Y respectively. If we define the function $f: X \to Y$ by $f(a) = q$ and $f(b) = p$ then f is continuous. Now U_2 is an IF α-open set in Y but $f^{-1}(U_2) = \langle x, (a/.2, b/.6), (a/.1, b/.3) \rangle$ is not an IF α-open set as $f^{-1}(U_2) \not\subseteq int(cl(int f^{-1}(U_2))) = 0_\sim$. Hence f is IFcontinuous but not IF α-irresolute mapping.

Example 3.3: Let $X = \{a, b\}$, $Y = \{p, q\}$ be two non empty sets and $G_1 = \langle x, (a/.5, b/.5), (a/.4, b/.3) \rangle$, $G_2 = \langle x, (a/.5, b/.35), (a/.4, b/.5) \rangle$ and $U_1 = \langle y, (p/.5, q/.6), (p/.2, q/.3) \rangle$. Then the family $\tau_1 = \{0_\sim, 1_\sim, G_1\}$ and $\tau_2 = \{0_\sim, 1_\sim, U_1\}$ are IFTS's on X and Y respectively. If we define the function $f: X \to Y$ by $f(a) = q$ and $f(b) = p$ then f is IF α-irresolute but not IFcontinuous as $f^{-1}(U_1) = \langle x, (a/.6, b/.5), (a/.3, b/.2) \rangle \notin \tau_1$.

Theorem 3.4: Every intuitionistic fuzzy strongly α-continuous mapping is IF α-irresolute.

Proof: Let (X, τ_1) and (Y, τ_2) be two IFTS's and $f: (X, \tau_1) \to (Y, \tau_2)$ is intuitionistic fuzzy strongly α-continuous. Then $f^{-1}(A)$ is IF

α-open set in X for each IFSO set A in Y. Now each IF α-open set is IFS-open set. Hence $f^{-1}(A)$ is IF α-open set in X for each IF α-open set A in Y.

But the converse is not true as shown in the following example.

Example 3.5: Let $X = \{a, b\}$, $Y = \{p, q\}$ be two non empty sets and

$$G_1 = < x, (a/.4, b/.2), (a/.5, b/.5) >$$
$$G_2 = < x, (a/.5, b/.4), (a/.4, b/.4) >$$
$$U_1 = < y, (p/.4, q/.5), (p/.4, q/.4) >$$
$$U_2 = < y, (p/.2, q/.4), (p/.5, q/.5) >$$
$$U_3 = < y, (p/.4, q/.5), (p/.3, q/.4) >$$

Then the family $\tau_1 = \{0_{\sim}, 1_{\sim}, G_1, G_2\}$ and $\tau_2 = \{0_{\sim}, 1_{\sim}, U_1, U_2\}$ are IFTS's on X and Y respectively. Let $f: X \to Y$ be a mapping, defined by $f(a) = q$ and $f(b) = p$. Then f is IF α-irresolute mapping but not IF strongly α-continuous. As U_3 is an IFS-open set but not IF α-open set and $f^{-1}(U_3) = < x, (a/.5, b/.4), (a/.4, b/.3) > \not\subseteq int(cl(int(f^{-1}(U_3)))) = G_1$, $f^{-1}(U_3)$ is not IF α-open set.

Theorem 3.6: Every intuitionistic fuzzy α-irresolute mapping is IFα-continuous.

Proof: It is obvious. But the converse is not true as shown by the following example.

Example 3.7: Let $X = \{a, b\}$, $Y = \{p, q\}$ be two non empty sets and

$$G_1 = < x, (a/.1, b/.1), (a/.5, b/.6) >$$
$$G_2 = < x, (a/.5, b/.6), (a/.2, b/.4) >$$
$$U_1 = < y, (p/.6, q/.5), (p/.4, q/.2) >$$

Then the family $\tau_1 = \{0_{\sim}, 1_{\sim}, G_1, G_2\}$ and $\tau_2 = \{0_{\sim}, 1_{\sim}, U_1\}$ are IFTS's on X and Y respectively. Let $f: X \to Y$ be a mapping, defined by $f(a) = q$ and $f(b) = p$. Then f is α-continuous mapping but not α-irresolute as $U_2 = < y, (p/.7, q/.7), (p/.2, q/.1) >$ is IF α-open set in Y but $f^{-1}(U_2) = < x, (a/.7, b/.7), (a/.1, b/.2) >$ is not IF α-open set in X since $f^{-1}(U_2) \not\subseteq int(cl(int(f^{-1}(U_2)))) = G_2$.

Remark 3.8: The relationships of intuitionistic fuzzy continuity, intuitionistic fuzzy α-continuity, intuitionistic fuzzy strongly α-continuity with intuitionistic fuzzy α-irresolute mappings are as follows.

IF strongly α-continuity ⇒ IF α-irresolute ⇒ IF α-continuity
⇑
IF-continuity

where "IF" means "intuitionistic fuzzy". However the converses are not true in general.

Next we prove two properties of α-open sets:

Theorem 3.9: $A \in$ IF α (X) if and only if for every IFP $c(\gamma, \beta) \in A$ there exists a IF α-open set O such that $c(\gamma, \beta) \in O, \subset A$.

Proof: If $A \in$ IF α (X) then we may take $O = A$ for every $c(\gamma\alpha, \beta) \in A$.

Conversely, let A be any subset of X satisfying the given condition. We have $A = \cup_{c(\gamma, \beta) \in A} [c(\gamma, \beta)] \subset \cup_{c(\gamma, \alpha) \in A} O \subset A$, and $A = \cup_{c(\gamma, \beta) \in A} O$. Hence $A \in$ IF α (X).

Theorem 3. 10: If $A \in$ IF α (X) and $B \in$ IF α (Y) then $A \times B \in$ IF α $(X \times Y)$, where (X, τ_1) and (Y, τ_2) be two IFTS's.

Proof: We have
$$\begin{aligned} A \times B &\subset int_X(cl_X(int_X(A))) \times int_Y(cl_Y(int_Y(B))) \\ &= int_{X \times Y}(cl_X(int_X(A)) \times cl_Y(int_Y(B))) \\ &= int_{X \times Y}(cl_{X \times Y}(int_X(A) \times int_Y(B))) \\ &= int_{X \times Y}(cl_{X \times Y}(int_{X \times Y}(A \times B))) \end{aligned}$$

Consequently, $A \times B \in$ IF α $(X \times Y)$.

Theorem 3.11: Let (X, τ_1) and (Y, τ_2) be IFTS's and $f: (X, \tau_1) \rightarrow (Y, \tau_2)$. Then the following conditions are equivalent.

(*a*) f is intuitionistic fuzzy α-irresolute mapping.

(*b*) for every IFP $c(\gamma, \beta) \in X$ and for every $B \in$ IF α (Y) containing $f(c(\gamma, \beta))$ there exist a IFS $A \in$ IF α (X) such that $c(\gamma, \beta) \in A \subseteq f^{-1}(B)$.

(*c*) for every IFP $c(\gamma, \beta) \in X$ and for every $B \in$ IF α (Y) containing $f[c(\gamma, \beta)]$ there exist a IFS $A \in$ IF α (X) such that $c(\gamma, \beta) \in A$ and $f(A) \subseteq B$.

(*d*) for every IF α-closed set B in $Y, f^{-1}(B)$ is IF α-closed set in X.

(*e*) for every $B \in$ IF α $(Y), f^{-1}(B) \subset int(cl(int(f^{-1}(B))))$.

(*f*) for every IF α-closed set B in Y, $f^{-1}(B) \supset cl(int(cl(f^{-1}(B))))$.

***Proof:* (a) $\Rightarrow$ (b):** Let $c(\gamma, b) \in X$ and $B \in$ IF $a(Y)$ such that $f[c(\gamma, \beta)] \in B$. Since f is intuitionistic fuzzy α-irresolute, $f^{-1}(B) \in$ IF $\alpha(X)$ and $c(\alpha, \beta) \in f^{-1}(B)$. Put $A = f^{-1}(B)$. Then $c(\gamma, \beta) \in A$ and $f(A) \subset B$. Hence $c(\gamma, \beta) \in A \subseteq f^{-1}(B)$.

(b) $\Rightarrow$ (a): Let $B \in$ IF $\alpha(Y)$ such that $c(\gamma, \beta) \in f^{-1}(B)$. Then $c(\gamma, \beta) \in f^{-1}(B)$. Then $f(cl(\gamma, \beta)) \in f(f^{-1}(B)) \subset B$. By (*b*) there exist a IFS $A \in$ IF $\alpha(X)$ such that $c(\gamma, \beta) \in A \subset f^{-1}(B)$. Hence by Theorem 3.9, $f^{-1}(B) \in$ IF $a(X)$.

(b) $\Rightarrow$ (c): Obvious.

(a) $\Rightarrow$ (d): Obvious.

(c) $\Rightarrow$ (e): Let $c(\gamma, \beta)$ be *IFP* of X and $B \subset IF\ \alpha(Y)$ and $c(\gamma, \beta) \in f^{-1}(B)$. Then by (*c*), there exists an IFαO set A in X containing $c(\gamma, \beta)$ such that $f(A) \subseteq B$. Therefore we obtain that $c(\gamma, \beta) \in A \subseteq int(cl(int(A))) \subseteq int(cl(int(f^{-1}(B))))$ and hence $f^{-1}(B) \subseteq int\{cl[int[f^{-1}(B)]\}$ for each IFαO set B of Y.

(e) $\Rightarrow$ (a): Obvious.

(d) $\Leftrightarrow$ (f): Follows from the fact that an intuitionistic fuzzy set A is IFα-closed set in an intuitionistic fuzzy topological space (X, τ) if and only if $A \supseteq cl(int(cl(A)))$.

Theorem 3.12: A mapping $f: X \to Y$ is intuitionistic fuzzy α-irresolute if and only if for every $B \subset Y$, $f^{-1}(\alpha\text{-}int\ B) \subset \alpha\text{-}int\ f^{-1}(B)$.

Proof: Let B be an IFS of Y. Since α-*int* B is IFαO set, $f^{-1}(\alpha\text{-}int\ B)$ is an IFαO set and we have $f^{-1}(\alpha\text{-}int\ B) \subseteq \alpha\text{-}int\ f^{-1}(\alpha\text{-}int\ B) = \alpha\text{-}int\ f^{-1}(B)$.

Conversely, let B be any intuitionistic fuzzy α-open set in Y then $\alpha\text{-}int\ B = B$ and $f^{-1}(B) = f^{-1}(\alpha\text{-}int\ B) \subseteq \alpha\text{-}int\ f^{-1}(B)$, which shows that $f^{-1}(B)$ is an IFαO set.

Theorem 3.13: Let $f: X \to Y$ be one-to-one and onto mapping. f is intuitionistic fuzzy α-irresolutemapping if and only if for every $A \subset X$, $\alpha\text{-}int(f(A)) \subset f(\alpha\text{-}int(A))$.

Proof: Let $A \subset X$. Since $\alpha\text{-}int(f(A)) \in$ IF $\alpha(Y)$ and f is intuitionistic fuzzy α-irresolute mapping, $f^{-1}(\alpha\text{-}int\ f(A)) \in$ IF $\alpha(X)$. Now $f^{-1}(f(A)) = A$, therefore by Theorem 3.12, we have $f^{-1}(\alpha\text{-}int\ f(A)) \subseteq$

(α-$int\, f^{-1}(f(A))) = \alpha$-$int\, A$. This implies that $f(f^{-1}(\alpha\text{-}int\, f(A))) \subseteq f(\alpha\text{-}int\, A)$ since f is on to α-$int\, f(A) = f(f^{-1}(\alpha\text{-}int\, f(A))) \subseteq f(\alpha\text{-}int\, A)$.

Conversely, let $B \in$ IF α (Y). Then α-$int\, B = B$. From hypothesis, $f(\alpha\text{-}int\, f^{-1}(B)) \supseteq \alpha\text{-}int\, (f(f^{-1}(B))) = \alpha\text{-}int\, B = B$. This implies that $f^{-1}(f(\alpha\text{-}int\, f^{-1}(B))) \supseteq f^{-1}(B)$. Since f is one-to-one, we have α-int $f^{-1}(B) \supseteq f^{-1}(B)$.

Hence $f^{-1}(B) = \alpha\text{-}int\, f^{-1}(B)$ and $f^{-1}(B) \in$ IF α (X). Therefore f is intuitionistic fuzzy α-irresolute mapping.

Theorem 3.14: Let $f: (X, \tau_1) \to (Y, \tau_2)$. Then f is IF α-irresolute mapping if and only if for every IFP $c(\gamma, \beta)$ of Y and for each IFO set G of Y such that $c(\gamma, \beta) \in int\,(cl\, G)$, $f^{-1}(G \cup c(\gamma, \beta))$ is IF α-open in X.

Proof: Let f is IF α-irresolute and for every IFP $c(\gamma, \beta)$ of Y and for each IFO set G of Y such that $c(\gamma, \beta) \in int\,(cl\, G)$. So $(G \cup c(\gamma, \beta)) \subseteq int\,(cl\, G)$ and $G \cup c(\gamma, \beta)$ is an IF α-open set of Y and hence $f^{-1}(G \cup c(\gamma, \beta))$ is IF α-open set in X.

Conversely, let $H \in$ IF α (Y). Then there exists a IF α-open set G of Y such that $G \subseteq H \subseteq int\,(cl\,(G))$ and for each IFP $c(\gamma, \beta) \in H$, $f^{-1}(G \cup c(\gamma, \beta))$ is IF α-open set in X. Since $f^{-1}(H) = \cup_{c(\gamma, \beta) \in H} (G \cup c(\gamma, \beta))$ is IF α-open set in X, f is IF α-irresolute.

The following theorem can be proved easily.

Theorem 3.15: If $f: X \to Y$ and $g: Y \to Z$ are IF α-irresolute mappings then $g\, o\, f$ is IF α-irresolute.

Definition 3.16: Let (X, τ) is an IFTS and N be an IFS in X. N is said to be an IF α-neighbourhood of an IFP $c(\gamma, \beta)$ in X if there exists an IFα-open set G in X such that $c(\gamma, \beta) \in G \subseteq N$. N is said to be an IF α-neighbourhood of an VIFP $c(\beta)$ if $\mu_{N(C)} = 0$ and there exists an IF α-open set G in X such that $c(\beta) \in \subseteq N$.

Definition 3.17: An IFTS (X, τ) is said to be an IF $\alpha - T_0$ space iff for every pair of distinct IF points (or VIF points) c, d in X, there exists an IF α-open set U in X such that $c \in U$ and $d \notin U$ or $d \in U$ and $c \notin U$.

For $c(\gamma, \beta)$ or $d(\theta, \delta)$ we discard the cases $\gamma = 1$, $\beta = 0$ or $\theta = 1$ or $\delta = 0$ and similarly for $c(\beta)$ or $d(\delta)$ the case $\beta = 0$ or $\delta = 0$.

Definition 3.18: An IFT space (X, τ) is said to be IF $\alpha - T_1$ space iff for every pair of distinct IF points (or VIF points) c, d in X, there exist IF–open sets U, V in X such that $c \in U, d \notin U$ and $d \in V, c \notin V$.

Definition 3.19: An IFT space (X, τ) is said to be IF $\alpha - T_2$ space iff for every pair of distinct IF points (or VIF points) c, d in X, there exist IF–open sets U, V in X such that $c \in U, d \in V, U \cap V = 0_{\sim}$.

Remark 3.20: The relationships between various IF α-topological spaces are as follows:

IF $\alpha - T_2$ space $\Rightarrow$ IF $\alpha - T_1$ space $\Rightarrow$ IF $\alpha - T_0$ space.

Theorem 3.21: Let f: $(X, \tau_1) \rightarrow (Y, \tau_2)$ is one-one IF α-irresolute mapping and (Y, τ_2) be IF $\alpha - T_i$ $(i = 0, 1, 2)$ space, then (X, τ_1) is also an IF $\alpha - T_i$ $(i = 0, 1, 2)$ space.

Proof: (*a*) Case $i = 2$. Let x_1 and x_2 be two distinct points of X so that $f(x_1) = y_1$ and $f(x_2) = y_2$ are distinct points of Y as f is one-one. Since (Y, τ_2) is IF $\alpha - T_2$ space, the distinct points y_1 and y_2 of Y there exist IF α–open sets U, V in Y such that $y_1 = f(x_1) \in U$, $y_2 = f(x_2) \in V$, $U \cap V = 0_{\sim}$. Therefore $x_1 \in f^{-1}(U)$, $x_2 \in f^{-1}(V)$, $f^{-1}(U) \cap f^{-1}(V) = 0_{\sim}$.

Since f is IF α-irresolute and U, V being IF α-open sets, $f^{-1}(U)$ and $f^{-1}(V)$ are IF α-open sets such that $x_1 \in f^{-1}(U)$, $x_2 \in f^{-1}(V)$, $f^{-1}(U) \cap f^{-1}(V) = 0_{\sim}$. Hence (X, τ_1) is also an IF $\alpha - T_2$ space.

(*b*) The proof for the case $i = 0, 1$ follows immediately.

REFERENCES

[1] Atanasov, K., "Intuitionistic Fuzzy Sets", **VII**, ITKR, Session, Sofia, (1983) (in Bulgaria).

[2] Atanasov, K., "Intuitionistic Fuzzy Sets", *Fuzzy sets and systems,* **20**, (1986), 87–96.

[3] Bhaumik, R.N., and Bhattacharjee, A., "Strongly α-Continuous Function in Intuitionistic Fuzzy Topological Space". *Proc. Nat. Sem. On recent development in Mathematics and its applications*, November 14-15, 2008.

[4] Coker, D., "An Introduction to IFT Spaces", *Fuzzy Sets and Systems,* **88 (1)**, (1997) 81–89.

[5] Coker, D., and Demirci, M., "On IF Points, Notes on IFS", **1 (2)**, (1995) 79–84.

[6] Gurcay H., Coker D., and Es. A.H., "On Fuzzy Continuity in Intuitionistic Fuzzy Topological Spaces", *J. Fuzzy Math*, **5**, (1997), 365–378.

[7] Hanafy, I.M., "Completely Continuous Functions in Intuitionistic Fuzzy Topological Spaces", *Czechoslovak Math. J.*, **53 (4)**, (2003), 793–803.

[8] Prasad, R., Thakur, S.S. and Saraf, R.K., "Fuzzy α-Irresolute Mappings", *The Journal of Fuzzy Mathematics*, **2**, No. 2, (1994), 335–339.

[9] Zadeh, L.A., "Fuzzy Sets, Information and Control", **8**, (1965), 338–353.

24

ON ALEXANDROFF FUZZIFYING TOPOLOGY AND INFORMATION QUANTA

R.N. Bhaumik[1] **and Md. Azad Hussain**[2]

[1]*Department of Mathematics, Tripura University, Surryamaninagar, Tripura, India-799130*
bhaumik_r_n@yahoo.co.in

[2]*Women's Polytechnic, Hapania, Tripura (w)*
azad_hussain07@rediffmail.com

Abstract: *The aim of this paper is to investigate the relationships between Alexandroff fuzzifying topology and fuzzy approximation space in the context of fuzzy topological structures. A fuzzy preorder relation induces an Alexandroff fuzzifying topology and it turns out to be a specialization order in fuzzifying topological space. The framework of information quanta introduced by Pagliani and Chakroborty distinguishes two levels of information structures, namely property system (the primary level) and information quantum relational system (the secondary level). With the help of these ideas, we introduce the notion of fuzzy property system and information quantum fuzzy relational system by fuzzy relation. There is a bijective correspondence between property system and finite fuzzy approximation space (i.e. their respective information quantum fuzzy relational system). The fuzzy property system in terms of fuzzy symmetric topological space is also characterized.*

Keywords: *Fuzzy preorder, Alexandroff fuzzifying topology, Fuzzifying neighbourhood system, Fuzzy approximation space, Specialization order.*

Mathematics Subject Classification (2000): 54A40.

1. INTRODUCTION

After introduction of the fuzzy topology by C.L. Chang (3), many authors discussed various aspects of fuzzy topology. The aim of this paper is to investigate the relationships between Alexandroff fuzzifying topology and information quanta introduced by Pagliani and Chakraborty (9, 10). A fuzzy preorder relation induces an Alexandroff fuzzifying topology and a fuzzifying topology induces a fuzzy preorder relation, which turns out to be a specialization order in fuzzifying topology.

The theory of rough sets was proposed by Pawlak (11) as an extension of set theory for the study of intelligent system characterized by insufficient and incomplete information. A key notion in Pawlak rough set model is an equivalence relation and the equivalence classes are the building blocks for the construction of the lower and upper approximations. It is seen that under a crisp reflexive and transitive relation, the pair of lower and upper approximation operators is just a pair of interior and closure operators of a topology (7, 15).

In (5, 8) the concept of fuzzy rough sets was proposed by replacing crisp binary relations by fuzzy relations on the universe. In this paper, we discuss concept of theory of information quanta developed by Pagliani and Chakraborty (9, 10) with the help of fuzzy approximation space. We present analogous definition of property system (the primary level) and Information Quantum Relational System IQRS (the secondary level) by fuzzy relation. In section 4, we prove that if (X, E) is a finite fuzzy approximation space and $(X, \Im_E)$ is the Alexandroff fuzzifying topological space, then (*i*) (X, E) is the IQFRS induced by $(X, \Im_E)$ and (*ii*) $(X, \Im_E)$ is a finite fuzzy approximation topological space.

2. PRELIMINARIES

Let X be a non-empty universal set. The set of all subsets (resp. all fuzzy subsets) of X will be denoted by $\mathcal{P}(X)$ or 2^X[resp. $\boldsymbol{F}(X)$ or I^X]. A fuzzy subset $R \in \boldsymbol{F}(X \times X)$ is referred to as a binary relation on X. For $(x, y) \in X \times X$, $R(x, y)$ is interpreted as the degree to which x is less than y. A fuzzy relation R on X is called:

1. reflexive if $R(x, x) = 1$, for each $x \in X$;
2. symmetric if $R(x, y) = R(y, x)$ for all $(x, y) \in X \times X$;
3. $\wedge$-transitive if $R(x, z) \geq \vee_{y \in X} [R(x, y) \wedge R(y, z)$, for all $x, y]$ $z \in X$.

In this paper sup–min transitivity fuzzy relation are considered, together with standard t-norm. A fuzzy relation R on X is called:

(i) fuzzy preorder if it is a reflexive and $\wedge$-transitive fuzzy relation.

(ii) fuzzy equivalence if it is a symmetric and fuzzy preorder.

Definition 2.1 (17): Let X be universal set. A map $\Im: \mathcal{P}^X \to I$ is called a fuzzifying topology on X if it satisfies the following conditions

(FOA1) $\Im(f) = \Im(X) = 1$;

(FOA2) $\Im(V_1 \cap V_2) \geq \Im(V_1) \wedge \Im(V_2), \forall V_1, V_2 \in \mathcal{P}^X$;

(FOA3) $\Im(\cup_{j\in J} V_j) \geq \wedge_{j\in J} \Im(V_j), \forall (V_j)_{j\in J} \leq \mathcal{P}^X$;

The pair $(X, \Im)$ is called a fuzzifying topological space.

If a fuzzifying topology $\Im$ satisfies the condition (SOA2): $\Im\left(\bigcap_{j\in J} V_j\right) \geq \wedge_{j\in J} \Im(V_j), \forall (V_j)_{j\in J} \leq \mathcal{P}^X$, then $(X, \Im)$ is called a Alexandroff fuzzifying topological space.

It is to be noted that in (4), Chattopadhyay, Hazra, Samanta generalized the concept of fuzzifying topology.

The concept of neighbourhood systems introduced by Ying, Zhang and *Xu* is as follows:

Definition 2.2 (17): Let X be a universal set and $\Im: \mathcal{P}^X \to I$ be a fuzzifying topology on X. For all $x \in X$, the neighbourhood system of x, denoted by $N_x: P^X \to I$, is defined as

$$N_x(V) = \vee_{x\in U \subseteq V} \Im(U), \forall V \in P^X.$$

$N_x(V)$ is open and the set $N = (N_x: x \in X)$ is called a fuzzifying neighbourhood system.

Lemma 2.3 (17): $\Im(V) = \wedge_{x\in V} N_x(V), \forall V \in \mathcal{P}^X$.

Definition 2.4 (18): A generalized neighbourhood system on X is a set $N = (N_x, x \in X)$, where $N_x: \mathcal{P}^X \to I$ satisfies the following conditions:

(GN1) $N_x(X) = 1, N_x(\phi) = 0$;

(GN2) $N_x(V_1) > 0 \Rightarrow x \in V_1$;

(GN3) $N_x(V_1 \cap V_2) = N_x(V_1) \wedge N_x(V_2)$;

(GN4) $N_x(V_1) = \vee_{x\in V_1 \subseteq V_2} \wedge_{y\in V_2} N_y(V_2); \forall V_1, V_2 \in \mathcal{P}^X$.

The fuzzifying neighbourhood system on X defined in Definition 2.2 is a generalized neighbourhood system on X. Zhang and *Xu* (18) proved that there is a one-to-one correspondence between fuzzifying topologies and generalized neighbourhood system on X.

3. ALEXANDROFF FUZZIFYING TOPOLOGY AND SPECIALIZATION ORDER

In this section, we focus on the problem how to obtain an Alexandroff fuzzifying topology by a fuzzy relation which is motivated by the approximation operator in the theory of fuzzy rough sets.

Definition 3.1: Let R be a fuzzy preorder relation on a non-empty set X.

(*i*) The upper fuzzifying approximation operator is a map R^*: $\mathcal{P}^X \to F(X)$ defined by

$$R^*(V)(x) = \vee_{y \in V} R(x, y),\ \forall V \in \mathcal{P}^X,\ \forall x \in X.$$

(*ii*) The lower fuzzifying approximation operator is a map R_*: $\mathcal{P}^X \to F(X)$ defined by

$$R_*(V)(x) = \wedge_{y \notin V} (1 - R(x, y)).$$

If $V = \{y\} \in \mathcal{P}^X$ for some $y \in X$, we have $R^*(V)(x) = R(x, y)$ for each $x \in X$.

If $V = X - \{y\} \in \mathcal{P}(X)$ for some $y \in X$, then $R_*(V)(x) = 1 - R(x, y)$ for each $x \in X$.

Theorem 3.2: Let R be a fuzzy preorder relation on X. Then

(*i*) $R_*(X) = 1_X$, $R_*(\phi) = 0_x$;

(*ii*) $R_*(\cap_{j \in J} V_j) = \wedge_{j \in J} R_*(V_j)$;

(*iii*) $R_*(\cup_{j \in J} V_j) = \wedge_{j \in J} R_*(V_j)$.

Proof: The proof is trivial.

From the above, it is seen that lower fuzzifying approximation operator forms a topology.

Definition 3.3: Let N_x^R: $2^X \to I$ be a map defined by lower fuzzifying approximation operator R_* as follows $N_x^R(V) = R_*(V)(x) = \wedge_{y \notin V} (1 - R(x, y))$, $\forall x \in X$, $\forall V \in 2^X$.

Theorem 3.4: Let R be a fuzzy preorder on X. Then the system $N^R = (N_x^R : x \in X)$ has the properties

$$(N1)\ N_x^R(X) = 1,\ N_x^R(\phi) = 0;$$

$(N2)\ N_x^R(V) > 0 \Rightarrow x \in V;$

$\rightarrow (N3)\ N_x^R(\cap_{j\in J} V_j) = \wedge_{j\in J} N_x^R(V_j);$

$(N4)\ N_x^R(V) = \vee_{x\in U\subseteq V} \wedge_{y\in U} N_y^R(U)$

Proof: The proof is straightforward.

The set $N^R = (N_x^R: x\in X)$ is called a Alexandroff fuzzifying neighbourhood system on X.

Theorem 3.5: Let R be fuzzy preorder relation on X. A map $\Im_R: 2^X \rightarrow I$ defined by

$$\Im_R(V) = \wedge_{x\in V} N_x^R(V),\ \forall\ V\in 2^X$$

is an Alexandroff fuzzifying topology on X.

Proof: (FOA1) Clearly $\Im_R(X) = 1$, $\Im_R(\phi) = 1$.

(FOA3) For any $(V_j) \subseteq 2^X$, $\Im_R(\cup_{j\in J} V_j) = \wedge_{x\in\cup_{j\in J} V_j} N_x^R(\cup_{j\in J} V_j)$ $= \wedge_{j\in J} \wedge_{x\in V_j} N_x^R(\cup_{j\in J} V_j) \geq \wedge_{j\in J} \wedge_{x\in V_j} N_x^R(V_j)$ (by Theorem 3.4) $=$ $\wedge_{j\in J} \Im_R(V_j)$, by definition of $\Im_R$.

(SOA2) For any $(V_j) \subseteq 2^X$, $\Im_R(\cap_{j\in J} V_j) = \wedge_{x\in\cap_{j\in J} V_j} N_x^R(\cap_{j\in J} V_j)$

$= \wedge_{x\in\cap_{j\in J} V_j} \wedge_{j\in J} N_x^R(V_j)$ by $N3$

$= \wedge_{j\in J} \wedge_{x\in\cap_{j\in J} V_j} N_x^R(V_j) \geq \wedge_{j\in J} \wedge_{x\in V_j} N_x^R(V_j) = \wedge_{j\in J} \Im_R(V_j)$.

Hence the theorem. The pair $(X, \Im_R)$ is an Alexandroff fuzzifying topological space.

Definition 3.6: Let $\Im: 2^X \rightarrow I$ be a fuzzifying topology on X. $R_\Im$ is a fuzzy relation, defined by $R_\Im(x, y) = \wedge_{(x,y)\in V\times(X-V)} (1 - \Im(V))$, $\forall\ (x, y) \in X \times X$.

Theorem 3.7: If $\Im: 2^X \rightarrow I$ is a fuzzifying topology on X, then $R_\Im(x, y) = \wedge_{(x,y)\in V\times(X-V)} (1 - \Im(V))$, $\forall (x, y) \in X \times X$ is a fuzzy preorder relation on X.

Proof: Clearly $R_\Im(x, x) = 1$, $\forall x \in X$. Hence $R_\Im$ is reflexive fuzzy relation on X.

For all $x, y, z, \in X$ and for all $V, G, H, F \in 2^X$ we have

$$1 - R_\Im(x, y) = 1 - \wedge_{(x,y)\in V\times(X-V)} (1 - \Im(V))$$
$$= 1 - \wedge_{(x,y)\in V\times(X-V)} (1 - \wedge_{z\in V} N_z(V))$$

$$= 1 - 1 + \vee_{(x, y) \in V \times (X - V)} \wedge_{z \in v} N_z(V)$$

$$= \vee_{(x, y) \in V(X - V)} \wedge_{z \in V} N_z(V)$$

Now $(1 - R_{\Im}(x, y)) \vee (1 - R_{\Im}(y, z)) = (\vee_{(x, y) \in G \times (X - G)} \wedge_{s \in G} N_s(G)) \vee (\vee_{(y, z) \in H \times (X - H)} \wedge_{t \in H} N_t(H))$ and $1 - R_{\Im}(x, z) = \vee_{(x, z) \in F \times (X - F)} \wedge_{r \in F} N_r(F)$.

To prove $R_{\Im}$ is $\wedge$-transitive, i.e. $R_{\Im}(x, y) \wedge R_{\Im}(y, z) \leq R_{\Im}(x, z)$, it is sufficient to show the following inequality.

$(\vee_{(x, y) \in G \times (X - G)} \wedge_{s \in G} N_s(G)) \vee (\vee_{(y, z) \in H \times (X - H)} \wedge_{t \in H} N_t(H)) \geq \vee_{(x, z) \in F \times (X - F)} \wedge_{r \in F} N_r(F)$.

Now for any $(x, z) \in F \times (X - F)$ and $y \in F$, we have $\wedge_{r \in F} N_r(F) \leq \vee_{(y, z) \in H \times (X - H)} \wedge_{t \in h} N_t(H)$, since $z \notin F$ in this case, otherwise $y \notin F$. Also for $x \in F$, we have $\wedge_{r \in F} N_r(F) \leq \vee_{(x, y) \in G \times (X - G)} \wedge_{s \in G} N_s(G)$. So we have, $\vee_{(x, z) \in F \times (X - F)} \wedge_{r \in F} N_r(F) \leq (\vee_{(x, y) \in G \times (X - G)} \wedge_{s \in G} N_s(G)) \vee (\vee_{(y, z) \in H \times (X - H)} \wedge_{t \in h} N_t(H))$. Hence the theorem.

Then we call $R_{\Im}$ as a specialization order.

Theorem 3.8: If R is a reflexive fuzzy relation on X, then

(*i*) $R_{\Im_R} \geq R$ in general.

(*ii*) $R_{\Im_R} = R$ if and only if R is fuzzy preorder.

Proof: The proof is easy.

The following theorem shows that Alexandroff fuzzifying topology is the largest specialization fuzzifying topology generated by the specialization order.

Theorem 3.9: Let $\Im$ be a fuzzifying topology on X and $\Im_{R\Im}$ be a specialization fuzzifying topology generated by specialization order $R_{\Im}$. Then

(*i*) $\Im_{R_{\Im}} \geq \Im$ in general.

(*ii*) $\Im_{R_{\Im}} = \Im$ if and only if $\Im$ is Alexandroff fuzzifying topology.

Proof: (*i*) By Theorem 3.5, $\Im_R(V_x) = \wedge_{x \in V} N_x^R(V) = \wedge_{x \in V} \wedge_{y \notin V} (1 - R(x, y)) = 1 - \vee_{x \in V} \vee_{y \in (X - V)} R(x, y) = 1 - \vee_{(x, y) \in V \times (X - V)} R(x, y)$. Now $\Im_{R\Im}(V) = 1 - \acute{U}_{(x, y) \in V \times (X - V)} R_{\Im}(x, y) = 1 - \vee_{(x, y) \in V \times (X - V)} \wedge_{(x, y) \in G \times (X - G)} (1 - \Im(G))$ (by Definition 3.6) $= \wedge_{(x, y) \in V \times (X - V)} \vee_{(x, y) \in G \times (X - G)} \Im(G) \geq \Im(V)$

(*ii*) Let $\Im$ be Alexandroff fuzzifying topology. To Show $\Im_{R_\Im} = \Im$, we have to show that $\Im_{R_\Im}(V) \leq \Im(V)$ for any $V \in 2^X$. For any $t \in I - \{1\}$, we have $\Im_{R_\Im}(V) = \wedge_{(x, y) \in V \times (X - V)} \vee_{(x, y) \in G \times (X - G)} \Im(G) > t$. For any $y \in X - V$, $x \in V$, there exists $G_y^x \in 2^X$, such that $x \in G_y^x$, $y \notin G_y^x$ and $\Im(G_y^x) > t$. Let $G_y = \cup_{x \in Vh}$, then $y \notin D$, $V \subseteq G_y$ and $\Im(G_y) \geq$

$\wedge \quad x \quad \in \quad V$

$\Im(G_y) \geq t$. Clearly, $V = \cap_{y \in X - V} G_y$. So $\Im(V) \geq \wedge_{y \in X - V} \Im(G_y) \geq t$, since $\Im$ is Alexandroff fuzzifying topology. Thus $t < \Im_{R_\Im}(V) \Rightarrow t \leq \Im(V)$. Hence $\Im_{R_\Im}(V) \leq \Im(V)$.

Conversely, let $\Im_{R_\Im} = \Im$, then by Theorem 3.7 and Theorem 3.5, $\Im$ is Alêxandroff fuzzifying topology.

Let $A(X)$ be the set of all Alexandroff fuzzifying topology on X and $\Re(X)$ be the set of all fuzzy preorders on X.

Theorem 3.10: There is a one-to-one correspondence between Alexandroff fuzzifying topology on a set X and fuzzy preorders on X.

Proof: Let $\theta: \Re(X) \to A(X)$ be the map defined by $\theta(R) = \Im_R$, $\forall R \in \Re(X)$. By Theorem 3.8 and Theorem 3.9, θ is a one-to-one correspondence from $\Re(X)$ to $A(X)$.

Definition 3.11 (17): Let $(X, \Im)$ be a fuzzifying topological space. The closure $Cl(V)$ of $V \subset X$ is defined as $Cl(V)(x) = \wedge_{x \notin U \supseteq V} (1 - F(U))$, $\forall U \subset X$ and $x \in X$.

Recall that for any topological space (X, τ) we can convert the relation of set inclusion on τ into preorder called specialization order, defined on elements of X: $x \preccurlyeq y$ iff $x \in Cl(y)$ or equivalently, $x \preccurlyeq y$ iff $Cl(x) \in Cl(y)$ where Cl is the closure operator induced by τ. Following this idea and definition 3.6 we call R_3 is a specialization order.

For any $(x, y) \in X \times X$ and a real number $t \in (0, 1)$, the condition $t \leq R_3(x, y)$ can be stated that in t-level, $x \preccurlyeq y$ iff $x \in Cl(y)$, or equivalently, in t-level, $x \preccurlyeq y$ iff $Cl(x) \leq Cl(y)$.......... (1).

Lemma 3.12 (17): $Cl(V)(x) = 1 - N_x(V)$.

By lemma 2.3 and lemma 3.12, the condition $t \leq R_3(x, y)$ can also be stated that in t-level, $x \preccurlyeq y$ iff $\Im(U_y) \leq \Im(V_x)$.......... (2), where $V_x \in 2^X$ containing x and $U_y \in 2^X$ containing y.

A fuzzifying topology τ on X is said to be generated by $\Im: 2^X \to I$ if τ is the coarsest fuzzifying topology on X containing $\Im$ and a fuzzifying topology δ on X with the property of $\Im \leq \delta$ implies $\tau \leq \delta$. Then $\Im$ is said to be a subbase of τ.

Theorem 3.13: If (X, τ) is any fuzzifying topological space and τ is generated by $\Im: 2^X \to I$, then $R_{\Im}(x, y) = R_\tau(x, y)$ (i.e. $R_{\Im} = R_\tau$).

Proof: Let $R_{\Im}(x, y) = \wedge_{(x, y) \in V \times (X - V)} (1 - \Im(V)), \forall (x, y) \in X \times X$. Then $R_{\Im}$ is a fuzzy relation on X. If $\Im_{R_{\Im}}$ is an Alexandroff fuzzifying topology on X induced by $R_{\Im}$, then for any $(x, y) \in X \times X$ and $V \in 2^{X'}$ we have $\Im_{R_{\Im}}(V) \geq \Im(V)$ (Theorem 3.9) i.e. $\Im_{R_{\Im}} \geq \Im$ and so $\tau \leq \Im_{R_{\Im}}$, since τ is the coarsest fuzzifying topology on X containing $\Im$.

Now, for all $(x, y) \in X \times X$, $R_{\Im}(x, y) = \wedge_{(x, y) \in V \times (X - V)} (1 - \Im(V)) \geq \wedge_{(x, y) \in V \times (X - V)} (1 - \Im(V)) = R_\tau(x, y)$ i.e. $R_{\Im} \geq R_t$. Also for all $(x, y) \in X \times X$, $R_t(x, y) = \wedge_{(x, y) \in V \times (X - V)} (1 - \Im(V)) \geq \wedge_{(x, y) \in V \times (X - V)} (1 - \Im_{R\Im}(V))$ (since $\tau \leq \Im_{R_{\Im}}$) $= \wedge_{(x, y) \in V \times (X - V)} (1 - \Im(V))$ (since $\Im_{R_{\Im}}$ is a Alexandroff fuzzifying topology) $= R_{\Im}(x, y)$ i.e. $R_\tau \geq R_{\Im}$.

Therefore, $\forall (x, y) \in X \times X$, $R_{\Im}(x, y) = R_\tau(x, y)$.

4. RELATIONSHIPS BETWEEN ALEXANDROFF FUZZIFYING TOPOLOGY AND INFORMATION QUANTA

In this section the concepts of the theory of property system and information quantum relational system developed by Pagliani and Chakraborty (9, 10) are defined with the help of fuzzy relation.

In property system, the objects are directly related with their properties.

Definition 4.1: A triple $(X, M, \Vdash)$, where X is a finite set of objects and M is a finite set of observable properties, $\Vdash$ is a fuzzy relation from X into M such that for all $x \in X$ there exists $m \in M$ satisfying $\Vdash(x, m) > 0$ and for all $m \in M$ there exists $x \in X$ satisfying $\Vdash(x, m) > 0$, is called a fuzzy property system.

If we consider the set of objects rather than single object, we get secondary level information structures as follows:

Definition 4.2: Let S be a property system and $Q_x = \{x' : \forall m \in M$ if $x \Vdash m$ then $x' \Vdash m\}$ for all $x \in X$, then there exists a fuzzy relation ρ defined by $\rho(x, x') = \min\{\Vdash(x, m) > 0, \Vdash(x', m) > 0\}$ induced by S

iff $(x, x') \in \rho$ iff $x' \in Q_x$. The system (X, ρ) is called an information quantum fuzzy relational system (IQFRS), ρ is called an information quantum fuzzy relation and Q_x is referred to as minimum quantum of perception.

The central concept of the theory of fuzzy rough set is the fuzzy approximation space. Let (X, E) be a fuzzy approximation space. If $(x, x') \in E$ for all $x, x' \in X$, then x, x' are regarded as indiscernible under the knowledge given above. Thus we can not distinguish x which is a "granulae" or "perception parcel" from those objects consisting x by means of gathered information (8, 9).

As any finite fuzzifying topological space is an Alexandroff fuzzifying topological space, so we focus on finite fuzzifying topological space which can be considered as a particular case of property system $(X, X, R_{\Im})$.

Theorem 4.3: If $(X, \Im)$ is a finite fuzzifying topological space and (X, ρ) is its IQFRS, then $\rho = R_{\Im}$.

Proof: By the definition of IQFRS, $Q_x = \{y\colon \forall N_x^R \in \Im_R, \forall V \in \mathcal{P}(X)$ if $x \in N_x^R(V)$ then $y \in N_x^R(V)\} = \{y\colon \acute{A}_R(V_y) \leq \Im_R(V_x)\} = \{y : y \in \Im_R(V_x)\}$. By definition $(x, y) \in \rho$ iff $y \in \Im_R(V_x)$ iff $x\, R_{\Im}\, y$, by (2). Hence $\rho = R_{\Im}$.

Thus in a finite fuzzifying topological space $(X, \Im)$ the information quantum fuzzy relation ρ induced by its property system coincides with specialization order $R_{\Im}$ of $(X, \Im)$.

In fuzzy approximation space, basic granules of knowledge can be represented by similarity classes for each element in X (14). The size of the support of every similarity classes is used to measure the granularity of the class. Thus the similarity class for $x \in X$, denoted by E_x, is a fuzzy set in X defined by the membership function $E_x(y) = E(x, y)$.

Definition 4.4: Let λ be a fuzzy subset of X and β be the family of all similarity classes E_x of $x \in X$. A fuzzy topological space $(X, \Im)$ with interior operation induced by similarity relation E, Int $(\lambda) = \vee(E_x \in \beta; E_x \leq \lambda;\ \beta$ is a subbase of $\Im)$, is called a fuzzy approximation topological space.

Theorem 4.5: If (X, E) is a finite fuzzy approximation space and $(X, \Im_E)$ is the Alexandroff fuzzifying topological space, then (X, E) is the IQFRS induced by $(X, \Im_E)$.

Proof: It follows from the Theorem 4.3.

Theorem 4.6: If (X, E) is a finite fuzzy approximation space and $(X, \Im_E)$ is the Alexandroff fuzzifying topological space induced by (X, E), then $(X, \Im_E)$ is a finite fuzzy approximation topological space.

Proof: It follows from the fact that $\Im_E(V_x)$ the fuzzy similarity class of E which contains x and is the smallest open set containing x.

Thus for any finite fuzzifying topological space we can consider fuzzy approximation spaces of the primary and the secondary level. By Theorem 3.10, there is a bijective correspondence between the levels of fuzzy approximation spaces. They provide same amount of information though they defined in different way.

Definition 4.7 [12]: A fuzzy topological space $(X, \Im)$ is said to be fuzzy symmetric iff $x\, q\, Cl\,(y)$ implies that $y\, q\, Cl\,(x)$ for all $x, y, \in X$ and Cl is the closure operator induced by $\Im$.

Theorem 4.8: Every fuzzy approximation topological space $(X, \Im)$ is fuzzy symmetric.

Proof: Let $R_\Im$ be the specialization order of $\Im$. Then $R_\Im$ is a fuzzy similarity relation, since $(X, \Im)$ is fuzzy approximation space. And so $(x, y) \in R_\Im \Rightarrow (y, x) \in R_\Im$. Then by (1), we have $x\, q\, Cl\,(y) \Rightarrow y\, q\, Cl\,(x)$ for all $x, y \in X$.

It is well known that, approximation topological space are characterized by the following clopen sets property: every closed set is open and every open set is closed (13). A fuzzy topological space $(X, \Im)$ is called zero dimensional if $\Im$ has a base clopen fuzzy sets. Every fuzzy approximation topological space is zero dimensional.

Theorem 4.9: A fuzzy topological space $(X, \Im)$ is a fuzzy approximation topological space if and only if $(X, \Im)$ is Alexandroff fuzzifying topological space and fuzzy symmetric space.

Proof: Let $(X, \Im)$ be fuzzy approximation topological space. Then by Theorem 4.8, it is symmetric. Since $(X, \Im)$ has clopen fuzzy set property, therefore $(X, \Im)$ is Alexandroff fuzzifying topological space.

Conversely, let $(X, \Im)$ be an Alexandroff fuzzifying topological space and fuzzy symmetric. Now we have to show that for any $\lambda \leq X$

is fuzzy open iff it is fuzzy closed. Let λ be fuzzy open. Then $\lambda = \vee_{x \in \lambda} \Im_R(V_x)$. Since specialization order $R_\Im$ of $\Im$ is fuzzy symmetric, $\Im_R(V_x) = \{y \in X: (x, y) \in R\}$. Also we know that in Alexandroff fuzzifying topological spaces, a fuzzy subset μ is closed iff $\mu = \vee_{x \in \mu} \Im_R(V_x)$. Thus λ is fuzzy closed. Similarly, we can prove that if λ is closed then it is open. Hence the theorem.

Corollary 4.10: A finite fuzzifying topological space $(X, \Im)$ is a fuzzy property system iff it is fuzzy symmetric.

ACKNOWLEDGEMENT

The authors gratefully acknowledge their thanks to the referee for his valuable suggestions.

REFERENCES

[1] P. Alexandroff, "Discrete Raume", *Mat. Sb. (N.S.)*, **2**, (1937), 501–519.

[2] R.N. Bhaumik and Md. Azad Hussain, "Fuzzy Alexandroff Space", *J. Tri. Math. Soc.*, **9**, (2007), 65–70.

[3] C.L. Chang, "Fuzzy Topological Spaces", *J. Math. Anal Appl.*, **24**, (1968), 182–190.

[4] K.C. Chattopadhyay, R.N. Hazra, S.K. Samanta, "Gradation of Openness: Fuzzy Topology, Fuzzy Sets and Systems", **49**, (1992), 237–2421.

[5] D. Dubois, H. Prade, "Rough Fuzzy Set and Fuzzy Rough Set", *Int. J. Gen. System*, **17**, (1990), 191–209.

[6] D. Dubois. H. Prade, "Putting Fuzzy Sets and Rough Sets Together", in: R. Slowinski (ed.), *Intelligent Decision support*, Kluwer Academic, Dordrecht, 1992, pp. 203–232.

[7] J. Kortelainen, "On the Relationship Between Modified Sets, Topological Spaces and Rough Sets", *Fuzzy Sets and Systems*, **61**, (1994), 91–95.

[8] A. Nakamura, "Fuzzy Rough Sets", *Note on Multiple–Valued Logic in Japan*, **9 (8)**, (1988), 1–8.

[9] P. Pagliani and M. Chakraborty, "Information Quanta and Approximation Spaces I: Non-Classical Approximation Operators", *Gr C 2005 IEEE International Conference on Granular Computing.*

[10] P. Pagliani, M. Chakraborty, "Information Quanta and Approximation Spaces II: Generalised Approximation Space", *Gr C 2005 IEEE International Conference on Granular Computing.*

[11] Z. Pawlak, "Rough Sets", *Internat. J. Comput Inform. Sci.*, **11**, (1982), 341–356.

[12] M.E. EL–Shaffi, "Some Applications of Generalized Closed Sets in Fuzzy Topological Spaces", *Kyungpook Math. J.*, **45**, (2005), 13 – 19.

[13] H. Rasiowa, "Algebraic Models of Logics", University of Warsaw, 2001.

[14] R. Slowinski and D.Vanderpooten, "A Generalized Definition of Rough Approximations Based on Similarity", *IEEE Trans. On knowledge and Data Engineering*, **12 (2)**, (2000), 331 – 336.

[15] Y.Y. Yao, "Two Views of the Theory of Rough Sets in Finite Universes", *Internat. J. Approx. Reasoning.*, **15**, (1996), 291 – 317.

[16] Y.Y. Yao, "Combination of Rough and Fuzzy Sets Based on L-Level Sets, Rough Set and Data Mining: Analysis for Imprecise Data", T.Y. Lin, and N. Cercone (Ed.), Kluwer. Academic, Boston, MA, pp. 301 – 321, 1997.

[17] M. Ying, "A New Approach for Fuzzy Topology (I)", *Fuzzy Sets and Systems*, **39**, (1991), 303 – 321.

[18] D. Zhang and L. Xu, "Categories Isomorphic to FNS", *Fuzzy Sets and Systems*, **104**, (1999), 373 – 380.

25

AN APPROACH TO ESTIMATE THE CRITERIA WEIGHTS FROM INTUITIONISTIC PREFERENCE RELATIONS

Debashree Guha and Debjani Chakraborty

Department of Mathematics, IIT-Kharagpur
Kharagpur-721302, India

Abstract: *The purpose of this paper is to introduce a new method for measuring attribute weights in decision analysis based on intuitionistic preference relation. A similarity measure of intuitionistic fuzzy sets is used to determine the degrees of importance or to estimate the weights of the attributes. The proposed importance-assessing method is a valuable tool in a variety of multi-attribute evaluation problems due to its simplicity, ease of interpretation. The method is also illustrated with the help of an example.*

1. INTRODUCTION

The theory of fuzzy sets introduced by Zadeh in 1965 [14] has achieved successful applications in various fields. An increasing number of publications devoted to this field prove that the interest in this theory is constantly rising. The concept of fuzzy set was further generalized and out of several higher order fuzzy sets, the concept of intuitionistic fuzzy sets (IFSs) proposed by Atanassov [1 – 5] in 1986 is found to be highly useful to deal with vagueness. IFSs are characterized by two functions, expressing degree of belongingness and the degree of non-belongingness respectively. Bustince and Burillo [6] pointed out that the notion of vague set [7] is the same as that of IFSs. Over the last decades a lot of research works were done over IFSs theory. Here we will not give detailed literature review about those research works and one can refer the papers [10, 12, 13] and the references therein.

Now in modeling a decision problem for instance an evaluation problem it is seen that the experts can not say very definitely that one criterion is better than the other. An uncertainty or hesitation may be included in their responses due to subjective estimation and perception, incomplete knowledge, the complexity of studied systems, etc. It is realized that such indeterminacy, (i.e. hesitation) related to human cognitive processes, such as thinking, reasoning etc. can be well captured by intuitionistic preference information. In such situation intuitionistic preference relation is a powerful tool [12, 13] to express the experts' intuitionistic preference information over criteria and/or alternatives (more detailed explanation is given in Section 2.2). Szmidt and Kacprzyk [11], and *Xu* [12] investigated the group decision making problems with intuitionistic preference relations.

Now in this respect how to calculate the weight or importance for the different criteria from this intuitionistic preference information of the experts is an essential part of a typical group decision making problem. In view of this *Z. Xu* [13] developed a method for estimating criteria weights from consistent and inconsistent intuitionistic preference relations [12] based on linear programming models. In this paper our main objective is to investigate the approach to estimate criteria weights from intuitionistic preference relations. In order to do so, the rest of this paper is organized as follows: In Section 2, we review some basic concepts. In order to compute the similarity for IFSs we have used our proposed similarity measure [8] in this work. Section 2.1 briefly describes the similarity measure for IFSs. In Section 3, we develop a method for estimating criteria weights from intuitionistic preference relations based on the similarity measure and finally, one example is given.

2. PRELIMINARIES

2.1 Similarity Measure Between two IFSs

Definition 1: Let us consider, $A, B \in IFSs\ (X)$, where $X = (x_1, x_2, \ldots, x_n)$. Then A and B are denoted as follows:

$$A = \sum_{i=1}^{n} \left[\mu_A(x_i), \nu_A(x_i)\right]/x_i \quad \text{and} \quad B = \sum_{i=1}^{n} \left[\mu_B(x_i), \nu_B(x_i)\right]/x_i$$

where $0 \le \mu_A(x_i) \le 1 - \nu_A(x_i)$ and $0 \le \mu_B(x_i) \le 1 - \nu_B(x_i)$.

Then the degree of similarity between two A and B is evaluated with the help of the following function:

$s^p(A, B) =$

$$1 - \frac{1}{\sqrt[p]{n}} \sum_{i=1}^{n} \frac{\sqrt[p]{\left|\delta_A(x_i) - \delta_B(x_i)\right|^p + \left|\alpha_A(x_i) - \alpha_B(x_i)\right|^p}}{1 + \sqrt[p]{\left|\delta_A(x_i) - \delta_B(x_i)\right|^p + \left|\alpha_A(x_i) - \alpha_B(x_i)\right|^p}};$$

for $1 \le p < \infty$. ... (1)

Where $\delta_A(x_i) = \mu_A(x_i) + [1 - \mu_A(x_i) - \nu_A(x_i)]\, \mu_A(x_i)$ and $\alpha_A(x_i) = \nu_A(x_i) + [1 - \mu_A(x_i) - \nu_A(x_i)]\, \nu_A(x_i)$.

The above proposed similarity measure is more reasonable than the existing similarity measures (8).

2.2 The Concept of Intuitionistic Preference Relation in Decision Making

In practice it is realized that that sometimes it is not possible for an expert to decide very definitely about the degree to which one alternative and/or criteria are better than others. The expert is unable to express his/her preferences accurately. In such cases, it may happen that the expert provides his/her preferences for alternatives to a certain degree, but he/she is no so sure about it [12]. Under this circumstances expert's preference values over the criteria and/or alternatives is more suitably represented by intuitionistic fuzzy values rather than exact numerical values or linguistic variables [9]. With this point of view *Xu* [12] introduced the notion of intuitionistic preference relation as follows:

Definition 2: An intuitionistic preference relation R on the set X is represented by a matrix $R = (r_{ij})_{n \times n} \subset X \times X$ with $r_{ij} = < (x_i, x_j), \mu(x_i, x_j), \nu(x_i, x_j) >$ for all $i, j = 1, 2, ..., n$. For convenience, we let $r_{ij} = (\mu_{ij}, \nu_{ij})$, for all $i, j = 1, 2, ..., n$ where r_{ij} is the intuitionistic fuzzy value, composed by the certainty degree μ_{ij} to which x_i is preferred to x_j, and the certainty degree ν_{ij} to which x_i is non-preferred to x_j, and

$\pi_{ij} = 1 - \mu_{ij} - \nu_{ij}$ is interpreted as the uncertainty degree to which x_i is preferred to x_j. Furthermore μ_{ij} and ν_{ij} satisfy the following characteristics:

$0 \leq \mu_{ij} + \nu_{ij} \leq 1$, $\mu_{ji} = \nu_{ij}$, $\nu_{ji} = \mu_{ij}$, $\mu_{ii} = \nu_{ii} = 0.5$ for all $i, j = 1, 2. ..., n$.

3. An Approach to Estimate the Criteria Weights Based on Intuitionistic Preference Relation

In this section we propose a method to find the criteria weights based on intuitionistic preference relation which can be described step wise as follows:

Step 1: For a multiple criteria decision making problem, let n numbers of criterion $C_1, C_2, ..., C_n$ are set by the expert where the weights are unknown. As discussed in Section 2.2, the expert provides his/her intuitionistic preference for each pair of criteria and constructs intuitionistic preference relation as follows:

$$R = (r_{ij})_{n \times n} \text{ where } r_{ij} = (\mu_{ij}, \nu_{ij})_{n \times n}, 0 \leq \mu_{ij} + \nu_{ij} \leq 1;$$
$$\mu_{ji} = \nu_{ij}, \nu_{ji} = \mu_{ij}, \mu_{ii} = \nu_{ii} = 0.5 \text{ for all } i, j = 1, 2, ..., n.$$

(*from Definition 2*)

Therefore for n criteria the intuitionistic preference relation $R = (r_{ij})_{n \times n}$ given by the expert is of the following form:

$$R = \begin{array}{c} \\ C_1 \\ C_2 \\ \vdots \\ C_i \\ \vdots \\ C_n \end{array} \begin{array}{c} \begin{array}{ccccccc} C_1 & & C_2 & & C_j & \cdots & C_n \end{array} \\ \begin{bmatrix} (0.5, 0.5) & (\mu_{12}, \nu_{12}) & \cdots & (\mu_{1j}, \nu_{1j}) & \cdots & (\mu_{1n}, \nu_{1n}) \\ (\mu_{21}, \nu_{21}) & (0.5, 0.5) & \cdots & (\mu_{2j}, \nu_{2j}) & \cdots & (\mu_{2n}, \nu_{2n}) \\ \vdots & \vdots & \cdots & \vdots & & \vdots \\ (\mu_{i1}, \nu_{i1}) & (\mu_{i2}, \nu_{i2}) & \cdots & (\mu_{ij}, \nu_{ij}) & \cdots & (\mu_{in}, \nu_{in}) \\ \vdots & \vdots & \cdots & \vdots & & \vdots \\ (\mu_{n1}, \nu_{n1}) & (\mu_{n1}, \nu_{n1}) & \cdots & (\mu_{nj}, \nu_{nj}) & \cdots & (0.5, 0.5) \end{bmatrix} \end{array}$$

Step 2: Among the n criteria, consider the pair (C_i, C_j), according to the expert if criteria C_i is strongly important than C_j, then the corresponding intuitionistic fuzzy value given by the expert is

$(r_{ij}) = (\mu_{ij}, \mu_{ij}) = (1, 0)$; where $\mu_{ij} = 1$ implies that the certainty degree to which C_i is preferred to C_j is 1, and the certainty degree to which C_i is non-preferred to C_j is 0. In view of this following definition can be introduced:

Definition 3: If the criteria C_i is *strongly important* than C_j, then the intuitionistic preference relation for the pair of criteria (C_i, C_j) is $(r_{ij}) = (1, 0) = r$.

Step 3: In this step compare each entry $r_{ij} = (\mu_{ij}, \nu_{ij})$ of R with $(1, 0)$ with help of the proposed similarity measure (Eq. 1).

For instance: if for the pair of criterion (C_1, C_j) the intuitionistic preference value given by the expert be $r_{1j} = (\mu_{1j}, \nu_{1j})$, then calculate the similarity between $r_{1j} = (\mu_{1j}, \nu_{1j})$ and $(1, 0)$ using Eq. (1). In this way for each pair of criteria (C_i, C_j) we can calculate $2 \times {}^nc_2$ similarity values and then it is expressed in a concise manner in the form of a matrix as follows:

$$R = \begin{array}{c} \\ C_1 \\ C_2 \\ \vdots \\ C_i \\ \vdots \\ C_n \end{array} \begin{array}{c} \begin{array}{cccccc} C_1 & C_2 & \cdots & C_j & \cdots & C_n \end{array} \\ \left[\begin{array}{cccccc} - & S_{12}^p & \cdots & S_{1j}^p & \cdots & S_{1n}^p \\ S_{21}^p & - & \cdots & S_{2j}^p & \cdots & S_{2n}^p \\ \vdots & \vdots & \cdots & \vdots & & \vdots \\ S_{i1}^p & S_{i2}^p & \cdots & S_{ij}^p & \cdots & S_{in}^p \\ \vdots & \vdots & \cdots & & \vdots & \vdots \\ S_{n1}^p & S_{n2}^p & \cdots & S_{nj}^p & \cdots & - \end{array}\right] \end{array}$$

Where $S_{ij}^p = s^p(r_{ij}, r)$ for $i \neq j$; $1 \leq p < \infty$.

Therefore the degree of importance of criteria C_i over the other criterion is defined by

$$d(c_i^p) = \min_{\substack{1 \leq j \leq n \\ i \neq j}} \left(S_{ij}^p\right) \text{ for } 1 \leq p < \infty \qquad \text{w (2)}$$

Therefore the weight vector of the criteria $C_1, C_2, \ldots, C_p$ is defined by $d(c_1^p), d(c_2^p), \ldots, d(c_n^p)$ for $1 \leq p \leq \infty$ respectively.

4. NUMERICAL EXAMPLE

Let us consider a multiple criteria decision making problem [13], where the expert gives his response depending upon five criteria C_i ($i = 1, 2, ..., 5$). Now the decision maker compares each pair of criteria C_i and C_j and provides his/her intuitionistic preference value $r_{ij} = (\mu_{ij}, \nu_{ij})$ composed by the certainty degree μ_{ij} to which C_i is preferred to C_j and the certainty degree ν_{ij} to which C_i is non-preferred to C_j degree, and then constructs the following intuitionistic preference relation

$$R = (r_{ij})_{5\times 5} = \begin{bmatrix} (0.5, 0.5) & (0.6, 0.3) & (0.4, 0.2) & (0.7, 0.2) & (0.4, 0.5) \\ (0.3, 0.6) & (0.5, 0.5) & (0.5, 0.3) & (0.6, 0.1) & (0.3, 0.6) \\ (0.2, 0.4) & (0.3, 0.5) & (0.5, 0.5) & (0.6, 0.2) & (0.4, 0.5) \\ (0.2, 0.7) & (0.1, 0.6) & (0.2, 0.6) & (0.5, 0.5) & (0.3, 0.6) \\ (0.5, 0.4) & (0.6, 0.3) & (0.5, 0.4) & (0.6, 0.3) & (0.5, 0.5) \end{bmatrix}$$

Using Eq. (1), calculate the similarity measure for $p = 1$; and finally applying the formulae (2) we will get the weight of the attributes as follows:

$$d(c_1) = 0.487,\ d(c_2) = 0.464,\ d(c_3) = 0.469,\ d(c_4) = 0.403 \text{ and } d(c_1) = 0.515.$$

ACKNOWLEDGEMENT

The first author gratefully acknowledges the financial support provided by the Council of Scientific and Industrial Research, India [Award No. 9/81 (712)/08-EMR-I).

REFERENCE

[1] K. Atanassov, "Intuitionistic Fuzzy Sets", *Fuzzy Sets and Systems*, **20**, (1986), 87–96.

[2] K. Atanassov, "More on Intuitionistic Fuzzy Sets", *Fuzzy Sets and Systems*, **33**, (1989), 37–45.

[3] K. Atanassov, "New Operations Defined Over the Intuitionistic Fuzzy Sets", *Fuzzy Sets and Systems*, **61**, (1994), 137–142.

[4] K. Atanassov, "Operators Over Interval Valued Intuitionistic Fuzzy Sets", *Fuzzy Sets and Systems*, **64**, (1994), 159–174.

[5] K. Atanassov, "Intuitionistic Fuzzy Sets: Theory and Applications", Physica-Verlag, Wyrzburg, (1999).

[6] H. Bustince, and P. Burillo, "Vague Sets are Intuitionistic Fuzzy Sets", *Fuzzy Sets and Systems*, **79**, (1996), 403–405.

[7] W.L. Gau, and D.J. Buehrer, "Vague Sets", *IEEE Trans. Systems Man Cybernet*, **23**, (1993), 610–614.

[8] Debashree Guha, and Debjani Chakraborty, "Theoretical Development of Similarity Measure Between Intuitionistic Fuzzy Sets and its Applications in Multiple Attribute Decision Making", *International Journal of fuzzy Mathematics*, To appear (2009).

[9] F. Herrera, L. Martinez, and P.J. Sanchez, "Managing Non-Homogeneous Information in Group Decision Making", *European Journal of Operational Research*, **166**, (2005), 115–132.

[10] Y. Li, David L. Olson, and Z. "Qin, Similarity Measures Between Intuitionistic Fuzzy (Vague) Sets: A Comparative Analysis", *Pattern Recognition Letters*, **28**, (2007), 278–285.

[11] E. Szmidt, and J. Kacprzyk, "Using Intuitionistic Fuzzy Sets in Group Decision Making", *Control and Cybernetics*, **31**, (2002), 1037–1053.

[12] Z.S. *Xu*, "Intuitionistic Preference Relations and their Application in Group Decision Making", *Information Sciences*, **177**, (2007), 2363–2379.

[13] Z.S. *Xu*, "A Method for Estimating Criteria Weights from Intuitionistic Preference Relations", in B.Y. Cao (Ed.) *Fuzzy Information and Engineering*, (ICFIE), ASC **40**, 503–512, Springer. **36**, (2007), 179–203.

[14] L.A. Zadeh, "Fuzzy Sets, Information and Control", **8**, (1965), 338–353.

26

ON $\mathcal{G}$-COVERING DIMENSION OF L-TOPOLOGICAL SPACES

Dalip Singh Jamwal[1] and Renu[2]

[1]*Department of Mathematics, University of Jammu, Jammu-180006, India. E-mail: dalipsj@yahoo.com*

[2]*Department of Mathematics, University of Jammu, Jammu-180006, India. E-mail: renu ju@yahoo.com*

Abstract: *In this paper the concept of $\mathcal{G}$-covering dimension of L-topological spaces is introduced and a characterization of $\mathcal{G}$-covering dimension is obtained. Also a relationship between $\mathcal{G}$-covering dimension and covering dimension of L-topological spaces is established and various properties of $\mathcal{G}$-covering dimension are studied.*

Keywords: *Fuzzy grill, $\mathcal{G}$-order, $\mathcal{G}$-covering dimension, L-topological spaces.*

2000 Subject Classification: *Primary 54A40, Secondary 54F45.*

1. INTRODUCTION AND PRELIMINARIES

The concept of grill was introduced by *G*. Choquet [2] in 1947. It has been seen from literature that the concept of grill is a powerful supporting tool, like nets and filters in dealing with many topological concepts. According to Choquet, a grill $\mathcal{G}$ on a set X is a nonempty collection $\mathcal{G}$ of nonempty subsets of X satisfying two conditions: (*i*) $U \in \mathcal{G}$ and $U \subseteq V \subseteq X \Rightarrow V \in \mathcal{G}$; and (*ii*) $U, V \subseteq X$ and $U \cup V \in \mathcal{G}$ or $V \in \mathcal{G}$.

The idea of fuzzy grill for fuzzy topological spaces was intiated by Azad [1] basically for the study and characterization of proximities in a fuzzy setting. For definition of *L*-topological space, see [4].

According to [4], all L-fuzzy topological spaces of [5] are L-topological spaces. Throughout this paper, L is a Boolean algebra. In this paper, by an L-topological space (X, δ), we mean X is a set and is an L-topology on X. Let (X, δ) be an L-topological space. A subfamily $\mathcal{A}$ of δ is said to be an open cover of (X, δ) if $\vee \mathcal{A} = 1$.

We first recall the following definition of a fuzzy grill from [1].

Let X be a non empty set. A *fuzzy grill* on X is a non empty collection $\mathcal{G}$ of L^X satisfying three conditions:

(*i*) $\underline{0} \notin \mathcal{G}$;

(*ii*) $U \in \mathcal{G}$, $V \in L^X$ and $U \leq V \Rightarrow V \in \mathcal{G}$; and

(*iii*) $U, V \in L^X$ and $U \vee V \in \mathcal{G} \Rightarrow U \in \mathcal{G}$ or $V \in \mathcal{G}$.

Now we recall the definition of order of a family of L-fuzzy sets and covering dimension of L-topological spaces as defined in [3].

Let $\Lambda \neq \phi$ and $\mathcal{A} = (A_\lambda : \lambda \in \Lambda)$ be a family of L-fuzzy subsets of a set X. The order of $\mathcal{A}$ is defined as under:

Case I: When $A_\lambda = \underline{0}$ for atleast one value of λ in Λ. Then the order of $\mathcal{A}$ is the largest non negative integer n for which there exists a subset M of Λ with $n + 1$ elements such that $\wedge_{\lambda \in M} A_\lambda \neq \underline{0}$, or is ∞ if there is no such largest integer.

Case II: When $A_\lambda = \underline{0}$ for all $\lambda \in \Lambda$. Then the order of $\mathcal{A}$ A is -1.

The Covering Dimension of an L-topological space (X, δ) is the least integer n such that every finite open covering of (X, δ) has an open refinement of order not exceeding n, or is ∞ if there is no such integer. The covering dimension of an L-topological space (X, δ) is denoted by dim (X, δ).

For basic concepts and results used in this paper, we refer [4, 5].

2. $\mathcal{G}$-COVERING DIMENSION

We begin by defining the concept of $\mathcal{G}$-order of a family of L-fuzzy subsets of a set X, which is a generlization of the concept of order of a family of L-fuzzy subsets.

Definition 2.1: Let $\Lambda \neq \phi$ and $\mathcal{A} = (A_\lambda : \lambda \in \Lambda)$ be a family of L-fuzzy subsets of a set X and $\mathcal{G}$ be a fuzzy grill on X. Then $\mathcal{G}$-order of $\mathcal{A}$ is defined as under:

Case I: When $A_\lambda \neq \underline{0}$ for atleast one value of λ in Λ. Then the $\mathcal{G}$-order of A is the largest nonnegative integer n for which there exists a subset M of Λ with $n + 1$ elements such that $\wedge_{\lambda \in M} A_\lambda \notin \mathcal{G}$, or is ∞ if there is no such largest integer n.

Case II: When $A_\lambda = \underline{0}$ for all $\lambda \notin \Lambda$. Then the $\mathcal{G}$-order of $\mathcal{A}$ is -1.

We now define the concept of $\mathcal{G}$-covering dimension of an L-topological space (X, δ), which is a generalization of covering dimension of an L-topological space.

Definition 2.2: Let (X, δ) be an L-topological space and $\mathcal{G}$ be a fuzzy grill on X. Then the *$\mathcal{G}$-covering dimension* of (X, δ), denoted by $\mathcal{G}$-dim (X, δ), is the least integer n such that for every finite open covering $\mathcal{U} = (U_1, U_2, \ldots, U_k)$ of (X, δ), there exists a subfamily $\mathcal{V} = (V_\alpha : \alpha \in \wedge)$ of δ of $\mathcal{G}$-order not exceeding n such that $(\wedge_{\alpha \in \wedge} V\alpha)' \notin \mathcal{G}$ and for each $V\alpha \in \mathcal{V}$, $\exists U_i \in \mathcal{U}$ such that $V_\alpha \leq U_i$. If there exists no such integer n, then $\mathcal{G}$-dim $(X, \delta) = \infty$.

We first obtain the following Characterization of $\mathcal{G}$-covering dimension.

Theorem 2.3: Let (X, δ) be an L-topological space and $\mathcal{G}$ be a fuzzy grill on X. Then the following statements about (X, δ) are equivalent:

(*a*) $\mathcal{G}$-dim $(X, \delta) \leq n$.

(*b*) For every finite open covering $\mathcal{U} = (U_1, U_2, \ldots, U_k)$ of (X, δ), there is a finite subfamily $\mathcal{V} = (V_1, V_2, \ldots, V_k)$ of δ of $\mathcal{G}$-order not exceeding n such that $(\vee_{i=1}^{k} V_i)' \notin \mathcal{G}$ and $V_i \leq U_i, \forall\, i = 1, 2, \ldots, k$;

(*c*) If $\mathcal{U} = (U_1, U_2, \ldots, U_{n+2})$ is finite open covering of (X, δ), there is a subfamily $\mathcal{V} = (V_1, V_2, \ldots, V_{n+2})$ of δ such that $V_i \leq U_i, \forall\, i, \wedge_{i=1}^{n+2} V_i \notin \mathcal{G}$ and $\left(\wedge_{i=1}^{n+2} V_i\right)' \notin \mathcal{G}$.

Proof: (*a*) $\Rightarrow$ (*b*): Suppose that $\mathcal{G}$-dim $(X, \delta) \leq n$ and let $\mathcal{U} = (U_1, U_2, \ldots, U_k)$ be a finite open cover of (X, δ).

Case I: When some $U_i = \underline{1}$. Then we can write $(U_1, U_2, \ldots\ldots, U_k) = (U_1, U_2, \ldots\ldots, U_{i-1}, U_{i+1}, \ldots\ldots, U_k)$. Clearly, $(V_1, V_2, \ldots\ldots, V_k) = (\underline{0}, \underline{0}, \ldots\ldots, \underline{0}, \underline{1}, \ldots\ldots, \underline{0})$ is a finite subfamily of of $\mathcal{G}$-order $= 0 \leq n$ such that $(\vee_{i=1}^{k} V_i)' = \wedge_{i=1}^{k} V_i' = \underline{0} \notin \mathcal{G}$ and $V_i \leq U_i, \forall\, i = 1, 2, \ldots\ldots, k$.

Case II: When no $U_i = \underline{1}$. Since $\mathcal{G}$-dim $(X, \delta) \leq n$, for each finite open covering $\mathcal{U} = (U_1, U_2, \ldots\ldots, U_k)$ of (X, δ), there exists a subfamily $\mathcal{W} = (W_\alpha : \alpha \in \Lambda)$ of δ of $\mathcal{G}$-order not exceeding n such that $(\vee_{\alpha\in\Lambda} W_\alpha)' \notin \mathcal{G}$ and for each $W_\alpha \in \mathcal{W}$, $\exists$ some $U_i \in \mathcal{U}$ such that $W_\alpha \leq U_i$. For each $i = 1, 2, \ldots\ldots, k$, let A_i be the join of those members Wa of $\mathcal{W}$ associated with U_i such that $W_\alpha \leq U_i$. Then each A_i is open and $A_i \leq U_i$. Let $\mathcal{W}_1 = (A_1, A_2, \ldots\ldots, A_k)$. For each $i = 1, 2, \ldots\ldots, k$, we define.

$$V_i = \begin{cases} A_i, & \textit{where } A_i \in \mathcal{W}_1 \textit{ and } A_i \not< A_j, \forall A_j \in \mathcal{W}_1 \\ \underline{0}, & \textit{if } A_i < A_j, \textit{ for some } A_j \in \mathcal{W}_1 \end{cases}$$

Let $\mathcal{V} = (V_1, V_2, \ldots\ldots, V_k)$. Then $\mathcal{V}$ is a subfamily of δ such that $V_i \leq U_i, \forall\, i = 1, 2, \ldots\ldots, k$ and $(\vee_{i=1}^{k} V_i)' = (\vee_{i=1}^{k} A_i)' = (\vee_{\alpha\in\Lambda} W_\alpha)' \notin \mathcal{G}$.

Next, we shall show that $\mathcal{G}$-order of $\mathcal{V}$ does not exceed n. For this, let $\mathcal{M} = (V_{\alpha 1}, V_{\alpha 2}, \ldots\ldots, V_{\alpha\, n+2})$ be any subfamily of $\mathcal{V}$ having $n + 2$ members.

Case I: If some $V_{\alpha i} = \underline{0}$, then clearly $\wedge \mathcal{M} = \underline{0} \notin \mathcal{G}$.

Case II: When all the members of $\mathcal{M}$ are non zero, then we have to show $\wedge \mathcal{M} \notin \mathcal{G}$. Suppose, contrary that $\wedge \mathcal{M} \notin \mathcal{G}$. Then $\wedge_{i=1}^{n+2} V_{\alpha i} \neq \underline{0}$. Let $\wedge_{i=1}^{n+2} V_{\alpha i} = B$. Then B is a non zero member of L^X. We claim that $B < V_{\alpha i}$, $\forall\, i = 1, 2, \ldots\ldots, n + 2$. For suppose, $B = V_{\alpha i}$ for some i. Then it means $V_{\alpha i}$ < all the remaining members of $\mathcal{M}$, which is a contradiction to the construction of $\mathcal{V}$. Hence, $B < V_{\alpha i}$, $\forall\, i = 1, 2, \ldots\ldots, n + 2$. But $B < V_{\alpha i}$, $\forall\, i = 1, 2, \ldots\ldots, n + 2$ implies B is less than $n + 2$ non-zero elements of $\mathcal{W}_1$. This is a contradiction because $\mathcal{G}$-order of $\mathcal{W}$ does not exceed n. Hence, $\wedge\mathcal{M} = \underline{0} \notin \mathcal{G}$. This proves that $\mathcal{G}$-order of $\mathcal{V}$ does not exceed n.

(b) ⇒ (c): Let $\mathcal{U} = (U_1, U_2, \ldots, U_{n+2})$ be a finite open covering of (X, δ). Then there exists a subfamily $\mathcal{V} = (V_1, V_2, \ldots, V_{n+2})$ of δ such that $V_i \le U_i$, $\forall\, i$, and $\mathcal{G}$-order of $\mathcal{V}$ does not exceed n, i.e. meet of atmost $n + 1$ members of $\mathcal{V}$ belongs to $\mathcal{G}$, so $\wedge_{i=1}^{n+2} V_i \notin \mathcal{G}$ and $\left(\vee_{i=1}^{n+2} V_i\right)' \notin \mathcal{G}$.

(c) ⇒ (b): Let $\mathcal{U} = (U_1, U_2, \ldots, U_k)$ be a finite open covering of (X, δ). We can assume that $k > n + 1$. Let $G_i = U_i$, if $i \le n + 1$ and $G_{n+2} = \vee_{i=n+2}^{k} U_i$. Then $(G_1, G_2, \ldots, G_{n+2})$ is an open covering of (X, δ) and so, by the given condition, there exists a subfamily $(H_1, H_2, \ldots, H_{n+2})$ of δ such that $H_i \le G_i, \forall i$, $\wedge_{i=1}^{n+2} H_i \notin \mathcal{G}$ and $\left(\wedge_{i=1}^{n+2} H_i\right)' \notin \mathcal{G}$. Let $W_i = H_i$, if $i \le n + 1$ and $W_i = U_i \wedge H_{n+2}$, if $i > n + 1$. Then $\mathcal{W} = (W_1, W_2, \ldots, W_k)$ is a subfamily of δ such that each $W_i \le U_i$.

Next, we shall show that $(\vee_{i=1}^{k} W_i)' \notin \mathcal{G}$. For this, we first observe that

$$
\begin{aligned}
(\vee_{i=1}^{k} W_i)' &= \wedge_{i=1}^{k} W_i' \\
&= \left[\wedge_{i=1}^{n+1} W_i'\right] \wedge \left[\wedge_{i=n+2}^{k} W_i'\right] \\
&= \left[\wedge_{i=1}^{n+1} H_i'\right] \wedge \left[\wedge_{i=n+2}^{k} \left(U_i \wedge H_{n+2}\right)'\right] \\
&= \left[\wedge_{i=1}^{n+1} H_i'\right] \wedge \left[\wedge_{i=n+2}^{k} \left(U_i' \vee H_{n+2}'\right)\right] \\
&= \left[\wedge_{i=1}^{n+1} H_i'\right] \wedge \left[\left(\wedge_{i=n+2}^{k} U_i'\right) \vee H_{n+2}'\right] \\
&= \left[\left(\wedge_{i=1}^{n+1} H_i'\right) \wedge \left(\wedge_{i=n+2}^{k} U_i'\right)\right] \vee \left[\wedge_{i=1}^{n+2} H_i'\right] \\
&= \left[\left(\wedge_{i=1}^{n+1} H_i\right)' \wedge \left(\vee_{i=n+2}^{k} U_i\right)'\right] \vee \left[\wedge_{i=1}^{n+2} H_i'\right] \quad ---(*)
\end{aligned}
$$

Now suppose $(\vee_{i=1}^{k} W_i)' \in \mathcal{G}$. Then from (*), either $\left(\vee_{i=1}^{n+1} H_i\right)' \wedge (\vee_{i=n+2}^{k} U_i)' \in \mathcal{G}$ or $\wedge_{i=1}^{n+2} H_i' \in \mathcal{G}$. But $\wedge_{i=1}^{n+2} H_i' = \left(\vee_{i=1}^{n+2} H_i\right)' \notin \mathcal{G}$ and so $\left(\vee_{i=1}^{n+2} H_i\right)' \wedge (\vee_{i=n+2}^{k} U_i)' \in \mathcal{G}$. One can show that $\left(\vee_{i=1}^{n+1} H_i\right)' \wedge (\vee_{i=n+2}^{k} U_i)' \le \left(\vee_{i=1}^{n+2} H_i\right)'$. Thus $\left(\vee_{i=1}^{n+1} H_i\right)' \wedge (\vee_{i=n+2}^{k} U_i)' \in \mathcal{G}$ implies $\left(\vee_{i=1}^{n+2} H_i\right)' \in \mathcal{G}$, which is not true. Therefore, $(\vee_{i=1}^{k} W_i)' \notin \mathcal{G}$

Also, $\wedge_{i=1}^{n+2} W_i = \wedge_{i=1}^{n+1} W_i \vee W_{n+2} = \wedge_{i=1}^{n+1} H_i \vee (U_{n+2} \wedge H_{n+2}) = \left(\vee_{i=1}^{n+2} H_i\right) \wedge U_{n+2} \notin \mathcal{G}$ because $\wedge_{i=1}^{n+2} H_i \notin \mathcal{G}$. Therefore, $\wedge_{i=1}^{n+2} W_i \notin \mathcal{G}$.

If there exists a subset B of $(1, 2, \ldots\ldots, k)$ with $n + 2$ elements such that $\wedge_{i \in B} W_i \in \mathcal{G}$, let the members of $\mathcal{W}$ be renumbered to give a family $\mathcal{P} = (P_1, P_2, \ldots\ldots P_k)$ such that $\wedge_{i=1}^{n+2} P_i \notin \mathcal{G}$. By applying the above construction to $\mathcal{P}$, we obtain a subfamily $\mathcal{C} = (C_1, C_2, \ldots\ldots, C_k)$ of δ such that each $C_i \leq P_i$, $\forall i$ and $\wedge_{i=1}^{n+2} C_i \notin \mathcal{G}$. Clearly, if B_1 is a subset of $(1, 2, \ldots\ldots, k)$ with $n + 2$ elements such that $\wedge_{i \in B_1} P_i \notin \mathcal{G}$, then $\wedge_{i \in B_1} C_i \notin \mathcal{G}$. Thus by a finite number of repetitions of this process, we obtain a finite subfamily $(V_1, V_2, \ldots\ldots V_k)$ of δ of $\mathcal{G}$-order not exceeding n such that each $V_i \leq U_i$.

Theorem 2.4: If (X, δ) is an L-topological space and $\mathcal{G}$ and $\mathcal{J}$ are two fuzzy grills on X such that $\mathcal{G} \subseteq \mathcal{J}$, then $\mathcal{G}$-dim $(X, \delta) \leq \mathcal{J}$-dim (X, δ).

Proof: It is sufficient to show that if $\mathcal{J}$-dim $(X, \delta) \leq n$, then $\mathcal{G}$-dim $(X, \delta) \leq n$. Suppose $\mathcal{J}$-dim $(X, \delta) \leq n$. Let $\mathcal{U} = (U_1, U_2, \ldots\ldots, U_k)$ be a finite open cover of (X, δ). Then there exists a subfamily $\mathcal{V} = (V\alpha: \alpha \in \Lambda)$ of δ of $\mathcal{J}$-order not exceeding n such that each V_α is less than or equal to some U_i and $(\vee_{\alpha \in \Lambda} V_\alpha)' \notin \mathcal{J}$. Since $\mathcal{G} \subseteq \mathcal{J}$, so $(\vee_{\alpha \in \Lambda} V_\alpha)' \notin \mathcal{J}$. Clearly, $\mathcal{G}$-order of $\mathcal{V} = (V_\alpha: \alpha \in \Lambda)$ does not exceed n and each V_α is less than or equal to some U_i. Therefore, $\mathcal{G}$-dim $(X, \delta) \leq n$. Hence, $\mathcal{G}$-dim $(X, \delta) \leq \mathcal{J}$-dim (X, δ).

Next, to obtain a relationship between $\mathcal{G}$-covering dimension and covering dimension of L-topological spaces, we first state a characterization of covering dimension of L-topological spaces proved in [3].

Theorem 2.5: Let (X, δ) be an L-topological space. Then following statements about (X, δ) are equivalent:

(*a*) dim $(X, \delta) \leq n$;

(*b*) For every finite open cover $(U_1, U_2, \ldots\ldots, U_k)$ of (X, δ), there is an open cover $(V_1, V_2, \ldots\ldots, V_k)$ of (X, δ) of order not exceeding n such that $V_i \leq U_i$, for $i = 1, 2, \ldots\ldots, k$;

(*c*) If $(U_1, U_2, \ldots, U_{n+2})$ is an open cover of (X, δ), there is an open cover $(V_1, V_2, \ldots, V_{n+2})$ of (X, δ) such that $V_i \le U_i$ and $\wedge_{i=1}^{n+2} V_i = \underline{0}$.

Theorem 2.6: (Relation between $\mathcal{G}$-dim (X, δ) and dim (X, δ))

For any fuzzy grill $\mathcal{G}$ on X, $\mathcal{G}$-dim $(X, \delta) \le$ dim (X, δ).

Proof: Suppose dim $(X, \delta) \le n$. Then by Theorem 2.5, for any finite open covering $\mathcal{U} = (U_1, U_2, \ldots, U_{n+2})$ of (X, δ) there is an open cover $\mathcal{V} = (V_1, V_2, \ldots, V_{n+2})$ of (X, δ) such that $V_i \le U_i$, $\forall i$ and $\wedge_{i=1}^{n+2} V_i = \underline{0}$

Clearly, $\wedge_{i=1}^{n+2} V_i = \underline{0} \notin \mathcal{G}$ and $\left(\wedge_{i=1}^{n+2} V_i\right)' = \underline{1}' = \underline{0} \notin \mathcal{G}$. Therefore, $\mathcal{G}$-dim $(X, \delta) \le n$.

Theorem 2.7: If $\mathcal{G} = L^X - (\underline{0})$, then $\mathcal{G}$-dim $(X, \delta) =$ dim (X, δ).

Proof: By Theorem 2.6, it is sufficient to show that if $\mathcal{G}$-dim $(X, \delta) \le n$, then dim $(X, \delta) \le n$. Suppose $\mathcal{G}$-dim $(X, \delta) \le n$ and $\mathcal{U} = (U_1, U_2, \ldots, U_k)$ is a finite open cover of (X, δ). Then there exists a subfamily $\mathcal{V} = (V_\alpha: \alpha \in \Lambda)$ of δ of $\mathcal{G}$-order not exceeding n such that $(\vee_{\alpha\in\Lambda} V_\alpha)' \notin \mathcal{G}$ and for each $V_\alpha \in \mathcal{V}$, $\exists U_i \in \mathcal{U}$ such that $V_\alpha \le U_i$. But $\mathcal{G} = L^X - (\underline{0})$ implies $(\vee_{\alpha\in\Lambda} V_\alpha)' = \underline{0}$ and so $\vee_{\alpha\in\Lambda} V_\alpha = \underline{1}$. Thus $\mathcal{V} = (V_\alpha: \alpha \in \Lambda)$ is an open refinement of $\mathcal{U}$ of order not exceeding n. Therefore, dim $X \le n$.

We now prove the following theorem regarding the restriction of a fuzzy grill.

Theorem 2.8: If $\mathcal{G}$ is a fuzzy grill on X and $A \subseteq X$, then $\mathcal{G}|A = \{u|A: u \in \mathcal{G}$ and $u|A \ne \underline{0}\}$ is a fuzzy grill on A.

Proof: Given, $\mathcal{G}$ is a fuzzy grill on X. We have to show $\mathcal{G}|A$ is a fuzzy grill on A.

(*i*) Let $u \in \mathcal{G}|A$ and $u \le v$, where $v \in L^A$. We have to show that $v \in \mathcal{G}|A$, i.e. $v = w|A$, where $w \in \mathcal{G}$. Since $u \in \mathcal{G}|A$, we have $u = u_1|A$, where $u_1 \in \mathcal{G}$.

Define $w: X \to L$ by $w(x) = \begin{cases} v(x), & \text{if } x \in A \\ u_1(x), & \text{if } x \in A \end{cases}$.

Clearly, $u_1(x) \le w(x)$, $\forall x \in X$ and so $u_1 \le w$. Since $u_1 \in \mathcal{G}$, we have $w \in \mathcal{G}$. Also $w|A = v$ and so $v \in \mathcal{G}|A$.

(*ii*) Let $u, v \in L^A$ be such that $u \vee v \in \mathcal{G}|A$. We have to show that $u \in \mathcal{G}|A$ or $v \in \mathcal{G}|A$, i.e $u = u_1|A$ or $v = v_1|A$ for some $u_1 \in \mathcal{G}$ or $v_1 \in \mathcal{G}$.

Since $u \vee v \in \mathcal{G}|A$, we have $u \vee v = w|A$ for some $w \in \mathcal{G}$.

Define $u_1, v_1 : X \to L$ by $u_1(x) =$

$$\begin{cases} 1, & \text{if } x \notin A \\ u(x), & \text{if } x \in A \end{cases} \quad \text{and} \quad v_1(x) = \begin{cases} 1, & \text{if } x \notin A \\ v(x), & \text{if } x \in A \end{cases}.$$

Let $w_1 = u_1 \vee v_1$. Then $w \leq w_1$ and so $w_1 \in \mathcal{G}$. Again, $w_1 = u_1 \vee v_1 \in \mathcal{G}$ implies $u_1 \in \mathcal{G}$ or $v_1 \in \mathcal{G}$. If $u_1 \in \mathcal{G}$, then $u_1|A = u$ implies $u \in \mathcal{G}|A$, and if $v_1 \in \mathcal{G}$, then $v_1|A = v$ implies $v \in \mathcal{G}|A$. Hence, $\mathcal{G}|A$ is a fuzzy grill on A.

Theorem 2.9: (Subset Theorem) If (X, δ) is an L-topological space, $\mathcal{G}$ is a fuzzy grill on X and A is a closed subspace of (X, δ), then $(\mathcal{G}|A)$-dim $(A, \delta|A) \leq \mathcal{G}$-dim (X, δ).

Proof: It is sufficient to show that if $\mathcal{G}$-dim $(X, \delta) \leq n$, then $(\mathcal{G}|A)$-dim $(A, \delta|A) \leq n$. Suppose $\mathcal{G}$-dim $X \leq n$. Let $\mathcal{U} = (U_1, U_2, ..., U_k)$ be a finite open cover of $(A, \delta|A)$.

Then for each U_i, there exists an L-fuzzy open set V_i in (X, δ) such that $V_i|A = U_i$, $\forall i$. Since $(A, \delta|A)$ is a closed subspace of (X, δ), $\mathcal{X}_A$ is an L-fuzzy closed set in (X, δ) and so $\mathcal{X}'_A$ is an L-fuzzy open set in (X, δ). One can show that $\mathcal{V} = (V_1, V_2, ..., V_k, \mathcal{X}'_A)$ is an open cover of (X, δ). Since $\mathcal{G}$-dim $(X, \delta) \leq n$, there is a subfamily $\mathcal{W} = (W_\alpha : \alpha \in \Lambda)$ of δ $\mathcal{G}$-order not exceeding n such that $(\vee_{\alpha \in \Lambda} W_\alpha)' \notin \mathcal{G}$ and for each $W_\alpha \in \mathcal{W}$, $\exists V \in \mathcal{V}$ such that $W_\alpha \leq V$. Let $\mathcal{W}_1 = (\mathcal{W}_\alpha|A : \alpha \in \wedge$ and $W_\alpha|A \neq 0)$. Clearly, $\mathcal{W}_1$ is a subfamily of $\delta|A$ such that for each $W_\alpha|A \in \mathcal{W}_1$, $\exists U_i \in \mathcal{U}$ such that $W_\alpha|A \leq U_i$.

We claim that $(\mathcal{G}|A)$-order of $\mathcal{W}_1$ does not exceed n. For this, let M be any subset of $\wedge$ with $n + 2$ elements. Then $\vee_{\alpha \in M}(W_\alpha|A) = (\wedge_{\alpha \in M} W_\alpha)|A \notin \mathcal{G}|A$ because $\wedge_{\alpha \in M} W_\alpha \notin \mathcal{G}$. Hence, $(\mathcal{G}|A)$-dim $A \leq \mathcal{G}$-dim X.

Next, we shall show that the above theorem is not true for arbitrary subspaces. For this, we first state a theorem proved in [3].

Theorem 2.10: Let (X, T) be an ordinary topological space and $(\delta = \chi_U : U \in T)$. Then dim $(X, \delta) =$ dim (X, T).

Example 2.11: We know that if (X, T) is an ordinary topological space and Y is a closed subspace of (X, T), then dim $Y \leq$ dim X. But this result is not true for arbitrary subspaces in ordinary topological spaces. We also know that if $\mathcal{G} = L^X - (\underline{0})$ is a fuzzy grill on an L-topological space (X, δ), then $\mathcal{G}$-dim (X, δ) = dim (X, δ).

Let (X, T) be an ordinary topological space and Y be any subspace of X such that dim $Y \not\leq$ dim X. Let $(\delta = \mathcal{X}_U: U \in T)$. Then by Theorem 2.9 and Theorem 2.10, it follows that $(\mathcal{G}|Y)$-dim $(Y, \delta|Y)$ = dim $(Y, \delta|Y)$ = dim $(Y, T|Y) \not\leq$ dim (X, T) = $\mathcal{G}$-dim (X, δ). Therefore, $(\mathcal{G}|Y)$-dim $(Y, \delta|Y) \not\leq \mathcal{G}$-dim (X, δ).

Theorem 2.12: Let (X, δ_1) and (Y, δ_2) be two L-topological spaces, $\mathcal{G}$ be a fuzzy grill on X. If $f: (X, \delta_1) \to (X, \delta_2)$ is a homeomorphism, then $\mathcal{G}$-dim $(X, \delta_1) = f(\mathcal{G})$-dim (Y, δ_2).

Proof: For images and inverse images of L-fuzzy sets, see page 437 of (4). It can be easily checked that $f(\mathcal{G}) = [f^{\to}(u) | u \in \mathcal{G}]$ is a fuzzy grill on Y, where $f^{\to}(u) \in L^Y$ is defined by $f^{\to}(u)(y) = \vee [u(x): x \in X, f(x) = y]$, $\forall y \in L$.

Let $\mathcal{G}$-dim $(X, \delta_1) \leq n$ and $\mathcal{V} = (V_1, V_2, ..., V_k)$ be any finite open cover of (Y, δ_2). Since $f: (X, \delta_1) \to (Y, \delta_2)$ is continuous, $V_i f$ is open in (X, δ_1), $\forall i = 1, 2, ..., k$. Also, since $V_i f = f^{\leftarrow}(V_i)$, $f^{\leftarrow}(V_i)$ is open in (X, δ_1), $\forall i = 1, 2, ..., k$. Thus $\mathcal{U} = [f^{\leftarrow}(V_i) | i = 1, 2, ..., k]$ is a finite family of open sets in (X, δ_1). Since $\vee_{i=1}^{k} f^{\leftarrow}(V_i) f^{\leftarrow}(\vee_{i=1}^{k} V_i) = f^{\leftarrow}(\underline{1}) = \underline{1}$, we find that $\mathcal{U}$ is an open cover of (X, δ_1). But $\mathcal{G}$-dim $(X, \delta_1) \leq n$ and so there is a finite subfamily $\mathcal{U}_1 = (U_1, U_2, ..., U_k)$ of δ_1 of $\mathcal{G}$-order not exceeding n such that $(\vee_{i=1}^{k} U_i)' \notin \mathcal{G}$ and $U_1 \leq f^{\leftarrow}(V_i)$, $\forall i = 1, 2, ..., k$.

Further, since $f^{-1}: (Y, \delta_2) \to (X, \delta_1)$ is continuous, $U_i f^{-1}$ is open in (Y, δ_2), $\forall i = 1, 2, ..., k$. But we know that $U_i f^{-1} = (f^{-1})^{\leftarrow} U_i = f^{\to}(U_i)$ and so $\mathcal{V}_1 = [f^{\to}(U_i) | i = 1, 2, ..., k]$ is a finite subfamily of δ_2. Also, $U_i \leq f^{\leftarrow}(V_i) \Rightarrow f^{\to}(U_i) \leq f^{\to}[f^{\leftarrow}(V_i)] \Rightarrow f^{\to}(U_i) \leq v_i$ $\forall i = 1, 2, ..., k$. Also, $[\vee_{i=1}^{k} f^{\to}(U_i)]' = [f^{\to}(\vee_{i=1}^{k} U_i)]' = f^{\to}[(\vee_{i=1}^{k} U_i)'] \notin f(\mathcal{G})$ because $(\vee_{i+1}^{k} U_i)' \notin \mathcal{G}$.

Finally, we shall show that $f(\mathcal{G})$-order of $\mathcal{V}_1$ does not exceed n. For this, let

$\mathcal{W} = [f^{\rightarrow}(U_{i1}), f^{\rightarrow}(U_{i2}), \ldots\ldots f^{\rightarrow}\left(U_{i_{n+2}}\right)]$ be any subfamily of $\mathcal{V}_1$ with $(n+2)$ members. Since $\mathcal{G}$-order of $\mathcal{U}_1$ does not exceed n, and so $\wedge_{t=1}^{n+2} U_{it} = \underline{0} \Rightarrow f^{\rightarrow}\left(\wedge_{t=1}^{n+2} U_{it}\right) = f^{\rightarrow} = (\underline{0}) \Rightarrow \wedge_{t=1}^{n+2} f^{\rightarrow}(U_{it}) = f^{\rightarrow} = (\underline{0})$ $= \underline{0}$ and so $f(\mathcal{G})$-order of $\mathcal{V}_1$ does not exceed n. That is $f(\mathcal{G})$-dim $(Y, \delta_2) \leq n = \mathcal{G}$-dim (X, δ_1). Similarly, it can be shown that $\mathcal{G}$-dim $(X, \delta_1) \leq f(\mathcal{G})$-dim (Y, δ_2). With these estimates it follows that $\mathcal{G}$-dim $(X, \delta_1) = f(\mathcal{G})$-dim (Y, δ_2).

REFERENCES

[1] Azad K. K. "Fuzzy Grills and a Characterization of Fuzzy Proximity", *J. Math. Anal. Appl.*, **79**, 13 – 17, 1981.

[2] Choquet G., "Sur Les Notions de Filtre et. de. Grille", *Comptes Rendus Acad. Sci. Paris*, **224**, (1947), 171 – 173.

[3] Jamwal D.S. and Shakeel Ahmed, "On Covering Dimension and Finitistic Spaces in *L*-topology", *The Journal of Fuzzy Mathematics*, **14 (1)**, 207 – 222, 2006.

[4] Hohle U. and Rodabaugh S.E., "Mathematics of Fuzzy Sets, Logic, Topology and Measure Theory", *the Handbook of fuzzy sets series*, **3**, (1999), Kluwer Academic Publishers (Dordrecht).

[5] Liu Y.M. and Luo M.K., "Fuzzy Topology", World Scientific Pub. Singapore, (1998 in English).

27

ON FINITISTIC FUZZY SUBSETS, FINITISTIC AND COMPLETELY FINITISTIC SPACES IN L-TOPOLOGY

Dalip Singh Jamwal[1], Shakeel Ahmed[2] and Ashwani Kumar[3]

[1]Department of Mathematics, University of Jammu, Jammu, India, E-mail: dalipsj@yahoo.com

Abstract: *In this paper, we have defined the concepts of finitistic L-fuzzy subsets and completely finitistic spaces of L-topological spaces and proved some of their properties. We have proved some results which give relationship between finitistic spaces and finitistic L-fuzzy subsets. We have also discussed some more properties, examples and counter examples of finitistic spaces.*

Keywords: *Finitistic Spaces, Completely finitistic spaces, Finitistic L-fuzzy subsets, L-topological spaces.*

2000 Subject Classification: *Primary 54A40, Secondary 03E72.*

1. INTRODUCTION AND PRELIMINARIES

The concept of finitistic spaces in topological spaces was introduced in 1960 by R.G. Swan [7]. The purpose of considering these spaces was to simplify and extend Borel's proof [1] of the classical P.A. Smith fixed point theorems for spheres. Of course, Swan did not name these as finitistic. The term 'finitistic' was used by Bredon in his book [6] and it has since then become a firmly established term.

According to Bredon ([6], p. 133), a topological space X is said to be finitistic if each open cover of X has a finite-dimensional open refinement. According to Pears ([2], p. 111), the terms finite-dimensional open refinement and finite order open refinement have same meaning. We also use the term 'finite order refinement' in place of the term 'finite-dimensional open refinement'. In topological spaces, the study of finitistic spaces became interesting with the basic paper of Deo-Tripathi [10] in which they proved a characterization for non finitistic paracompact spaces. Several works on finitistic spaces in topological spaces appeared after [10], one can see [8], [9], [11], [12], [13].

We first recall the definition of order of a family of subsets of a set X and the covering dimension of a topological space.

The *order* ([2], p. 111) of a family $(A_\lambda : \lambda \in \Lambda)$ of subsets, not all empty, of some set X is the largest integer n for which there exists a subset M of Λ with $n + 1$ elements such that $\cap_{\lambda \in M} A_\lambda$ is nonempty, or is ∞ if there is no such largest integer. A family of empty subsets has order -1.

One can see the definition of L-topological space in [14]. According to [14], all L-fuzzy topological spaces defined in [15] are L-topological spaces. Throughout this paper, L is a complete lattice. Some additional conditions on L are taken when they are needed.

Let (X, δ) be an L-topological space. A subfamily $\mathcal{A} \subset \delta$ is said to be an open cover of (X, δ) if $\vee \mathcal{A} = \underline{1}$. Let $A \in L^X$. Then a subfamily $\mathcal{B} \subset \delta$ is said to be an open cover of A if $\vee \mathcal{B} \geq A$ (page 187 of [15]). Throughout this paper, we have used these covers for L-topological spaces and L-fuzzy subsets, respectively.

Now, we recall the definition of order of a family of L-fuzzy subsets of a set X and finitisticness of L-topological spaces defined in [5].

Definition 1.1: Let $\Lambda \neq \emptyset$ and $\mathcal{A} = (A_\lambda : \lambda \in \Lambda)$ be a family of L-fuzzy subsets of a set X. Then order of A is defined as under:

Case I: When $A_\lambda = \underline{0}$ for atleast one value of λ in Λ. Then order of $\mathcal{A}$ is the largest nonnegative integer n for which there exists a subset M of Λ having $n + 1$ elements such that $\wedge_{\lambda \in M} A\lambda \neq \underline{0}$, or is ∞ if there is no such largest integer n.

Case II: When $A_\lambda = \underline{0}$ for all $\lambda \in \Lambda$. Then order of $\mathcal{A}$ is -1.

Definition 1.2: An L-topological space (X, δ) is said to be finitistic if each open cover of (X, δ) has a finite order open refinement.

We now state the results proved in [5], which are used in this paper.

Theorem1.3: A topological space (X, T) is finitistic if and only if the L-topological space (X, δ) is finitistic, where $\delta = (\chi_U: U \in T)$ and L is a complete lattice having atleast two elements.

Theorem 1.4: Every closed subspace of a finitistic L-topological space is finitistic.

Theorem 1.5: If (X, δ) is a weakly induced finitistic L-topological space, where $L = (0, 1)$, then $(X, [\delta])$ is finitistic. Converse is not true.

Theorem 1.6: Let $L = (0, 1)$ and (X, T) be an ordinary topological space. Then $(X, \omega_L(T))$ is finitistic implies (X, T) is finitistic, where ω_L: *Top* $\rightarrow$ *L-Top* is the Lowen Functor.

Remark 1.7: In L-topology, continuous image of a finitistic space need not be finitistic.

Let $(X, \mathcal{D})$ be an ordinary discrete topological space, where X is an infinite set. Then clearly $(X, \mathcal{D})$ is finitistic. Let T be another topology on X such that (X, T) is not finitistic. Then by Theorem 1.6, (X, δ_1) is finitistic and (X, δ_2) is not finitistic, where $\delta_1 = \{\chi_U: U \in \mathcal{D}\}$ and $\delta_2 = \{\chi_V: V \in T\}$. Define $I: X \rightarrow X$ as $I(x) = x, \forall x \in X$. Then clearly $I: (X, \delta_1) \rightarrow (X, \delta_2)$ is a continuous and onto. It means (X, δ_2) is continuous image of (X, δ_1). Here (X, δ_1) is finitistic but (X, δ_2) is not finitistic.

In remaining two sections, we have defined the concepts of finitistic L-fuzzy subsets and completely finitistic spaces in L-topology and studied their some of the basic properties. These sections also contain some more properties, examples and counter examples of finitistic spaces of L-topological spaces.

2. WE START THIS SECTION BY DEFINING THE CONCEPT OF COMPLETELY FINITISTICNESS OF L-TOPOLOGICAL SPACES.

Definition 2.1: An L- topological space (X, δ) is said to be completely finitistic if every subspace of (X, δ) is finitistic.

Theorem 2.2: A topological space (X, T) is completely finitistic if and only if the L- topological space (X, δ) is completely finitistic, where $\delta = \{\chi_U : U \in T\}$ and L is any complete lattice with atleast two elements.

Proof: Let $(Y, \delta|_Y)$ be any subspace of (X, δ). Then clearly $(Y, T|_Y)$ is subspace of (X, T). It is easy to show that $U \in T|_Y \Leftrightarrow \chi_U \in \delta|_Y$. First, suppose (X, T) is completely finitistic. Let $(Y, \delta|_Y)$ be any subspace of (X, δ) and $\mathcal{U} = (\chi_{U\lambda} : \lambda \in \Lambda)$ be an open cover of $(Y, \delta|_Y)$. Then $\vee\{\chi_{U\lambda} : \lambda \in \Lambda\} = \underline{1}|_Y \Rightarrow \chi U_{(U\lambda, \lambda \in \Lambda)} = \underline{1}|_Y \Rightarrow \cup\{U_\lambda : \lambda \in \Lambda\} = Y$. Thus $\mathcal{V} = \{U_\lambda : \lambda \in \Lambda\}$ is an open cover of $(Y, T|_Y)$. Since $(Y, T|_Y)$ is finitistic, therefore, $\mathcal{V}$ has a finite order open refinement, say $\mathcal{V}_1 = \{V_\alpha : \alpha \in \Delta\}$. Then clearly $\mathcal{U}_1 = \{\chi_{V\alpha} : \alpha \in \mathcal{V}_1\}$ is a finite order open refinement of $\mathcal{U}$. Hence $(Y, \delta|_Y)$ is finitistic. It means (X, δ) is completely finitistic.

Conversly, suppose (X, δ) is completely finitistic. Let $(Y, T|_Y)$ be any subspace of (X, T) and $\mathcal{U} = \{U_\lambda : \lambda \in \wedge\}$ be any open cover of $(Y, T|_Y)$. Then $\cup\{U_\lambda : \lambda \in \Lambda\} = Y \Rightarrow \mathcal{X}\cup_{\{U\lambda : \lambda \in \Lambda\}} = \mathcal{X}Y \Rightarrow \vee\{\mathcal{X}U_\lambda : \lambda \in \Lambda\} = \underline{1}|_Y$. Now (2.1) and (2.2) implies that $\mathcal{V} = \{\mathcal{X}U_\lambda : U_\lambda \in \mathcal{U}\}$ is an open cover of $(Y, \delta|_Y)$. Since $(Y, \delta|_Y)$ is finitistic, therefore $\mathcal{V}$ has a finite order open refinement, say $\mathcal{V}_1 = \{\chi_{V\alpha} : \alpha \in \Delta\}$. Then it can be easily seen that $\mathcal{U}_1 = (V_\alpha : \mathcal{X}_{V\alpha} \in \mathcal{V}_1)$ is a finite order open refinement of $\mathcal{U}$. Hence $(Y, T|_Y)$ is finitistic. It means (X, T) is completely finitistic.

Remark 2.3: In topological spaces, a closed subspace of a finitistic space is finitistic. This result is also true in L-topological spaces (see Theorem 1.4). We also know that any subspace of a finitistic topological space need not be finitistic. We now give an example to show that any subspace of a finitistic L-topological space need not be finitistic.

Example 2.4: let (X, T) be an ordinary finitistic space and $(Y, T|_Y)$ be a subspace of (X, T) which is not finitistic. Let (X, δ) be the L-toplogical space, where $\delta = \{\chi_U : U \in T\}$ and L is any complete lattice with atleast two elements. Then (X, δ) is finitistic (by Theorem 1.3). Again by Theorem 1.3, $(Y, \delta|_Y)$ is subspace of (X, δ) which is not finitistic.

Next, we define the concept of finitisticness of L-fuzzy subsets.

Definition 2.5: Let (X, δ) be an L-topological space and A be an L-fuzzy subset of X. A is said to be finitistic if each open cover of A in (X, δ) has a finite order open refinement.

Theorem 2.6: Let (X, δ) be an L-topological space. Then every L-fuzzy subset A of X is finitistic if and only if each L-fuzzy open subset of (X, δ) is finitistic.

Proof: Suppose each L-fuzzy open subset of (X, δ) is finitistic. Let A be any L-fuzzy subset of X. Let $\mathcal{U} = \{U_\lambda : \lambda \in \Lambda\}$ be any open cover of A. Then $\vee\{U_\lambda : \lambda \in \Lambda\}$ is an L-fuzzy open subset in (X, δ). Since $\vee\{U_\lambda : \lambda \in \Lambda\}$ is finitistic and $\mathcal{U}$ is an open cover of $\vee\{U_\lambda : \lambda \in \Lambda\}$, therefore $\mathcal{U}$ has a finite order open refinement, say $\mathcal{V}$. Clearly $\mathcal{V}$ is also open cover of A. Hence A is finitistic. Converse is trivial.

Theorem 2.7: Let (X, δ) be an L-toplogical space. If each L-fuzzy open subset of (X, δ) is finitistic, then (X, δ) is completely finitstic.

Proof: Let $(Y, \delta|_Y)$ be any subspace of (X, δ). Let $\mathcal{U} = \{U_\alpha : \alpha \in \Delta\}$ be any open cover of $(Y, \delta|_Y)$. Then, $\forall U_\alpha \in \mathcal{U}$, there is some $V_\alpha \in \delta$ such that $U_\alpha = V_\alpha|Y$. Clearly $\vee V_\alpha$ is L-fuzzy open subset of (X, δ) and $\mathcal{V} = \{V_\alpha : U_\alpha = V_\alpha|Y\}$ is an open cover of $\vee V_\alpha$. Since $\vee V_\alpha$ is finitistic, therefore $\mathcal{V}$ has a finite order open refinement, say $\mathcal{V}_1 = \{W_\beta : \beta \in \Delta_1\}$. Clearly $\mathcal{U}_1 = \{W_\beta| Y : W_\beta \in \mathcal{V}_1\}$ is a finite order open refinement of $\mathcal{U}$. Hence $(Y, \delta|_Y)$ is finitistic. It means (X, δ) is completely finitistic.

Theorem 2.8: Let (X, δ) be an L-topological space. If $' : L \to L$ is a complementary opertion, then every L-fuzzy closed subset of (X, δ) is finitistic.

Proof: Let A be any L-fuzzy closed subset of (X, δ). Let $\mathcal{U} = \{U_\alpha : \alpha \in \Delta\}$ be any open cover of A. Since $' : L \to L$ is a complementary opertion, therefore $\mathcal{U}_1 = \mathcal{U} \cup (A')$ is an open cover of (X, δ). Since (X, δ) is finitistic, therefore $\mathcal{U}_1$ has a finite order open refinement, say $\mathcal{V}_1 = \{V_\lambda : \lambda \in \Lambda\}$, in which case $\mathcal{V} = \{V_\lambda \wedge A : \in \Lambda\}$ is a finite order open refinement of $\mathcal{U}$ because

$$V_\lambda \wedge A = \begin{cases} V_\lambda, & \text{if } V_\lambda \nleq A' \\ \underline{0}, & \text{if } V_\lambda \leq A \end{cases}$$

Hence A is finitistic.

Theorem 3.1: An L-topological space (X, δ) is finitistic if and only if each basic open cover of (X, δ) has a finite order open refinement.

Proof: Suppose each basic open cover of (X, δ) has a finite order open refinement. Let $\mathcal{U} = \{U_\lambda : \lambda \in \Lambda\}$ be any open cover of (X, δ).

Let $\mathcal{B}$ be any base of (X, δ). Then each U_λ is join of a subfamily of $\mathcal{B}$. Let $\mathcal{V}$ be the union of all such subfamilies of $\mathcal{B}$. Then clearly $\mathcal{V}$ is a basic open cover of (X, δ). By the given condition, $\mathcal{V}$ has a finite order open refinement, say $\mathcal{V}_1$. Then clearly $\mathcal{V}_1$ is also finite order open refinement of $\mathcal{U}$. Hence (X, δ) is finitistic.

Sufficiency part is trivial.

Definition 3.2: A family $\mathcal{A}$ of L-fuzzy subsets of a set X is said to have finite meet property (FMP) if meet of any finite subfamily of $\mathcal{A}$ is nonzero.

Theorem 3.3: If each family of closed subsets of an L-topological space (X, δ) which has (FMP) has nonzero meet, then (X, δ) is finitistic.

Proof: Let every collection of closed subsets of (X, δ) with (FMP) has nonzero meet. We have to show that (X, δ) is finitistic. Let $\mathcal{U} = (U_\lambda: \lambda \in \Lambda)$ be any open cover of (X, δ). Then clearly $\vee (U_\lambda : \lambda \in \Lambda) = \underline{1} \Rightarrow [\vee (U_\lambda: \lambda \in \Lambda)]' = \underline{1}' \Rightarrow \wedge (U'_\lambda: \lambda \in \Lambda) = \underline{0}$. Thus $(U'_\lambda: \lambda \in \Lambda)$ is a collection of closed subsets of (X, δ) with zero meet and by hypothesia this collection doesnot have (FMP). Hence there exists a finite subfamily, say $(U'_{\lambda 1}, U'_{\lambda 2}, \cdots, U'_{\lambda K})$ of $(U'_\lambda: \lambda \in \Lambda)$ such that $\wedge (U'_{\lambda 1}, U'_{\lambda 2}, \cdots, U'_{\lambda K}) = \underline{0} \Rightarrow [\wedge (U'_{\lambda 1}, U'_{\lambda 2}, \cdots, U'_{\lambda K})]' = \underline{0}' \Rightarrow \vee (U_{\lambda 1}, U_{\lambda 2}, \cdots, U_{\lambda K}) = \underline{1}$. Then clearly $\mathcal{V} = (\underline{0}, U_{\lambda 1}, U_{\lambda 2}, \cdots, U_{\lambda K})$ is a finite order open refinement of $\mathcal{U}$. Hence (X, δ) is finitistic.

Theorem 3.4: Let (X, δ) be an L-topological space such that weight $\omega(\delta)$ of (X, δ) is finite. Then (X, δ) is finitistic.

Note 3.5: In L-topology, continuous image of a finitistic space need not be finitistic (see Remark 1.7.). Following is a case in which continuous image of a finitistic space is finitistic in L-topology.

Theorem 3.6: In L-topology, continuous image of a finitistic space of finite weight is finitistic.

Proof: Follows by Theorem 3.4 and Theorem 7.1.5 of [15].

Theorem 3.7: Let (X, δ) be an L-topological space. Then $[X, (\delta)]$ is finitistic if and only if $[X, crs\,(\delta)]$ is finitistic.

Proof: It follows from the definition of (δ) that $crs\,(\delta) = [\chi_U: \in (\delta)]$. Hence $[X, (\delta)]$ is finitistic if and only if $[X, crs\,(\delta)]$ is finitistic follows by Theorem 1.3.

Theorem 3.8: If (X, δ) is weakly induced L-topological finitistic space and meet of any finite number of nonzero elements of L is nonzero, then $[X, crs(\delta)]$ is finitistic.

Proof: We know if (X, δ) is weakly induced L-topological space and meet of any finite number of nonzero elements of L is nonzero, then $[X, (\delta)]$ is finitistic. By this result and Theorem 3.7, $[X, crs(\delta)]$ is finitistic.

Example 3.9: Here we give an example to show that converse of Theorem 3.8 need not be true.

Let X be any non-empty set. Let $L = (0, 1)$. Let $\delta = (\underline{\alpha}: \alpha \in L)$. Then clearly (X, δ) is an L-topological space. We first show that (X, δ) is weakly induced. Let $\underline{\alpha} \in \delta$ and $a \in L$.

Case I: When $\alpha < a$. Then

$$\begin{aligned}(\underline{\alpha})_{(a)} &= [x: \underline{\alpha}(x) \nleq a] \\ &= [x: \underline{\alpha}(x) > a] \\ &= \emptyset \in (\delta).\end{aligned}$$

Case II: When $\alpha > a$. Then

$$\begin{aligned}(\underline{\alpha})_{(a)} &= [x: \underline{\alpha}(x) \nleq a] \\ &= [x: \underline{\alpha}(x) > a] \\ &= X \in (\delta).\end{aligned}$$

It means $(\underline{\alpha})_a \in (\delta)$, $\forall \underline{\alpha} \in \delta$ and $\forall a \in L$. Hence by ([15], Theorem 4.2.17), (X, δ) is weakly induced. Here $(\delta) = (\emptyset, X)$. It means $crs(\delta) = (\chi_\emptyset, \chi_X) = (\underline{0}, \underline{1})$. Clearly $[X, crs(\delta)]$ is finitistic and meet of any finite number of nonzero elements of L is nonzero but (X, δ) is not finitistic because the open cover

$$\mathcal{U} = (a_n: \forall n \in N, \text{ where } a_n = 1 - 1/n) \text{ of } (X, \delta)$$

has no finite order open refinement. (Here N is the set of all natural numbers).

Theorem 3.10: Let (X, δ) be any L-topological space. Then $[X, \iota_0^*(\delta)]$ is finitistic.

Proof: Definition of $\iota_0^*(\delta)$ can be seen on page 188 of [15]. Let $U \in \delta$. Then we know $U_{(0)} = [x \in X: U(x) \geq 0] = X$. It means every subbasic open subset of $[X, \iota_0^*(\delta)]$ is X. It means $X, \iota_0^*(\delta)$ is indiscrete space. Therefore it is finitistic.

Next, we give an example of an infinite dimensional finitistic L-topological space.

Example 3.11: Let X be any nonempty set. Let $Y = N$ (The set of all natural numbers). Let $L = P(Y)$. Here $0 = \emptyset$ and $1 = N$. Let $\delta = (\underline{A}: A \in L$ and $X \setminus A$ is finite$] \cup (\underline{0})$. It can be checked that (X, δ) is an L-topological space and it is finitistic. It can be shown that dim $(X, \delta) = \infty$.

Theorem 3.12: Let (X, δ_1) and (Y, δ_2) be two L-topological spaces. Then $(X \cup Y, \delta_1 \oplus \delta_2)$ is finitistic if and only if (X, δ_1) and (Y, δ_2) are finitistic. (Here $X \cap Y = \emptyset$.

Proof:

Necessity: Suppose $(X \cup Y, \delta_1 \oplus \delta_2)$ is finitistic. Let $\mathcal{U} = (V_\lambda: \lambda \in \Lambda)$ be any open cover of (X, δ_1). $\forall V_\lambda \in \mathcal{U}$, define $R_\lambda = V_\lambda$ on X and $R_\lambda = \underline{1}$ on Y. Let $\mathcal{A}$ be the family of all these R_λ's. Then clearly $\mathcal{A}$ is an open cover of $(X \cup Y, \delta_1 \oplus \delta_2)$. Since $(X \cup Y, \delta_1 \oplus \delta_2)$ is finitistic, therefore $\mathcal{A}$ has a finite order open refinement, say $\mathcal{A}_1 = (U_\alpha: \alpha \in \Delta)$. Then clearly $\mathcal{U}_1 = (U_\alpha|_X: U_\alpha \in \mathcal{A}_1)$ is a finite order open refinement of $\mathcal{U}$. Hence (X, δ_1) is finitistic. Similarly we can show that (Y, δ_2) is finitistic.

Sufficiency: Suppose both (X, δ_1) and (Y, δ_2) are finitistic. Let $\mathcal{U} = (U_\lambda: \lambda \in \Lambda)$ be any open cover of $(X \cup Y, \delta_1 \oplus \delta_2)$. Then clearly $\mathcal{U}_X = (U_\lambda|_X: U_\lambda \in \mathcal{U})$ and $\mathcal{U}_Y = (U_\lambda|_Y: U_\lambda \in \mathcal{U})$ are open covers of (X, δ_1) and (Y, δ_2), respectively. Since both (X, δ_1) and (Y, δ_2) are finitistic, therefore $\mathcal{U}_X$ and $\mathcal{U}_Y$ have finite order open refinements, say $\mathcal{V}_X = (V_\alpha: \alpha \in \Delta_1)$ and $\mathcal{V}_Y = (W_\beta: \beta \in \Delta_2)$. $\forall V_\alpha \in \mathcal{V}_X$, define $R_\alpha = V_\alpha$ on X and $R_\alpha = \underline{0}$ on Y and $S_\beta = W_\beta$ on Y and $S_\beta = \underline{0}$ on X. Let $\mathcal{A}$ and $\mathcal{B}$ be the families of these R_α's and S_β's respectively. Then clearly $\mathcal{U}_1 = \mathcal{A} \cup \mathcal{B}$ is a finite order open refinement of $\mathcal{U}$. Hence $(X \cup Y, \delta_1 \oplus \delta_2)$ is finitistic.

Example 3.13: (See definition of L-fuzzy quasi-uniformity and related terms in Chapter 13 of [15]). Let $X = (x)$ and $L = (0, 1)$. Here we show that $[X, \delta (D)]$ need not be finitistic. Here $\delta (D)$ is the L-topology generated by the L-fuzzy quasi-uniformity D. Let $D \subset \epsilon (L^X)$ be an L-fuzzy quasi-uniformity on X such that $I \in D$ where $I: L^X \to L^X$ is identity function. We show $\delta (D) = L^X$. Clearly $\delta (D) \subset L^X$.

Let $A \in L^X$. Then $A^\circ = \vee\,(C \in L^X\colon \exists f \in D$ such that $f(C) \leq A)$. Since $I \in D$ and $I(A) = A \leq A$, therefore $I(A) \leq A$. It means $A \in [C \in L^X\colon \exists f \in D$ such that $f(C) \leq A]$. Then clearly $\vee\,[C \in L^X\colon \exists f \in D$ such that $f(C) \leq A] = A$

$$\Rightarrow A^\circ = A \Rightarrow A\ \delta\,(D)$$

$$\Rightarrow L^X \subset \delta\,(D). \qquad \text{...}\ (3.1)$$

From (3.1) and (3.2), $\delta\,(D) = L^X$.

Then clearly $[X, \delta\,(D)]$ is not finitistic because the open cover $\mathcal{U} = (a_n\colon n \in N$ and $a_n = 1 - 1/n)$ of $[X, \delta\,(D)]$ has no finite order open refinement.

REFERENCES

[1] A. Borel, "Nouvelle Demonstration d'um Theorem" *de P.A. Smith, Commt. Math. Helv.*, **28**, (1955), 27–39.

[2] A. R. Pears, "Dimension Theory of General Spaces", Cambridge University Press (1975).

[3] C.L. Chang, "Fuzzy Topological Spaces", *J. Math. Anal. Appl.*, **24**, (1968), 182–190.

[4] D.S. Jamwal and Shakeel Ahmed, "On Covering Dimension of *L*-fuzzy Subsets and *L*-topological Spaces", *J. Fuzzy Math.*, No. 2, **12**, (2004), 331–339.

[5] D.S. Jamwal and Shakeel Ahmed, "On Covering Dimension and Finitistic Spaces in *L*-topology", *J. Fuzzy Math.*, **14**, No. 1, (2006), 207–222.

[6] G.E. Bredon, "Introduction to Compact Transformation Groups", Academic Press (1972).

[7] R.G., Swan, "A New Nethod of Fixed Point Theory", *Comm. Math. Helv.*, **34**, (1960), 1–16.

[8] S. Deo, "Topology of Finitistic Spaces and Related Topics", *Bull. Allahabad Math. Soc.*, **2**, (1987), 31–61.

[9] S. Deo and A.R. Pears, "A Completely Finitistic Space is Finite Dimensional", *Bull. London Math. Soc.*, **17**, (1985), 49–51.

[10] S. Deo and H.S. Tripathi, "Compact Lie Group Action on Finitistic Spaces, Topology", **21**, (1982), 393–399.

[11] S. Deo and J.S. Andotra, "On Completely Finitistic Spaces", *Bull. Cal. Math. Soc.*, **80**, (1988), 173–177.

[12] S. Deo and M. Singh, "On Certain Construction in Finitistic Spaces", *Int. J. Math. and Math. Math. Soc.*, **6**, (1983), 477–482.

[13] S. Deo and D.S. Jamwal, "Finitistic and Locally Finitistic Spaces", *Indian J. Math.*, **27**, (1985), 183–194.

[14] U.H¨ohle and S.E. Rodabaugh, "Mathematics of Fuzzy Sets: Logic, Topology and Measure Theory", *The Handbook of Fuzzy Sets Series*, **3**, (1999), Kluwer Academic Publisher (Dordecht).

[15] Y.M. Liu and M.K. Luo, "Fuzzy Topology", World Scientific Pub., Singapore (1997).

28

SOME GENERALISED CLASSES OF FUZZY UNIFORM SPACES

D. K. Mitra and D. Hazarika

Department of Mathematical Sciences
Tezpur University, Napaam-784028, India
E-mail: dkmitra@tezu.ernet.in

Abstract: *Fuzzy semi-uniform and fuzzy semi-quasi-uniform spaces are introduced in the category* ***I-TOP*** *of fuzzy topological spaces. Fuzzy semi-quasi-uniform spaces are shown to be interior spaces. A necessary condition for a fuzzy semi-quasi-uniform space to be fuzzy topological is obtained. Various results on fuzzy semi-uniformly continuous functions are obtained. Every fuzzy semi-uniform space with countable base is shown to be fuzzy semi-pseudo-metrizable.*

Keywords: *fuzzy semi-uniformity, fuzzy semi-quasi-uniformity, fuzzy semi-pseudometrizability*

1 INTRODUCTION

Uniform spaces play a very important role as a bridge between metric spaces and general topological spaces. Many generalisations of uniform spaces, viz. quasi-uniform spaces [1], semi-uniform spaces, semi-quasi-uniform spaces, locally uniform spaces [10, 2] and locally quasi-uniform spaces [4] have been developed in general topology. Semi-quasiuniform spaces were introduced and a necessary and sucient condition under which semi-quasi-uniform spaces are topological was obtained in [8]. Closure spaces in relation to semi-quasi-uniform spaces have been studied and various results on semi-uniformly continuous and semi-pseudo-metrization have been obtained in [9].

Whereas quasi-uniform space have been developed in the fuzzy setting [3], other generalisations have not been developed in the fuzzy setting. In one of our previous work [7] we have developed fuzzy locally uniform spaces in the category *I-TOP*. In this paper we dene and study fuzzy semi-uniform spaces in terms of fuzzy semi-quasiuniform spaces in the same category.

Throughout the paper X is an arbitrary non-empty set. A fuzzy set A in X is a function from X to $I = (0, 1)$. In particular, $X(x) = 1$ and $\phi(x) = 0$ are fuzzy sets in X, denoted by $\underline{1}$ and $\underline{0}$ respectively. The collection of all fuzzy sets in X is denoted by I^X. A fuzzy point P in X is a fuzzy set in X such that $P(x) = \alpha$ if $x = x_0$, and $P(x) = 0$ if $x \neq x_0$, where $0 < \alpha < 1$. P is said to have support x_0 and value α, and is denoted by P^α_{x0}. Let A be a fuzzy set in X, then $P^\alpha_{x0} \subseteq A \Leftrightarrow \alpha \leq A(x_0)$. In particular, $P^\alpha_{x0} \subseteq P^\beta_y \Leftrightarrow x_0 = y, \alpha \leq \beta$. A fuzzy point P^α_{x0} is said to be in A, denoted by $P^\alpha_{x0} \in A$ iff $\alpha < A(x_0)$.

The collection of all points P^α_x, where $0 < \alpha \leq 1$ and $x \in X$ is denoted by $\mathbb{P}$. The collection of all fuzzy points P^α_x, with P^0_x, where $0 < \alpha < 1$ and $x \in X$ denoted by $\mathbb{P}_*$. $\mathbb{P}_0$ denotes all fuzzy points P^α_x, where $0 < \alpha < 1$ and $x \in X$.

Operations on fuzzy sets are taken as usual. For denitions of fuzzy topology and fuzzy uniformity we refer to [5].

2 FUZZY SEMI UNIFORMITY

Denition 2.1: Let $\text{int}: I^X \to I^X$ be a mapping on I^X.

Then int is called an interior operator on I^X, if it fullls the following conditions:

(IO1) $\text{int}(\underline{1}) = 1$.

(IO2) $\text{int}(A) \subseteq A, | \forall A \in I^X$.

(IO3) $\text{int}(A \cap B) = \text{int}(A) \cap \text{int}(B), \forall A, B \in I^X$.

I^X together with an interior operator 'int' shall be called as interior space.

An interior operator 'int' is said to be fuzzy topological interior operator if in addition it satises the following:

(IO4) $\text{int}[\text{int}(A)] = \text{int}(A); \forall A \in I^X$.

Denition 2.2: [5] A mapping d: $\mathbb{P}_* \times \mathbb{P}_* \to (0, \infty)$ is called continuous for membership grade iff for every fuzzy point $P_x^\alpha \in \mathbb{P}_0$ and $\varepsilon > 0$ there exists a $\delta > 0$ s.t. $|\beta - \alpha| < \delta$ implies $d(P_x^\alpha, P_y^\beta) < \varepsilon$.

Denition 2.3: A fuzzy semi-pseudo-metric on I^X is a mapping d: $\mathbb{P}_* \times \mathbb{P}_* \to (0, \infty)$ which is continuous for membership grade and for any $P_x^\alpha, P_y^\beta \in \mathbb{P}_*$ satisfies the following axioms:

(QM1) If $P_y^\beta \subseteq P_x^\alpha$, then $d(P_x^\alpha, P_y^\beta) = 0$.

(QM2) $d(P_x^\alpha, P_y^\beta) = d(P_y^{1-\beta}, P_x^{1-\alpha})$.

The above is a generalisation of fuzzy metric in the sense of Hu Cheng-Ming [5].

Denition 2.4: [5] Let $\mathcal{U}^*$ be the collection of subsets U of $\mathbb{P}_* \times \mathbb{P}_*$ satisfying the following conditions:

(c1) If $(P_x^\alpha, P_y^\beta) \in U$, then $(P_x^\alpha, P_y^\gamma) \in U$ for every $0 \leq \gamma < \beta$.

(c2) If $P_x^\alpha = \cup_\lambda P_x^{\alpha\lambda}$ then $P_y^{\beta^*} = \cup_\lambda P_y^{\beta^*_\lambda}$ for every $y \in X$, where $\beta^* = \sup [\beta | (P_x^\alpha, P_y^\beta) \in U]$, $\beta^*_\lambda = \sup [\beta_\lambda | (P_x^{\alpha\lambda}, P_y^{\beta\lambda}) \in U]$.

We shall require the following denitions and results from [5]:

For any $U, V \in \mathcal{U}^*$ the composition operator is given by

$U \circ V = [(P_x^\alpha, P_z^\gamma) | \exists P_y^\beta \in \mathbb{P}_*$ s.t. $(P_x^\alpha, P_y^\beta) \in V, (P_x^\beta, P_z^\gamma) \in U]$

For any $P_x^\alpha \in \mathbb{P}$ and $A \in I^X$:

$U(P_x^\alpha) = \cup [P_y^\beta | [(P_x^\alpha, P_y^\beta) \in U]$ and $U(A) = \cup [P_y^\beta | [(P_x^\alpha, P_y^\beta) \in U, P_x^\alpha \subseteq A]$

Further, $U(P_x^\alpha)$, $U(A)$ are in I^X

$U(\cup_\lambda P_x^{\alpha\lambda}) = \cup_\lambda U(P_x^{\alpha\lambda})$, $U(A) = \cup_{x \in X} U\{P[x, A(x)]\}$.

Moreover $U(\cup_\lambda A_\lambda) = \cup_\lambda U(A_\lambda)$.

The inverse of $U \in \mathcal{U}^*$ is,

$U^{-1} = [(P_x^\alpha, P_y^\beta) | (P_y^{1-\beta}, P_x^{1-\alpha}) \in U]$.

If $U^{-1} = U$, then U is said to be symmetric. Also $(U \circ V)^{-1} = V^{-1} \circ U^{-1}$.

Denition 2.5: The fuzzy set $\Delta\mathbb{P}_* = [(P_x^\alpha, P_x^\alpha) | P_x^\alpha \in \mathbb{P}_*]$ is called the diagonal of $\mathbb{P}_* \times \mathbb{P}_*$.

Denition 2.6: The nonvoid family $\mathcal{U} \subseteq \mathcal{U}^*$ of subsets of $\mathbb{P}_* \times \mathbb{P}_*$ is called a fuzzy semi-quasi-uniformity on I^X iff the following axioms are satised:

S_1: If $U \in \mathscr{U}$, then $(P_x^\alpha, P_x^\alpha) \in U$ for any $P_x^\alpha \in \mathbb{P}_*$, moreover if $P_x^\alpha \in \mathbb{P}_0$, there exits $\delta = \delta\,(P_x^\alpha) > 0$ such that $(P_x^\alpha, P_x^{\alpha+\delta}) \in U$.

S_2: If $U, V \in \mathscr{U}$, then $U \cap V \in \mathscr{U}$.

S_3: If $U \in \mathscr{U}$ and $U \subseteq V$, then $V \in \mathscr{U}$.

This generalizes the notion of fuzzy uniformity of Hu Cheng-Ming [5].

Denition 2.7: The nonvoid subfamily $\mathscr{B}$ of a fuzzy semi-quasi-uniformity $\mathscr{U}$ is called a base for $\mathscr{U}$ iff for any $U \in \mathscr{U}$, there is a $V \in \mathscr{B}$ such that $V \subseteq U$.

The nonvoid subfamily $\mathscr{B}$ of subsets of $\mathbb{P}_* \times \mathbb{P}_*$ is a base for some fuzzy semi-quasi-uniformity iff it satises axiom (S_1) and the following:

(S_4): If $U, V \in \mathscr{B}$, then there is a $W \in \mathscr{B}$ such that $W \subseteq U \cap V$.

Denition 2.8: The non void subfamily $\mathscr{S}$ of $\mathscr{U}^*$ is called a subbase for some fuzzy semi-quasi-uniformity if the collection of nonempty intersection of nitely many members of $\mathscr{S}$ is a base for some fuzzy semi-quasi-uniformity.

Denition 2.9: We shall say that a base $\mathscr{B}$ for a fuzzy semi-quasi-uniformity $\mathscr{U}$ on I^X is a fuzzy semi-uniformity $U^{-1} \in \mathscr{B}, \forall\, U \in \mathscr{B}$. $(I^X, \mathscr{U})$ is called a fuzzy semi-uniform space.

Clearly, fuzzy semi-uniformity generalizes the notion of fuzzy uniformity [5]. Also, the collection of symmetric members of $\mathscr{U}$ is a base for $\mathscr{U}$.

Denition 2.10: [6] We shall say that a base $\mathscr{B}$ for a fuzzy semi-quasi-uniformity $\mathscr{U}$ on I^X is a fuzzy local quasi-uniformity iff $\forall\, U \in \mathscr{B}$ and $\forall\, P_x^\alpha \in \mathbb{P}_*$ there exists $V \in \mathscr{U}$ such that $V \circ V\,(P_x^\alpha) \subseteq U\,(P_x^\alpha)$. If $\mathscr{U}$ is a fuzzy semi-quasi-uniformity which is fuzzy local quasi-uniformity, we call $\mathscr{U}$ to be a fuzzy local quasi-uniformity and $(I^X, \mathscr{U})$ is called a fuzzy locally quasi-uniform space.

Clearly, fuzzy local quasi-uniformity generalizes the notion of fuzzy uniformity [5].

Denition 2.11: The non void subfamily $\mathscr{S}$ of $\mathscr{U}^*$ is called a subbase for some fuzzy local quasi-uniformity if the collection of nonempty intersection of nitely many members of $\mathscr{S}$ is a base for some fuzzy local quasi-uniformity.

Proposition 2.1: [3] Let $i: I^X \to I^X$ be a fuzzy topological interior operator.

Then $\mathbb{F} = [V \in I^X \mid i(V) = V]$ is a fuzzy topology and $i(V)$ is the interior of V in (I^X, F).

Theorem 2.2: Let $(I^X, \mathcal{U})$ be a fuzzy semi-quasi-uniform space and $\mathcal{B}$ be any base of $\mathcal{U}$. Then the mapping int: $I^X \to I^X$ dened by, $\text{int}(A) = \cup [P_x^\alpha \mid \exists V \in \mathcal{B} \text{ s.t. } V(P_x^\alpha) \subseteq A]$, is an interior operator on I^X.

Proof: $(IO1)$ Clearly int $(\underline{1})$ and $(IO2)$ int $(A) \subseteq A$.

Finally, $(IO3)$ $\text{int}(A \cap B) = \text{int}(A) \cap \text{int}(B)$ is fullled by virtue of (S_2).

Clearly every fuzzy semi-quasi-uniformity generates an interior space.

Lemma 2.1: Let $(I^X, \mathcal{U})$ be a fuzzy semi-quasi-uniform space. Then int dened in Theorem 2.2 is a fuzzy topological interior operator under the following condition:

For any $U \in \mathcal{U}$ and $P_x^\alpha \in \mathbb{P}_*$, there exists $V \in \mathcal{U}$ such that to each $P_y^\beta \in V(P_x^\alpha)$ there corresponds $W \in \mathcal{U}$ with $W(P_y^\beta) \subseteq U(P_x^\alpha)$.

Proof: Let $(I^X, \mathcal{U})$ be a fuzzy semi-quasi-uniform space satises the given condition and $A \in I^X$. Let $P_x^\alpha \in \text{int}(A)$, then there exists $U \in \mathcal{U}$ such that $U(P_x^\alpha) \subseteq A$. Let $V \in \mathcal{U}$ such that for any $P_y^\beta \in V(P_x^\alpha)$ there is $W \in \mathcal{U}$ such that $W(P_y^\beta) \subseteq U(P_x^\alpha)$. By (S_1) we have $P_x^\alpha \in V(P_x^\alpha)$. But for $P_x^\alpha \in V(P_x^\alpha)$ we may choose $W \in \mathcal{U}$ such that $W(P_x^\alpha) \subseteq U(P_x^\alpha)$. This implies $P_x^\alpha \in \text{int}[\text{int}(A)]$ and since the other inclusion follows by $(IO2)$ in Theorem 2.2, we have $\text{int}(A) = \text{int}[\text{int}(A)]$.

Now by Proposition 2.1 we have the following:

Theorem 2.3: Every fuzzy semi-quasi-uniformity with the condition given in Lemma 2.1 generates a fuzzy topological space.

For any fuzzy semi-uniform space $(I^X, \mathcal{U})$, the collection $\mathcal{N}_{x_\alpha} = [V(x_\alpha) \mid V \in \mathcal{U}]$ is the neighborhood system at x_α in the generating interior space.

If the family $(\mathcal{N}_{x_\alpha} \mid x_\alpha \in \mathbb{P}_*)$ is a neighborhood system for some fuzzy topology $\mathbb{F}$, we say that $\mathbb{F}$ is the fuzzy topology generated by $\mathcal{U}$

Denition 2.12: Let $(I^X, \mathcal{U})$ and $(I^Y, \mathcal{V})$ be fuzzy semi-uniform spaces, a function $f: (I^X, \mathcal{U}) \rightarrow (I^Y, \mathcal{V})$ is fuzzy semi-uniformly continuous iff for every $V \in \mathcal{V}$, there exists $U \in \mathcal{U}$ such that $\hat{f}(U) \subseteq V$, where $\hat{f}(P_x^\alpha, P_y^\beta) = [f(P_x^\alpha), f(P_y^\beta)]$.

Now since $\hat{f}(U) \subseteq V$ implies $\hat{f}(U)[f(P_x^\alpha)] \subseteq V[f(P_x^\alpha)], \forall P_x^\alpha \in \mathbb{P}_*$ therefore we have the following:

Theorem 2.4: Fuzzy semi-uniformly continuous functions on fuzzy semi-uniform spaces are continuous with respect to the relative interior spaces.

Proof: Let $(I^X, \mathcal{U})$ and $(I^Y, \mathcal{V})$ be fuzzy semi-uniform spaces. Let $f: (I^X, \mathcal{U}) \rightarrow (I^Y, \mathcal{V})$ be fuzzy semi-uniform and $i_{\mathcal{U}}$ and $i_{\mathcal{V}}$ be the respectively interior operators induced by $\mathcal{U}$ and $\mathcal{V}$. For $P_x^\alpha \in \mathbb{P}_*$ and for each neighborhood N of $f(P_x^\alpha)$ in $(I_Y, i_{\mathcal{V}})$, we may choose $V \in \mathcal{V}$ and $U \in U$ so that $V[f(P_x^\alpha)] \subseteq N$ and $\tilde{f}(U) \subseteq V$ so that $f[U(P_x^\alpha)] = \tilde{f}(U)[f(P_x^\alpha)] \subseteq V[f(P_x^\alpha)] \subseteq N$.

Theorem 2.5: Composition of two fuzzy semi-uniformly continuous is fuzzy semi-uniformly continuous.

Proof: Straightforward.

Theorem 2.6: Every fuzzy local quasi-uniformity is a fuzzy semi-quasi-uniformity with condition given in Lemma 2.1.

Proof: Let $(I^X, \mathcal{U})$ be a fuzzy local quasi-uniformity. Then $(I^X, \mathcal{U})$ is a fuzzy semi-quasi-uniform space. Let $U \in \mathcal{U}$ and $P_x^\alpha \in \mathbb{P}_*$. Then there exits $V \in \mathcal{U}$ such that $V o V(P_x^\alpha) \subseteq U(P_x^\alpha)$. This implies for any $P_y^\beta \in \mathcal{V}(P_x^\alpha)$ there exits $W (= V) \in U$ such that $W(P_y^\beta) \subseteq U(P_x^\alpha)$.

Thus we have the following result:

Theorem 2.7: Every fuzzy local quasi-uniform space generates a fuzzy topological space.

Remark 2.8: The converse of the Theorem 2.7 is also true which was shown in (6).

The following results can now be concluded in view of results in (5).

Theorem 2.9: Every fuzzy semi-pseudo metric induces a fuzzy semi-uniformity.

Theorem 2.10: A fuzzy semi-uniform space $(I^X, \mathcal{U})$ is a fuzzy semi-pseudo metric iff $\mathcal{U}$ has a countable base.

REFERENCES

[1] P. Fletcher and W. Lindgren, "Quasi-uniform Spaces", Marcel Dekker INC., 1982.

[2] T.L. Hicks and S.M. Human, "A Note on Locally-Uniform Spaces", *J. Canad. Math. Bull.*, **19**, (1976), No. 4, 501–504.

[3] B. Hutton, "Uniformities of Fuzzy Topological Spaces", *J. Math. Anal. Appl.*, **58**, (1977), 559–571.

[4] W. Lindgren and P. Fletcher, "Locally Quasi-uniform Spaces with Countable Bases", *Duke Math. J.*, **41**, (1974), 231–240.

[5] Hu Cheng Ming, "Fuzzy Topological Spaces", *J. Math. Anal. Appl.* **110**, (1985), 141–178.

[6] D.K. Mitra and D. Hazarika, "Fuzzy Locally Quasi-uniform Spaces", *Submitted.*

[7] "Fuzzy Locally Uniform Spaces", *J. Fuzzy Math. To appear.*

[8] H.W. Pu and H.H. Pu, "Semi-Quasi-Uniform Spaces", *J. Portugal. Math.*, **33**, (1974), 177–184.

[9] Eduard Cech, "Topological Space, Interscience", New York, 1966.

[10] J. Williams, "Locally Uniform Spaces", *Trans. American Math. Soc.*, **168**, (1972), 435–469.

29

ON FUZZY $\breve{g}$–LOCALLY CLOSED SETS IN FUZZY BI TOPOLOGICAL SPACES

Anjan Mukherjee[1*] and Snigdhendu Bikas Chakraborty[2]

[1*]*Department of Mathematics, Tripura University Surjyamaninagar, Tripura-799130 Tripura, India, E-mail: anjan2002_m@yahoo.co.in*

[2]*Lalsingmura H.S. School Tripura, India, E-mail: snigdhendu2007@yahoo.co.in*

Abstract: *The aim of the paper is to introduce the notions of* $\breve{g}$ *-locally closed sets in fuzzy setting. We also define fuzzy pair wise* $\breve{G}$*-LC-continuous functions in fuzzy bi topological spaces.*

Keywords: *Fuzzy bi topological spaces, fuzzy* $\breve{g}$ *– locally closed sets, fuzzy pair wise* $\breve{G}$ *– LC – continuous functions.*

AMS subject classification: *54A40.*

1. INTRODUCTION AND PRELIMINARIES

The concept of fuzzy sets was first introduced by Zadeh (16) and the concept of fuzzy bi topological spaces was initiated by Kandil (6). The triplet (X, τ_1, τ_2) where X is a non–empty set and τ_1, τ_2 are two fuzzy topologies on X is called a fuzzy bi topological space. The concept of fuzzy generalized closed sets was first considered by Thakur and Malviya (14). In 1998 Maki, etc. (9) introduced the notions of fuzzy semi generalized closed sets, recently Saraf and Khanna (11) introduced the concepts of fuzzy generalized semi pre closed sets.

The notions of locally closed sets was first introduced by Kuratowski and Sierpienski (5). According to Bourbaki (2) a subset of a topological space is locally closed if it is the intersection of an open set and a closed set in the space. Levine in 1970 (7) introduced generalized closed (g-closed) sets. Since then several new notions are defined in terms of g-closed sets. Recently, Rajesh and Ekici (3, 4) has introduced and studied $\breve{g}$ -locally closed sets in topological spaces and in 2007 Rajesh and Ekici (3, 4) generalized this concept in bi topological spaces.

In this paper we define $\breve{g}$ -locally closed sets and other related concepts in fuzzy setting. We also study various properties of this concepts in fuzzy bi topological setting. We also define pair wise $\breve{G}$-LC-continuous functions in fuzzy bi topological spaces, and establish some results on them.

In 1980 PU P.M. and Liu Y.M. (10) introduced the concepts of quasi-coincidence and q-neighborhoods by which the extensions of functions in fuzzy setting can be effectively carried out, which can be very interesting. Two fuzzy sets λ and μ are said to be quasi-coincident (q-coincident in short) if there exits x in X such that $\lambda(x) + \mu(x) > 1$. The closure of a fuzzy sub set A of X, denoted by cl (A) is the intersection of all fuzzy super sets of A and the interior of A, denoted by int (A) is the union of all fuzzy open sub sets of A.

Definition 1.1: (1) A fuzzy set λ of the space X is called (*i*) a fuzzy regular open set of X if int cl (λ) = λ (*ii*) a fuzzy regular closed set of X if cl int (λ) = λ (*iii*) fuzzy semi-open set if λ < cl int (λ) and fuzzy semi closed if int cl (λ) < λ (13) λ is fuzzy pre open if λ < int cl (λ) and fuzzy pre closed if cl int (λ) < λ and λ is fuzzy semi pre open if λ < cl int cl (λ) and fuzzy semi pre closed if int cl int (λ) < λ.

We know that fuzzy semi interior s int (λ) [respectively p int (λ)] of fuzzy subset λ of X is the union of all fuzzy semi-open sets (fuzzy pre open sets) contained in λ. The fuzzy semi closure s cl (λ) [resp. p cl (λ)] is the intersection of all fuzzy semi closed (respectively fuzzy pre closed) sets that contain λ. Thakur and Singh (15) defined semi pre closure and semi pre interior of a fuzzy sub set λ of the fuzzy topological space (X, τ) as:

sp cl (λ) = inf (δ : δ > λ; δ is fuzzy semi pre closed in X)

sp int (λ) = sup (δ : δ < λ; δ is fuzzy semi pre open of X)

Definition 1.2 (14): A fuzzy sub set λ of (X, τ) is said to be (*i*) fuzzy generalized closed if cl $(\lambda) < \delta$ whenever $\lambda < \delta$ and δ is fuzzy open. (*ii*) fuzzy generalized semi closed if *s* cl $(\lambda) < \delta$, whenever $\lambda < \delta$ and δ is fuzzy open.

Definition 1.3: A fuzzy subset A of a fuzzy topological space (X, τ) is called (*i*) a fuzzy g^*-closed set if cl $(A) \leq U$ whenever $A \leq U$ and U is fuzzy g open set (generalized open) in (X, τ). The complement of a fuzzy g^*-closed set is fuzzy g^*-open set. (*ii*) fuzzy $\hat{g}$-closed if cl $(A) < U$ whenever $A < U$ and U is fuzzy semi open set in (X, τ). The complement of a fuzzy $\hat{g}$-closed is called fuzzy $\hat{g}$-open. (*iii*) fuzzy $^*g$-closed if cl $(A) < U$ whenever $A < U$ and U is fuzzy $\hat{g}$-open in (X, τ). The complement of fuzzy $^*g$-closed set is fuzzy $^*g$-open. (*iv*) fuzzy $^\#g$-semi closed (briefly $^\#gs$ closed) if *s* cl $(A) < U$ whenever $A < U$ and U is fuzzy $^*g$-open in (X, τ). The complement of a fuzzy $^\#gs$-closed set is called fuzzy $^\#gs$-open. (*v*) fuzzy $\hat{g}$-closed if cl $(A) < U$ whenever $A < U$ and U is fuzzy $^\#gs$-open in (X, τ). The complement of a fuzzy $\breve{g}$-closed set is called fuzzy $\breve{g}$-open.

Definition 1.4: The intersection of all fuzzy $\breve{g}$-closed sets containing the set A is called $\breve{g}$-closure of A and is denoted by $\breve{g}$ – cl (A).

Throughout the paper (X, τ_1, τ_2) and (Y, σ_1, σ_2) denote fuzzy bi topological spaces.

2. $\breve{g}$-LOCALLY CLOSED SETS IN FUZZY BI TOPOLOGICAL SPACES

In this section we define and introduce (τ_1, τ_2)-fuzzy locally closed (briefly flc), sets and discuss their basic characteristics and the relationship among themselves.

Definition 2.1: A subset A of a fuzzy bi topological space (X, τ_1, τ_2) is called (τ_1, τ_2)-fuzzy locally closed (flc) if $A = G \cap F$, where G is τ_1-fuzzy open and F is τ_2-fuzzy closed in (X, τ_1, τ_2).

The collection of all (τ_1, τ_2)-flc sets of (X, τ_1, τ_2) will be denoted by (τ_1, τ_2)-fLC (X).

Definition 2.2: A fuzzy subset A of a fuzzy bi topological space (X, τ_1, τ_2) is called (τ_1, τ_2)-fuzzy $\breve{g}$-locally closed $[(\tau_1, \tau_2) - f\breve{g} - \text{lc}]$ if $A = G \cap F$, where G is τ_1-fuzzy $\breve{g}$-open and F is τ_2-fuzzy-$\breve{g}$-closed in fuzzy bi topological space (X, τ_1, τ_2).

Definition 2.3: A fuzzy subset A of a fuzzy bi topological space (X, τ_1, τ_2) is called (τ_1, τ_2)-fuzzy $\breve{g}$-locally closed* $[(\tau_1, \tau_2) - f\breve{g}\text{ lc}^*]$ if there exits a τ_1-fuzzy $\breve{g}$ open set G and τ_2-fuzzy closed set F of (X, τ_1, τ_2) such that $A = G \cap F$.

Definition 2.4: A fuzzy subset A of a fuzzy bi topological space (X, τ_1, τ_2) is called (τ_1, τ_2)-fuzzy $\breve{g}$ locally closed** if there exits a τ_1 fuzzy open set G and τ_2 fuzzy $\breve{g}$ closed set F of (X, τ_1, τ_2) such that $A = G \cap F$.

The collection of all $(\tau_1, \tau_2) - f\breve{g}$ lc [respectively $(\tau_1, \tau_2) - f\breve{g}$ lc*, $(\tau_1, \tau_2) f\breve{g}$ lc**] sets of (X, τ_1, τ_2) will be denoted by $(\tau_1, \tau_2)f - \breve{G}LC$ (X) [respectively $(\tau_1, \tau_2)f - \breve{G}LC^*(X)$, $(\tau_1, \tau_2)f - \breve{G}LC^{**}(X)$]

Theorem 2.5: If A be a fuzzy sub set of a fuzzy bitopological space (X, τ_1, τ_2).

(*i*) if $A \in (\tau_1, \tau_2) - fLC(X)$ then $A \in (\tau_1, \tau_2)f - \breve{G}LC^*(X)$ and $A \in (\tau_1, \tau_2)f - \breve{G}LC^{**}(X)$.

(*ii*) if $A \in (\tau_1, \tau_2)f - \breve{G}LC^*(X)$ then $A \in (\tau_1, \tau_2)f - \breve{G}LC(X)$.

(*iii*) if $A \in (\tau_1, \tau_2)f - \breve{G}LC^{**}(X)$ then $A \in (\tau_1, \tau_2)f - \breve{G}LC(X)$.

Proof: Proof is obvious from the definitions.

Theorem 2.6: If A be a fuzzy sub set of a fuzzy bitopological space (X, τ_1, τ_2) then the following are equivalent

(*i*) $A \in (\tau_1, \tau_2)f - \breve{G}LC^*(X)$.

(*ii*) $A = G \cap [\tau_2 - \text{cl}(A)]$ for some fuzzy $\tau_1 - \breve{g}$ open set G.

(*iii*) fuzzy $\tau_2 - \text{cl}(A) - A$ is $\tau_1 - f\ \breve{g}$-closed set.

(*iv*) $A \cup [X - \tau_2 - \text{cl}(A)]$ is fuzzy $\tau_1\ \breve{g}$-open set.

***Proof:* (*i*) => (*ii*):** Let A $\in (\tau_1, \tau_2)f - \breve{G}LC^*(X)$ then there exits a fuzzy $t_1 - g$ open set G and fuzzy τ_2-closed set F in (X, τ_1, τ_2) such that $A = G \cap F$. Since $A \subset G$ and $A \subset [\tau_2 - \text{cl}(A)]$, we have $A \subset G \cap [\tau_2 - \text{cl}(A)]$, on the other hand since $\tau_2 - \text{cl}(A) \subset F$, $\Rightarrow G \cap [\tau_2 - \text{cl}(A)] \subset G \cap F = A$. Therefore $A = G \cap [\tau_2 - \text{cl}(A)]$.

(*ii*) => (*i*): Since G is fuzzy $\tau_1 - \breve{g}$-open and fuzzy $\tau_2 - \text{cl}(A)$ is fuzzy τ_2-closed, we have $G \cap [\tau_2 - \text{cl}(A)] \in (\tau_1, \tau_2)f - \breve{G}LC^*(X)$.

(*iii*) => (*iv*): Let $F = [\tau_2 - \text{cl}(A)] - A$, then by (*iii*) F is $\tau_2 - \breve{g}$-closed. Now $X - F = A \cup [X - \tau_2 - \text{cl}(A)]$, since $(X - F)$ is fuzzy $t_1 - \breve{g}$-open, we obtain $A \cup [X - \tau_2 - \text{cl}(A)]$ is fuzzy $\tau_1 - \breve{g}$ open.

(iv) => (iii): Let $G = A \cup [X - \tau_2 - \text{cl}(A)]$, this implies that $X - G$ is fuzzy τ_1 g-closed and $X - G$ = fuzzy $\tau_2 - \text{cl}(A) - A$. Hence $\tau_2 - \text{cl}(A) - A$ is fuzzy $\tau_1 - \breve{g}$-closed.

(iv) => (ii): $G = A \cup [X - \tau_2 - \text{cl}(A)]$, therefore $G \cap \tau_2 - \text{cl}(A) = A$, hence $A = G \cap [\tau_2 - \text{cl}(A)]$ for some fuzzy $\tau_1 - \breve{g}$-open set G.

(ii) => (iv): Let $A = G \cap \tau_2 - \text{cl}(A)$ for some fuzzy $\tau_1 - \breve{g}$-open set G. Then $A \cup [X - \tau_2 - \text{cl}(A)] = G$ which is fuzzy $\tau_1 - \breve{g}$-open.

Theorem 2.7: A and B be any two fuzzy sub sets of a fuzzy bi topological space (X, τ_1, τ_2). If $A \in (\tau_1, \tau_2) f - \breve{G}LC(X)$ and B is fuzzy τ_1 $\breve{g}$-open or fuzzy τ_2 $\breve{g}$-closed then $A \cap B \in (\tau_1, \tau_2) f - \breve{G}LC(X)$.

Proof: Let $A \in (\tau_1, \tau_2) f - \breve{G}LC(X)$ this implies that $A = G \cap F$, where G is fuzzy $\tau_1 - \breve{g}$-open and F is fuzzy τ_2 $\breve{g}$-closed in (X, τ_1, τ_2). Now $A \cap B = (G \cap B) \cap F$.

Case (i): If B is fuzzy $\tau_1 - \breve{g}$-open then $G \cap B$ is also fuzzy τ_1 $\breve{g}$-open and F is fuzzy τ_2 $\breve{g}$-closed in (X, τ_1, τ_2) hence $A \cap B \in (\tau_1, \tau_2) f - \breve{G}LC(X)$.

Case (ii): If B is fuzzy τ_2 $\breve{g}$-closed, then $A \cap B = G \cap (B \cap F)$ where G is fuzzy τ_1 $\breve{g}$-open and $B \cap F$ is fuzzy τ_2 $\breve{g}$-closed in (X, τ_1, τ_2). Hence $A \cap B \in (\tau_1, \tau_2) f - \breve{G}LC(X)$. It is clear that the intersection of any two $(\tau_1, \tau_2) - f$ lc sets is again a $(\tau_1, \tau_2) - f$ lc set.

Theorem 2.8: If $A, B \in (\tau_1, \tau_2) f - \breve{G}LC^*(X)$ then $A \cap B \in (\tau_1, \tau_2) f - \breve{G}LC^*(X)$.

Proof: $A, B \in (\tau_1, \tau_2) f - \breve{G}LC^*(X)$, then there exits fuzzy τ_1 $\breve{g}$-open sets P and Q such that $A = P \cap [\tau_2 - \text{cl}(A)]$ and $B = Q \cap [\tau_2 - \text{cl}(B)]$ since $P \cap Q$ is fuzzy τ_1 $\breve{g}$-open and $A \cap B = (P \cap Q) \cap [(\tau_2 - \text{cl}(A)] \cap [\tau_2 - \text{cl}(B)]$, hence $A \cap B \in (\tau_1, \tau_2) f - \breve{G}LC^*(X)$.

Theorem 2.9: If $A, B \in (\tau_1, \tau_2) f - \breve{G}LC^{**}(X)$ then $A \cap B \in (\tau_1, \tau_2) f - \breve{G}LC^{**}(X)$.

Proof: Proof is same as Theorem 2.8.

Theorem 2.10: If $A \in (\tau_1, \tau_2) f - \breve{G}LC^*(X)$ and B is either τ_2 closed or fuzzy τ_1 open subset of (X, τ_1, τ_2) then $A \cap B \in (\tau_1, \tau_2) f - \breve{G}LC^{**}(X)$.

Proof: Let $A \in (\tau_1, \tau_2) f - \breve{G}LC^*(X)$. This implies that $A = G \cap F$ where G is fuzzy τ_1-open and F is fuzzy $\tau_2 - \breve{g}$-closed. If B is fuzzy τ_1 open then $B \cap G$ is also fuzzy τ_1 open, therefore $A \cap B \in (\tau_1, \tau_2)$

$f - \breve{G}LC^{**}(X)$. Now if B is fuzzy τ_2 closed then $B \cap F$ is also fuzzy τ_2 $\breve{g}$-closed. Therefore $A \cap B \in (\tau_1, \tau_2) f - \breve{G}LC^{**}(X)$.

3. FUZZY PAIR WISE $\breve{G}$-LC CONTINUOUS FUNCTIONS

In this section we define fuzzy pair wise $\breve{G} - LC$ continuous functions in fuzzy bi topological locally closed space and study some basic properties of these functions.

Definition 3.1: A function f from a fuzzy bi topological space (X, τ_1, τ_2) to another fuzzy bi topological space (Y, σ_1, σ_2). $f: (X, \tau_1, \tau_2) \rightarrow (Y, \sigma_1, \sigma_2)$ is called fuzzy pair wise LC-continuous (respectively fuzzy pair wise $\breve{G}LC$ continuous, fuzzy pair wise $\breve{G}LC^*$ continuous, fuzzy pair wise $\breve{G}LC^{**}$ continuous) if $f^{-1}(V) \in (\tau_1, \tau_2) - f\,LC\,(X)$ [respectively $f^{-1}(V) \in (\tau_1, \tau_2) f - \breve{G}LC(X)$, $f^{-1}(V) \in (\tau_1, \tau_2) f - \breve{G}LC^*(X)$, $f^{-1}(V) \in (\tau_1, \tau_2) f - \breve{G}LC^{**}(X)$] for every fuzzy σ_i closed set V of fuzzy bi topological space (Y, σ_i, σ_j) $i\,j = 1, 2$ and $i \neq j$.

Definition 3.2: A function $f: (X, \tau_1, \tau_2) \rightarrow (Y, \sigma_1, \sigma_2)$ is fuzzy pair wise $\breve{G}LC$ irresolute (respectively fuzzy pair wise $\breve{G}LC^*$ irresolute, fuzzy pair wise $\breve{G}LC^{**}$ irresolute) if $f^{-1}(V) \in (\tau_1, \tau_2) f - \breve{G}LC(X)$ [respectively $f^{-1}(V) \in (t_1, t_2) f - \breve{G}LC^*(X)$, $f^{-1}(V) \in (\tau_1, \tau_2) f - \breve{G}LC^{**}(X)$] for every $V \in (\sigma_1, \sigma_2) f\,\breve{G}LC(Y)$ [respectively $V \in (\sigma_1, \sigma_2) f - \breve{G}LC^*(Y)$, $V \in (\sigma_1, \sigma_2) f - \breve{G}LC^{**}(Y)$].

Theorem 3.3: $f: (X, \tau_1, \tau_2) \rightarrow (Y, \sigma_1, \sigma_2)$ be a function then the following conditions are satisfied

(*i*) If f is fuzzy pair wise LC continuous then it is both fuzzy pair wise $\breve{G}LC^*$ continuous and fuzzy pair wise $\breve{G}LC^{**}$ continuous.

(*ii*) If f is fuzzy pair wise $\breve{G}LC^*$ continuous, then it is fuzzy pair wise $\breve{G}LC$ continuous.

(*iii*) If f is fuzzy pair wise $\breve{G}LC^{**}$ continuous, then it is fuzzy pair wise $\breve{G}LC$ continuous.

(*iv*) If f is fuzzy pair wise $\breve{G}LC$ irresolute, then it is fuzzy pair wise $\breve{G}LC$ continuous.

(v) If f is fuzzy $\breve{g}$ pair wise $\breve{G}LC$ irresolute, then it is fuzzy pair wise $\breve{G}LC^*$ continuous.

Proof: Proof is obvious from the definition of the functions.

Theorem 3.4: f: $(X, \tau_1, \tau_2) \rightarrow (Y, \sigma_1, \sigma_2)$ and g: $(Y, \sigma_1, \sigma_2) \rightarrow (Z, \eta_1, \eta_2)$ are two fuzzy pair wise functions then the following conditions are satisfied:

(*i*) If f is fuzzy pair wise $\breve{G}LC$ irresolute and $g \circ f$ is fuzzy pair wise $\breve{G}LC$ continuous, then $g \circ f$ is fuzzy pair wise $\breve{G}LC$ continuous.

(*ii*) If f and g are fuzzy pair wise $\breve{G}LC$ irresolute then $g \circ f$ is fuzzy pair wise $\breve{G}LC$ irresolute.

(*iii*) If f is fuzzy pair wise $\breve{G}LC^*$ irresolute and g is fuzzy pair wise $\breve{G}LC^*$ continuous, then $g \circ f$ is fuzzy pair wise $\breve{G}LC^*$ continuous.

(*iv*) If f is fuzzy pair wise $\breve{G}LC^{**}$ irresolute and g is fuzzy pair wise $\breve{G}LC^{**}$ continuous, then $g \circ f$ is fuzzy pair wise $\breve{G}LC^{**}$ continuous.

Proof: Proof is obvious from the definitions.

Remark 3.5: The composition of fuzzy pair wise $\breve{G}LC$ continuous functions need not be fuzzy pair wise $\breve{G}LC$ continuous.

REFERENCES

[1] Azad K. K., "On Fuzzy Semi Continuity, Fuzzy Almost Continuity and Fuzzy Weakly Continuity", *J. Math Anal. Appl.*, **82**, (1981), 14–32.

[2] Bourbaki N., "General Topology Part 1 Reading MA", *Addison-Wesley* 1996.

[3] Ekici E. and Rajesh N., " $\breve{g}$-locally Closed Sets", *Kochi J. Math* **2**, (2007), 1–9.

[4] Ekici E. and Rajesh N., "On $\breve{g}$-locally Closed Sets in Bitopological Spaces". *Bulletin of Allahabad Mathematical Society*, **22**, (2007) 11–20.

[5] Kuratowski C. and Sierpinski N., "Surles Difference Deux Ensembles Ferms", *Tohuku Math J*, **20**, (1921), 22–25.

[6] Kandil A., "Biproximities and Fuzzy Topological Space", *Simon Stevin*, **63**, (1989), 45-46.

[7] Levine N., "Generalized Closed Sets in Topology", *Rend. Cir. Mat. Palermo*, **19**, (1970), 89–96.

[8] Mukherjee M.N. and Singha S.P., "On Some Near Fuzzy Continuous Functions Between Fuzzy Topological Spaces". *Fuzzy Sets and Systems*, **34**, (1990), 245–254.

[9] Maki H., Fukutaka T., Kojima M., "Generalized Closed Sets in Fuzzy Topological Spaces", *Meeting on topo. Space Theory and Appl.*, (1998) 23–26.

[10] PU P.M. and Liu Y.M., "Fuzzy Topology I-Neighbourhood Structure of a Fuzzy Point and Moore" *Smith convergence. J. Math Anal. Appl.*, **76**, (1980), 571–599.

[11] Saraf R.K., and Khanna M., "Fuzzy Generalized Semi-Pre-Closed Sets", *Journal Tri. Math. Soc.*, **3**, (2001), 59–68.

[12] Saraf R.K. and Caldas M., "Contra Fuzzy Continuity in Fuzzy Topology". *Journal Tri. Math. Soc.*, **7**, (2005), 1–6.

[13] Shahna B., "On Fuzzy Strongly Semi Continuity and Fuzzy Pre Continuity", *Fuzzy Sets and Systems*, **44**, (1991), 303–308.

[14] Thakur S.S., and Malviya R., "Fuzzy Ge-irresolute Mapping, *Proc. Math. Soc. BHU*, **2**, (1995), 184–186.

[15] Thakur S.S. and Singh S., "On Fuzzy Semi Pre Open Sets and Fuzzy Semi Pre Continuity", *Fuzzy Sets and Systems*, **98**, (1998), 383–391.

[16] Zadeh. L.A., "Fuzzy Sets", *Inform. and Control.*, **8**, (1965), 338–353.

30

LOWER AND UPPER MINIMAL CONTINUOUS FUZZY MULTIFUNCTIONS

Anjan Mukherjee[1] **and Rupak Sarkar**[2]

[1]*Department of Mathematics, Tripura University, Surjyamaninagar, Agartala-799130, Tripura, India.*
E-mail: anjan2002_m@yahoo.co.in

[2]*Department of Mathematics, National Institute of Technology, Agartala-799055, Jirania, Tripura, India.*
E-mail: rupak.math@rediffmail.com

Abstract: *Fuzzy minimal structures and fuzzy minimal space were introduced by Alimohammady and Roohi in the following way: A family m_X of fuzzy sets in X is called a fuzzy minimal structure (briefly m-structure) on X if 0_X and 1_X belong to m_X and (X, m_X) is a fuzzy minimal space. The members of m_X are called fuzzy m-open sets and their complements are called fuzzy m-closed sets. The purpose of this paper is to introduce the lower and upper minimal continuous fuzzy multifunction in the sense of Papageorgiou. Some basic properties and characterization theorems are also studied.*

Keywords: *Fuzzy minimal structure, fuzzy m-open (closed) functions, fuzzy m-continuity, fuzzy multifunction.*

AMS Subject Classification: *54A40.*

1. INTRODUCTION

Using fuzzy semiopen sets (3), fuzzy preopen sets (9), fuzzy α-open sets (10), fuzzy β-open sets (8, 11) and fuzzy δ-open sets (4) several authors introduced and studied various types of modifications of fuzzy open functions in fuzzy topological spaces.

Fuzzy minimal structures and fuzzy minimal space were introduced by Alimohammady and Roohi in (1, 2) in the following way: A family

m_X of fuzzy sets in X is called a fuzzy minimal structure (briefly m-structure) on X if 0_X and I_X belong to m_X and (X, m_X) is a fuzzy minimal space. The members of m_X are called fuzzy m-open sets and their complements are called fuzzy m-closed sets.

In Section 2, we introduce the notion of fuzzy lower m-continuous (fuzzy upper m-continuous) multifunction in the sense of Papageorgiou (7). We also establish some characteristic theorems and basic properties of the lower (upper) m-continuous fuzzy multifunction. We introduce the notion of fuzzy lower (upper) m-neighbourhood and characterize the fuzzy lower (upper) m-continuous multifunction via fuzzy lower (upper) m-neighbourhoods.

Let (X, τ) be a fuzzy topological space and λ a fuzzy subset of X. The fuzzy closure of λ and fuzzy interior of λ are denoted by Fcl (λ) and Fint (λ) respectively. Throughout the present paper (X, τ), (Y, Δ) denote fuzzy topological spaces.

Some results and definition are given for ready reference:

Definition 1.2: (1, 2) Let (X, m_X) be a fuzzy minimal space and λ be a fuzzy subset of X, the fuzzy m_X-closure of λ and the fuzzy m_X-interior of λ are defined as follow:

m_X-int $(\lambda) = \vee\,(\alpha: \alpha \leq \lambda, \alpha \in m_X)$

m_X-cl $(\lambda) = \wedge\,(\gamma: \lambda \leq \gamma, \gamma^C \in m_X]$.

Proposition 1.3: (1, 2) For two fuzzy sets a and λ in X.

(*a*) m_X-int $(\alpha) \leq \alpha$, and m_X-int $(\alpha) = \alpha$ if α is a fuzzy m-open set.

(*b*) $\alpha \leq m_X$-cl (α) and m_X-cl $(\alpha) = \alpha$ if α is a fuzzy m-closed set.

(*c*) m_X-int $(\alpha) \leq m_X$-int (λ) and m_X-cl $(\alpha) \leq m_X$-cl (λ) if $\alpha \leq \lambda$.

(*d*) m_X-int $(\alpha \wedge \lambda) = [m_X$-int $(\alpha)] \wedge [m_X$-int $(\lambda)]$ and $[m_X$-int $(\alpha)] \vee [m_X$-int $(\lambda)] \leq m_X$-int $(\alpha \vee \lambda)$.

(*e*) m_X-cl $(\alpha \vee \lambda) = [m_X$-cl $(\alpha)] \vee [m_X$-cl $(\lambda)]$ and m_X-cl $(\alpha \wedge 1) \leq [m_X$-cl $(\lambda)] \wedge [m_X$-cl $(\lambda)]$.

(*f*) m_X-int $[m_X$-int $(\alpha)] = m_X$-int (α) and m_X-cl $[m_X$-cl $(\lambda)] = m_X$-cl (λ).

(*g*) $[m_X$-cl $(\alpha)]^C = m_X$-int (α^C) and $[m_X$-int $(\alpha)]^C = m_X$-cl (α^C).

Definition 1.4: (5) A fuzzy minimal space (X, m_X) enjoys the property B if arbitrary union of fuzzy m-open sets is fuzzy m-open sets.

Proposition 1.5: (1, 2) For a fuzzy minimal structure m_X on a set X the following are equivalent:

(*a*) m_X has the property B.

(*b*) If m_X-int $(\alpha) = \alpha$ then $\alpha \in m_X$

(*c*) If m_X-cl $(\lambda) = \lambda$ then $\lambda^C \in m_X$

Definition 1.6: (1, 2) Let (X, m_X) and (Y, m_Y) be two fuzzy minimal space then a function $f: (X, m_X) \rightarrow (Y, m_Y)$ is fuzzy minimal continuous (briefly fuzzy m-continuous) if $f^{-1}(\lambda) \in m_X$ for every $\lambda \in m_Y$.

Definition 1.7: (1, 2) Suppose (X, m_X) and (Y, m_Y) be two fuzzy minimal space.

A function $f: (X, m_X) \rightarrow (Y, m_Y)$ is said to be

(*a*) fuzzy m-open if $f(\alpha) \in m_Y$ for any fuzzy m-open set in X.

(*b*) fuzzy m-closed if $f(\lambda)$ is a fuzzy m-closed set in (Y, m_Y) for any fuzzy m-closed λ in (X, m_X).

Definition 1.8: (6) For a fuzzy point x_p and a fuzzy set a in X, we write $x_p\ q\ \alpha$ to denote that x_p is *quasi-coincident* with α if there is at least one point $x \in X$ such that $p + \alpha(x) > 1$.

$x_p \not{q}\ \alpha$ means that x_p is not quasi coincident with a if $a(x) + b(x) \leq 1$.

For two fuzzy sets α and β in X, we write $\alpha\ q\ \beta$ to denote that α is *quasi-coincident* with β if there is at least one point $x \in X$ such that $\alpha(x) + \beta(x) > 1$.

$\alpha \not{q}\ \beta$ means that α is not quasi coincident with β if $\alpha(x) + \beta(x) \leq 1$.

2. LOWER AND UPPER *m*-CONTINUOUS FUZZY MULTIFUNCTIONS

Papageoregiou (7) introduced and investigated the concept of multifunction in fuzzy setting. According to (7) if (X, τ) be an ordinary topological space and (Y, σ) be a fuzzy topological space then $f: (X, \tau) \rightarrow (Y, \sigma)$ is a fuzzy multifunction if corresponding to each $x \in X$. $F(x)$ is a unique fuzzy set in Y. For a fuzzy multifunction $F: X \rightarrow Y$, The upper inverse F^+ and lower inverse F^- are defined as follows:

For any fuzzy set $\mu \in Y$, $F^+(\mu) = [x \in X: F(x) \leq \mu]$ and $F^-(\mu) = [x \in X: F(x)\ q\ \mu]$.

For a fuzzy multifunction $F: X \to Y$. $F^-(1 - \mu) = X - F^+(\mu)$ for any fuzzy set μ in Y.

In this section, we introduce concepts of lower and upper μ-continuous fuzzy multifunction.

Definition 2.1: Let (X, T) be an ordinary topological space and (Y, m_Y) be a fuzzy m-space. A fuzzy multi function $F: (X, T) \to (Y, m_Y)$ is said to be

(*a*) fuzzy lower m-continuous (briefly flmc) at a point $x_0 \in X$ if and only if for every fuzzy m-open set μ in Y with $x_0 \in F^-(\mu)$ there exists a T-open neighbourhood U of x_0 in X such that $U \subset F^-(\mu)$ i.e. $F(x_0)\ q\ U$ for all $x_0 \in U$.

(*b*) fuzzy upper m-continuous (briefly fumc) at a point $x_0 \in X$ if and only if for every fuzzy m-open set μ in Y with $x_0 \in F^+(\mu)$ there exists a T-open neighbourhood U of x_0 in X such that $U \subset F^+(\mu)$ i.e. $F(x) \leq \mu$ for all $x_0 \in U$.

(*c*) fuzzy lower (upper) m-continuous on (X, T) if and only if it is fuzzy lower (upper) m-continuous at each point $x_0 \in X$.

Theorem 2.2: A fuzzy multifunction $F: (X, T) \to (Y, m_Y)$ is flmc if and only if for any fuzzy m-open set U in Y, $F^-(U)$ is T-open in X.

Proof: Let F be flmc and let $x \in F^-(U)$ when U is any fuzzy m-open set in Y. By flmc of F there exists an open neighbourhood V of x such that $V \subset F^-(U)$. Then Int $V = V \subset$ Int $F^-(U)$ and hence $x \in$ Int $F^-(U)$ and consequently $F^-(U)$ is T-open in X.

Conversely let $x \in X$ and U be a fuzzy m-open set in Y such that $x \in F^-(U)$. Then $V = F^-(U)$ is an open neighbourhood of x such that $V \subset F^-(U)$. Hence f is flmc at x and consequently flmc on X.

Corollary 2.3: A fuzzy multifunction $F: (X, T) \to (Y, m_Y)$ is fumc if and only if for any fuzzy m-open set λ in Y, $F^+(\lambda)$ is T-open in X.

Proof: It follows from the fact that $F^-(1 - \lambda) = [F^+(\lambda)]^C$ for any fuzzy m-set λ in Y.

Definition 2.4: A fuzzy set a in a fuzzy m-space (Y, m_Y) is said to be a fuzzy lower (upper) m-nbd of a fuzzy set β in Y if there exists a fuzzy m-open set v in Y such that $\beta\ q\ v$ (resp. $\beta \leq v$) and m_X-cl $v\ \not{q}$ $(1 - \alpha)$.

We now characterize a flmc and fumc multifunctions via fuzzy lower (upper) m-nbds of a fuzzy set.

Theorem 2.5: A fuzzy multifunction F: $(X, T) \to (Y, m_Y)$ is flmc on X if and only if for each point $x_0 \in X$ and each fuzzy lower m-nbd M of $f(x_0)$, $F^-(M)$ is a open nbd of x_0.

Proof: Let F be flmc on X. Then for any $x_0 \in X$ and any fuzzy lower m-nbd M of $F(x_0)$, There exists a fuzzy m-open set v in Y such that $x_0 \in F^-(v)$ and m_X-cl $v \not q (1 - \mu)$ i.e. $F(x_0)\, q\, v$ and $m_Y -$ cl $v \leq M$.

As f is flmc and $F(x_0)\, q\, v$, there exists T-open nbd U of x_0 such that $U \subset F^-(m_X$-cl $v)$. Then $x_0 \in U \subset F^-(M)$ showing that $F^-(M)$ is a open nbd of x_0.

Conversely, let $x_0 \in X$ and v be any fuzzy m-open set in Y with $x_0 \in F^-(v)$. Then $v\, q\, F(x_0)$ and m_Y-cl $v \not q (1 - m_Y$-cl $v)$. Thus m_Y-cl v is a fuzzy lower m-nbd of $f(x_0)$. Then by the given condition $F^-(m_Y$-cl $v)$ is a open nbd of x_0.

Thus by the definition there exists $U \in T$ such that $x_0 \in U \subset F^-(m_Y$-cl $v)$ i.e. $F(x_U)\, q$ m_Y-cl v for all $x_U \in U$. Hence F is flmc on X.

Theorem 2.6: A fuzzy multifunction F: $(X, T) \to (Y, m_Y)$ is flmc on X if and only if for each point $x_0 \in X$ and each fuzzy upper m-nbd M of $f(x_0)$, $F^+(M)$ is a open nbd of x_0.

Proof: Similar to Theorem 2.5.

Definition 2.7: A fuzzy multifunction $F: X \to Y$ is called

(*a*) fuzzy lower almost minimal continuous (flamc, in short) at some point $x_0 \in X$ iff for every fuzzy m-open set V in Y with $x_0 \in F^-(V)$, there exists an open nbd U of x_0 such that $U \subset F^-(m_Y$-Int m_Y Cl $V)$.

(*b*) fuzzy upper almost minimal continuous (fuamc, in short) at $x_0 \in X$ iff for every fuzzy m-open set V in Y with $x_0 \in F^+(V)$, there exists an open nbd U of x_0 such that $U \subset F^+(m_Y$ Int m_Y Cl $V)$.

(*c*) flamc (fuamc) on X iff F respectively so at each point $x_0 \in X$.

Theorem 2.8: For a fuzzy multi function F: $(X, T) \to (Y, m_Y)$ the following are equivalent:

1. f is flamc

2. G is fuzzy m-open in $Y \Rightarrow F^-(G) \subset \text{Int } F^-(m_Y \text{ Int } m_Y\text{-Cl } G)$
3. G is fuzzy m-regularly open in $Y \Rightarrow F^-(G)$ is T-open in X.
4. G is fuzzy m-open in $Y \Rightarrow F^-(m_Y \text{ Int } m_Y \text{ Cl } G)$ is T-open in X.

Proof: (1) → (2): Let G be a fuzzy m-open set in Y and $x \in F^-(G)$.

Then there exists an open nbd U of x such that $U \subset F^-(m_Y \text{ Int } m_Y\text{-Cl } G)$. Then $x \in U \subset \text{Int } F^-(m_Y \text{ Int } m_Y \text{ Cl } G)$. Hence $F^-(G) \subset \text{Int } F^-(m_Y \text{ Int } m_Y \text{ Cl } G)$.

(2) → (3): G being fuzzy m-regular open, $G = m_Y \text{ Int } m_Y \text{ Cl } G$ and hence by (2), $F^-(G) \subset \text{Int } F^-(m_Y \text{ Int } m_Y \text{ Cl } G) = \text{Int } F^-(G)$. Consequently $F^-(G)$ is T-open in X.

(3) → (4): Obvious, because for a fuzzy m-open set G, m_Y-Int m_Y-Cl G is fuzzy m-regularly open.

(4) → (1): For any arbitrary $x \in X$ and arbitrary fuzzy m-open set G in Y with $x \in F^-(m_Y \text{ Int } m_Y \text{ Cl } G) = U$ (say) is T-open in X.

Now $G \subset m_Y \text{ Int } m_Y \text{ Cl } G \Rightarrow x \in U$ and so (1) follows.

Remark 2.9: For a fuzzy multifunction $F: X \to Y$ the following are equivalent

1. F is flamc.
2. V is fuzzy m-regularly closed in $Y \Rightarrow F^+(V)$ is T-closed in X.

Definition 2.10: A fuzzy set A in a f minimal space (Y, m_Y) is fuzzy m-semi open iff there exists a fuzzy m-open set U in Y so that $U \le A \le m_Y$-Cl U or A in (Y, m_Y) is fuzzy m-semi open iff $A \le m_Y \text{ Cl } m_Y \text{ Int } A$.

Theorem 2.11: A fuzzy multifunction $F: X \to Y$ is flamc (fuamc) iff for any fuzzy m-semi open set A in Y, $\text{Cl } [F^+(A)] \subset F^+(m_Y \text{ Cl } A)$ [resp. $\text{Cl } [F^-(A)] \subset F^-(m_Y \text{ Cl } A)$.

Proof: Let F be flamc and A be fuzzy m-semi open in Y. Then $A \le m_Y \text{ Cl } m_Y \text{ int } A = B$ (say) where B is fuzzy m-regularly closed in Y.

Then by Remark 2.9, $F^+(B)$ is T-closed in X and thus $\text{Cl } F^+(A) \subset \text{Cl } F^+(B) = F^+(B) = F^+(m_Y \text{ Cl } m_Y \text{ Int } A) \subset F^+(m_Y \text{ Cl } A)$.

Conversely, since every fuzzy m-regularly closed set is fuzzy m-semi open for any fuzzy m-regularly closed set A in Y.

We have $\text{Cl } [F^+(A)] \subset F^+(m_Y \text{ Cl } A) = F^+(A)$.

Then $F^+(A)$ is T-closed in X, hence F is flamc.

REFERENCES

[1] M. Alimohammady, M. Roohi; "Fuzzy Minimal Structure and Fuzzy Minimal Vector Spaces", *Chaos, Solitons and Fractals*, **27 (3)**, (2006), 599–605.

[2] M. Alimohammady, M. Roohi; "Compactness in Fuzzy Minimal Spaces", *Chaos, Solitons and Fractals*, **28 (4)**, (2006), 906–912.

[3] K.K. Azad; "On Fuzzy Semi Continuity, Fuzzy Almost Continuity and Fuzzy Weakly Continuity", *J. Math. Anal. Appl.*, **82**, (1981), 14–32.

[4] S. Ganguly and S. Saha; "A Note on δ-continuity and δ-connected Sets in Fuzzy Set Theory", *Simon Stevin*, **62**, (1988), 127–141.

[5] A. Mukherjee; "Fuzzy Minimal Structure, Fuzzy *m*-regularity, Fuzzy *m*-normality and Fuzzy Bitopological Spaces", *Presented the paper in AA-BHU-09*, 19–22, March, 2009.

[6] P. Pao-Ming, L. Ying-Ming; "Fuzzy Topology *I*, Neighbourhood Structure of a Fuzzy Point and Moore Smith Convergence", *J. Math. Anal. Appl.*, **76 (2)**, (1980), 571–599.

[7] N.S. Papageorgiou, "Fuzzy Topology and Fuzzy Multifunction", *Journal of Math. Anal. Appl.*, **109 (2)**, (1985), 397–425.

[8] J.H. Park, B.Y. Lee, "Fuzzy Semi Pre Open and Fuzzy Semi Pre Continuous Maps", *Fuzzy Sets and Systems*, **67**, (1994), 359–364.

[9] M.K. Singal and N. Prakash; "Fuzzy Pre Open Sets and Fuzzy Pre Separation Axioms", Fuzzy Sets and Systems, **44**, (1991), 345–351.

[10] M.K. Singal and Niti Rajvansi, "Fuzzy Alpha Sets and Alpha Continuous Maps", *Fuzzy Sets and Systems*, **48**, (1992), 383–390.

[11] S.S. Thakur and Surendra Singh, "On Fuzzy Semi Pre Open Sets and Fuzzy Semi Pre Continuity", *Fuzzy Sets and Systems*, 98, (1998), 383–391.

[12] L.A. Zadeh, "Fuzzy Sets", *Information and Control*, **8**, (1965), 338–353.

31

FRACTIONAL BINARY OPERATIONS

Anjan Mukherjee[1] and Subrata Bhowmik[2*]

[1]*Department of Mathematics, Tripura University*
Suryamaninagar, Tripura-799130, India
E-mail: anjan2002_m@yahoo.co.in
[2*]*E-mail: subrata_bhowmik_math@rediffmail.com*

Abstract: *We are all familiar with algebraic operations with crisp numbers. The purpose of this paper is to introduce a new type of operation on a set, which is the extension of the binary operation concept on a set—called fractional binary operation. It is seen that the structures like fractional groupoid, fractional semi-group can be developed. Lastly an application is to be shown in topology.*

Keywords: *Groupoid, semi-group, monoid, group, partial order.*

Mathematics Subject Classification 2000: *54A99*

1. INTRODUCTION

We know binary operations on a non-empty set, here we compose two elements of the set by a rule or a mapping called a binary operation on the set. We can think that we are composing two elements of the set, but due to an interruption in the middle of the process of completing composing them, the composition process have not done completely, as much as we give the time to the composition process, we get more developed/superior result, like we are going to mix two things in a mixture grinder, but the power connection cuts off before mixing them completely—this idea motivated us to think of binary operations fractionally, i.e. we are thinking a new type of operation on a set which is the extension of the binary operation like membership value gradation of a point in a fuzzy set – we call this new operation on a non-empty set "fractional binary operation".

Here we have taken the set G equipped with a binary operation o; we are thinking to compose $a, b \in G$ fractionally with respect to the operation o. So fractional composition can be done from zero fractional level to fractional level 1, fractional level 1 mean complete composition, i.e. aob. Now what will be the value if we compose a and b with respect to o at fractional level zero, so we think to equip G with a monoid structure with respect to o and assign the value of fractional composition a and b with respect to o at zero level be the identity element e of the monoid (G, o). Now what will be the value of fractional composition with respect to o at fractional level between 0 and 1, so we need a order relation to show as greater as we take the fractional level we get more superior/developed result.

All these ideas will be described in the next articles with some properties and with an application in general topology.

Before going to our main topic here is some pre-requisite definitions:

Definition 1.1 (Nanda[2]): Let G be a non-empty set, a relation "$\leq$" on G is said to be a partial order relation on G if "$\leq$" is reflexive, antisymmetric and transitive. The set G with a partial order relation "$\leq$" on G is called a partially ordered set to be denoted by $(G, \leq)$.

Definition 1.2 (Nanda[2]): Let G be non-empty set, a mapping

$$o : G \times G \to G$$

$$(a, b) \mapsto aob$$

is called a binary operation on G. The set G with a binary operation o is called a groupoid to be denoted by (G, o).

Definition 1.3 (Nanda[2]): A groupoid (G, o) is said to be a semi-group if the binary operation is associative, i.e.

$$ao\,(boc) = (aob)\,oc \;\forall a, b, c \in G$$

Definition 1.4 (Nanda[2]): A semi-group (G, o) is said to be a monoid if there exists an element $e \in G$, called the identity element, such that

$$aoe = eoa \;\forall a \in G$$

It can be easily prove that the identity element e is unique in the semi-group.

2. FRACTIONAL BINARY OPERATIONS

Definition 2.1: An ordered groupoid (semi-group, monoid) is a groupoid (repectively semi-group, monoid) (G, o) with a partial order "$\leq$" on G with the property that for $a, b, c, d \in G, a \leq c$ and $b \leq d \Rightarrow a o b \leq c o d$.

Let us denote the ordered groupoid (semi-group, monoid) by $(G, o, \leq)$.

Now let us introduce our main definition of this paper:

Definition 2.2: Let $(G, o, \leq)$ be a ordered monoid, then a fractional binary operation $\odot$ on $(G, o, \leq)$ is a mapping

$$\odot : G \times G\,[0, 1] \to G$$
$$(a, b, r) \mapsto a \odot_r b$$

such that: $\forall a, b \in G$

(*i*) $a \odot_0 b = e$, the identity element of the monoid (G, o),

(*ii*) $0 \leq r_1 \leq r_2 \leq 1 \Rightarrow a \odot_{r1} b \leq a \odot_{r2} b$,

(*iii*) $a \odot_1 b = a o b$.

The ordered monoid $(G, o, \leq)$ with a fractional binary operation $\odot$, we call it a fractional groupoid and let us denote this fractional groupoid by $(G, o, \leq, \odot)$.

Example 2.3: In the ordered monoid $(\boldsymbol{R}^+, +, \leq)$, where $\leq$ is the natural ordering in set $\boldsymbol{R}^+$ of non-negative reals, we define the mapping

$$\odot : \boldsymbol{R}^+ \times \boldsymbol{R}^+ \times [0, 1] \to \boldsymbol{R}^+$$
$$(a, b, r) \mapsto a \odot_r b = r\,(a + b)$$

then $\odot$ is a fractional binary operation on $(\boldsymbol{R}^+, +, \leq)$ and also $(\boldsymbol{R}^+, +, \leq, \odot)$ is a fractional groupoid $\odot$.

Theorem 2.4: In a fractional groupoid $(G, o, \leq, \odot)$

$$e \odot_r e = e, \text{ for every } r \in (0, 1) \text{ and } e \leq a, \forall a \leq G$$

Proof: Since $e = e \odot_0 e \leq e \odot_r e \leq e \odot_1 e = e o e = e$ and

$$e = a \odot_0 e \leq a \odot_1 e = a o e = a. \odot$$

Theorem 2.5: In a fractional groupoid $(G, o, \leq, \odot)$, if (G, o) is a group, G is a singleton set.

Proof: Since $e \leq a \leq a$, $e \leq a^{-1} \Rightarrow e \leq aoe \leq aoa^{-1} = e$, i.e. $a = e$.

Theorem 2.6: In a fractional groupoid $(G, o, \leq, \odot)$, for $a, b \in G$, a and b both are not simultaneously e, we define $< a, b > = [a \odot_r b : r \in (0, 1)]$. Let T be a t-norm and S be a t-co-norm. We can define the binary operations $\ominus$ and $\ominus'$ on $< a, b >$ as $(a \odot_{r1} b) \ominus (a \odot_{r2} b) = a \odot_{T(r1, r2)} b$ and $(a \odot_{r1} b) \ominus' (a \odot_{r2} b) = a \odot_{S(r1, r2)} b$ $\forall r \in [0, 1]$ if $0 \leq r_1 < r_2 \leq 1 \Rightarrow a \odot_{r1} b < a \odot_{r2} b$, then $(< a, b >, \ominus, \leq)$ and $(< a, b >, \ominus', \leq)$ both are commutative ordered monoids. (Here comutativity is in binary operation sense).

Proof: We consider the binary operation $\ominus$ only, the other one has similar proof. The binary operation $\ominus$ is well defined.

We see that: For $a \odot_{r1} b$, $a \odot_{r2} b$, $a \odot_{r3} b$, $a \odot_{r4} b \in < a, b >$

(*i*) $(a \odot_{r1} b) \ominus [(a \odot_{r2} b) \ominus (a \odot_{r3} b)] = (a \odot_{r1} b) \ominus [a \odot_{T(r2, r3)} b]$

$$= a \odot_{T[r1, T(r2, r3)]} b = a \odot_{T[T(r1, r2) r3)]} b$$

$$= [a \odot_{T(r1, r2)} b] \ominus (a \odot_{r3} b)$$

$$= [(a \odot_{r1} b) \ominus (a \odot_{r2} b)] \ominus (a \odot_{r3} b)$$

(*ii*) We see that $aob = a \odot_1 b \in < a, b >$ and $(a \odot_{r1} b) \ominus (aob) = (a \odot_{r1} b) \ominus (a \odot_1 b) = a \odot_{T(r1, 1)} b = a \odot_{r1} b = a \odot_{T(1, r1)} b = (a \odot_{r1} b) \ominus (a \odot_1 b) = (a \odot_{r1} b) \ominus (aob)$

so, aob is the identity element of $< a, b >$ with respect to the binary operation $\ominus$. Hence $(< a, b >, \ominus)$ is a monoid.

Again we see that:

(*iii*) $(a \odot_{r1} b) \ominus (a \odot_{r2} b) = a \odot_{T(r1, r2)} b = a \odot_{T(r2, r1)} b = (a \odot_{r2} b) \ominus (a \odot_{r1} b)$

If with respect to the partial order $\leq$ on $< a, b >$ induced from G, let $a \odot_{r1} b \leq a \odot_{r2} b$ and $a \odot_{r3} b \leq a \odot_{r4} b$ then obviously $r_1 \leq r_2$ and $r_3 \leq r_4$ (if $r_2 < r_1$ then $a \odot_{r2} b < a \odot_{r1} b$, which is not possible) and so $T(r_1, r_3) \leq T(r_2, r_4)$ and hence $a \odot_{T(r1, r3)} b \leq a \odot_{T(r2, r4)} b$, i.e. $(a \odot_{r1} b) \ominus (a \odot_{r3} b) \leq (a \odot_{r2} b) \ominus (a \odot_{r4} b)$.

Hence $(< a, b >, \ominus, \leq)$ is a commutative ordered monoid.

For the binary operation $\ominus'$ on $< a, b >$, it can be easily shown that $e = a \odot_0 b \in < a, b >$ is the identity element $\odot$.

Cor 2.7: Let us define U: $[0, 1] \times [0, 1] \to [0, 1]$ as

$$U(r, s) = \begin{cases} r+s, & if \quad r+s \le 1 \\ r+s-1, & if \quad r+s > 1 \end{cases}$$

Now we can define the binary operation $\varnothing$ on $<a, b>$ as: $(a \odot_{r1} b) \varnothing (a \odot_{r2} b) = a \odot_{U(r1, r2)} b$, $\forall r \in (0, 1)$, then $(<a, b>, \varnothing)$ is a commutative group.

Proof: Proof is similar that of the Theorem 2.6, so we leave the proof. Here $aob \in <a, b>$ is the identity element and inverse element of $a \odot_r b \in <a, b>$ is $a \odot_{1-r} b \in <a, b>$ with respect to the binary operation $\varnothing$ on $<a, b> \odot$.

Theorem 2.8: For fractional groupoid $(G, o, \le,)$ for $0 \le r \le 1$ we define $G(r) = (a \odot_r b$: $a, b \in G)$. We define a binary operation $\otimes$ on $G(r)$ as $\forall a, b, c, d \in G$ $(a \odot_r b) \otimes (c \odot_r d) = (aoc) \odot_r (bod)$, if it is will defined then $[G(r), \otimes]$ is a monoid and if (G, o) is a commutative then so also $[G(r), \otimes]$.

Proof: The proof is very simple, here $e = e \odot_r e \in G(r)$ identity element of the monoid.

Definition 2.9: A fractional groupoid $(G, o, \le, \odot)$ will be called a fractional semi-group if $\forall a, b, c \in G$ and for every $r \in (0, 1)$

$$a \odot_r (boc) = (aob) \odot_r c$$

Example 2.10: The fractional groupoid $(\boldsymbol{R}^+, +, \le, \odot)$ in Example 2.3 is also a fractional semi-group $\odot$.

Theorem 2.11: For fractional semi-group $(G, o, \le, \odot)$

(*i*) (G, o) is commutative iff $a \odot_r b = b \odot_r a \quad \forall a, b \in G$

(*ii*) $\forall a, b \in G$ with $a \le b \Rightarrow a \odot_r e \le b \odot_r e$ iff

$$\forall a, b, c, d \in G \text{ with } a \le c \text{ and } b \le d \Rightarrow a \odot_r b \le c \odot_r d$$

Proof: (*i*) If (G, o) is commutative then $\forall a, b \in G$ $a \odot_r b = a \odot_r (boe) = (aob) \odot_r e = (boa) \odot_r e = b \odot_r (aoe) = b \odot_r a$.

Converse follows directly.

(*ii*) Let $\forall a, b \in G$ with $a \le b \Rightarrow a \odot_r e \le b \odot_r e$

Let $a, b, c, d \in G$ with $a \le c$ and $b \le d$.

Now $a \odot_r b = a \odot_r (boe) = (aob) \odot_r e \le (cod) \odot_r e = c \odot_r (doe) = c \odot_r d$. Converse follows directly.

3. AN APPLICATION TO TOPOLOGY

Definition 3.1: A topological fractional groupoid is a topological space (G, τ) where $(G, o, \leq, \odot)$ is a fractional groupoid such that the mapping

$$G \times G\,[0, 1] \to G$$
$$(a, b, r) \mapsto a \odot_r b$$

is a continuous mapping. Let us denote the topological fractional groupoid as $(G, \tau, o, \leq, \odot)$.

Example 3.2: Let us consider the product space $\boldsymbol{R}^+ \times \boldsymbol{R}^+$ with the topology τ induced by the usual topology on $\boldsymbol{R} \times \boldsymbol{R}$ ($\boldsymbol{R}^+$ is the set of non-negative reals).

Let us take

(*i*) the order relation $\leq$ on $\boldsymbol{R}^+ \times \boldsymbol{R}^+$ defined as $(x_1, y_1) \leq (x_2, y_2)$ iff $x_1 \leq x_2$ and $y_1 \leq y_2$ therefore $(\boldsymbol{R}^+ \times \boldsymbol{R}^+, \leq)$ is a partially order set.

(*ii*) the binary operation + on $\boldsymbol{R}^+ \times \boldsymbol{R}^+$ defined as $(x_1, y_1) + (x_2, y_2) = (x_1 + x_2, y_1 + y_2)$ therefore $(\boldsymbol{R}^+ \times \boldsymbol{R}^+, +, \leq)$ is a ordered gropoid.

(*iii*) the mapping $\odot$: $(\boldsymbol{R}^+ \times \boldsymbol{R}^+) \times (\boldsymbol{R}^+ \times \boldsymbol{R}^+) \times [0, 1] \to \boldsymbol{R}^+ \times \boldsymbol{R}^+$ $[(x_1, y_1), (x_2, y_2)\, r] \mapsto [r\,(x_1 + x_2), r\,(y_1 + y_2)]$ therefore $(\boldsymbol{R}^+ \times \boldsymbol{R}^+, +, \leq, \odot)$ is a fractional groupoid and hence $(\boldsymbol{R}^+ \times \boldsymbol{R}^+, \tau, +, \leq, \odot)$ a topological fractional groupoid $\odot$.

Theorem 3.3: For a topological fractional groupoid $(G, \tau, o, \leq, \odot)$, (G, τ) is a path connected and hence a connected topological space.

Proof: For each $a \in G$ we define the mapping

$$[0, 1] \to G$$
$$r \mapsto a \odot_r e$$

then this is a continuous mapping and hence a path from e to a, i.e. every point is connected with e by a continuous path and so (G, τ) is a path connected and hence a connected topological space $\odot$.

Let us denote the parametric above path in the theorem for $a \in G$ as $\sigma_t(a) = a \odot_t e$, $0 \leq t \leq 1$.

We define the binary operation "·" on $\Sigma = [\sigma_t(a) : a \in G]$ as $\sigma_t(a)\, \sigma_t(b) = \sigma_t(aob)$ and we can easily deduce the following theorem:

Theorem 3.4: $(\Sigma, \cdot)$ is a monoid $\odot$.

Here the identity element is the one point mapping $\sigma_l(e)$.

We can also define a order relation $\leq$ on Σ induced from the partially ordered set $(G, \leq)$ as $\sigma_l(a) \leq \sigma_l(b)$ iff $a \leq b$ and we can easily deduce the following theorem:

Theorem 3.5: $(\Sigma, ., \leq)$ is a ordered monoid.

Again we define a mapping

$$O : \Sigma \times \Sigma \times (0, 1) \rightarrow \Sigma$$

$[\sigma_l(a), \sigma_l(b), r] \mapsto \sigma_l(a)\, O_r \sigma_l(b) = \sigma_l(a \odot_r b)$ and we have

Theorem 3.6: $(\Sigma, ., \leq, O)$ is a fractional groupoid and if $(G, o, \leq, \odot)$ is a fractional semi-group, then $(\Sigma, ., \leq, O)$ is also a fractional semi-group.

REFERENCES

[1] Dugundji J., "Topology", University Book Stall, New Delhi, (1989).

[2] Nanda S., "Topics in Algebra", Allied Publishers Pvt. Ltd., New Delhi (2005).

[3] Zimmermann, H.J., "Fuzzy Set Theory and its Applications", Allied Pub. Ltd., New Delhi (1991).

32

PAIRWISE SEMI PARACOMPACTNESS IN FUZZY BITOPOLOGICAL SPACE

Anjan Mukherjee[1] and Bishnupada Debnath[2]

[1]*Department of Mathematics, Tripura University, Suryamaninagar*
Agartala, Pin-799130, Tripura, India
E-mail: anjan 2002_m@yahoo.co.in
[2]*E-mail: debnath_bishnupada@rediffmail.com*

Abstract: *The purpose of this paper is to study the concept of pairwise semi paracompactness in fuzzy bitopological spaces. Some characterizations of pairwise fuzzy semi paracompactness are also studied. We also introduce two types of pairwise fuzzy semi paracompactness viz,* p_1*-fuzzy semi paracompact and* p_2*- fuzzy semi paracompact spaces and then disscuss the relation between them.*

Key words: *Pairwise fuzzy semi paracompact spaces, pairwise fuzzy semi regular spaces, pairwise fuzzy semi normal spaces, etc.*

AMS Subject Classificatin: 54A 40.

1. INTRODUCTION AND PRELIMINARIES

The concept of bitopological spaces was initiated by J.C. Kelly [5] whereas the notion of fuzzy bitopological spaces was first introduced by A. Kandil [4]. A non-empty set X equipped with two arbitrary fuzzy topologies τ_1 and τ_2 is called a fuzzy bitopological space which is denoted by (X, τ_1, τ_2).

A fuzzy point (7) is a fuzzy set which is denoted by x_t and defined by

$$x_t = t\ (0 < t \leq 1), \quad \text{for, } y = x,$$
$$= 0, \quad \text{for, } y \neq x.$$

In 1981, K.K. Azad [1], introduced fuzzy semi-open Set and fuzzy semi closed set in the following way:

A fuzzy set $\lambda \in I^X$ of a fuzzy space (x, z) is called (*i*) fuzzy semi open it then omits a $\nu \in \tau$ such that $\nu \leq \lambda \leq \mathrm{cl}(\nu)$ and (*ii*) fuzzy semi closed if there omits a $e \in \tau$ such that $int(\beta) \leq \lambda \leq \beta$. In a fuzzy topological space, S.S. Kumar [6], defined (τ_i, τ_j)-fuzzy semi open set and (τ_i, τ_j)-fuzzy semi closed set as follows:

A fuzzy set $\lambda \in I^X$ of a fuzzy bitopological space $(X333, \tau_1, \tau_2)$ is called (*i*) a (τ_i, τ_j)-fuzzy semi open (briefly, (τ_i, τ_j)-*fso*) set of x if 7 a $\gamma \in \tau_i$ such that $\gamma \leq \lambda \leq \tau_j - cl(\nu)$ and (*ii*) a (τ_i, τ_j)-fuzzy semi closed (briefly, (τ_i, τ_j)-*fso*) set of x if 7 a τ_i-fuzzy closed set β such that $\tau_j - int(\beta) \leq \lambda \leq \beta$. The concepts of fuzzy semi closed mappings [1] and fuzzy irresolute mappings [7] were introduced by Azad and Mukherjee, Sinha in the following way:

Let $f: (X, \tau) \to (\gamma, a)$ be a mapping from a fuzzy space to an another fuzzy space Y, f is said to be (*i*) fuzzy semi closed mapping if $f(\beta)$ is a fuzzy semi closed set in Y for each fuzzy closed set β of x. (*ii*) fuzzy irresolute mapping if $f^{-1}(\lambda)$ is fuzzy semi open in x for each fuzzy semi open set λ in Y.

Let $S(\tau)$ be the collection of all fuzzy semi-open sets in X. We use p to denote pairwise, e.g. pairwise fuzzy semi compact can be written as p-fuzzy semi-compact. The intersection of all fuzzy semi-closed sets containing a fuzzy subset $\lambda \in I^X$ of a fuzzy bitopological space (X, τ_1, τ_2) is called fuzzy semi-closure (1) of λ and is denoted by FSCl (λ) and it's complement denoted by FSInt (λ) is called fuzzy semi-inerior of λ. The fuzzy semi-closure and fuzzy semi-interior of λ with respect to $S(\tau_i)$ are denoted by $\mathrm{FSCl}_i(\lambda)$ and $\mathrm{FSInt}_i(\lambda)$ respectively. The fuzzy closure of λ and fuzzy interior of λ with respect to τ_i are denoted by $\mathrm{FCl}_i(\lambda)$ and $\mathrm{FInt}_i(\lambda)$ respectively.

2. PAIRWISE FUZZY SEMI PARACOMPACT SPACES

In this section we define and study two types of pairwise fuzzy semi paracompactness namely p_1-fuzzy semi paracompact and p_2-fuzzy semi paracompact spaces and the relation between them are established.

Definition 2.1: A family A of fuzzy subsets of a fuzzy bitopological space (X, τ_1, τ_2) is called $\tau_1\tau_2$-fuzzy semi-open if $A \subset S(\tau_1) \cup S(\tau_2)$.

If, in addition $A \cap S(\tau_1) \neq 1_\phi$ and $A \cap S(\tau_2) \neq 1_\phi$, then A is called pairwise-fuzzy semi-open (simply, p-fuzzy semi-open).

Definition 2.2: If $\mathcal{P}$ and $\mathcal{R}$ are $\tau_1\tau_2$-fuzzy semi-open covers of a fuzzy bitopological space (X, τ_1, τ_2), then $\mathcal{P}$ is called a refinement of $\mathcal{R}$ if each $U \in P \cap S(\tau_i)$ is contained in some $V \in R \cap S(\tau_i)$, $i = 1, 2$.

Definition 2.3: A family $\mathcal{U}$ of fuzzy subsets of a fuzzy space (X, τ) is locally finite in $(X, S(\tau))$ if for each $x_t \in X$, $\exists$ a fuzzy semi open set A such that $x_t \in A$ and A intersects at most finitely many members of $\mathcal{U}$.

Definition 2.4: A collection $\mathcal{U}$ of fuzzy subsets of a fuzzy bitopological space (X, τ_1, τ_2) is called p_1-locally finite if $\mathcal{U} \cap S(\tau_i)$ is locally finite in $(X, S(\tau_i))$, $i = 1, 2$.

Definition 2.5: A collection $\mathcal{U}$ of fuzzy subsets of a fuzzy bitopological space (X, τ_1, τ_2) is called p_2-locally finite if $\mathcal{U} \cap S(\tau_i)$ is locally finite in $(X, S(\tau_j))$, $i \neq j$, $i, j = 1, 2$.

Definition 2.6: A fuzzy bitopological space (X, τ_1, τ_2) is called p_1-fuzzy semi-paracompact (respectively, p_2-fuzzy semi-paracompact) if every p-fuzzy semi-open cover of X has a p_1-locally (respectively, p_2-locally) finite $\tau_1\tau_2$-fuzzy semi-open refinement.

Remark 2.7: p_1-fuzzy semi-paracompactness does not imply p_2-fuzzy semi-paracompactness and p_2-fuzzy semi-paracompactness does not imply p_1-fuzzy semi-paracompactness.

Definition 2.8: A fuzzy subset G of a fuzzy space (X, τ_1, τ_2) is called τ_1–fuzzy semi paracompact (resp. τ_2-fuzzy semi paracompact) relative to X if every fuzzy semi-open cover of G by members of $S(\tau_1)$ (resp. $S(t_2)$) has a locally finite (in X) refinement by members of $S(\tau_1)$ (resp. $S(\tau_2)$).

Theorem 2.9: Let $\{A_\alpha: \alpha \in \Lambda\}$ and $\{B_\beta: \beta \in \Gamma\}$ be two p-fuzzy semi-open covers of a fbts (X, τ_1, τ_2), then any common refinement of $\{A_\alpha\}$ and $\{B_\beta\}$ is also a refinement of $\{A_\alpha \cap B_\beta\}$.

Proof: Let the p-fuzzy semi-open covering $\{C_\nu: \nu \in \Lambda\}$ refines both $\{A_\alpha\}$ and $\{B_\beta\}$, then for each C_ν, $(\exists A_\alpha: C_\nu \subset A_\alpha)$ and $(\exists B_\beta: C_\nu \subset B_\beta)$ so that $C_\nu \subset A_\alpha \cap B_\beta$ and $\{C_\nu\} \subset \{A_\alpha \cap B_\beta\}$. Hence $\{C_\nu\}$ is a refinement of $\{A_\alpha \cap B_\beta\}$.

Theorem 2.10: For fbts (X, τ_1, τ_2) the following statements are equivalent:

(i) (X, τ_1, τ_2) is p_1-fuzzy semi-paracompact

(ii) Each τ_1-fuzzy semi closed proper subsets of X is τ_2-fuzzy semi paracompact relative to X and each τ_2-fuzzy semi closed proper subsets of X is τ_1-fuzzy semi paracompact relative to X.

Proof: (i) ⇒ (ii): Let $G \neq 1_\phi$ and $G \neq 1_X$ be a τ_1-fuzzy semi closed subset of X and $\mathcal{P} = \{U_\alpha : \alpha \in \Lambda\}$ be a τ_2-fuzzy semi open cover of G. Then, $1_X - G$ is τ_1-fuzzy semi open subset of X and $\mathfrak{R} = \{U_\alpha : \alpha \in \Lambda$, an index set$\} \cup (1_X - G)$ is p_1-fuzzy semi open cover of X. But (X, τ_1, τ_2) is p_1- fuzzy semi-paracompact, then $\mathfrak{R}$ has a p_1-locally finite $\tau_1\tau_2$-fuzzy semi-open refinement, say $\{U'_\alpha : \alpha \in \Lambda\} \cup \{V_\beta : \beta \in \Gamma$, an index set$\}$, where, $U'_\alpha \in S(\tau_2)$ and $V_\beta \in S(\tau_1)$ for each $\alpha \in \Lambda$ and $\beta \in \Gamma$. Thus $\{U'_\alpha : \alpha \in \Lambda\}$ is a $S(\tau_2)$-locally finite τ_2-fuzzy semi-open refinement of $\mathfrak{R}$ for G. Hence G is τ_2-fuzzy semi paracompact relative to X.

(ii) ⇒ (i): Let $\mathcal{P} = \{U_\alpha : \beta \in \Gamma\} \cup \{V_\alpha : \alpha \in \Lambda\}$ be a p-fuzzy semi-open cover of X, where, $U_\beta \in S(\tau_1)$ and $V_\alpha \in S(\tau_2)$ for each $\alpha \in \Lambda$ and $\beta \in \Gamma$. Now we have two cases:

Case I: If $\cup \{V_\alpha : \alpha \in \Lambda\} = 1_X$, choose $\beta_0 \in \Gamma$ such that $U_{\beta 0} \neq 1_\phi$. Since $\{V_\alpha : \alpha \in \Lambda\}$ is τ_2-fuzzy semi open cover of τ_1-fuzzy semi-closed proper subset $X - \beta_0$, therefore it has a τ_2-fuzzy semi open $S(\tau_2)$-locally finite refinement $\{V'_\alpha : \alpha \in \Lambda\}$. Hence $\{U_{\beta 0}\} \cup \{V' : \alpha \in \Lambda\}$ is a $\tau_1\tau_2$-fuzzy semi-open p_1-locally finite refinement of $\mathcal{P}$.

Case II: If $\cup \{V_\alpha : \alpha \in \Lambda\} \neq 1_X$, then $F = 1_X - \cup \{V_\alpha : \alpha \in \Lambda\}$ is a τ_2-fuzzy semi-closed proper subset of X and $F \subseteq \{U_\beta : \beta \in \Gamma\}$. Hence there is a $S(\tau_1)$-locally finite τ_1-fuzzy semi-open refinement of $\{U_\beta : \beta \in \Gamma\}$, say $\{U'_\beta : \beta \in \Gamma\}$. If $\cup \{U'_\beta : \beta \in \Gamma\} = 1_X$ there is nothing to prove. But if $\cup \{U'_\beta : \beta \in \Gamma\} \neq 1_X$, then $1_X - \cup \{U'_\beta : \beta \in \Gamma\}$ is a τ_1-fuzzy semi-closed proper subset contained in $\cup \{V_\alpha : \alpha \in \Lambda\}$, so there is a τ_2-locally finite τ_2-fuzzy semi-open refinement $\{V'_\alpha : \alpha \in \Lambda\}$ of $\{V_\alpha : \alpha \in \Lambda\}$. Therefore, $\{U'_\beta : \beta \in \Gamma\} \cup \{V'_\alpha : \alpha \in \Lambda\}$ is a $\tau_1\tau_2$-fuzzy semi-open p_1-locally finite refinement of $\mathcal{P}$. Hence the theorem is proved.

Theorem 2.11: For fbts (X, τ_1, τ_2) the following statements are equivalent:

(i) (X, τ_1, τ_2) is p_2-fuzzy semi-paracompact.

(*ii*) If F is any τ_i-fuzzy semi-closed proper subsets of X then any τ_j-fuzzy semi-open cover of F has a τ_j-fuzzy semi-open $S(\tau_i)$-locally finite (in X) refinement, $i \neq j$, $i, j = 1, 2$.

Proof: Proof is similar to the proof of Theorem 2.10.

Theorem 2.12: For fbts (X, τ_1, τ_2) the following statements are equivalent:

(*i*) (X, τ_1, τ_2) is p_1-fuzzy semi-paracompact and p_2-fuzzy semi-paracompact.

(*ii*) Every p_1-fuzzy semi-open cover of X has a $S(\tau_1)$-locally finite $\tau_1\tau_2$-fuzzy semi-open-refinement and every p-fuzzy semi-open cover of X has a $S(\tau_2)$-locally finite $\tau_1\tau_2$-fuzzy semi-open-refinement.

Proof: (i) $\Rightarrow$ (ii): Let $\mathcal{U}$ be a p-fuzzy semi-open cover of X. Then $\exists A \in \mathcal{U} \cap S(\tau_1)$ and $B \in \mathcal{U} \cap S(\tau_2)$ such that $A \neq 1_\phi$ and $B \neq 1_\phi$. Since, (X, τ_1, τ_2) is p_2-fuzzy semi-paracompact, $\exists$ *a* p_2-locally finite $\tau_1\tau_2$-fuzzy semi-open-refinement $\mathcal{P}$ of $\mathcal{U}$. Since (X, τ_1, τ_2) is p_1 fuzzy-semi-paracompact, the p-fuzzy semi-open cover $\mathcal{P}$ (add A to the cover $\mathcal{P}$ if $\mathcal{P} \cap S(\tau_1) = 1_\phi$ and add B to the cover $\mathcal{P}$ if $\mathcal{P} \cap S(\tau_2)=1_\phi$) has a p_1-locally finite $\tau_1\tau_2$-fuzzy semi-open-refinement $\mathcal{R}$. Then $(\mathcal{P} \cap S(\tau_1)) \cup (R \cap S(\tau_2))$ is a $S(\tau_2)$-locally finite $\tau_1\tau_2$-fuzzy semi-open-refinement (replace $\mathcal{P} \cap S(\tau_1)$ by (A) if $\mathcal{P} \cap S(\tau_1) = 1_\phi$) of $\mathcal{U}$ and $(\mathcal{P} \cap S(\tau_2)) \cup (R \cap S(\tau_1))$ is a $S(\tau_1)$-locally finite $\tau_1\tau_2$-fuzzy semi-open-refinement (replace $\mathcal{P} \cap S(\tau_2)$ by (B) if $\mathcal{P} \cap S(\tau_2) = 1_\phi$) of U.

(ii) $\Rightarrow$ (i): Similar to the above technique.

Definition 2.13: In a fbts (X, τ_1, τ_2), $S(\tau_1)$ is said to be fuzzy semi-regular with respect to $S(\tau_2)$ if for each fuzzy point x_t in X and each τ_1-fuzzy semi-closed set F such that $x_t \notin F$, there are τ_1-fuzzy semi-open set U and τ_2-fuzzy semi-open set V such that $x_t \in U$ and $F \subseteq V$ and $U \cap V = 1_\phi$.

Definition 2.14: A fbts (X, τ_1, τ_2) is said to be p-fuzzy semi-regular if $S(\tau_i)$ is fuzzy semi-regular with respect to $S(\tau_j)$, $i \neq j$, $i, j = 1, 2$.

Definition 2.15: In a fbts (X, τ_1, τ_2), $S(\tau_1)$ is said to be fuzzy semi-normal with respect to $S(\tau_2)$ if for every τ_1-fuzzy semi-closed set F_1 and each τ_2-fuzzy semi-closed set F_2 such that there are τ_2-fuzzy semi-open set U and τ_1-fuzzy semi-open set V such that $F_1 \subseteq U$ and $F_2 \subseteq V$ and $U \cap V = 1_\phi$.

Definition 2.16: A fbts (X, τ_1, τ_2) is said to be p-fuzzy semi-normal if $S(\tau_i)$ is fuzzy semi-normal with respect to $S(\tau_j)$, $i \neq j$, $i, j = 1, 2$.

Theorem 2.17: Let $f: (X, \tau_1, \tau_2) \to (Y, \sigma_1, \sigma_2)$ be a fuzzy semi-closed map and $A \subset Y$ be a fuzzy subset such that $f^{-1}(A) \subset U$, where U is p-fuzzy semi-open set in X. Then $\exists$ a p-fuzzy semi-open set $V \supset A$ in Y such that $f^{-1}(V) \subset U$.

Proof: Let us suppose f be a fuzzy semi closed map from a fbts (X, τ_1, τ_2) into an another fbts (Y, σ_1, σ_2). Let $A \subset Y$ be a fuzzy subset such that $\exists$ *a* p-fuzzy semi-open set U in X for which $f^{-1}(A) \subset U$. Suppose, $V = 1_Y - f(1_X - \text{U})$. Since, $f^{-1}(A) \subset U$, it follows that $A \subset V$ and V is p-fuzzy semi-open because ϕ is fuzzy semi-closed map and $f^{-1}(V) = 1_X - f^{-1}(f(1_X - \text{U})) \subset 1_X - (1_X - U) = U$. Hence, the theorem is complete.

Theorem 2.18: If $f: (X, \tau_1, \tau_2) \to (Y, \sigma_1, \sigma_2)$ be a fuzzy irresolute semi-closed surjection map and X is p-fuzzy semi normal. Then Y is also p-fuzzy semi normal.

Proof: Let X is p-fuzzy semi normal and $f: (X, \tau_1, \tau_2) \to (Y, \sigma_1, \sigma_2)$ be a fuzzy irresolute semi-closed surjection mapping and let A and B be two disjoint p-fuzzy semi closed sets in Y. Since f is fuzzy irresolute, so $f^{-1}(A)$ and $f^{-1}(B)$ are also disjoint and p-fuzzy semi closed in X. Since X is p-fuzzy semi normal, $\exists$ p-fuzzy semi open sets U and V such that $f^{-1}(A) \subset U$ and $f^{-1}(B) \subset V$ and $U \cap V = 1_\phi$. As f is fuzzy semi-closed map, so by **Theorem 2.17,** $\exists$ two p-fuzzy semi-open sets U_A and V_B containing respectively A and B such that $f^{-1}(U_A) \subset U$ and $f^{-1}(V_B) \subset V$ and so U_A and V_B are disjoint p-fuzzy semi-open sets in Y and hence Y is p-fuzzy semi normal.

Definition 2.19: 'A' fbts (X, τ_1, τ_2) is said to be p-fuzzy semi-Lindelof if each p-fuzzy semi-open cover of X has a countable fuzzy subcover.

Definition 2.20: A fuzzy subset in a fbts (X, τ_1, τ_2) is said to be p-fuzzy semi-Lindeiof relative to (X, τ_1, τ_2) if every p-fuzzy semi-open cover of A, by the members in $S(\tau_1) \cup S(\tau_2)$ has a countable fuzzy subcover.

Definition 2.21: Let λ be a fuzzy subset of a fbts (X, τ_1, τ_2). Then λ is said to be p-fuzzy semi-dense if $SCl_1\lambda = SCl_2\lambda = 1_X$ and for every p-fuzzy semi-open (in X) p_2-locally finite countable cover $\mathcal{U}$ of λ we have,

$$SCl_1 \cup \{A: A \in \mathcal{U} \cap S(\tau_1)\} \subseteq SCl_2 \cup \{A: A \in \mathcal{U} \cap S(\tau_1)\}$$

or

$$SCl_2 \cup \{A: A \in \mathcal{U} \cap S(\tau_2)\} \subseteq SCl_1 \cup \{A: A \in \mathcal{U} \cap S(\tau_2)\}$$

Theorem 2.22: If a p-fuzzy semi-regular space (X, τ_1, τ_2) is p_2-fuzzy semi-paracompact having a p-fuzzy semi-dense subset λ which is p-fuzzy semi-Lindelof relative to X, then X is p-fuzzy semi-Lindelof space.

Proof: Let $\mathcal{U}$ be a p-fuzzy semi-open cover of X. Then $\exists$ two non-empty sets $V_0, U_0 \in \mathcal{U}$ such that $V_0 \in S(\tau_1)$ and $U_0 \in S(\tau_2)$. Since (X, τ_1, τ_2) is p-fuzzy semi-regular and p_2-fuzzy semi-paracompact, $\mathcal{U}$ has a p_2-locally-finite $\tau_1\tau_2$-fuzzy semi-open refinement $\mathcal{P}$ such that for each $A \in \mathcal{P} \cap S(\tau_1)$, $SCl_2 A$ is contained in some $B \in \mathcal{P} \cap S(\tau_1)$ and for each $A \in \mathcal{P} \cap S(\tau_2)$, $SCl_1 A$ is contained in some $B \in \mathcal{P} \cap S(\tau_2)$. Now P is p-fuzzy semi-open cover of λ by fuzzy semi-open sets in $S(\tau_1) \cup S(\tau_2)$ [add V_0 to $\mathcal{P}$ if $\mathcal{P} \cap S(\tau_1) = 1_\phi$ and add U_0 to $\mathcal{P}$ if $\mathcal{P} \cap S(\tau_2) = 1_\phi$]. Since λ is p-fuzzy semi-Lindelof relative to X, $\exists$ a countable subcover $\mathcal{P}'$ of $\mathcal{P}$ which covers λ. We may assume that $\mathcal{P}'$ is p-fuzzy semi-open. But λ is p-fuzzy semi-dense in X, so we have two cases to consider:

Case I: $SCl_1 \cup \{A: A \in \mathcal{P}' \cap S(\tau_1)\} \subseteq SCl_2 \cup \{A: A \in \mathcal{P}' \cap S(\tau_1)\}$: In this case we have:

$$1_X = SCl_1 \lambda \subseteq SCl_1 \cup \{A: A \in \mathcal{P}'\}$$
$$= (SCl_1 \cup \{A: A \in \mathcal{P}' \cap S(\tau_1)\}) \cup (SCl_1 \cup \{A: A \in \mathcal{P}' \cap S(\tau_2)\})$$
$$\subseteq (SCl_2 \cup \{A: A \in \mathcal{P}' \cap S(\tau_1)\}) \cup (SCl_1 \cup \{A: A \in \mathcal{P}' \cap S(\tau_2)\})$$
$$= (\cup \{SCl_2 A: A \in \mathcal{P}' \cap S(\tau_1)\}) \cup (\cup \{SCl_1 A: A \in \mathcal{P}' \cap S(\tau_2)\})$$

Case II: $SCl_2 \cup \{A: A \in \mathcal{P}' \cap S(\tau_2)\} \subseteq SCl_1 \cup \{A: A \in \mathcal{P}' \cap S(\tau_2)\}$. In this case we have:

$$1_X = SCl_2 \lambda \subseteq SCl_2 \cup \{A: A \in \mathcal{P}'\}$$
$$= (SCl_2 \cup \{A: A \in \mathcal{P}' \cap S(\tau_1)\}) \cup (SCl_2 \cup \{A: A \in \mathcal{P}' \cap S(\tau_2)\})$$
$$\subseteq (SCl_2 \cup \{A: A \in \mathcal{P}' \cap S(\tau_1)\}) \cup (SCl_1 \cup \{A: A \in \mathcal{P}' \cap S(\tau_2)\})$$
$$= (\cup \{SCl_2 A: A \in \mathcal{P}' \cap S(\tau_1)\}) \cup (\cup \{SCl_1 A: A \in \mathcal{P}' \cap S(\tau_2)\})$$

Now in each case, we have $X \subseteq (\cup \{SCl_2 A: A \in \mathcal{P}' \cap S(\tau_1)\}) \cup (\cup \{SCl_1 A: A \in \mathcal{P}' \cap S(\tau_2)\})$. For each $A \in \mathcal{P}' \cap S(\tau_i)$ choose one element $B_A \in \mathcal{U}$ such that $SCl_j A \subseteq B_A$ $(i \neq j;\ i, j = 1, 2)$. Then $\{B_A : A \in \mathcal{P}'\}$ is a countable subcover of $\mathcal{U}$ for X. Hence X is p-fuzzy semi-Lindelof.

REFERENCES

[1] Azad K.K., "On Fuzzy Continuity, Fuzzy Almost Continuity and Fuzzy Semi Continuity", *Jour. Math. and Appl.*, **82**, (1981), 14–32.

[2] Crossley S.G. and Hildbrand S.K., "Semi-closure", *Texas J. Sci.*, **22**, (1971), 99–112.

[3] Dorset Ch., Semi Compactness, "Semi Separation Axioms, and Product Spaces", *Bull. Malaysian Math. Soc.*, **4 (2)**, (1981), 21–28.

[4] Kandil A., Nouth A.A. and El-Sheikh S.A., "On Fuzzy Bitopological Soaces", *Fuzzy Sets and Systems*, **74**, (1995), 353–363.

[5] Kelly J.C., "Bitopological Spaces", *Proc. London Math. Soc.*, **13**, (1963), 71–89.

[6] Kumar Sampat, "Semiopen Set's, Semi Continuity and Semiopen Mappings in Fuzzy Bitopological Space," *Fuzzy Sets and Systems*, **64**, (1994), 421–426.

[7] Mukherjee M.N. and Sinha S.P, "Irresolute and Almost Open Functions Between Fuzzy Topological Spaces, Fuzzy Sets and Systems, 29, (1989), 381–388.

[8] Mustafa Jamal M., "Semi Compactness and Semi Lindelofness in Bitopological Spaces", *Abhath Al-Yarmouk: Basic Sci and Eng.*, **13 (2)**, (2004), 309–314.

[9]. Pao-Ming P. and Ying-Ming L., "Fuzzy Topology I: Neighborhood Structure of a Fuzzy Point and Moore Smith convergence", *J. Math. Anal. and Appl.*, **76 (2)**, (1980), 571–599.

[10] Zadeh L.A., "Fuzzy Sets", *Information and Control*, **8**, (1965), 338–353.

33

PAIRWISE WEAKLY FUZZY δ-SEMI PRE IRRESOLUTE MAPPINGS

Anjan Mukherjee[1] and Runu Dhar[2]

[1]*Department of Mathematics, Tripura University Suryamaninagar, Agartala, Tripura west-799130, India, E-mail: anjan2002_m@yahoo.co.in*

[2]*Department of Mathematics, BBM College, Agartala, Tripura west-799004, India, E-mail: mr.runu@rediffmail.com*

Abstract: *The aim of this paper is to introduce pairwise weakly fuzzy δ-semi pre irresolute mapping in fuzzy bitopological spaces and study its properties and its relationship with other mappings. Lastly some applications are shown with the help of the above concept.*

Keywords: *Fuzzy bitopology, (i, j)-semiopen set, (i, j)-semiclosed set, pairwise fuzzy semi continuity, pairwise fuzzy irresolute mapping, pairwise weakly fuzzy δ-semi pre irresolute mapping.*

AMS Subject Classification: *54A40.*

1. INTRODUCTION

The notion of fuzzy bitopological space was introduced by Kandil (6). He established that fuzzy bitopological space is naturally a richer structure than that of fuzzy topological space. It is of much use and importance to generalize a bitopological notion and their implications in a fuzzy bitopological situation. The concept of weakly pairwise irresolute mapping is introduced Dubey et.al. (4). The concept of weakly fuzzy δ-semi pre irresolute mapping in fuzzy topological space was introduced by the authors (7). In this paper we generalize the above concept in fuzzy bitopological space. Throughout the present study, the space X, Y and Z always represent fuzzy bitopological spaces (X, τ_1, τ_2), (Y, τ_1, τ_2) and (Z, Γ_1, Γ_2).

2. PRELIMINARIES

Some results and definitions are given for ready references.

Let (X, τ) be a fuzzy topological space. For a subset A of a space (X, τ), cl (A) and int (A) denote the closure of A and the interior of A, respectively.

Definition 2.1: (5) A subset S of a fuzzy topological space (X, τ) is said to be fuzzy δ-open if for each fuzzy point $x_p \in S$ there exists a fuzzy regular open set W such that $x_p \in W \leq S$. The complement of δ-open set is called δ-closed set. Every fuzzy δ-open set is the union of fuzzy regular open sets.

Definition 2.2: (3) Let A be a fuzzy subset of a fuzzy topological space (X, τ). Then A is called fuzzy δ-preopen if $A \leq \text{int}\,(\delta\,\text{cl} A)$. The complement of a fuzzy δ-preopen set is called fuzzy δ-preclosed.

Every fuzzy open set is fuzzy preopen and every fuzzy preopen set is fuzzy δ-pre open. Every fuzzy open set is fuzzy δ-open, every fuzzy δ-open set is fuzzy semiopen and every fuzzy open set is fuzzy semi preopen and every fuzzy semi preopen set is fuzzy δ-semi preopen.

Definition 2.3: A fuzzy subset λ in a fuzzy topological space X is called

(*i*) (1) fuzzy semiopen if $\lambda \leq \text{clint}\,(\lambda)$

(*ii*) (2) fuzzy preopen if $\lambda \leq \text{int}\,[\text{cl}\,(\text{int}\,\lambda)]$

(*iii*) (13) fuzzy semi preopen if there exists a fuzzy preopen set μ such that $\mu \leq \lambda\ \text{cl}\ \mu$.

Definition 2.4: (10) A fuzzy subset γ in (X, τ) is said to be fuzzy δ-semi preopen if there exists a fuzzy δ-preopen set μ such that $\mu \leq \gamma \leq \delta\text{-cl}\,(\mu)$ or equivalently $\gamma \leq \delta\text{-cl}\,(\text{int}\ \delta\text{-cl}\ \gamma)$.

Definition 2.5: (1) A fuzzy subset A of a fuzzy topological space (X, τ) is called

(*i*) a fuzzy regular open set of (X, τ) if intcl $(A) = A$ and

(*ii*) a fuzzy regular closed set of (X, τ) if clint $(A) = A$.

Definition 2.6: The set $(x \in X: x \in U \leq A$ for some regular open set U of $X)$ is called the δ-interior of A and is denoted by $\text{int}_\delta(A)$.

Definition 2.7: (8) A fuzzy subset A of a fuzzy topological space (X, τ) is said to be fuzzy δ-semiopen if $A \leq \text{cl int}_\delta(A)$.

(7) The complement of a fuzzy δ-semiopen set is called fuzzy δ-semiclosed set.

Definition 2.8: (7) A mapping $f: (X, \tau) \rightarrow (Y, \sigma)$ from a fts (X, τ) to another fts (Y, σ) is said to be weakly fuzzy δ-semi pre irresolute mapping if $f^{-1}(\alpha) \in f\delta\text{spo}(X)$ for each $\alpha \in \text{fspo}(Y)$, where fδspo (X) [respectively fδpo (X)] denotes the family of all δ-semi preopen (respectively fuzzy δ-preopen) sets of X.

Definition 2.9: In a fuzzy bitopolgical space X a fuzzy subset A is (i, j)-fuzzy semiopen (11) if there exists a τ_i-fuzzy open set B such that $B \leq A \leq \tau_j\text{-cl}(B)$, $i \neq j$; $i, j = 1, 2$ [τ_j-cl (B) denotes the closure of B with respect to τ_j]. [Every τ_i-fuzzy open set is (i, j)-fuzzy semiopen but not conversely and any union of (i, j)-fuzzy semiopen set is (i, j)-fuzzy semi interior of A and we denote it by (i, j)-sint (A). Thus, A is (i, j)-fuzzy semiopen iff $A = (i, j)$-fuzzy semiopen. The union of all (i, j)-fuzzy semiopen sets contained in A is called the (i, j)-fuzzy semi interior of A and we it by (i, j)-sint (A). Thus, A is (i, j)-fuzzy semiopen iff $A = (i, j)$-sint (A), $i \neq j$; $i, j = 1, 2$. Complement of an (i, j)-fuzzy semiopen set is known as (i, j)-fuzzy semiclosed. The intersection of all (i, j)-fuzzy semiclosed sets containing A is called (i, j)-fuzzy semiclosure of A and is denoted by (i, j)-scl (A). Further, $A \leq (i, j)\text{-scl}(A) \leq \tau_i\text{-cl}(A)$, also $A \leq B$ implies $(i, j)\text{-scl}(A) \leq (i, j)\text{-scl}(B)$. Infact, A is (i, j)-fuzzy semiclosed iff $A = (i, j)$-scl (A), $i \neq j$; $i, j = 1, 2$.

Definition 2.10: A mapping $f: (X, \tau_1, \tau_2) \rightarrow (Y, \tau_1, \tau_2)$ is pairwise fuzzy semi continuous (11) [resp. pairwise fuzzy irresolute (12)] if the inverse image of each τ_i-open set [resp. (i, j)-fuzzy semiopen set] in Y is (i, j)-fuzzy semiopen set in X, $i \neq j$; $i, j = 1, 2$.

Definition 2.11: (9) For any fuzzy bitopological spaces (X, τ_1, τ_2) and (Y, τ_1, τ_2), a mapping $f: X \rightarrow Y$ is said to be pairwise fuzzy weakly irresolute if for each $x_\alpha \in X$ and each (i, j)-fuzzy semiopen set E containing $f(x_\alpha)$ there is an (i, j)-fuzzy semiopen set H in X such that $x_\alpha H$ and $f(H) \leq (i, j)\text{-scl}(E)$, $i \neq j$ and $i, j = 1, 2$.

3. PAIRWISE WEAKLY FUZZY δ-SEMI PRE IRRESOLUTE MAPPING

In this section the concept of weakly fuzzy δ-semi pre irresolute mapping is introduced. Some characterization theorems and its basic properties are also studied.

Definition 3.1: A subset A of a fuzzy bitopological space (X, τ_1, τ_2) is said to be (i, j)-δ-semiopen if $A \leq \tau_j\text{-cl}\ [\tau_i\text{-int}_\delta\ (\mathrm{A})]$. The complement of (i, j)-δ-semiopen set is called (i, j)-δ-semiclosed.

Definition 3.2: Let A be a subset of a fuzzy bitopological space (X, τ_1, τ_2). Then

(*i*) the intersection of all (i, j)-δ-semiclosed sets containing A is called the (i, j)-δ-semi closure of A and is denoted by $(i, j)\text{-scl}_\tau\ (A)$.

(*ii*) the union of all (i, j)-δ-semiopen sets contained in A is called the (i, j)-δ-semi interior of A and is denoted by $(i, j)\text{-sint}_\delta\ (A)$.

Definition 3.3: A subset A of a fuzzy bitopological space (X, τ_1, τ_2) is said to be (i, j)-δ-semi preopen if $A \leq [(i, j)\text{-spint}_\delta\ (i, j)\text{-spcl}_\delta\ (A)]$. The complement of (i, j)-δ-semi preopen set is called (i, j)-δ-semi preclosed.

Definition 3.4: Let A be a subset of a fuzzy bitopological space (X, τ_1, τ_2). Then

(*i*) the intersection of all (i, j)-δ-semi preclosed sets containing A is denoted by $(i, j)\text{-spcl}_\delta\ (A)$.

(*ii*) the union of all (i, j)-δ-semi preopen sets contained in A is called the (i, j)-δ-semi preinterior of A and is denoted by $(i, j)\text{-spint}_\delta\ (A)$.

Definition 3.5: For any fuzzy bitopological spaces (X, τ_1, τ_2) and (Y, τ_1, τ_2), a mapping $f: X \to Y$ is said to be pairwise weakly fuzzy δ-semi pre irresolute if for each x_p in X and each (i, j)-δ-semi preopen set V containing $f(x_p)$, there is an (i, j)-δ-semi preopen set U in X such that $x_p \in U$ and $f(U) \leq (i, j)\text{-spcl}_\delta\ (V)$, $i \neq j$ and $i, j = 1, 2$.

Theorem 3.6: For any mapping $f: (X, \tau_1, \tau_2) \to (Y, \tau_1, \tau_2)$ the following conditions are equivalent:

(*i*) For any subset A of Y, $(i, j)\text{-spcl}_\delta\ [f^{-1}\ (i, j)\text{-spint}_\delta\ (i, j)\ \text{spcl}_\delta\ (A)] \leq f^{-1}\ [(i, j)\text{-spcl}_\delta\ (A)]$.

(*ii*) For any (i, j)-δ-semi preopen set G in Y, (i, j)-spcl$_\delta$ $[f^{-1}(G)] \leq f^{-1}[(i, j)$-spcl$_\delta(G)]$.

(*iii*) For any (i, j)-δ-semi preopen set H in G in Y, (i, j)-spcl$_\delta$ $[f^{-1}(i, j)]$-spint$_\delta(H) \leq f^{-1}(H)$, where $i \neq j$ and $i, j = 1, 2$.

Proof: (i) ⇒ (ii): Suppose that G is any (i, j)-δ-semi preopen set in Y. Then, by (*i*), (i, j)-spcl$_\delta$ $[f^{-1}(i, j)$-spint$_\delta$ (i, j) spcl$_\delta(G)] \leq f^{-1}[(i, j)$-spcl$_\delta(G)]$. Since G is (i, j)-δ-semi preopen, G $\leq [(i, j)$-spint$_\delta(i, j)$-spcl$_\delta(G)]$. Consequently, (i, j)-spcl$_\delta[f^{-1}(G)] \leq f^{-1}[(i, j)$-spcl$_\delta(G)]$.

(ii) ⇒ (iii): For any (i, j)-δ-semi preclosed set H in Y, (i, j)-spint$_\delta$ (H) is (i, j)-δ-semi preopen set in Y. Therefore, by (*ii*), (i, j)-spcl$_\delta$ $[f^{-1}(i, j)$-spint$_\delta(H)] \leq f^{-1}[(i, j)$-spcl$_\delta(i, j)$-spint$_\delta(H)]$. Since H is (i, j)-δ-semi preclosed, (i, j)-spcl$_\delta[(i, j)$-spint$_\delta(H)] \leq H$. Therefore, (i, j)-spcl$_\delta[f^{-1}(i, j)]$-spint$_\delta(H)] \leq f^{-1}(H)$.

(iii) ⇒ (i): Let A be any subset of Y. Let $H = (i, j)$-spcl$_\delta(A)$. Then for the (i, j)-δ-semi preclosed set H, by (*iii*), (i, j)-spcl$_\delta$ $[f^{-1}(i, j)$-spint$_\delta(i, j)$-spcl$_\delta(A)] \leq f^{-1}[(i, j)$-spcl$_\delta(A)]$.

Theorem 3.7: A mapping f: $(X, \tau_1, \tau_2) \rightarrow (Y, \tau_1, \tau_2)$ is pairwise weakly fuzzy δ-semi pre irresolute if and only if for any (i, j)-δ-semi preopen set V in Y, $f^{-1}(V) \leq (i, j)$-spint$_\delta$ $\{f^{-1}[(i, j)$-spcl$_\delta(V)]$, $i \neq j$ and $i, j = 1, 2$.

Proof: Suppose that f is pairwise weakly fuzzy δ-semi pre irresolute and let V be any (i, j)-δ-semi preopen set in Y. Then for any $x_p \in X$ with $x_p \in f^{-1}(V)$, then there exists some (i, j)-δ-semi preopen set U in X such that $x \in U$ and $f(U) \leq [(i, j)$-spcl$_\delta(V)]$. Hence $x_p \in U \leq f^{-1}[(i, j)$-spcl$_\delta(V)]$. Cosequently, $x_p \in (i, j)$-spint$_\delta$ $[f^{-1}(i, j)$-spcl$_\delta$ $(V)]$ and $f^{-1}(V) \leq (i, j)$-spint$_\delta[f^{-1}(i, j)$-spcl$_\delta(V)]$.

Conversely, let $x_p \in X$ and V be any (i, j)-δ-semi preopen set in Y with $f(x_p)$ V. Then by hypothesis $f^{-1}(V) \leq (i, j)$-spint$_\delta[f^{-1}(i, j)$-spcl$_\delta$ $(V)]$. Put $U = (i, j)$-spint$_\delta$ $[f^{-1}$ (i, j)-spcl$_\delta$ $(V)]$. Then (i, j)-δ-semi preopen subset U is such that $x_p \in U \leq f^{-1}[(i, j)$-spcl$_\delta(V)]$. Therefore, $f(U) \leq (i, j)$-spcl$_\delta(V)$.

Theorem 3.8: A mapping f: $(X, \tau_1, \tau_2) \rightarrow (Y, \sigma_1, \sigma_2)$ is pairwise weakly fuzzy δ-semi pre irresolute if and only if for any subset A of Y, (i, j)-spcl$_\delta[f^{-1}(i, j)$-spint$_\delta[(i, j)$-spcl$_\delta(A)] \leq f^{-1}[(i, j)$-spcl$_\delta(A)]$, $i \neq j$ and $i, j = 1, 2$.

Proof Sufficiency: Let A be any subset of Y and $x_p \in X$ be such that $x_p \notin f^{-1}[(i,j)\text{-spcl}_\delta(A)]$. Then $f(x_p) \notin f^{-1}[(i,j)\text{-spcl}_\delta(A)]$ and so there exists some (i,j)-δ-semi preopen set W in Y such that $f(x_p)$ W and $WCA = \Phi$. Since f is pairwise weakly fuzzy δ-semi pre irresolute, there exists some (i,j)-δ-semi preopen set U in X such that $x \in U$ and $f(U) \leq (i,j)\text{-spcl}_\delta(W)$. Further, $W \cap (i,j)\text{-spcl}_\delta(A) = \Phi$ and (i,j)-$\text{spcl}_\delta[Y\text{-}(i,j)\text{-spcl}_\delta(A)] = [Y\text{-}(i,j)\text{-spint}_\delta[(i,j)\text{-spcl}_\delta(A)]$. Therefore, $f(U) \leq \{Y\text{-}(i,j)\text{-spint}_\delta[(i,j)\text{-spcl}_\delta(A)]\}$ and hence, $f(U)$ $(i,j)\text{-spint}_\delta$ $(i,j)\text{-spcl}_\delta(A) = \Phi$. Consequently, $U \cap f^{-1}[(i,j)\text{-spint}_\delta(i,j)\text{-spcl}_\delta(A)] = \Phi$. It follows that $x \notin (i,j)\text{-spcl}_\delta$ $[f^{-1}(i,j)\text{-spint}_\delta(i,j)\text{-spcl}_\delta(A)]$. Hence $(i,j)\text{-spcl}_\delta[f^{-1}(i,j)\text{-spint}_\delta[(i,j)\text{-spcl}_\delta(A)] \leq f^{-1}[(i,j)\text{-spcl}_\delta(A)]$.

Necessity: Let $x_p \in X$ and V be any fuzzy (i,j)-δ-semi preopen set in Y with $f(x_p) \in V$. Then $V \cap [Y - (i,j)\text{-spcl}_\delta(V)] = \Phi$. Therefore, $f(x_p) \notin (i,j)\text{-spcl}_\delta[Y - (i,j)\text{-spcl}_\delta(V)]$ and hence $x_p \notin f^{-1}[(i,j)\text{-spcl}_\delta$ $[Y - (i,j)\text{-spcl}_\delta(V)]$. Now, $[Y - (i,j)\text{-spcl}_\delta(V)] \leq (i,j)\text{-spint}_\delta(i,j)$-$\text{spcl}_\delta[Y - (i,j)\text{-spcl}_\delta(V)]$ and by hypothesis, $(i,j)\text{-spcl}_\delta\{f^{-1}(i,j)$-$\text{spint}_\delta(i,j)\text{-spcl}_\delta[Y - (i,j)\text{-spcl}_\delta(V)]\} \leq f^{-1}(i,j)\text{-spcl}_\delta[Y - (i,j)\text{-spcl}_\delta$ $(V)]$. Therefore, $x_p \notin \{(i,j)\text{-spcl}_\delta[f^{-1}[Y - (i,j)\text{-spcl}_\delta(V)]]\}$. Therefore, there exists some (i,j)-δ-semi preopen set U in X such that $x_p \in U$ and $U \cap f^{-1}[(i,j)\text{-spcl}_\delta(i,j)] = \Phi$. Consequently, $U \leq X$ –

$f^{-1}[Y - (i,j)\text{-spcl}_\delta(V)] = f^{-1}[(i,j)\text{-spcl}_\delta(V)]$. Therefore, it follows that $f(U) \leq (i,j)\text{-spcl}_\delta(V)$.

Theorem 3.9: Let (X, τ_1, τ_2) and (Y, σ_1, σ_2) be fuzzy bitoplogical spaces and let $f: (X, \tau_1, \tau_2) \rightarrow (Y, \sigma_1, \sigma_2)$. Then the following statements are equivalent:

(*i*) The mapping f is pairwise weakly fuzzy δ-semi pre irresolute.

(*ii*) For each $A \leq Y$, $(i,j)\text{-spcl}_\delta[f^{-1}(i,j)\text{-spint}_\delta(i,j)\text{-spcl}_\delta(A)] \leq f^{-1}[(i,j)\text{-spcl}_\delta(A)]$.

(*iii*) For each (i,j)-δ-semi preopen set G in Y, $(i,j)\text{-spcl}_\delta[f^{-1}(G)] \leq f^{-1}[(i,j)\text{-spcl}_\delta(G)]$.

(*iv*) For each (i,j)-δ-semi preopen set G in Y, $(i,j)\text{-spcl}_\delta\{f^{-1}[(i,j)$-$\text{spint}_\delta(H)]\} \leq f^{-1}(H)$.

(*v*) For each (i,j)-δ-semi preopen set G in Y, $f^{-1}(G) \leq \{(i,j)\text{-spint}_\delta$ $f^{-1}[(i,j)\text{-spcl}_\delta(G)]\}$ where, $i \neq j$ and $i, j = 1, 2$.

Proof: The proof follows from the Theorems 3.7, 3.8 and 3.9 and hence omitted.

Theorem 3.10: For a fuzzy bitopological spaces (X, τ_1, τ_2), (Y, σ_1, σ_2) and (Z, Γ_1, Γ_2) if the mapping $f: (X, \tau_1, \tau_2) \to (Y, \sigma_1, \sigma_2)$ and $g: (Y, \sigma_1, \sigma_2) \to (Z, \Gamma_1, \Gamma_2)$ are pairwise weakly fuzzy δ-semi pre irresolute, then $gof: (X, \tau_1, \tau_2) \to (Z, \Gamma_1, \Gamma_2)$ is pairwise weakly fuzzy δ-semi pre irresolute.

Proof: Let $x_p \in X$ and W be any (i, j)-δ-semi preopen subset of Z such that $(gof)(x_p) \in W$. Since g is pairwise weakly fuzzy δ-semi pre irresolute, there exists an (i, j)-δ-semi preopen set V in containing $f(x_p)$ such that $V \le g^{-1}[(i, j)\text{-spcl}_\delta(W)]$. Further f being pairwise weakly fuzzy δ-semi pre irresolute, there exists an (i, j)-δ-semi preopen set U in X such that $x_p \in U \le f^{-1}[(i,j)\text{-spcl}_\delta(V)]$. Thus, $x_p \in U \le f^{-1}[(i,j)\text{-spcl}_\delta\{g^{-1}[(i,j)\text{-spcl}_\delta(W)]\}$. But g being pairwise weakly fuzzy δ-semi pre irresolute, $(i, j)\text{-spcl}_\delta[g^{-1}(W)] \le g^{-1}[(i, j)\text{-spcl}_\delta(W)]$. Therefore, $x_p \in U \le (gof)^{-1}[(i, j)\text{-spcl}_\delta(W)]$. Consequently gof is pairwise weakly fuzzy δ-semi pre irresolute.

Lemma 3.11: In a fuzzy bitopological space (X, τ_1, τ_2), if A is (i, j)-δ-semi preopen and B is biopen, then $A \cap B$ is (i, j)-δ-semi preopen, $i \neq j$ and $i, j = 1, 2$.

Theorem 3.12: For fuzzy bitopological spaces (X, τ_1, τ_2) and (Y, σ_1, σ_2), let $f: (X, \tau_1, \tau_2) \to (Y, \sigma_1, \sigma_2)$ be pairwise weakly fuzzy δ-semi pre irresolute A be a biopen subset of X. Then $f/A : A \to Y$ is pairwise weakly fuzzy δ-semi pre irresolute.

Proof: Let $x_p \in A$ and let V be an (i, j)-δ-semi preopen set in Y containing $f(x_p)$. Then f being pairwise weakly fuzzy δ-semi pre irresolute, there exists an (i, j)-δ-semi preopen set U in X such that $x_p \in U \le f^{-1}[(i,j)\text{-spcl}_\delta(V)] \cap A = (f/A)^{-1}[(i, j)\text{-spcl}_\delta(V)]$ and so by Lemma 3.11, $U \cap A$ is (i, j)-δ-semi preopen set in A. It follows that f/A is pairwise weakly fuzzy δ-semi pre irresolute.

Theorem 3.13: Let $f: (X, \tau_1, \tau_2) \to (Y, \sigma_1, \sigma_2)$ be a mapping and $g: X \to X \times Y$ be the graph mapping of f, given by $g(x_p) = [x_p, f(x_p)]$ for $x \in X$. If $g: X \to X \times Y$ is pairwise weakly fuzzy δ-semi pre irresolute, then f is pairwise weakly fuzzy δ-semi pre irresolute.

Proof: Let $x_p \in X$ and V be an (i, j)-δ-semi preopen set containing $f(x_p)$ in Y. Then $X \times Y$ is (i,j)-δ-semi preopen set in $X \times Y$ containing

$g(x_p)$. Since g is pairwise weakly fuzzy δ-semi pre irresolute, there exists an (i, j)-δ-semi preopen set U containing x_p in X such that $g(U) \leq (i, j)\text{-spcl}_\delta(X \times Y) \leq X \times \leq (i, j)\text{-spcl}_\delta(V)$. Since g is the graph mapping of f, we have $f(U) \leq (i, j)\text{-spcl}_\delta$. This shows that f is pairwise weakly fuzzy δ-semi pre irresolute.

REFERENCES

[1] K.K. Azad, "On Fuzzy Semi Continuity, Fuzzy Almost Continuity and Fuzzy Weakly Continuity", *J. Math. Anal. Appl.*, **82**, (1981), 14–32.

[2] A.S. Bin Sahana, "On Fuzzy Strong Semi Continuity and Fuzzy Pre Continuity", *Fuzzy Sets and Systems*, **44**, (1991), 303–308.

[3] M. Caldas, Sjafari and R.K. Saraf, "Fuzzy $(\delta, P) - T_1$ Topological Spaces", *J. Tri. Math. Soc.*, **9**, (2007), 1–4.

[4] K.K. Dubey, O.S. Panwar and R.K. Tiwari, "On Weakly Pairwise Irresolute Mappings", *Bull. Cal. Math. Soc.* **82**, (1990), 250.

[5] S. Ganguly and S. Saha, "A Note on δ-continuity and δ-connected Sets in Fuzzy Set Theory", *Simon Stevin*, **62**, (1988), 127–141.

[6] A. Kandil, "On Biproximities and Fuzzy Bitopological Spaces", Simon Stevin, **63**, (1989), 45–66.

[7] A. Mukherjee and R. Dhar, "On Weakly Fuzzy δ-semi Pre Continuous Mapping and Weakly Fuzzy δ-semi Pre Irresolute Mapping", (*to appear in Int. J.Fuzzy .Math.*) **18 (1)**, 2010.

[8] A. Mukherjee and S. Debnath, "On δ-semiopen Seta in Fuzzy Setting", *J. Tri. Math. Soc.*, **8**, (2006), 51–54.

[9] R.K. Saraf and Seema Mishra, "On Pairwise Fuzzy Weakly Irresolute Mappings", *J. Tri. Math. Soc.*, 6, (2004), 11–14.

[10] S.S. Thakur and R.K. Khare, "Fuzzy Semi δ-pre Open Sets and Fuzz Semi δ-pre Continuous Mappings", *Stud. Cercet. Stint. Ser. Met. Univ. Bacau.*, 2004.

[11] S.S. Thakur and R Malviya, "On Semiopen Sets and Semi Continuity in Fuzzy Bitopological Spaces", *Fuzzy Sets and Systems*, **79**, (1996), 251.

[12] S.S. Thakur and R. Malviya, "On Pairwise Fuzzy Irresolute Mapping", *Math. Bohemica*, **121**, (1996), 273.

[13] S.S. Thakur, S. Singh, "On Fuzzy Semi Preopen Sets and Fuzzy Semi Pre Continuity", *Fuzzy Sets and Systems*, **98**, (1998), 383–391.

[14] N.V. Veliko, "On *H* Closed Topological Spaces", *Amer Math Soc Transl*, **78**, (1968), 103–118.

34

A STUDY ON INDUCED FUZZY TOPOLOGICAL SPACE GENERATED BY *m*-RLSC FUNCTIONS

A.K. Saha and D. Bhattacharya

National Institute of Technology, Agartala
Barjala, Jirania-799055 Tripura, India

Abstract: *Cardinal extension of RLSC functions to be called m-RLSC function using the concept of regular G_m-subset has been studied. Its various properties has been investigated. Using this class of functions a generalised induced fuzzy topological space (fts) to be called m-induced fts has been introduced and its various properties has been examined. The newly introduced induced fts includes many of the induced fts found in the literature.*

AMS Classification*: 54 A 40.*

1. INTRODUCTION

The aim of the present note is to study the cardinal extension of the regular lower semi-continuous (RLSC) functions introduced by E.P. Lane (1). The natural tool for the proposed extension is the class of regular G_m-subset introduced by J. Mack (2). Using this class of generalised functions a fts called m-induced fts has been defined and studied its various properties.

Definition 1.1: J. Mack (2) introduced the concept of regular G_m-subset as follows:

For an infinite cardinal m, a subset A in a topological space X is called a regular G_m-subset provided it is the intersection of at most m

closed sets whose interiors contain A. Thus A is regular G_m-subset if $A = \cap_{a \leq m} F_a$ where each F_a is closed in X and $A \subseteq$ int (F_a). Equivalently A is regular G_m-subset if $A = \cap_{a \leq m} G_a = \cap_{a \leq m}$ cl (G_a), where each G_a is open in X.

Complement of a regular G_m-subset is called a regular F_s-subset, i.e. a subset F is regular F_s-subset if there is a family (F_a) of closed sets of cardinality at most m contained in F such that $F = \cup_{a \leq m} F_a = \cup_{a \leq m}$ int (F_a). If $m = \aleph_0$, then the familiar term regular G_δ-subset is used, i.e. A subset B of X is called a regular G_δ-subset if $B = \cap_i G_i = \cap_i \mathrm{cl}_X G_i$ for $i \in N$, where each G_i is open in X.

E. P. Lane (1) defined the regular lower semi-continuous (RLSC) function as follows:

A function $f: X \rightarrow R$ is RLSC if for each real number r, the Lebesgue set $L_r(f) = [x : f(x) > r]$ is a regular G_δ-subset of X, i.e. $[x: f(x) > r]$ is regular F_σ-subset.

Generalising the concept of RLSC function we define m-RLSC functions using regular G_m subset as follows:

Definition1.2: A function $f: X \rightarrow R$ is called m-RLSC if for each real number r, the Lebesgue set $L_r(f) = [x : f(x) \leq r]$ is regular G_δ-subset of X, i.e. $[x : f(x) > r]$ is a regular F_σ-subset of X.

2. UNION AND INTERSECTION PROPERTIES OF REGULAR G_m (RESP. REGULAR F_s) SUBSETS

The proofs of the following theorems are straightforward.

Theorem 2.1: The intersection (resp. union) of a family of regular G_δ-subsets (resp. regular F_σ-subsets) with cardinality at most m is again a regular G_δ-subset (resp. regular F_σ-subset).

Theorem 2.2: The union (intersection) of two regular G_δ-subsets (resp. regular F_σ-subsets) is again a regular G_δ-subset (resp. regular F_σ-subset).

3. PROPERTIES OF NEWLY DEFINED m-RLSC FUNCTION

Theorem 3.1: The pointwise supremum of a family of m-RLSC functions of cardinality at most m is again a m-RLSC function.

Proof: Let (f_α: $\alpha \in \Lambda$) be a family of m-RLSC functions of cardinality at most m. Again, let $g(x) = \sup [f_\alpha(x), \alpha \in \Lambda]$, from which it follows that $g(x) \geq f_\alpha(x)$ for all $\alpha \in \Lambda$ and for given $\varepsilon > 0$, there exists at least one $\alpha \in \Lambda$ such that $f_\alpha(x) = g(x) - \varepsilon$, $x \in X$. Let us consider the set $C = [x \in X: g(x) \leq r, r \in R]$. Since f_α is m-RLSC, the set $H_{\alpha, r} = [x \in X: f_\alpha(x) \leq r]$ is a regular G_δ-subset of X for each $r \in R$ and $\alpha \in \Lambda$. Hence, $\cap_{\alpha \leq m} H_{\alpha, r}$ is also a regular G_δ-subset of X. For $y \in \cap_{\delta \leq m} H_{\alpha, r}$, we have $f_\alpha(y) \leq r$ for each $\alpha \in \Lambda$.

Now for given $\varepsilon > 0$, however small, there exists $\alpha \in \Lambda$ such that $g(y) < f_\alpha(y) + \varepsilon$, i.e. $g(y) < r + \varepsilon$. Since $\varepsilon > 0$ is arbitrary, so $g(y) \leq r$. Thus $\cap_{\alpha \leq m} H_{\alpha, r} \subseteq [x \in X: g(x) \leq r, r \in R]$.

If possible, let $x_0 \in [x \in X: g(x) \leq r] - \cap_{\delta \leq m} H_{\alpha, r}$. Then for at least one α, $f_\alpha(x_0) > r$, but $g(x_0) \leq r$ which is a contradiction. Thus $[x \in X: g(x) \leq r, r \in R] = \cap_{\delta \leq m} H_{\alpha, r}$. Which follows that g is m-RLSC function.

Theorem 3.2: If $f_1, f_2, \ldots\ldots\ldots f_n$ are m-RLSC functions on a topological space X, then the function $h = \inf(f_i: i = 1, 2, \ldots\ldots\ldots n)$ is also an m-RLSC function.

Proof: Since $h = \inf(f_i: i = 1, 2, \ldots\ldots\ldots n)$, we have, $h(x) \leq f_i(x)$ for all $x \in X$, and for given $\varepsilon > 0$, there exists at least one f_j such that $f_j(x) < h(x) + \varepsilon$, $x \in X$. Let us consider the set $A = [x \in X: h(x) \leq r]$, $r \in R$. Since f_j is m-RLSC function, the set $H_{j,r} = [x \in X: f_j(x) \leq r]$ is a regular G_δ-subset of X for each $r \in R$. Hence $\cup_j H_{j,r}$, $j = 1, 2, \ldots\ldots\ldots n$, is also a regular G_δ-subset of X. For $y \in A$, we have, $h(y) \leq r$. Now, for given $\varepsilon > 0$, however small, we have, $f_j(y) < h(y) + \varepsilon$, for some j, i.e. $f_j(y) < r + \varepsilon$. Since ε is arbitrary so $f_j(y) \leq r$, i.e. $A \subseteq \cup_j H_{j,r}$.

Again, if possible, let $x_0 \in \cup_j H_{j,r} - A$. Then for some f_j, $f_j(x_0) \leq r$ but $h(x_0) > r$, which is a contradiction, i.e. $A = [x \in X: h(x) \leq r] = \cup_j H_{j,r}$. It follows that h is m-RLSC function. Hence finite infima of m-RLSC functions is also m-RLSC.

Remark: It may be seen that the arbitrary suprema and countable infima of m-RLSC functions are not m-RLSC functions.

Definition 3.3: *A* function $f: X \rightarrow R$ is called m-RUSC, if for each real number r, the Lebesgue set $L^r(f) = [x: f(x) \geq r]$ is a regular G_δ-subset of X or equivalently $[x: f(x) < r]$ is a regular F_σ-subset.

Theorem 3.4: The function $f: X \to R$ is m-RLSC iff $-f$ is m-RUSC function, where $(-f)(x) = -f(x)$.

Proof: Let $f: X \to R$ be an m-RLSC function, i.e. $[x : f(x) \leq r]$ is regular G_m-subset of X for each $r \in$ R. Now, let us consider the set $[x: (-f)(x) \geq r]$. Here $[x: (-f)(x) \geq r] = [x: -f(x) \geq r] = [x: f(x) \leq -r]$, which is a regular G_m-subset of X, as f is m-RLSC. Thus $-f$ is m-RUSC function.

Theorem 3.5: If f and g are two m-RLSC functions then so is $f + g$.

Proof: Let us consider the set $[x : f(x) + g(x) \leq \lambda]$.

Now, $[x : f(x) + g(x) \leq \lambda] = \cap \{[x : f(x) \leq \lambda - t] \cup [x : g(x) \leq t]\}$, $t \in D$, where D is a set dense in R having cardinality at most m. Now, $A_t = [x : f(x) \leq \lambda - \text{t}]$ and $B_t = [x : g(x) \leq t]$ are both regular G_m-subset for each $t \in D$ and hence $A_t \cup B_t$ is regular G_m and since $t \in D$, a set having cardinality atmost m, so $\cap (A_t \cup B_t)$ is also regular G_m-subset.

Hence, $f + g$ is m-RLSC function.

Theorem 3.6: If f be an m-RLSC function and $a > 0$ be any real number, then af also is an m-RLSC function.

Proof: $f: X \to R$ be an m-RLSC function, i.e. $[x : f(x) \leq r]$ is regular G_m-subset of X. Let us consider the set $[x \in X: af(x) \leq t]$.

Now, $[x \in X: af(x) \leq t] = [x \in X: f(x) \leq t/a] = [x \in X: f(x) \leq \lambda]$, $\lambda = t/a$, which is a regular G_m-subset of X, since f is m-RLSC. Hence, af is m-RLSC function.

Theorem 3.7: If f and g are two m-RLSC functions then so is f.g.

Proof: Let us consider the set $[x \in X: f(x) + g(x) \leq \lambda]$.

Now, $[x \in X: f(x) + g(x) \leq \lambda] = \cap \{[x \in X: f(x) \leq t] \cup [x \in X: g(x) \leq \lambda/t]\}$, $t \in D$, where D is a set dense in R having cardinality at most m. Now, $A_t = [x \in X: f(x) \leq t]$ and $B_t = [x \in X: g(x) \leq \lambda/t]$ are both regular G_δ-subset for each $t \in M$ and hence $A_t \cup B_t$ is regular G_δ and since $t \in D$, a set having cardinality atmost m, so $\cap (A_t \cup B_t)$ is also regular G_δ-subset.

Hence, f.g. is m-RLSC function.

Remark: If f is m-RLSC and g is m-RUSC functions then f/g is m-RLSC function.

Theorem 3.8: The characteristic function of a regular F_s-subset is m-RLSC.

Proof: Let A be a regular F_s-subset, i.e. $A = \cup_{a \leq m} F_a = \cup_{a \leq m}$ int (F_a), where each F_a is closed. Now, the characteristic function μ_A of the set A is defined as

$$\mu_A(x) = 1 \text{ if } x \in A = 0 \text{ if } x \in X - A.$$

We have to show that μ_A is m-RLSC, i.e. to show that $[x : \mu_A(x) \leq r]$ is a regular G_m-subset for each $r \in R$.

Case 1: If $r < 0$, then $[x: \mu_A(x) \leq r] = \phi$, which is a regular G_m-subset, since ϕ is open and $\phi = \cap\phi = \cap \operatorname{cl} \phi$.

Case 2: If $0 \leq r < 1$, then $[x : \mu_A(x) \leq r] = X - A$, which is a regular G_m-subset, being the complement of a regular F_s-subset A.

Case-3: If $r \geq 1$, then $[x: \mu_A(x) \leq r] = X$, which is a regular G_m-subset, since $X = \cap X = \cap \operatorname{cl} X$. Hence the theorem.

4. m-INDUCED FUZZY TOPOLOGICAL SPACES

Definition 4.1: A family $\mathfrak{F} \subseteq I^X$, is called a m-fuzzy topology on X if

(*i*) $0, 1 \in \mathfrak{F}$

(*ii*) $\operatorname{Inf}(f_i, i = 1, 2, ...n) \in \mathfrak{F}, f_i \in \mathfrak{F}$

(*iii*) $\operatorname{Sup}(f_a)_{a \in \Lambda} \in \mathfrak{F}$, where each $f_a \in \mathfrak{F}$, for $a \in \Lambda$ and $|\Lambda| \leq m$.

Theorem 4.2: Let (X, T) be a topological space. The family m-RL (X) of all m-RLSC functions from X to I forms an m-fuzzy topology on X.

Proof: Clearly, **0** and **1** are m-RLSC functions and the remaining conditions (*ii*) and (*iii*) follows from Theorems 3.1 and 3.2.

Definition 4.3: The m-fuzzy topology formed by the family m-RL (X) of all m-RLSC functions from a topological space (X, T) to I is called m-induced fuzzy topology on X and is denoted by $m(T)$ and the space $[X, m(T)]$ is called m-induced fuzzy topological space. The members of $m(T)$ are called m-open sets.

Gillman and Henriksen (3), in 1954 defined a topological space to be pseudo-discrete or for brevity a P-space if it is a Tychonoff space and each prime ideal in the ring $C(X)$ of all continuous real valued functions on X is maximal. They obtained a number of equivalent conditions for P-spaces in the class of Tychonoff spaces. One of the important of such condition is given by

P-Space: A tychonoff space X is a P-space if every G_δ-subset in X is open.

In their paper 'On Cardinal Extension of P-topology' S.K. Acharyya, K.C. Chattopadhyay and Saral Datta (4) introduced the concept of P_α-space as follows:

P_α-Space: A T_1-space X is called a P_α-space if the intersection of any family $\mathcal{B}$ of open sets in X with $|\mathcal{B}| \leq \aleph_\alpha$ is open. A point p in X is called a P_α-point if the nbd. System of p is closed under the intersection of $\aleph_\alpha$ many members of it. Evidently X is a P_α-space iff each of its point is a P_α-point.

In this paper we use the term P_m-space for such cardinal extension of P-space, i.e. a T_1-space is called a P_m-space if the intersection of any family $\mathcal{B}$ of open sets in X with $|\mathcal{B}| \leq m$ is open. Trivially every P_m-space is a P-space.

AN EXAMPLE OF m-INDUCED FUZZY TOPOLOGICAL SPACE

Example 4.4: Let us construct a topological space as follows:

Let S be a uncountable non-discrete space. In S, all points are isolated except for a distinguished point s, *a* nbd. of s being any set containing s whose complement has a cardinality at most m. So S is a P_m-space. The closed set (s) is not regular G_m-subset since (s) is an uncountable intersection of its nbds. which are open and closed and so regular G_m-subset.

Again let us consider *a* nbd. G_a of s. Then G_a, being a clopen subset is regular G_m-subset and hence $S - G_a$ is a regular F_s-subset. Let us define f_a: $S \to R$ for $a \in \Lambda$ by

$$f_a(x) = 1 \quad \text{if } x \in S - G_a$$
$$= 0 \quad \text{if } x \in G_a$$

So each f_a is m-RLSC, being the characteristic function of a regular F_s-subset. Now we consider the family $\mathfrak{F}$ consisting of all f_a's and $\mathbf{1}$ $(= \mu_S)$ [$\mathbf{0}$ $(= \mu_\phi)$ can be obtained from f_a when $G_a = S$]. Clearly all the members of $\mathfrak{F}$ are m-RLSC functions. Now let us construct a family $\mathfrak{F}^*$ of functions as follows:

$\mathfrak{F}^* = (rf_a : f_a \in \mathfrak{F}, 0 \le r \le 1)$. So $\mathfrak{F}^*$ is a collection of m-RLSC functions on S such that

(*i*) $\mathbf{0} (= \mu_\phi), \mathbf{1} (= \mu_S) \in \mathfrak{F}^*$

Considering a family (f_β) of m many members of $\mathfrak{F}^*$. Then $\vee_\beta r_\beta f_\beta = f$, say is given by

$$f(x) = \vee_\beta r_\beta, \quad \text{if } x \in S - \cap_a G_a$$
$$= 0, \quad \text{if } x \in \cap_a G_a$$

Since S is a P_m-space, so every G_a is open and hence $\cap_a G_a$ is open set containing s. So $f \in F$.

(*ii*) Lastly, $r_1 f_1, r_2 f_2, \dots\dots\dots, r_n f_n \in \mathfrak{F}^*$. Then $\wedge_i r_i f_i = g$ (say), is given by,

$$g(x) = \wedge_i r_i, \quad \text{if } x \in S - \cup G_i$$
$$= 0, \quad \text{if } x \in \cup G_i.$$

Which clearly belongs to $\mathfrak{F}^*$.

Thus we see that$(S, \mathfrak{F}^*)$ forms an m-induced fuzzy topological space.

Theorem 4.5: A fuzzy subset λ in an m-induced fts $[X, m(T)]$ is m-open iff for each $r \in I$, the strong r-cut $\sigma_r(\lambda)$ is regular F_s-subset in the topological space (X, T).

Proof: Let $[X, m(T)]$ be an m-induced fts of the topological space (X, T). Let $\lambda \in m(T)$ i.e. $\lambda \in m - RL(X)$.

Now, $\lambda \in m - RL(X)$ iff for each $r \in I$, the set $[x \in X : \lambda(x) \le r]$ is regular G_m-subset of X, i.e. iff the set $[x \in X : \lambda(x) > r]$ is regular F_s-subset of X, i.e. iff $\sigma_r(\lambda) = [x \in X : \lambda(x) > r]$ is regular F_s-subset of (X, T).

Lemma 4.6: A function $f : (X, T) \to (R, \sigma_1)$, where $\sigma_1 = [(r, \infty) : r \in R]$ is m-RLSC function iff the inverse image of an open subset of (R, σ_1) is regular F_s-subset in (X, T).

Proof: Let $f : (X, T) \to (R, \sigma_1)$ be m-RLSC. Hence by definition of m-RLSC function

$f^{-1}(r, \infty) = [x \in X : f(x) > r] = X - [x \in X : f(x) \le r]$ is regular F_s-subset of X and conversely if

$f^{-1}(r, \infty) = [x \in X : f(x) > r]$ is regular F_s-subset of X then $[x \in X : f(x) \le r] = X - [x \in X : f(x) > r]$ is a regular G_m-subset of X and hence f is a m-RLSC function.

Definition 4.7: Let $f: [X, m(T_1)] \rightarrow [Y, m(T_2)]$ be a mapping between two m-induced fuzzy topological spaces. Then f is called fuzzy m-continuous if the inverse image of an m-open fuzzy subset of $m(T_2)$ is an m-open fuzzy subset of $m(T_1)$.

Theorem 4.8: If $f: X \rightarrow Y$ is continuous then the inverse image of a regular G_m (regular F_s) subset of Y is a regular G_m (regular F_s) subset of X.

Proof: Let $B = \cap_{a \leq m} G_a = \cap_{a \leq m} \text{cl}(G_a)$ be a regular G_m-subset of Y, where each G_a is open in Y.

Let $f^{-1}(G_a) = V_a$, $a \leq m$, then each V_a is open in X and let $A = \cap_{a \leq m} V_a$. It remains to show that

$$A = \cap_{a \leq m} V_a = \cap_{a \leq m} \text{cl}(V_a)$$

Clearly, $\cap_{a \leq m} V_a \subseteq \cap_{a \leq m} \text{cl}(V_a)$ ... *(i)*

Again let $p \in \cap_{a \leq m} \text{cl}(V_a)$, so $p \in \text{cl}(V_a)$ for each a. Now, by the continuity of f

$f(p) \in f[\text{cl}_X(V_a)] \subseteq \text{cl}_Y(G_a)$, i.e. $f(p) \in \cap_a \text{cl}_Y(G_a) = \cap_a G_a$.

i.e. $p \in f^{-1}(\cap_a G_a) = \cap_a f^{-1}(G_a) = \cap_a V_a$.

i.e. $\cap_{a \leq m} \text{cl}(V_a) \subseteq \cap_{a \leq m} V_a$ ... *(ii)*

Hence, $A = \cap_{a \leq m} V_a = \cap_{a \leq m} \text{cl}(V_a)$ is a regular G_m-subset of X such that $A = f^{-1}(B)$.

Definition 4.9: A mapping $f: X \rightarrow Y$ is said to be a s-continuous if the inverse image under f of any regular F_s-subset of Y is regular F_s-subset of X.

Clearly, every continuous mapping is an s-continuous mapping.

Theorem 4.10: A mapping $f: [X, m(T_1)] \rightarrow [Y, m(T_2)]$ is fuzzy m-continuous iff

$$f: (X, T_1) \rightarrow (Y, T_2) \text{ is } s\text{-continuous.}$$

Proof: Let $f: [X, m(T_1)] \rightarrow [Y, m(T_2)]$ be a fuzzy m-continuous mapping and let B be a regular F_s-subset of (Y, T_2).

Now, $f^{-1}(B) = [x \in X: f(x) \in B]$

$= [x \in X: \mu_B f(x) = 1]$, where μ_B is the characteristic function of the crisp set B.

$= [x \in X: \mu_B f(x) > r, 0 \leq r < 1]$

$= \{x \in X: [f^{-1}(\mu_B)](x) > r, 0 \leq r < 1\}$

$$= \sigma_r [f^{-1} (\mu_B)]$$

Now μ_B, being the characteristic function of a regular F_s-subset of Y, is an m-RLSC function. So, $\mu_B \in m (T_2)$. Again, f: $[X, m (T_1)] \rightarrow [Y, m (T_2)]$ being a fuzzy m-continuous mapping $f^{-1} (\mu_B) \in m (T_1)$.

Hence by Theorem 4.5, $f^{-1} (B) = \sigma_r [f^{-1} (\mu_B)]$ is a regular F_s-subset of X, i.e. f: $(X, T_1) \rightarrow (Y, T_2)$ is s-continuous.

Conversely, let f: $(X, T_1) \rightarrow (Y, T_2)$ be an s-continuous mapping and β be an m-open fuzzy subset of $[Y, m (T_2)]$. Now, for $0 < r < 1$,

$$\begin{aligned} \sigma_r [f^{-1} (\beta)] &= \{x \in X: [f^{-1} (\beta)] (x) > r\} \\ &= \{x \in X: \beta [f(x)] > r\} \\ &= (\beta f)^{-1} (r, 1) \\ &= f^{-1} \beta^{-1} (r, 1) \\ &= f^{-1} [\beta^{-1} (r, 1)] \\ &= f^{-1} [\sigma_r (b)] \end{aligned}$$

Now, since $\beta \in m (T_2)$ i.e. $\beta \in m\,\mathrm{RL}\,(Y)$. So $\sigma_r(\beta) = [y \in Y: \beta (y) > r]$ is a regular F_s-subset of Y, i.e. $\sigma_r(\beta) = \beta^{-1} (r, 1) = [y \in Y: \beta (y) > r]$ is a regular F_s-subset of Y. Finally, f: $(X, T_1) \rightarrow (Y, T_2)$, being an s-continuous mapping, $\sigma_r [f^{-1} (b)] = f^{-1} [b^{-1} (r, 1)]$ is a regular F_s-subset of X. Hence, $f^{-1} (\beta) \in m (T_1)$ and thus f is fuzzy m-continuous.

REFERENCES

[1] Lane E.P., "Weak C Insertion of a Continuous Function", *Notices, Trans. Amer. Math. Soc.*, **26**, (1979), A-231.

[2] Mack J.E., "Countably Paracompactness, Weak Normality Properties", *Trans. Amer. Math. Soc.*, **148**, (1970), 256–272.

[3] Gillman L. and Henriksen M., "Concerning Rings of Continuous Functions", *Trans. Amer. Math. Soc,* **77**, (1954), 360–372.

[4] Acharya S.K., Chattopadhyay K.C. and Datta S., "On Cardinal Extension of P-Topology", *Journal of the Ramanujan Inst. Math. Sci.*, 51–63.

35

ON A CLASS OF GENRALIZED DIFFERENCE SEQUENCE SPACES RELATED TO ℓ_p SPACE DEFINED BY MODULUS FUNCTION

Binod Chandra Tripathy[1] and Stuti Borgohain[2]

Mathematical Sciences Division Institute of Advanced Study in Science and Technology, Paschim Boragaon, Garchuk, GUWAHATI-781035, ASSAM, INDIA,

E-mail: tripathybc@rediffmail.com, tripathybc@yahoo.com[1]

E-mail: stutiborgohain@yahoo.com[2]

Abstract: *The sequence space $m(f, \varphi, \Delta^n_m, p)^F$ of fuzzy real numbers for $0 < p < \infty$, is introduced. Some properties of this sequence space like solidness, symmetricity, convergence-free, etc. are studied. We obtain some inclusion relations involving this sequence space.*

Keywords: *Modulus function; symmetric space; solid space; convergence free; metric space; completeness.*

AMS Classification No: *40A05, 40A25, 40A30; 40C05.*

1. INTRODUCTION

The concept of fuzzy set theory was introduced by Zadeh [11]. Later on different classes of sequences of fuzzy numbers have been discussed by many others.

Kizmaz [1] studied the notion of difference sequence spaces at the initial stage. Tripathy, Esi and Tripathy [9] further generalized this notion as follows. Let $m \geq 1$ and $n \geq 1$ be integers, then

The work of the second author was financially supported by a Department of Science and Technology (GoI) project no. SR/WOS-A/MS-07/2008

$$Z(\Delta_m^n) = [x = (x_k): (\Delta_m^n x_k) \in Z], \text{ for } Z = \ell_\infty, c \text{ and } c_0.$$

The notion of modulus function was introduced by Nokano [2], defined as f: $[(0, \infty) \to (0, \infty)]$ is called a modulus if, $f(x) \geq 0$, for each x, $f(x) = 0$ if and only if $x = 0$, $f(x + y) \leq f(x) + f(y)$, f is increasing and continuous from the right at 0.

Later on it was studied from sequence space point of view by Ruckle [4] and many others.

Let X be a sequence space. Then the sequence space $X(f)$ is defined as,

$$X(f) = \{x = (x_k): [f(|x_k|) \in X\},$$

for a modulus f.

Sargent [5] introduced the crisp set sequence spacc $m(\phi)$ in crisp set and studied some properties of this space. Later on it was studied from the sequence space point of view and some matrix classes were characterized with one member as $m(\phi)$ by Tripathy [7], Tripathy and Sen [10] and others. In this article we introduced the space $m(f, \varphi, \Delta_m^n, p)^F$ of fuzzy real numbers defined by modulus function.

Throughout the article w^F, ℓ^F, ℓ_∞^F represent the classes of *all*; *absolutely summable*, *bounded* sequences fuzzy real numbers respectively.

2. DEFINITIONS AND BACKGROUND

A fuzzy real number X is a fuzzy set on R, i.e. a mapping X: $R \to I$ $(= [0, 1])$ associating each real number t with its grade of membership $X(t)$. In this article we shall consider $R(I)$ the class of all *upper semi-continuous*, *normal*, *convex* fuzzy real numbers.

Let D be the set of all closed and bounded intervals $X = \left(X_1^\alpha, X_2^\alpha\right)$ and $Y = \left(Y_1^\alpha, Y_2^\alpha\right)$. Define $d: R(I) \times R(I) \to R$ by, $\bar{d}(X, Y) = \sup\limits_{0<\alpha\leq 1} \max\left(\left|X_1^\alpha - Y_1^\alpha\right|, \left|X_2^\alpha - Y_2^\alpha\right|\right)$ for $X, Y \in R(I)$. It is well known that $\left(R(I), \bar{d}\right)$ is a complete metric space.

A sequence space E is said to be *solid* if $(Y_n) \in E$, whenever $(X_n) \in E$ and $|Y_n| \leq |X_n|$, for all $n \in N$. Let $X = (X_n)$ be a sequence, then $S(X)$ denotes the set of all permutations of the elements of (X_n) i.e. $S(X) = \{[X_{\pi(n)}]: \pi$ is a permutation on $N\}$. A sequence space E is said to be *symmetric* if $S(X) \subset E$ for all $X \in E$. A sequence space E is said to be *convergence free* if $(Y_n) \in E$ whenever $(X_n) \in E$ and $X_n = 0$ implies $Y_n = 0$. A sequence space E is said to be *monotone* if E contains the canonical pre-images of all its step spaces.

Lemma: A sequence space E is solid implies that E is monotone.

Let $\wp_s$ be the class of all subsets of N those do not contain more than s number of elements. Throughout (ϕ_n) is a non-decreasing sequence of positive real numbers such that $n\phi_{n+1} \leq (n+1)\phi_n$ for all $n \in N$.

The space $m(\phi)$ introduced by Sargent (5). In this article we introduce the following difference sequence space:

$$m\left(f, \varphi, \Delta_m^n, p\right)^F =$$

$$\left\{X = (X_k) : \sup_{s \geq 1, \sigma \in \wp_s} \frac{1}{\varphi_s} \sum_{k \in \sigma} \left[f\left(\bar{d}\left(\Delta_m^n X_k, \bar{0}\right)\right)\right]^p < \infty\right\},$$

for $0 < p < \infty$.

3. MAIN RESULTS

In this section we prove some results involving all these sequence spaces.

Theorem 1 (a): The sequence space $m\left(f, \varphi, \Delta_m^n, p\right)^F$ is a complete metric space with the metric,

$$\eta(X, Y) =$$

$$\sum_{r=1}^{mn} \bar{d}(X_r, Y_r) + \sup_{s \geq 1, \sigma \in \wp_s} \frac{1}{\varphi_s} \sum_{k \in \sigma} \left(f\left(\bar{d}\left(\Delta_m^n X_k, \Delta_m^n Y_k\right)\right)\right)^p,$$

for X, Y $m\left(f, \varphi, \Delta_m^n, p\right)^F$ and $0 < p < 1$.

(b) The sequence space $m\left(f, \varphi, \Delta_m^n, p\right)^F$ is a complete metric space with the metric,

$$\eta(X,Y) =$$

$$\sum_{r=1}^{mn} \bar{d}(X_r, Y_r) + \sup_{s \geq 1, \sigma \in \wp_s} \frac{1}{\varphi_s} \left(\sum_{k \in \sigma} \left(f\left(\bar{d}\left(\Delta_m^n X_k, \Delta_m^n Y_k \right)\right)\right)^p \right)^{p^{-1}},$$

for X, Y $m\left(f, \varphi, \Delta_m^n, p\right)^F$ and $1 \leq p < \infty$.

Proof (a): Clearly, $m\ (f, \phi, \Delta_{mp}^n, p)^F$ is a metric space with the above defined metric h. We have to prove that it is a complete metric space.

Let $(X^{(i)})$ be a Cauchy sequence in $m\left(f, \varphi, \Delta_m^n, p\right)^F$ where $X^{(i)} = \left(x_n^{(i)}\right)_{n=1}^{\infty}$.

Let $\varepsilon > 0$ be given. Then there exits a positive integer $n_0 = n_0\ (\varepsilon)$ such that

$$\eta\left(X^{(i)}, X^{(j)}\right) < \varepsilon, \text{ for all } i, j \geq n_0.$$

By the definition of η, we get:

$$\sum_{r=1}^{mn} \bar{d}(X_r, Y_r) + \sup_{s \geq 1, \sigma \in \wp_s} \frac{1}{\varphi_s} \sum_{k \in \sigma} \left(f\left(\bar{d}\left(\Delta_m^n X_k, \Delta_m^n Y_k \right)\right)\right)^p < \varepsilon,$$

for all $i, j \geq n_0$ ... (1)

Which implies that, $\sum_{r=1}^{mn} \bar{d}\left(X_r^{(i)}, X_r^{(j)}\right) < \varepsilon$, for all $i,\ j \geq n_0$.

$$\Rightarrow \bar{d}\left[X_r^{(i)}, X_r^{(j)}\right] < \varepsilon, \text{ for all } i,\ j \geq n_0, r = 1, 2, 3 \ldots mn.$$

Hence $(X_r^{(i)})$ is a Cauchy sequence in $R\ (I)$, so it is convergent in $R\ (I)$, by the completeness property of $R\ (I)$, for $r = 1, 2, 3 \ldots mn$.

Let, $\lim_{i \to \infty} X_r^{(i)} = X_r$ for $r = 1, 2, 3 \ldots mn$ ... (2)

Also, $\sup_{s \geq 1, \sigma \in \wp_s} \frac{1}{\varphi_s} \sum_{k \in \sigma} \left(f\left(\bar{d}\left(\Delta_m^n X_k^{(i)}, \Delta_m^n X_k^{(j)} \right)\right)\right)^p < \varepsilon$, for all $i,\ j \geq n_0$ ... (3)

For $s = 1$ and σ varying over $\wp_s$, we get,

$$f\left(\bar{d}\left(\Delta_m^n X_k^{(i)}, \Delta_m^n X_k^{(j)}\right)\right) < \left(\varepsilon\varphi_1\right)^{p^{-1}} = f(\varepsilon) \text{ (say), for all } i, j \geq n_0.$$

$$\Rightarrow \bar{d}\left[\Delta_m^n X_k^{(i)}, \Delta_m^n X_k^{(j)}\right] < \varepsilon \text{, using the continuity of } f.$$

Which implies that $\left(\Delta_m^n x_k^{(i)}\right)$ is a Cauchy sequence in $R(I)$ and so is convergent in $R(I)$.

Let, $\lim\limits_i \Delta_m^n X_k^{(i)} = Y_k$ (say), in $R(I)$, for each $k \in N$.

We have to prove that, $\lim\limits_i X^{(i)} = X$ and $X \in m\left(f, \varphi, \Delta_m^n, p\right)^F$.

For $k = 1$, we get, from (2), $\lim\limits_{i \to \infty} X_{mn+1}^{(i)} = X_{mn+1}$ for $m \geq 1, n \geq 1$.

Hence by induction we have, $\lim\limits_{i \to \infty} X_k^{(i)} = X_k$, for each $k \in N$. Also, $\lim\limits_i \Delta_m^n X_k^{(i)} = \Delta_m^n X_k$, for each $k \in N$.

Taking $j \to \infty$ and fixing i, we get, from (4),

$$\sup_{s \geq 1, \sigma \in \wp_s} \frac{1}{\varphi_s} \sum_{k \in \sigma} \left(f\left(\bar{d}\left(\Delta_m^n X_k^{(i)}, \Delta_m^n X_k\right)\right)\right)^p < \varepsilon \text{, for } i \geq n_0.$$

Hence we get,

$$\sum_{r=1}^{mn} \bar{d}\left(X_r^{(i)}, X_r\right) + \sup_{s \geq 1, \sigma \in \wp_s} \frac{1}{\varphi_s} \sum_{k \in \sigma} \left(f\left(\bar{d}\left(\Delta_m^n X_k^{(i)}, \Delta_m^n X_k\right)\right)\right)^p$$

$$< \varepsilon + \varepsilon = 2\varepsilon \text{, for all } i \geq n_0.$$

which implies that, $\eta\,[X^{(i)}, X] < 2\varepsilon$, for all $i \geq n_0$, i.e. $\lim\limits_i X^{(i)} = X$.

Now, we show that $X \in m\left(f, \varphi, \Delta_m^n, p\right)^F$.

We know that, $\eta\,(X, \theta) \leq \eta\,[X^{(n)}, X] + \eta\,[X^{(n)}, \theta] < \varepsilon + M$, for all $n \geq n_0(\varepsilon)$.

Which implies that $X \in m\left(f, \varphi, \Delta_m^n, p\right)^F$.

Hence $\in m\left(f, \varphi, \Delta_m^n, p\right)^F$ is a complete metric space.

(*b*) This part can be proved by following similar techniques.

Result 1: The sequence space $m\left(f, \varphi, \Delta_m^n, p\right)^F$ is not solid in general, for $0 < p < \infty$.

Proof: The result follows from the following example:

Example 1: Let $m = 2$, $n = 4$, $p = 2$. Let, $X_k = \bar{k}$, for all $k \in N$ and $\phi_s = s$ for all $s \in N$.

Let $M(x) = |x|$, for all $x \in (0, \infty)$. Then, we have, $\bar{d}\left(\Delta_2^4 X_k, \bar{0}\right) = 0$, for all $k \in N$.

Hence, we have, $\sup\limits_{s \geq 1, \sigma \in \wp_s} \frac{1}{s} \sum\limits_{k \in \sigma} \left(f\left(\bar{d}\left(\Delta_2^4 X_k, \bar{0}\right)\right)\right)^2 < \infty$,

which implies that, $(X_k) \in m\left(f, \varphi, \Delta_m^n, p\right)^F$.

Consider the sequence (α_k) of scalars defined by, $\alpha_k = 1$, for all k even and $\alpha_k = 0$, otherwise. Then, $\left(\alpha_k X_k\right) = \left(X_1, \bar{0}, X_3, \bar{0}, X_5, \bar{0}, ...\right)$

Which implies that, $\sup\limits_{s \geq 1, \sigma \in \wp_s} \frac{1}{s} \sum\limits_{k \in \sigma} \left(f\left(\bar{d}\left(\Delta_2^4 \alpha_k X_k, \bar{0}\right)\right)\right)^2 = \infty$

Which shows that, $(a_k X_k) \notin m\,(f, s, \Delta_2^4, 2)$.

Hence $m\left(f, \varphi, \Delta_m^n, p\right)^F$ is not solid in general, for $0 < p < \infty$.

Result 2: The sequence space $m\left(f, \varphi, \Delta_m^n, p\right)^F$ is not symmetric in general, for $0 < p < \infty$.

Proof: The result follows from the following example:

Example 2: Let $m = 1$, $n = 1$, $p = 2^{-1}$ and $M(x) = x^2$, for all $x \in (0, \infty)$.

Let $\phi_s = s$, for all $s \in N$. Let $X_k = \bar{k}$, for all $k \in N$.

Then, $\bar{d}\left(\Delta X_k, \bar{0}\right) = 1$, for all $k \in N$. Hence $(X_k) \in m\left(f, s, \Delta, 2^{-1}\right)$

Let (Y_k) be a rearrangement of (X_k) such that $(Y_k) = (X_1, X_2, X_4, X_3, X_9, X_5, X_{16}, X_6, X_{25}.....)$. Then $\bar{d}\left(\Delta Y_k, \bar{0}\right) \approx k - (k-1)^2 \approx k^2$, for all $k \in N$.

Which shows that, $\sup\limits_{s \geq 1, \sigma \in \wp_s} \frac{1}{s} \sum\limits_{k \in \sigma} \left(f\left(\bar{d}\left(\Delta Y_k, \bar{0}\right)\right)\right)^{2^{-1}} = \infty$.

Hence, $(Y_k) \notin m\left(f, s, \Delta, 2^{-1}\right)$

Thus, $m\left(f, \varphi, \Delta_m^n, p\right)^F$ is not symmetric in general, for $0 < p < \infty$.

Result 3: The sequence space $m\left(f, \varphi, \Delta_m^n, p\right)^F$ is not convergence-free, in general.

Proof: The result follows from the following example:

Example 3: Let $m = 4$, $n = 1$, $p = 2^{-1}$. Let $M(x) = x^4$, for all $x \in (0, \infty)$. Let $\phi_s = s$ for all $s \in N$. Consider the sequence (X_k) defined as follows:

$$X_k(t) = \begin{cases} 1 + kt & \text{for} \quad t \in \left(-k^{-1}, 0\right) \\ 1 - kt & \text{for} \quad t \in \left(0, k^{-1}\right) \\ 0 & \text{otherwise} \end{cases}$$

Then, $\Delta_4 X_k(t) =$

$$\begin{cases} 1 + k(k+4)(2k+4)^{-1} t & \text{for} \quad t \in \left\{-(2k+4)\left[k(k+4)\right]^{-1}, 0\right\} \\ 1 - k(k+4)(2k+4)^{-1} t & \text{for} \quad t \in \left\{0, (2k+4)\left[k(k+4)\right]^{-1}\right\} \\ 0 & \text{otherwise} \end{cases}$$

Such that, $\bar{d}\left(\Delta_4 X_k, \bar{0}\right) =$

$$(2k+4)\left[k(k+4)\right]^{-1} = 2(k+1)^{-1} + 4\left[k(k+4)\right]^{-1}.$$

We have, $\sup\limits_{s \geq 1, \sigma \in \wp_s} \dfrac{1}{s} \sum\limits_{n \in \sigma} \left(f\left(\bar{d}\left(\Delta_4 X_n, \bar{0}\right)\right)\right)^{2^{-1}} < \infty$.

Thus $(X_n) \in m\left(f, s, \Delta_4, 2^{-1}\right)$

Now, let us take another sequence (Y_k) such that,

$$Y_k(t) = \begin{cases} 1 + t\left(k^2\right)^{-1} & \text{for } t \in \left(-k^2, 0\right) \\ 1 - t\left(k^2\right)^{-1} & \text{for } t \in \left(0, k^2\right) \\ \bar{0}, & \text{otherwise} \end{cases}$$

So that,

$$\Delta_4 Y_k(t) = \begin{cases} 1 + t\left(2k^2 + 8k + 16\right)^{-1} & \text{for } t \in \left[-\left(2k^2 + 8k + 16\right), 0\right] \\ 1 - t\left(2k^2 + 8k + 16\right)^{-1} & \text{for } t \in \left(0, 2k^2 + 8k + 16\right) \\ 0 & \text{otherwise} \end{cases}$$

But, $\left[\bar{d}\left(\Delta_4 Y_k, \bar{0}\right)\right] = \left(2k^2 + 8k + 16\right)$, for all $k \in N$.

Which implies that, $\sup\limits_{s \geq 1, \sigma \in \wp_s} \frac{1}{s} \sum\limits_{k \in \sigma} \left(f\left(\bar{d}\left(\Delta_4 Y_k, \bar{0}\right)\right)\right)^{2^{-1}} = \infty$.

Thus, $(Y_n) \notin m\left(f, s, \Delta_4, 2^{-1}\right)^F$,

Hence $m\left(f, \varphi, \Delta_m^n, p\right)^F$ is not convergence-free, in general.

Now we shall state some results. The proofs of the results are easy, so omitted.

Proposition 1: $m\left(f, \varphi, \Delta_m^n\right)^F \subseteq m\left(f, \varphi, \Delta_m^n, p\right)^F$, for $1 \leq p < \infty$.

Proposition 2: $m\left(f, \varphi, \Delta_m^n, p\right)^F \subseteq m\left(f, \psi, \Delta_m^n, p\right)^F$ if and only if $\sup\limits_{s \geq 1} \varphi_s \left(\psi_s\right)^{-1} < \infty$, for $0 < p < \infty$.

Corollary 1: $m\left(f, \varphi, \Delta_m^n, p\right)^F = m\left(f, \psi, \Delta_m^n, p\right)^F$ if and only if $\sup\limits_{s \geq 1} \left(\eta_s\right) < \infty$ and $\sup\limits_{s \geq 1} \left(\eta_s^{-1}\right) < \infty$, where $\eta_s = \varphi_s \left(\psi_s\right)^{-1}$, for $0 < p < \infty$.

Proposition 3: $\ell_p\left(f, \Delta_m^n\right)^F \subseteq m\left(f, \varphi, \Delta_m^n, p\right)^F \subseteq \ell_\infty\left(f, \Delta_m^n\right)^F$, for $1 \leq p < \infty$.

Proposition 4: $m\left(M, \varphi, \Delta_m^n, p\right)^F = \ell_p\left(M, \Delta_m^n\right)^F$ if and only if $\sup\limits_{s \geq 1} \left(\varphi_s\right) < \infty$ and $\sup\limits_{s \geq 1} \left(\varphi_s^{-1}\right) < \infty$.

Corollary 2: $m\left(f, \varphi, \Delta_m^n, p\right)^F = \ell_\infty\left(f, \Delta_m^n\right)^F$ if $\lim\limits_{s \to \infty} \varphi_s s^{-1} > 0$, for $0 < p < \infty$.

REFERENCE

[1] KIZMAZ H., "On Certain Sequence Spaces", *Canad. Math. Bull.*, **24 (2)**, (1981), 169–176.

[2] NAKANO, H., "Concave Modulus", *J. Math. Soc. Japan*, **5**, (1953), 29–49.

[3] RATH, D. and TRIPATHY, B.C., "Characterization of Certain Matrix Operators", *J Orissa Math Soc*, 1989, **8**, 121–134.

[4] RUCKLE W.H., "FK Spaces in Which the Sequence of Coordinate Vectors is Bounded". *Canad. J. Math.*, **25**, (1973), 973–978.

[5] SARGENT, W.L.C., "Some Sequence Spaces Related to Spaces", *J. London Math. Soc*, **35**, 1960, 161–171.

[6] TRIPATHY, B.C., "Matrix Maps on the Power Series Convergent on the Unit Disc", *J. Analysis.*, 1998, **6**, 27–31.

[7] TRIPATHY, B.C., "A Class of Difference Sequences Related to the *p*-Normed Space", *Demonstratio Math*, (2003), **36 (4)**, 867–872.

[8] TRIPATHY, B.C. and ESI, A., "A New Type of Difference Sequence Spaces", *International Journal of Science and Technology*, **1 (1)**, (2006), 11–14.

[9] TRIPATHY, B.C., ESI, A. and TRIPATHY, B.K., "On a New Type of Generalized Difference Cesàro Sequence Spaces", *Soochow J. Math.*, **31**, (2005), 333–340.

[10] TRIPATHY, B.C. and SEN, M., "On a Class of Sequences Related to the *p*-Normed Space", *Journal of Beijing University of Technology*. 2005, **31**, 112–115.

[11] ZADEH L.A., "Fuzzy Sets", *Information and control*, **8**, (1965), 338–353.

36

A VARIABLE LEAD-TIME PERIODIC REVIEW INVENTORY SYSTEM IN THE PRESENCE OF FUZZINESS AND RANDOMNESS

Oshmita Dey, Debjani Chakraborty

Department of Mathematics, Indian Institute of Technology, Kharagpur-721302, India

1. INTRODUCTION

With the evolution of inventory modeling, it has been noted that randomness alone fails to capture another kind of uncertainty, namely imprecision or fuzziness, which exists in any real life inventory situation. For instance, where fuzziness may arise in the statistical data itself, fuzzy randomness provides a more generalized framework to tackle issues of uncertainty. With this point of view, Dey and Chakraborty (6) developed a fuzzy random periodic review 'order up to *r*' inventory model assuming the lead-time to be a constant. However, with the successful implementation of the Japanese manufacturing philosophy of Just-In-Time (JIT), it has been seen that lead-time may be controlled by adopting any or all of the following means such as added manpower, renew equipment, work overtime, reset layout or choose better logistic etc. Besides reduced lead-time leads to advantages such as reduced safety stock levels, reduced loss due to stockout and provides a competitive advantage thereby yielding a smoother and more efficient supply chain. Developments regarding this issue may be found in (3), (4), (8), etc. and the references therein. In all of the work mentioned above however, the lead-time is assumed to comprise of several components (1). It is further assumed that each of these components can be crashed from their normal duration to a minimum duration by paying some extra amount of money. This additional amount is what is referred to as the crashing cost for each component. And the total crashing cost is assumed to be dependent on the crashing durations and the associated crashing cost for each component. Now for this approach to work, it is essential to have a complete knowledge regarding the components into which the lead-time can be broken, the normal duration of each component, the minimum duration to which each may be crashed and the crashing cost incurred for this purpose. But with the increasing complexity of real life inventory situations, it is very difficult to separate the lead-time into components and to acquire concrete and correct information regarding all these components. Therefore, it is suitable to adopt a more general approach in trying to reduce the lead-time, where even the aggregate information suffices to estimate the crashing cost (1). Keeping this in mind in this paper the crashing cost has been assumed to be a negative exponential function of the lead-time. The principal aim of this paper, therefore, is to extend the methodology proposed in the fuzzy random periodic review system (6) to include the lead-time as one of the control parameters.

2. METHODOLOGY

2.1 Model and Assumptions

Let the annual demand $\tilde{D}(\omega)$ be a fuzzy random variable (7) of the form $[(\tilde{D}1, \tilde{p}_1), (\tilde{D}_2, \tilde{p}_2), \ldots\ldots, (\tilde{D}_n, \tilde{p}_n)]$, where $\tilde{D}_i$'s are triangular fuzzy numbers of the form $(\underline{D}_i, D_i, \bar{D}_i)$ and $\tilde{p}_i$'s their corresponding probabilities, also triangular fuzzy numbers of the same form $(\underline{p}_i, p_i, \bar{p}_i)$ and $i = 1, 2, \ldots, n$ being the number of data considered. The lead-time demand and the lead-time plus one period's demand have been assumed to be connected to the annual demand through the durations of the lead-time and period as in (6). It is to be noted here that the same notations as in (6) have been followed here.

Now as explained earlier, the lead-time crashing cost has been assumed to be a negative exponential function of the lead-time (Ben-Daya and Raouf 1994) of the form:

Crashing cost $R(L) = \alpha e^{-\beta L}$

where the parameters α, β can be estimated easily even if only a few values of the crashing cost are known for given values of L.

Therefore, the total inventory cost is given as

$$\widetilde{C}(\omega)(r, L, T) = \frac{A}{T} + h\left[r - \tilde{D}(\omega)L - \frac{\tilde{D}(\omega)T}{2}\right] +$$

$$\frac{\pi}{T}\bar{M}\left(\tilde{D}_{L+T} - r\right)^{+} + \frac{\alpha}{T}e^{-\beta L} \qquad \text{... (1)}$$

where $\bar{M}\left(\tilde{D}_{L+T} - r\right)^{+}$ gives the expected shortage (6) during each period.

2.2 Defuzzification of the Fuzzy Expected Cost Function Using its Possibilistic Mean Value

Now since the total cost is a fuzzy random variable, therefore its expectation is a unique fuzzy number. For computational purposes, this unique fuzzy number is defuzzified using the concept of possibilistic mean value (2). Let, $E[\widetilde{C}(\omega)]$ denote the fuzzy expected cost function, then we calculate its possibilistic mean value, denoted by $\bar{M}(r, L, T)$. By definition,

$$\bar{M}(r, L, T) = \int_{0}^{1}\alpha\left\{E\left[C_{\alpha}^{-}(\omega)\right] + E\left[C_{\alpha}^{+}(\omega)\right]\right\}d\alpha \qquad \text{... (2)}$$

Therefore,

$$\bar{M}(r, L, T) = h\left[\begin{array}{c} r\sum_{i=1}^{n}\bar{M}(\tilde{p}_i) - \sum_{i=1}^{n}\bar{M}\left(\tilde{D}_i\,\tilde{p}_i\right)L - \\ \frac{1}{2}\sum_{i=1}^{n}\bar{M}\left(\tilde{D}_i\,\tilde{p}_i\right)T \end{array}\right] +$$

$$\sum_{i=1}^{n}\left[\begin{array}{c} \frac{A}{T} + \frac{\pi}{T}\bar{M}\left(\tilde{D}_{L+T} - r\right)^{+} + \\ \frac{\alpha}{T}e^{-\beta L} \end{array}\right]\bar{M}(\tilde{p}_i) \qquad \text{... (3)}$$

2.3 Determination of the Optimal Solution

For each period of review T selected, the values of r, L are evaluated from the following set of equations in each of the cases:

Situation 1: When r lies between $\underline{D}_{L+T}$ and D_{L+T}

$$\bar{M}(r, L, T) = h\begin{bmatrix} r\sum_{i=1}^{n}\bar{M}(\tilde{p}_i) - \sum_{i=1}^{n}\bar{M}(\tilde{D}_i\,\tilde{p}_i)\,L - \\ \frac{1}{2}\sum_{i=1}^{n}\bar{M}(\tilde{D}_i\,\tilde{p}_i)\,T \end{bmatrix} +$$

$$\sum_{i=1}^{n}\left\{\frac{\pi}{T}\left[\begin{matrix}\int_0^1 \alpha D^{+}_{L+T,\alpha}\,d\alpha + \int_{L(r)}^{1} \alpha D^{+}_{L+T,\alpha} \\ d\alpha - r\left[1-0.5L^2(r)\right]\end{matrix}\right]\right\}$$

$$\bar{M}(\tilde{p}_i) + \sum_{i=1}^{n}\left(\frac{A}{T} + + \frac{\alpha}{T}e^{-\beta L}\right)\bar{M}(\tilde{p}_i) \qquad \text{... (4)}$$

$$\frac{\partial \bar{M}}{\partial r} = h - \frac{\pi}{T}\left[1 - \frac{1}{2}L^2(r)\right] = 0 \qquad \text{... (5)}$$

$$\frac{\partial \bar{M}}{\partial L} = (\pi/T)\sum_{i=1}^{n}\bar{M}(\tilde{p}_i)\frac{\partial}{\partial L}\bar{M}\left(\tilde{D}_{L+T} - r\right)^{+}$$

$$-h\sum_{i=1}^{n}\bar{M}(\tilde{D}_i\,\tilde{p}_i) - (\alpha\beta/T)\,e^{-\beta L}$$

$$\sum_{i=1}^{n}\bar{M}(\tilde{p}_i) = 0 \qquad \text{... (6)}$$

Situation 2: When r lies between D_{L+T} and $\bar{D}_{L+T}$

$$\bar{M}(r, L, T) = h\begin{bmatrix} r\sum_{i=1}^{n}\bar{M}(\tilde{p}_i) - \sum_{i=1}^{n}\bar{M}(\tilde{D}_i\,\tilde{p}_i)\,L - \\ \frac{1}{2}\sum_{i=1}^{n}\bar{M}(\tilde{D}_i\,\tilde{p}_i)\,T \end{bmatrix} +$$

$$\sum_{i=1}^{n}\left\{\begin{bmatrix} \frac{A}{T}+\frac{\pi}{T}\begin{bmatrix} \int_{0}^{R(r)} \alpha D_{L+T,\alpha}^{+}\, d\alpha - \\ \frac{1}{2} rR^{2}(r) \end{bmatrix} + \\ \frac{\alpha}{T} e^{-\beta L} \end{bmatrix}\right\} \bar{M}(\tilde{p}_i) \quad \dots (7)$$

$$\frac{\partial \bar{M}}{\partial r} = h - \frac{\pi}{2T} R^2(r) = 0 \quad \dots (8)$$

$$\frac{\partial \bar{M}}{\partial L} = (\pi/T)\sum_{i=1}^{n} \bar{M}(\tilde{p}_i)\frac{\partial}{\partial L}\bar{M}\left(\tilde{D}_{L+T} - r\right)^{+}$$

$$-h\sum_{i=1}^{n} \bar{M}\left(\tilde{D}_i\, \tilde{p}_i\right) - (\alpha\beta/T)\, e^{-\beta L}$$

$$\sum_{i=1}^{n} \bar{M}(\tilde{p}_i) = 0 \quad \dots (9)$$

2.4 Safety Stock

The safety stock, defined as the net inventory just prior to the arrival of stock, is given by $\left[r - (L+1)\,\tilde{D}(\omega)\right]$. Since, for all practical purposes, no manufacturer wants to have a negative safety stock, a non-negativity criterion is imposed on it. Thus, the safety stock criterion is given by

$$SS = r\sum_{i=1}^{n} \bar{M}(\tilde{p}_i) - \sum_{i=1}^{n} \bar{M}\left(\tilde{D}_i \tilde{p}_i\right) \times (L+T) \geq 0 \dots (10)$$

Thus, the optimal period of review, the optimal target inventory level and the optimal lead-time T^*, r^*, L^*, respectively, are determined such that the total cost in the fuzzy sense is minimized. The non-negativity of the safety stock is verified at each step.

3. CONCLUDING REMARKS

A methodology has been proposed in this paper to incorporate the lead-time as one of the decision variables in the fuzzy random periodic review system. The proposed model may be especially applied to inventory situations characterized by lack of historical information as it is suitable to model such situations by means of a discrete fuzzy random variable.

REFERENCES

[1] Ben-Daya M., Raouf A., "Inventory Models Involving Lead Time as Decision Variable". *Journal of the Operational Research Society,* **45**, (1994); 579–582.

[2] Carlsson C., Fuller R., "On Possiblistic Mean Value and Variance of Fuzzy Numbers". *Fuzzy Sets and Systems* **122**, (2001); 315–326.

[3] Chang H.C., Yao J.S., Ouyang L.Y., "Fuzzy Mixture Inventory Model Involving Fuzzy Random Variable Lead Time Demand and Fuzzy Total Demand". *European Journal of Operational Research* **169**, (2006); 65–80.

[4] Chang H.C., Yao J.S., Ouyang L.Y., "Fuzzy Mixture Inventory Model with Variable Lead-Time Based on Probabilistic Fuzzy Set and Triangular Fuzzy Number". *Mathematical and Computer Modelling,* **39**, (2004); 287–304.

[5] Chang H.C., "Lead Time Reduction on the Periodic Review Inventory Model with Fuzzy Backorders". *Information Sciences,* **12**, (2001); 15–27.

[6] Dey O., Chakraborty D., "Fuzzy Periodic Review Inventory System with Fuzzy Random Variable Demand", *European Journal of Operational Research,* **198**, (2009): 113–120.

[7] Kwakernaak H., "Fuzzy Random Variables: Definitions and Theorems". *Information Sciences,* **15**, (1978): 1–29.

[8] Lin Y.J., "A Periodic Review Inventory Model Involving Fuzzy Expected Demand Short and Fuzzy Backorder Rate". *Computers and Industrial Engineering,* **54**, (2008); 666–676.

37

THE USE OF GENETIC ALGORITHM TO PRIORITY BASED FUZZY GOAL PROGRAMMING MODEL FOR PATROL MANPOWER DEPLOYMENT PROBLEM IN TRAFFIC MANAGEMENT SYSTEM: A CASE STUDY

Bijay Baran Pal[1], Somsubhra Gupta[2] and Papun Biswas[3]

[1]*Department of Mathematics, University of Kalyani, Kalyani-741235, W.B., India, E-mail: bbpal18@hotmail.com*

[2]*Department of Information Technology, JIS College of Engineering, Kalyani-741235, W.B., India, E-mail: gsomsubhra@gmail.com*

[3]*Department of Electrical Engineering, Narula Institute of Technology, Kolkata-700109, W.B., India, E-mail: papunbiswas@yahoo.com*

Abstract: *This paper demonstrates how the genetic algorithm (GA) method can be used in the process of solving a fuzzy goal programming (FGP) formulation of patrol manpower allocation problems to the road-segment areas of Metropolitan cities to deter traffic violations and accidents.*

In the model formulation of the problem, the different types of goals involved with the problem are first fuzzily described. Then, a pre-emptive priority based FGP model for achievement of the highest membership value (unity) of the membership goals defined for the fuzzy goals is developed for taking a proper decision in the deployment planning horizon.

In the solution process, an GA method is employed to the FGP model in an iterative manner for goal achievement on the basis of the assigned priorities in the decision making context.

In the decision process, the notion of Euclidean Distance Function is used to obtain an ideal point dependent solution under an appropriate priority structure and thereby to reach a satisfactory decision in the deployment planning situation.

A case example is solved to illustrate the approach.

Keywords: *Fractional Programming, Fuzzy Goal Programming, Genetic Algorithm, Membership Function, Traffic Management.*

1. INTRODUCTION

The rapid rate of development of society along with the upgrade of the technological aspects for production of wide variety of road-vehicles for use in daily life, traffic load is increasing in an alarming rate day-by-day in the major cities of a country.

As a result, accident frequency mainly in the major road-segments areas in most of the cities is increasing day to day throughout the world. It has become a great challenging problem to a traffic control department for enforcing traffic laws and thereby reducing accident rates. Here, it may be mentioned that although different effective measures like construction of new roads, expansion of existing roads, increasing of the road segments, etc. have been taken by most of the metropolitan cities with high traffic density, police departments are facing a lot of problems due to negligence of obeying traffic laws by the common people and proper implementation of traffic enforcement programme.

From the mid-1960s to early 1970s, a considerable amount of efforts was devoted to analytical models in [12, 13, 16, 17, 18] for allocation and deployment of police patrol units in the context of smoothing the traffic flow by enforcing traffic laws. The methodological aspect of the problem studied then was surveyed by Chaiken et. al. in [10] in 1972.

During the late 1970s, patrol manpower planning problems were studied in [7, 14, 21] and widely circulated in the literature. The general mathematical model within the framework of goal programming (GP) in [5, 11] for state patrol deployment problems were introduced by Lee et. al. in [19] in 1979. However, their model involves huge initial allocation to reach an optimal allocation, and hidden nonlinearity in deployment of patrol manpower problem was not discussed there.

To overcome the above difficulty, a gradient search technique within the framework of GP was studied by Taylor et. al. in [20] in 1985, where the fractional in nature of deployment policy goals have been taken into account.

Here, it may be mentioned that, in most of the problems studied previously, patrol manpower deployment activities depend on trial and error method and they are often imprecise in nature. As such, crisp definition of them and use of the conventional approaches create decision trouble to solve such practical multiobjective decision making (MODM) problems.

To overcome the above difficulty, FGP in [2] as an extension of fuzzy programming (FP) in [8] have been studied in (3) in the past.

In this article, a FGP formulation of patrol unit deployment planning problem with various performance achievement functions in fractional form is considered. A solution scheme based on GA in [6, 9] is introduced to reach a satisfactory decision on the basis of priorities of achieving the defined objectives in the decision making environment.

An illustrative case example of the metropolitan city Kolkata, India is solved.

A comparison of the solution with the solution of additive FGP model studied in [4] in the past is made to expound the potential use of the approach.

2. FGP PROBLEM FORMULATION

In a fuzzy decision making situation, description of the objectives as well as constraints can be made for taking proper decision, and that depends on the needs and desires of the decision maker (DM) in the decision environment.

Now, fuzzy goal description is presented in the following Section 2.1.

2.1 Description of Fuzzy Goal

Let b_k be the imprecise aspiration level of the k-th objective $F_k(X)$, $k = 1, 2, \ldots, K$.

Then, the fuzzy goals may appear in one of the forms

$$F_k(X) \gtrsim b_k \quad \text{and} \quad F_k(X) \lesssim b_k,$$

where $X (\geq 0)$ is the vector of decision variables, and where & and. indicate the fuzziness of $\geq$ and $\leq$ restrictions, respectively, in the sense of Zimmermann [8].

Now, in the field of FP, the fuzzy goals are characterized by their respective membership functions.

2.2 Characterization of Membership Function

Let t_{lk} and t_{uk} be the lower and upper-tolerance ranges, respectively, for achievement of the aspired level b_k of the kth fuzzy goal. Then, the membership function (MF), say $\mu_k(X)$, for the fuzzy goal $F_k(X)$ can be characterized as follows [3]:

For & type of restriction, $\mu_k(X)$, takes the form

$$\mu_k(X) = \begin{cases} 1 & \text{if } F_k(X) \geq b_k, \\ \dfrac{F_k(X) - (b_k - t_{lk})}{t_{lk}} & \text{if } b_k - t_{lk} \leq F_k(X) < b_k, \\ 0 & \text{if } F_k(X) < b_k - t_{lk}, \end{cases} \quad \ldots (2.1)$$

where $(b_k - t_{lk})$ represents the lower-tolerance limit for achievement of the stated fuzzy goal.

Again, for · type of restriction, $u_k(X)$ becomes

$$\mu_k(X) = \begin{cases} 1 & \text{if } F_k(X) \geq b_k, \\ \dfrac{(b_k + t_{uk}) - F_k(X)}{t_{lk}} & \text{if } b_k < F_k(X) \leq b_k + t_{uk}, \\ 0 & \text{if } F_k(X) > b_k + t_{uk}, \end{cases} \quad \ldots (2.2)$$

where $(b_k + t_{uk})$ represents the upper-tolerance limit for achievement of the stated fuzzy goal.

Now, construction of the FGP model is presented in the following Section 2.3.

2.3 Construction of FGP Model

The FGP model of the problem under a pre-emptive priority structure for achievement of the highest membership values (unity) can be presented as:

Find X so as to

$$\text{Minimize } Z = [P_1(d^-), P_2(d^-), \ldots, P_r(d^-), \ldots, P_R(d^-)] \text{ and}$$

satisfy

$$\frac{F_k(x) - (b_k - t_{\ell k})}{t_{\ell k}} + d_k^- - d_k^+ = 1$$

$$\frac{(b_k + t_{uk}) - F_k(x)}{t_{uk}} + d_k^- - d_k^+ = 1$$

$$d_k^-, d_k^+ \geq 0, \qquad k = 1, 2, \ldots, K, \qquad \ldots (2.3)$$

where Z represents the vector of R priority achievement function, and d_k^-, d_k^+ are the under and over-deviational variables of the kth goal. $P_r(d^-)$ is a linear function of the weighted under-deviational variables where $P_r(d^-)$ is of the form [3]:

$$P_r(d^-) = \sum_{k=1}^{K} w_{rk}^- d_{rk}^-, (k = 1, 2, \ldots, K; R \leq K),$$

where d_{rk}^- is renamed for d_k^- to represent it at the rth priority level, w_{rk}^- (≥ 0) is the numerical weight associated with d_{rk}^- and it represents the weight of importance of achieving the aspired level of the kth goal relative to other which are grouped at the rth priority level [3].

Now, the use of the Euclidean distance function for proper priority structure selection is presented in the Section 2.4.

2.4. Euclidean Distance Function for Priority Structure Selection

In the context of assigning priorities to the goals in the above FGP formulation, it may be mentioned here that the DM is always confused with that of assigning the proper priorities to the goals, because the goals are often conflict among themselves for achieving the aspired levels in the decision making environment. To overcome such a situation, the concept using the Euclidean distance function proposed in [15] for group decision analysis can be used for selection of proper priority structure under which most satisfactory decision can be obtained.

In the patrolman allocation situation, since the highest membership value of each fuzzy goal is unity, the ideal point would be a vector with each element equal to 1.

The *Euclidean Distance Function* can be presented as

$$D_p = \left[\sum_{k=1}^{K} \left(1 - \mu_k^p\right)^2 \right]^{1/2} \qquad \text{... (2.4)}$$

where μ_k^p represents the achieved membership value of the kth goal under the pth priority structure of the goal.

Here, it can easily be realized that the solution which is closest to the ideal point must correspond to:

$$\min_{p \in P} \{D_p\} = D_q \text{ (say)}, \qquad 1 \le q \le P$$

Then, qth priority structure would be considered as an appropriate one to arrive at the most satisfactory decision.

Now, the decision variables and different types of parameters involved with the problem are defined in the following Section 2.5.

2.5 Definition of Variables and Parameters

Decision Variables

(i) Independent Variable

x_{ij} = Deployment of patrol units to the road-segment area i during the shift j.

(ii) Dependent variable

AR_{ij} = Reduction of accident rate contributed by the road-segment area i during the shift j.

PC_{ij} = Number of physical-contacts made by a patrolman in the road-segment area i during shift j.

SC_{ij} = Number of sight-contacts made by a patrolman in the road-segment area i during shift j.

Fuzzy Resource Parameters

E = Estimated total number of patrol units required for the various road-segment areas in all the shifts of the time period.

E_i = Estimated total number of patrol units required in the road-segment area i in all the shifts of the time period.

F_j = Estimated total number of patrol units required during the shift j in all the road segment areas of the period.

AR = Total accident reduction level contributed by all the road-segment areas of the period.

r_i = Accident reduction level desired at the road-segment area i in all the shifts of the period.

PC = Goal level for the total physical contacts in all the road segment areas of the period.

p_i = Goal level for physical contacts at the road segment area i in all the shifts of the period.

SC = Goal level for the total sight contacts in all the road segment areas in all the shifts.

s_i = Goal level for sight contacts at the road segment area i in all the shifts of the period.

CE = Estimated total cash expenditure for deployment of the patrol units in all the road segment areas during all the shifts of the period.

Crisp Coefficients

C_{ij} = Estimated cost for deployment of a patrol unit to the road-segment area i during the shift j.

Now, for the defined variables and parameters, the algebraic structures of the fuzzy goals are described in the following Section 2.6.

2.6 Fuzzy Goal Description

(a) Patrol Manpower Requirement Goal

A certain level of manpower should be provided to deploy to the road-segment areas during different shifts of the period.

The fuzzy goal expression appears as:

$$\sum_{i=1}^{m}\sum_{j=1}^{n} X_{ij} \,.\, E \qquad \dots (2.5)$$

(*i*) ***Segment-wise allocation:*** Depending on the accident frequency and traffic density, a minimum number of patrol units to be deployed to each of the road-segment areas in all the shifts.

The goal expression takes the form

$$\sum_{j=1}^{n} X_{ij} \;\&\; E_i, \qquad i = 1, 2, \dots, m \qquad \dots (2.6)$$

(*ii*) ***Shift-wise allocation:*** For smoothing the traffic flow in all the road segment areas, a certain number of patrol units need be provided in each shift to all the road-segment areas. The goal expression appears as

$$\sum_{i=1}^{m} X_{ij} \;\&\; F_j, \qquad j = 1, 2, \dots\dots, n. \qquad \dots (2.7)$$

(b) Deployment Performance Goals

It is to be observed that the reduction of accident rate as well as the other performance measures of increase with the increase of patrol units to the road-segment areas. But, increase of them beyond a certain level does not give any additional effect and that depends on the traffic density and accident frequency of a road-segment-area.

In context of the above, a general mathematical expression for relationship with x_{ij} and each of the AR_{ij}, PC_{ij} and SC_{ij} can be defined as follows:

Performance Measure Function

The expression for measuring deployment performance appears as

$$y_{ij} = a_{ij} - \frac{b_{ij}}{x_{ij}}, \qquad \dots (2.8)$$

where $x_{ij} \geq 0$ and integer $\forall\, i$ and j, and where a_{ij} and b_{ij} are the estimated parameters.

In the expression (2.8), it is to be followed that when $x_{ij} = 0$, y_{ij} becomes infeasible. As such $y_{ij} = 0$ is taken into consideration as and when $x_{ij} = 0$.

Now, from the Expression in (2.8), it follows that the measure of performance becomes asymptotic for deployment after a certain number of patrol units, which is shown in the Fig. 1.

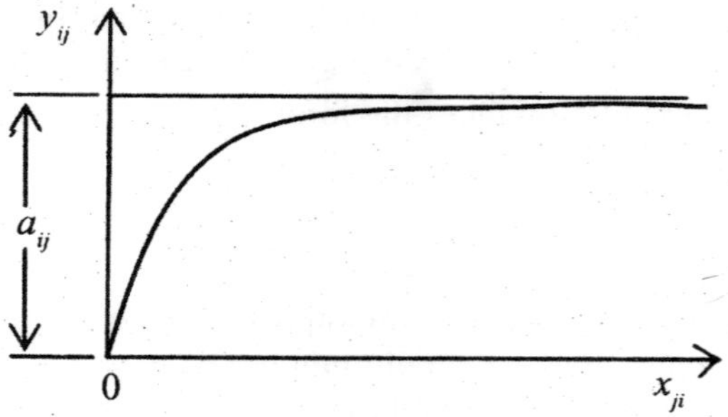

Fig. 1: *Deployment Efficiency Measure*

The potential use of the Expression in (2.8) in different phases of traffic operations activities is presented as follows.

(c) *Accident Reduction Goals*

The expression of accident rate reduction, which increases with the increase of patrol units, appears as

$$AR_{ij} = R_{ij} - \frac{r_{ij}}{x_{ij}} \qquad \text{... (2.9)}$$

where R_{ij} and r_{ij} ($R_{ij} > r_{ij}$) are the estimated parameters associated with AR_{ij} and x_{ij}.

(*i*) Segment-wise accident reduction goals. The reduction of accident rate at the *i*th road-segment in all the shifts appears as

$$\sum_{j=1}^{n}\left(R_{ij} - \frac{r_{ij}}{x_{ij}}\right) \& \, r_i, \qquad i = 1, 2, \ldots m \qquad \text{... (2.10)}$$

(*ii*) Total accident reduction goal. The goal expression for reduction of overall accident rate for all the road-segments areas during all the shifts of the period takes the form

$$\sum_{i=1}^{m}\sum_{j=1}^{n}\left(R_{ij} - \frac{r_{ij}}{x_{ij}}\right) \& \, AR \qquad \text{... (2.11)}$$

(d) Physical-Contact Goals

The most important activity of the patrol units is to make physical-contact to deter traffic violations and accidents. Physical-contact indicates the direct contact to the vehicles for the spot-checking of driving license, car registration, road tax clearance, etc.

Now, following the Expression (2.9), the expression for PC_{ij} takes the form

$$PC_{ij} = P_{ij} - \frac{p_{ij}}{x_{ij}}, \quad i = 1, 2,.., m; j = 1, 2, ..., n \qquad ... (2.12)$$

where P_{ij} and p_{ij} $(P_{ij} \geq p_{ij})$ are the estimated parameters.

Here, it is to be followed that the number of physical contacts increases with the increase of x_{ij}, but at a decreasing rate.

Then, the goal expressions can be defined as follows:

(*i*) ***Segment-wise physical contact goals:*** Since the physical-contact operations are made in terms of number of contacts and on the basis of traffic density, segment-wise physical-contact goal expression appears as

$$\sum_{j=1}^{n} \left(P_{ij} - \frac{p_{ij}}{x_{ij}} \right) \gtrsim p_i, \quad i = 1, 2, ..., m \qquad ... (2.13)$$

(*ii*) ***Total physical-contact goal:*** To measure the overall performance against deployment of patrol units, the goal expression for total physical-contact takes the form

$$\sum_{i=1}^{m} \sum_{j=1}^{n} \left(P_{ij} - \frac{p_{ij}}{x_{ij}} \right) \gtrsim PC \qquad ... (2.14)$$

(e) Sight-Contact Goals

Beyond physical-contact, sight-contact should be an integral and vital part to smooth running of the vehicle, preventing against violation of general traffic rules during crossing of a road-segment areas.

Now, similar to the physical-contact, the expression of SC_{ij} takes the form

$$SC_{ij} = S_{ij} - \frac{s_{ij}}{x_{ij}}, i = 1, 2, ..., m; j = 1, 2, ..., n \quad ... (2.15)$$

where S_{ij} and s_{ij} ($S_{ij} > s_{ij}$) are estimated parameters. The goal expressions can be defined as follows.

(*i*) ***Segment-wise sight-contact goals:*** For smoothing the traffic operations, the goal expression for segment-wise sight-contact appears as

$$\sum_{j=1}^{n} \left(S_{ij} - \frac{s_{ij}}{x_{ij}} \right) \& s_i, i = 1, 2, ..., m, \quad ... (2.16)$$

(*ii*) ***Total sight-contact goal:*** Similar to the total physical-contact goal, the goal expression here also appears as

$$\sum_{i=1}^{m} \sum_{j=1}^{n} \left(S_{ij} - \frac{s_{ij}}{x_{ij}} \right) \& SC \quad ... (2.17)$$

(*f*) Cash Expenditure Goal

An estimated amount of money should also be provided for incurring different costs for patrolling to the road-segment areas in different shifts of the period.

The goal expression appears as

$$\sum_{i=1}^{m} \sum_{j=1}^{n} c_{ij} x_{ij} \cdot CE \quad ... (2.18)$$

Now, it is to be observed that the traffic operational goals are fractional in form to overcome the difficulty associated with linear approximation approach, the GA approach as a goal satisfier (6, 9) for multiobjective decision analysis is introduced here in the solution search process.

The proposed GA method is presented in the following Section 3.

3. GA BASED SOLUTION METHOD

The basic steps of the GA procedure with the core functions adopted in the solution search process are presented via the following steps.

Step 1: Representation and initialization: Let V denote the binary coded representation of chromosome in a population as $V = (x_1, x_2, ..., x_n)$. The population size is defined by pop-size, and pop-size chromosomes are randomly initialized in its search domain.

Step 2: Fitness function: The fitness value of each chromosome is judged by the value of an objective function. The fitness function is defined as

$$\text{eval}\,(V_L) = (Z)_r = \left(\sum_{k=1}^{K} w_{rk}^{-}\, d_{rk}^{-}\right)_L, \qquad \text{... (3.1)}$$

where Z_r represents the rth priority factor of the goal achievement of the function Z in (2.3) and where the subscript L in (3.1) refers to the fitness value of the selected Lth chromosome, $L = 1, 2, ...,$ pop_size. The best chromosome with largest fitness value at each generation is determined as

$$V^* = \min\,\{\text{eval}\,(V_L) \mid L = 1, 2, ..., \text{pop\_size}\},$$

which depends on searching of the best value of an objective.

Step 3: Selection: The simple roulette-wheel scheme (6) is used for selecting two parents for mating purposed in the genetic search process.

Step 4: Crossover: The parameter P_c is defined as the probability of crossover. The simple arithmetic crossover operation (single point crossover) of a genetic system is applied here. The arithmetic crossover for two parents $V_1, V_2 \in S$ is defined as

$$X_1 = \alpha_1 V_1 + \alpha_2 V_2, \quad X_2 = \alpha_2 V_1 + \alpha_1 V_1,$$

for producing two offspring X_1 and X_2, where $\alpha_1 . \alpha_2 \geq 0$ with $\alpha_1 + \alpha_2 = 1$ always belong to S, and where S is a convex set.

Here, a chromosome is selected as a parent, if for a defined random number $r \in (0, 1)$, $r < P_c$ is satisfied.

Here, arithmetic crossover for two parents E_1, E_2 is defined as

$$X_1 = \alpha_1 E_1 + \alpha_2 E_2, X_2 = \alpha_2 E_1 + \alpha_1 E_1$$

for producing two offspring X_1 and X_2 where $\alpha_1, \alpha_2 \geqslant 0$ with $\alpha_1 + \alpha_2 = 1$

Step 5: Mutation: As in the conventional GA scheme, a parameter P_m of the genetic system is defined as the probability of mutation. The mutation operation is performed on a bit-by-bit basis, where for a random number $r \in (0, 1)$, a chromosome is selected for mutation provided that $r < P_m$.

Step 6: Termination: The execution of the whole process terminates when the fittest chromosome is identified at a certain generation number in the solution search process.

Now, the FGP formulation of the problem by defining the membership functions of the fuzzy goals and thereby solving the problem by using the proposed GA is demonstrated through the case example presented in the Section 4.

4. A DEMONSTRATIVE CASE EXAMPLE

The manpower deployment problem of the department of Traffic Police of the Eastern Metropolitan City Kolkata of India is considered to illustrate the potential use of the proposed model.

The required data were collected from the Annual Review Bulletin (2007) (1) published by the Kolkata Traffic Police Department. Some other relevant information provided in the Bulletins published in the few preceding years were also considered.

The City Calcutta (presently renamed: Kolkata) was the capital of the British Regime in colonial India and it is over 300 years old. It is one of the most densely populated cities in the world with a very high vehicle density.

In the patrolmen deployment situation, since the overall yearly performance is measured on the basis of day-to-day performance, the period under consideration for traffic operations is considered as day. The record shows that the most accident cases occur during the time interval 6–30 hrs. – 22.00 hrs. Generally, the four shifts: 6 am to 10 am, 10 am to 2 pm, 2 pm to 6 pm and 6 pm to 10 pm are arranged to deploy patrolmen to different road segment areas on the basis of traffic density and accident frequency. They are denoted by the shifts members 1, 2, 3 and 4, respectively. The four most accident-prone areas are: BT road, AJC road, Strand road and EM By-pass. They are designated by the road-segment areas: 1, 2, 3, 4, respectively.

An instance of a patrolman on duty in a road-segment area is shown in Fig. 2.

Fig. 2: *An Instance of a Patrolman on Duty*

Now, following the records, the different fuzzy aspiration levels (A^L) with their associated tolerance limits, upper-tolerance (U^T) and lower-tolerance (L^T) limits, as well as the crisp coefficients are presented in the following Tables 1-4.

Table 1: *Total Utilization Resource Data*

Resource	*E* (A^L, U^T)	*AR* (A^L, L^T)	*PC* (A^L, L^T)	*SC* (A^L, L^T)	*CE* (A^L, U^T)
Utilization levels	(35, 50)	(1000, 900)	(3.5, 3)	(2930, 2800)	(2780, 3000)

The fuzzy resource parameters *E*, *AR*, *PC*, *SC*, and *CE* are presented in the Section 2.5

Table 2: *Segment-Wise Resource Utilization Data*

Segment area (i)	E_i (A^L, U^T)	r_i (A^L, L^T)	p_i (A^L, L^T)	s_i (A^L, L^T)
1	(12, 9)	(1.15, 1.1)	(280, 250)	(790, 750)
2	(10, 8)	(1.02, 1)	(290, 250)	(770, 750)
3	(8, 6)	(0.97, 0.9)	(250, 200)	(700, 650)
4	(8, 7)	(0.87, 0.8)	(240, 200)	(670, 650)

Table 3: *Shift-Wise Resource Utilization Data*

Shift (j)	*1*	*2*	*3*	*4*
$F_j(A^L, L^T)$	(10, 8)	(12, 10)	(8, 6)	(9, 7)

Note: $F_j(.)$ = Shift-wise aspiration level and lower tolerance limit.

Table 4: *Data Description of C_{ij} (Cost in Rs.) for Deployment of Patrol Unit*

Shift (j) *Road segment (i)*	*1*	*2*	*3*	*4*
1	75	75	75	75
2	80	80	80	80
3	65	65	65	65
4	70	70	70	70

Now, regarding traffic operations, it is to be followed that there is no additional effect for deployment of more than 5 patrolmen in any road-segment area during any shift. As such, the range of x_{ij} is considered as

$$1 \leq x_{ij} \leq 5, \quad i = 1, 2, ..., m; \quad j = 1, 2, ..., n. \tag{4.1}$$

The different parameters involved with the traffic operations are estimated by using the conventional two point estimates and presented in the Table 5.

Table 5: *Parameters Values for AR_{ij}, PC_{ij}, SC_{ij}*

Road segment i	*Shift j*	*Accident Reduction (AR_{ij})*		*Physical Contacts (PC_{ij})*		*Sight Contacts (SC_{ij})*	
		a	*b*	*a*	*b*	*a*	*b*
1	1	0.26	0.14	77	60	220	175
1	2	0.49	0.25	106	82	295	225
1	3	0.21	0.12	62	52	169	154
1	4	0.34	0.20	90	69	266	206
2	1	0.24	0.14	76	60	217	174
2	2	0.39	0.17	100	77	289	221
2	3	0.20	0.10	60	50	162	150
2	4	0.32	0.20	87	66	261	205
3	1	0.23	0.12	71	57	194	169

Table Contd.

Table 5 Contd.

3	2	0.37	0.21	97	75	277	212
3	3	0.19	0.09	54	46	137	127
3	4	0.30	0.18	82	62	237	187
4	1	0.23	0.13	68	55	188	165
4	2	0.35	0.22	93	71	272	210
4	3	0.17	0.08	50	42	125	116
4	4	0.24	0.15	81	62	225	175

Now, using the data presented in the Tables 1–5, the membership goals for the defined fuzzy goals can be constructed by following the fuzzy goal Expressions in (2.1)–(2.2).

1. ***Patrolmen Allocation Goals:*** The membership goal expressions appear as:

   (i) Total patrolmen allocation goals: μ_1: $3.33 - 0.0667\,(x_{11} + x_{12} + x_{13} + x_{14} + x_{21} + x_{22} + x_{23} + x_{24} + x_{31} + x_{32} + x_{33} + x_{34} + x_{41} + x_{42} + x_{43} + x_{44}) + d_1^- - d_1^+ = 1$. In the same way, the membership goals for the other defined fuzzy goals are determined as follows:

   (ii) Segment-wise patrolmen allocation goals:

$$\mu_2 : 0.33\left(x_{11} + x_{12} + x_{13} + x_{14}\right) - 3 + d_2^- - d_2^+ = 1,$$

$$\mu_3 : 0.5\left(x_{21} + x_{22} + x_{23} + x_{24}\right) - 4 + d_3^- - d_3^+ = 1,$$

$$\mu_4 : 0.5\left(x_{31} + x_{32} + x_{33} + x_{34}\right) - 3 + d_4^- - d_4^+ = 1,$$

$$\mu_5 : \left(x_{41} + x_{42} + x_{43} + x_{44}\right) - 7 + d_5^- - d_5^+ = 1,$$

   (iii) Shift-wise patrolmen allocation goals:

$$\mu_6 : 0.5\left(x_{11} + x_{21} + x_{31} + x_{41}\right) - 4 + d_6^- - d_6^+ = 1,$$

$$\mu_7 : 0.5\left(x_{12} + x_{22} + x_{32} + x_{42}\right) - 5 + d_7^- - d_7^+ = 1,$$

$$\mu_8 : 0.5\left(x_{13} + x_{23} + x_{33} + x_{43}\right) - 3 + d_8^- - d_8^+ = 1,$$

$$\mu_9 : 0.5\left(x_{14} + x_{24} + x_{34} + x_{44}\right) - 3 + d_9^- - d_9^+ = 1,$$

2. ***Accident Reduction Goals***

   (i) Segment-wise accident reduction goals:

$$\mu_{10} : 0.67\left[0.2 - \left(\frac{0.14}{x_{11}} + \frac{0.25}{x_{12}} + \frac{0.12}{x_{13}} + \frac{0.2}{x_{14}}\right)\right] + d_{10}^- - d_{10}^+ = 1,$$

$$\mu_{11} : 50\left[0.15-\left(\frac{0.14}{x_{21}}+\frac{0.17}{x_{22}}+\frac{0.10}{x_{23}}+\frac{0.20}{x_{24}}\right)\right]+$$

$$d_{11}^{-}-d_{11}^{+}=1,$$

$$\mu_{12} : 14.28\left[0.19-\left(\frac{0.12}{x_{31}}+\frac{0.21}{x_{32}}+\frac{0.09}{x_{33}}+\frac{0.18}{x_{34}}\right)\right]+$$

$$d_{12}^{-}-d_{12}^{+}=1,$$

$$\mu_{13} : 14.28\left[0.19-\left(\frac{0.13}{x_{41}}+\frac{0.22}{x_{42}}+\frac{0.08}{x_{43}}+\frac{0.15}{x_{44}}\right)\right]+$$

$$d_{13}^{-}-d_{13}^{+}=1,$$

(ii) *Total accident reduction goal:* The goal expression takes the form:

$$\mu_{14} : 2\left[4.53-\left(\begin{array}{l}\frac{0.14}{x_{11}}+\frac{0.25}{x_{12}}+\frac{0.12}{x_{13}}+\frac{0.2}{x_{14}}+\frac{0.14}{x_{21}}+\\ \frac{0.17}{x_{22}}+\frac{0.10}{x_{23}}+\frac{0.20}{x_{24}}+\frac{0.12}{x_{31}}+\frac{0.21}{x_{32}}+\\ \frac{0.09}{x_{33}}+\frac{0.18}{x_{34}}+\frac{0.13}{x_{41}}+\frac{0.22}{x_{42}}+\frac{0.08}{x_{43}}+\\ \frac{0.15}{x_{44}}\end{array}\right)\right]+$$

$$d_{14}^{-}-d_{14}^{+}=1,$$

3. *Physical-Contact Goals*

(i) *Segment-wise physical-contact goals:*

$$\mu_{15} : 0.033\left[85-\left(\frac{60}{x_{11}}+\frac{82}{x_{12}}+\frac{52}{x_{13}}+\frac{69}{x_{14}}\right)\right]+d_{15}^{-}-d_{15}^{+}=1,$$

$$\mu_{16} : 0.025\left[73-\left(\frac{60}{x_{21}}+\frac{77}{x_{22}}+\frac{50}{x_{23}}+\frac{66}{x_{24}}\right)\right]+d_{16}^{-}-d_{16}^{+}=1,$$

$$\mu_{17} : 0.02\left[104 - \left(\frac{57}{x_{31}} + \frac{75}{x_{32}} + \frac{46}{x_{33}} + \frac{62}{x_{34}}\right)\right] - d_{17}^{-} - d_{17}^{+} = 1,$$

$$\mu_{18} : 0.025\left[92 - \left(\frac{55}{x_{41}} + \frac{71}{x_{42}} + \frac{42}{x_{43}} + \frac{62}{x_{44}}\right)\right] + d_{18}^{-} - d_{18}^{+} = 1,$$

(*ii*) *Total physical-contact goal*

$$\mu_{19} : 0.01\left[1254 - \left(\begin{array}{c}\frac{60}{x_{11}} + \frac{82}{x_{12}} + \frac{52}{x_{13}} + \frac{69}{x_{14}} + \frac{60}{x_{21}} + \\ \frac{77}{x_{22}} + \frac{50}{x_{23}} + \frac{66}{x_{24}} + \frac{57}{x_{31}} + \frac{75}{x_{32}} + \\ \frac{46}{x_{33}} + \frac{62}{x_{34}} + \frac{55}{x_{41}} + \frac{71}{x_{42}} + \frac{42}{x_{43}} + \\ \frac{62}{x_{44}}\end{array}\right)\right] +$$

$$d_{19}^{-} - d_{19}^{+} = 1,$$

4. *Sight-Contact Goals*

(*i*) *Segment-wise sight-contact goals:*

$$\mu_{20} : 0.025\left[200 - \left(\frac{175}{x_{11}} + \frac{225}{x_{12}} + \frac{154}{x_{13}} + \frac{204}{x_{14}}\right)\right] +$$

$$d_{20}^{-} - d_{20}^{+} = 1,$$

$$\mu_{21} : 0.05\left[179 - \left(\frac{174}{x_{21}} + \frac{221}{x_{22}} + \frac{150}{x_{23}} + \frac{205}{x_{24}}\right)\right] +$$

$$d_{21}^{-} - d_{21}^{+} = 1,$$

$$\mu_{22} : 0.02\left[130 - \left(\frac{169}{x_{31}} + \frac{277}{x_{32}} + \frac{127}{x_{33}} + \frac{187}{x_{34}}\right)\right] +$$

$$d_{22}^{-} - d_{22}^{+} = 1,$$

$$\mu_{23} : 0.05\left[160 - \left(\frac{165}{x_{41}} + \frac{210}{x_{42}} + \frac{116}{x_{43}} + \frac{175}{x_{44}}\right)\right] +$$

$$d_{23}^{-} - d_{23}^{+} = 1,$$

(ii) Total sight-contact goal:

$$\mu_{24}:0.007\left[3469-\left(\begin{array}{l}\frac{175}{x_{11}}+\frac{225}{x_{12}}+\frac{154}{x_{13}}+\frac{206}{x_{14}}+\frac{174}{x_{21}}+\\ \frac{221}{x_{22}}+\frac{150}{x_{23}}+\frac{205}{x_{24}}+\frac{169}{x_{31}}+\frac{277}{x_{32}}+\\ \frac{127}{x_{33}}+\frac{187}{x_{34}}+\frac{165}{x_{41}}+\frac{210}{x_{42}}+\frac{116}{x_{43}}+\\ \frac{175}{x_{44}}\end{array}\right)\right]+$$

$$d_{24}^{-}-d_{24}^{+}=1,$$

5. *Cash Expenditure Goal*

μ_{25}: $13.64 - 0.0045\,[75\,(x_{11} + x_{12} + x_{13} + x_{14}) + 80\,(x_{21} + x_{22} + x_{23} + x_{24}) + 65\,(x_{31} + x_{32} + x_{33} + x_{34}) + 70\,(x_{41} + x_{42} + x_{43} + x_{44})] + d_{25}^{-} - d_{25}^{+} = 1.$

Now, for the stated membership goals of the problem and the decision restriction given in (4.1), using the achievement function expression in (2.3), the executable FGP model under the four assigned priorities are obtained as:

Find $\{x_{ij} \mid i = 1, 2, 3, 4;\quad j = 1, 2, 3, 4\}$ so as to

Minimize:

$$Z = [P_1\,(0.067d_1^- + 0.33d_2^- + 0.5d_3^- + 0.5d_4^- + 0.5d_5^- + 0.5d_6^- + 0.5d_7^- + 0.5d_8^- + 0.5d_9^-),\ P_2\,(0.0045d_{10}^-),\ P_3\,(20d_{11}^- + 50d_{12}^- + 14.29d_{13}^- + 14.29d_{14}^- + 2d_{15}^-),\ P_4\,(0.033d_{16}^- + 0.025d_{17}^- + 0.02d_{18}^- + 0.025d_{19}^- + 0.01d_{20}^- + 0.025d_{21}^- + 0.05d_{22}^- + 0.025d_{23}^- + 0.05d_{24}^- + 0.0077d_{25}^-)] \tag{4.2}$$

and satisfy the defined membership goal expressions.

Then, the GA method presented in the Section 3 is employed to solve the problem. The achievement function in (4.2) here appears as the fitness function in the solution search process.

In the genetic search process, the following parameter values are introduced.

- probability of crossover $P_c = 0.8$

- probability of mutation $P_m = 0.08$
- population size = 100
- chromosome length = 30

The GA based program is designed in Programming Language C. The parametric values are determined from specific ranges for each of them in the search space. The execution is done in an Intel Pentium IV with 2.66 GHz. Clock-pulse and 1GB RAM.

The model solution under the three priority structures for goal achievement is presented in the Table 6.

Table 6: *Priority Structure Selection*

Run	*Priority Structure*	*Shift -> Road Segment*	*1*	*2*	*3*	*4*
	$P_1\left(\sum_{k=1}^{10} w_{rk}^- d_{rk}^-\right)$,	B. T Road	2	2	3	2
	$P_2\,(w_{10k}\, d_{10k}^-)$,	A.J.C Road	3	2	4	2
1	$P_3\left(\sum_{k=11}^{15} w_{rk}\, d_{rk}\right)$,	Strand Road	4	5	3	3
	$P_4\left(\sum_{K=16}^{24} w_{rk}\, d_{rk}^-\right)$	EM Bypass	2	4	4	2
	$P_2\left(\sum_{k=1}^{10} w_{rk}^- d_{rk}^-\right)$,	B. T Road	4	5	3	3
	$P_3\,(w_{10k}\, d_{10k}^-)$,	A.J.C Road	2	4	4	2
2	$P_1\left(\sum_{k=11}^{15} w_{rk}\, d_{rk}\right)$,	Strand Road	2	2	3	2
	$P_4\left(\sum_{K=16}^{24} w_{rk}\, d_{rk}^-\right)$	EM Bypass	3	2	4	2
	$P_3\left(\sum_{k=1}^{10} w_{rk}^- d_{rk}^-\right)+$	B. T Road	4	4	3	4

Table Contd.

Table 6 Contd.

3	$P_1(w_{10k}\, d^-_{10k}) +$ $P_2\left(\sum_{k=11}^{15} w_{rk}\, d_{rk}\right) +$ $P_4\left(\sum_{K=16}^{24} w_{rk}\, d^-_{rk}\right)$	A.J.C Road	3	5	4	2
		Strand Road	3	5	3	3
		EM Bypass	5	4	4	2

Now, using the *Euclidean Distance Function*, to the solutions under the three different priority structures, it is found that the priority structure under the Run-2 is the appropriate one.

Hence, the optimal solution corresponds to the Run-2 of the model execution process.

Note: If the additive FGP approach in [4] is used to the problem, where maximization of the sum of the defined MF μ_i such that $\mu_i \leq$ the associated membership function expression and $\mu_i \leq 1$ is considered, then using the software LINGO (var. 6.0) the obtained solution is presented in the Table 7.

Table 7: *Patrolmen Allocation Under the Additive FGP*

Shift / *Road segment area*	*6 am to 10 am*	*10 am to 2 pm*	*2 pm to 6 pm*	*6 pm to 10 pm*
B. T Road	1	4	5	2
A. J. C Road	3	5	1	1
Strand Road	5	1	1	1
EM By-pass	1	1	1	5

A comparison of the model solution with the result in the Table 7 shows that of better deployment plan is achieved here under the proposed approach in the decision making environment.

5. CONCLUSION

The main advantage of using the proposed solution approach is that the computational complexity due to involvement of inherent fractional objectives and the computational load for the use of linear approximation technique as involved in the traditional approaches can be avoided here with the use of GA scheme. Again for the use of the Euclidean distance function in the process of appropriate priority selection, a satisfactory solution can always be obtained on the basis of the needs and desires of the DM in the decision making environment.

In future study, the extension of proposed approach for proper patrol manpower deployment plan for traffic controlling in different emergency situations can be considered.

However, it is hoped that the solution approach presented here can contribute to future research works for patrolmen deployment planning in the current complex decision analysis for traffic management systems.

ACKNOWLEDGEMENT

The authors are thankful to the anonymous Referees and the Conference Chair, ICRFSC2009 for their helpful suggestions and comments which have lead to improve the quality of presentation of the paper. The authors are grateful to the Kolkata Traffic Police Department, Government of West Bengal, India for providing the data and various supports to carry out the research work.

REFERENCES

[1] Annual Review Bulletin, "Kolkata Traffic Police, Kolkata, Government of West Bengal", India, 2007.

[2] B.B. Pal and B.N. Moitra, "A Goal Programming Procedure for Solving Problems with Multiple Fuzzy Goals Using Dynamic Programming", *European Journal of Operational Research*, **144 (3)**, pp. 480–491, 2002.

[3] B.B. Pal, B.N. Moitra, and U. Maulik, "A Goal Programming Procedure for Fuzzy Multiobjective Linear Fractional Programming Problem". *Fuzzy Sets and Systems*, **139**, pp. 395–405, 2003.

[4] B.B. Pal, S. Gupta. A. Mukhopadhyay, P. Biswas, "An Application of Genetic Algorithm Method for Solving Patrol Manpower Deployment Problems Through Fuzzy Goal Programming in Traffic Management System: A Case Study", *in Proc. of International Conference on Operations Research Applications in Engineering and Management* (ICOREM 2009), pp. 253–283, 2009.

[5] C. Romero. "A Survey of Generalized Goal Programming", *European Journal of Operational Research*, **25 (2)**, pp. 183–191, 1986.

[6] D.E. Goldberg. "Genetic Algorithms in Search. Optimization and Machine Learning". Addison-Wesley. Reading, MA, 1989.

[7] D.G. Olson and G.P. Wright, "Models for Allocating Police Preventive Patrol Effort", *Operational Research Quarterly*, **26**, **(4)**, pp. 703–715, 1975.

[8] H.J. Zimmerman, "Fuzzy Sets, Decision Making and Expert Systems", Kluwer Academic Publishers, Boston, Dordrecht, Lamcaster. 1987.

[9] J.H. Holland, "Adaptation in Natural and Artificial Systems". University of Michigan Press. Ann Arbor, MI 1975.

[10] J.M.Chaiken and R.C. Larson, "Methods for Allocating Urban Emergency Units: A Survey", *Management Science*, **19**, **(4)**, pp. 110–130. 1972.

[11] J.P. Ignizio, "Goal Programming and Extensions", Lexington Books, Lexington MA, 1976.

[12] J.T. McEwen, Project Director. "Allocation of Patrol Manpower Resources in the Saint Louis Police Department", **I and II**, Report of the St. Louis Police Department, 1968.

[13] J.T. McEwen, "A Mathematical Model for Prediction of Police Patrol Workload", *Paper presented at TIMS/ORSA Joint Nation Meeting*, May 1968.

[14] P.J. Kolesar, K.L. Rider, T.B. Crabill and W.E. Walker. "A Quing Linear Programming Approach to Scheduling Police Patrol Cars", *Operations Research*. **23**, **(6)**, pp. 1045–1062, 1975.

[15] P.L. Yu, "A Class of Solutions for Group Decsion Problems", *Management Science*, **19 (8)**, pp. 936–946, 1973.

[16] R.C. Larson, "Models for the Allocation of Urban Police Patrol Forces", *MIT Operations Research Center Technical Report No. 44*, 1969.

[17] R.P. Shumate and R.F. Crowther, "Quantitative Methods for Optimizing the Allocation of Police Resources", *Journal of Criminal Law, Criminal Police Science*. **(57)**. pp. 197–206, 1966.

[18] R.F. Crowther and F. Richard. "The use of Computer System for Police Manpower Allocation in St. Louis Missouri", *Department of Police Administration*", Indiana University, 1974.

[19] S.M. Lee, S.L. Franz and J.A. Wyne, "Optimizing State Patrol Manpower Allocation", *Journal of the Operation Research Society*, **30 (10)**, pp. 885–896, 1979.

[20] W.B. Taylor, J.L. Moore, R.E. Clayton, D. Roseoe and R.R. Terry, "An Integer Nonlinear Goal Programming Model for the Deployment of State Highway Patrol Units", **31 (11)**, pp. 1335–1347, 1985.

[21] W. Chow. "An Optimization Model for Freeway Stationary Emergency Service Systems", *Transportation Science*, **10**, pp. 20–33, 1976.

38

A GENETIC ALGORITHM BASED GOAL PROGRAMMING APPROACH FOR SOLVING INTERVAL VALUED BILEVEL PROGRAMMING PROBLEMS

Bijay Baran Pal[1], Somsubhra Gupta[2] and Debjani Chakraborti[3]

[1]*Department of Mathematics, University of Kalyani*
Kalyani-741235, W.B., India.
E-mail: bbpal18@hotmail.com

[2]*Department of Information Technology*
JIS College of Engineering, Kalyani-741235, W.B., India
E-mail: gsomsubhra@gmail.com

[3]*Department of Mathematics, Narula Institute of Technology*
Kolkata-700109, W.B., India.
E-mail: debjani_333@yahoo.co.in

Abstract: *In this article, the efficient use of genetic algorithm (GA) for solving goal programming (GP) formulations of interval valued bilevel programming problems (IVBLPPs) in a hierarchical decision making organization is presented.*

In the proposed approach, first the individual best and least solutions of the objectives of both the decision makers (DMs) located at the two hierarchical decision levels are determined by using an GA method. Then, the target interval for achievement of each of the objectives as well as the target interval of the decision vector controlled by the upper-level DM are defined in the inexact decision making environment.

In the executable goal programming (GP) formulation of the problem, the defined interval valued goals are transferred into the standard goals in GP by using the interval arithmetic technique in interval programming (IP) approach.

In the solution process, both the aspects of minsum and minmax GP formulation are adopted to minimize the lower bounds of the regret intervals for goal achievement within the specified interval from the optimistic point of view and thereby distribution of proper decision powers to the DMs of the hierarchical levels.

In the GA based solution search process, the conventional Roulette wheel selection scheme, arithmetic crossover and random mutation are taken into consideration to reach a satisfactory decision.

In the decision process, the GA scheme is directly used to the achievement function for minimizing the deviational variables of the goals in the decision making situation.

A numerical example is solved to illustrate the proposed approach, and the obtained solution is compared with the conventional approach studied previously.

Keywords: *Bilevel Programming, Genetic Algorithm, Goal Programming, Interval Programming, Interval Valued Bilevel Programming.*

1. INTRODUCTION

Bilevel programming problem (BLPP) is a special case of Multilevel programming problem (MLPP) of a large hierarchical decision system. In a BLPP, two DMs are located at two different hierarchical levels, each independently control one set of decision variables for optimizing the individual objective function which often conflict each other in the decision making situation.

In a hierarchical decision process, the lowerlevel DM (the follower) executes his/her decision powers, after the execution of decision of the upper-level DM (the leader). Although the leader independently optimizes its own benefits, the decision may be affected by the reactions of the follower. As a consequence, decision deadlock arises frequently and the problem of distribution of proper decision powers is encountered in most of the practical hierarchical decision situations.

The concept of BLPP was first introduced by Candler and Townsley in (6) in 1982. Thereafter, various versions of bilevel programming (BLP) have been presented in (2, 3). During 1980s, several solution approaches have been developed in (2, 3, 6, 10, 20, 21, 25) for solving BLPPs as well as MLPPs in general from the view point of their potential applications to hierarchical decentralized decision systems as economic systems, warfare, network designs and is specially applicable to conflict resolution.

Multiobjective solution techniques for BLPPs with post optimality analysis on the objective values to obtain efficient solutions have been suggested by Wen and Hsu in (25). In their approach, three efficient compromise solutions, threat point, ideal point and ideal–threat point dependent solutions depending on the DM's preference have been provided. The use of the concept of membership function of fuzzy set theory to BLPPs as well as MLPPs for satisfactory decision was first introduced by Lai in (10). Thereafter, Lai's satisfactory solution concept was extended by Shih et. al in (20, 21). The basic concept of the fuzzy programming (FP) approaches in (20) is that there is a possibility of rejecting the solution again and again by the leader and re-evaluation of the problem with the elicited membership values of the membership functions is repeatedly introduced in the solution search process due to the conflict in nature of the objectives. Again, for the use of such approaches, decision deadlock frequently arises due to the paradox that the leader's decision power is dominated by that of the follower.

To overcome the above difficulties, the BLPPs in the framework of fuzzy goal programming (FGP) in (5) have been also studied by Pal et. al in (4, 15), and others in the past.

Now, it may be noted that although FP as well as FGP have been successfully implemented to the BLPPs as well as MLPPs, the main difficulty of using such approaches is that it may not always be possible for the DM to assigning the fuzzy aspiration levels to the objectives and thereby defining the tolerance ranges of goal achievement in the actual decision situations.

To overcome the above difficulty, IP approaches in (23, 24) have appeared as the prominent tools for solving multiobjective decision problems in imprecise environment. The IP problems have also been studied by Pal et. al. in (16, 17) in the recent past.

However, the methodological extension of IP problems is still at an early stage. Further, IP approach to hierarchical decision problems is yet to appear in the literature.

In this article, interval valued objectives of both the DMs located at two hierarchical decision levels are considered. In the objectives of the DMs, both the coefficients and the target values are taken as intervals. Then the interval valued objectives are transformed into objective goals in the standard GP formulation using interval arithmetic technique in (12) by introducing under- and over-deviational variables to each of them. In the model formulation of the problem, both the aspects of GP, *minsum* GP in (4) and *minmax* GP in (19) are taken into consideration to construct the achievement function for goal achievement of the interval goals defined in the decision situation.

In the solution process, an GA method in (7, 8) is introduced to the GP formulation of IVBLPP. The GA approach in solving multiobjective decision making problems has been investigated by various researchers in (1, 11, 20) in the past. In using the GA process, the goal satisficing philosophy in GP, expound by Simon in (22) in 1957, for linear goals defined for the interval objectives is taken into consideration to reach a satisfactory decision.

The proposed approach is illustrated by a numerical example and the model solution is compared with the solution approach studied previously.

2. PROBLEM FORMULATION

Let X_1, X_2 be the vectors of decision variables controlled by the leader and the follower, respectively, in the hierarchical decision system, and F_1, F_2 be the objectives to be optimized at the two decision levels.

Then, the IVBLPP can be constructed as:

Find $X(X_1, X_2)$ so as to:

$$\max_{X_1} F_1(X_1, X_2) = (c_{11L}, c_{11U})X_1 + (c_{12L}, c_{12U})X_2 \quad \text{(leader's problem)} \qquad \text{... (2.1)}$$

where, for given X_1, X_2 solves

$$\max_{X_2} F_2(X_1, X_2) = (c_{21L}, c_{21U})X_1 + (c_{22L}, c_{22U})X_2 \quad \text{(follower's problem)} \qquad \text{... (2.2)}$$

subject to

$$X \in S = [(X_1, X_2) \mid A_1X_1 + A_2X_2 = b, X \geq 0],$$
$$X_1, X_2 \in R^n, b \in R^m. \qquad \text{... (2.3)}$$

where, (c_{ijL}, c_{ijU}) $i, j = 1, 2$, are the vector of interval coefficients, L and U denote the lower and upper bounds, respectively, of the defined interval. A_1, A_2 are the constant matrices and b is a constant vector. It is assumed that the feasible region S ($\neq \Phi$) is bounded.

Now, using the rules of interval arithmetic operations in (9), the interval valued objectives in (2.1) and (2.2) can be successively expressed as :

$$\max_{X_1} F_1(X_1, X_2) = (c_{11L}X_1 + c_{12L}X_2, c_{11U}X_1 + c_{12U}X_2) \quad \text{(leader's problem)} \quad ... (2.4)$$

and

$$\max_{X_2} F_2(X_1, X_2) = (c_{21L}X_1 + c_{22L}X_2, c_{21U}X_1 + c_{22U}X_2) \quad \text{(follower's problem)} \quad ... (2.5)$$

Now, due to inexactness of the decision environment, it becomes generally difficult for a DM to assign the exact target levels for achievement of the objective. In such a case, interval valued target called the target interval for goal achievement is to be taken into account by the DM.

In the present decision situation, the individual best and least objective values can be reasonably taken into account as the upper and lower bounds of the target intervals for goal achievement.

Now, to determine the target intervals and thereby to make a satisfactory solution, an GA scheme is introduced here in the solution search process.

3. DESIGJN OF THE GA SCHEME

In the literature of the GAs, there is a number of schemes in (7, 8) for generation of new populations with the use of the different operators: selection, crossover and mutation. Here, the binary coded representation of a candidate solution called chromosome is considered to perform genetic operations in the solution search process. The conventional Roulette wheel selection scheme in (7), singlepoint crossover (11) and bit-by-bit mutation operations are adopted to generate offspring in new population in search domain defined in the decision making environment. The fitness score of a chromosome v (say) in evaluating a function, say, eval (Ev), based on maximization or minimization of an objective function defined on the basis of DMs' needs and desires in the decision making context.

Now, the model formulation of the problem is described in the Section 4.

4. INTERVAL VALUED MODEL FORMULATION

To introduce the target intervals to the objectives F_1 and F_2 as well as the decision vector X_1 of the leader, the individual best and least solutions of the objectives are determined by employing the proposed GA scheme. Let the individual best and least solutions of the leader be $(X_1^{lb}, X_2^{lb}; T_1U)$ and $(X_1^{ll}, X_2^{ll}; T_{1L})$ respectively, where

$$T_{1U} = \underset{(X_1, X_2) \in S}{\text{Max}} F_1(X_1, X_2)$$

and
$$T_{1L} = \underset{(X_1, X_2) \in S}{\text{Min}} F_1(X_1, X_2)$$

Similarly, let $(X_1^{fb}, X_2^{fb}; T_{2U})$ and $(X_1^{fl}, X_2^{fl}; T_{2L})$ be respectively the individual best and least solutions of the follower, where

$$T_{2U} = \underset{(X_1, X_2) \in S}{\text{Max}} F_2(X_1, X_2)$$

and
$$T_{2L} = \underset{(X_1, X_2) \in S}{\text{Min}} F_2(X_1, X_2)$$

Then, the objectives in (2.4) and (2.5) with target intervals can be successively presented as

$$(c_{11L}X_1 + c_{12L}X_2, c_{11U}X_1 + c_{12U}X_2) = (T_{1L}, T_{1U}),$$
(leader's problem) ... (4.1)

$$(c_{21L}X_1 + c_{22L}X_2, c_{21U}X_1 + c_{22U}X_2) = (T_{2L}, T_{2U}),$$
(follower's problem) ... (4.2)

Again, since the leader has a higher power of making decision, a certain relaxation of X_1^{lb} as a lower tolerance limit should be given for searching of a better decision by the follower.

Let. X_1^p $(X_1^{ll} < X_1^p < X_1^{lb})$ be the lower tolerance limit of the decision vector X_1 controlled by the leader.

Using the concept of mid-point interval arithmetic in (12), the control vector X_1 with target itervals can be obtained as

$$X_1 = (X_1^p, X_1^{lb}) \quad \text{... (4.3)}$$

Now, the GP formulation of the problem is discussed in the Section 4.1.

4.1 GP Formulation

To formulate the GP model of the problem, the interval valued goals in (4.1), (4.2) and (4.3) are to be transformed into the conventional form of goals of GP methodology by using the rules of interval arithmetic in IP in (9), and under- and over-deviational variables are introduced to each of them.

Now, from the expression in (4.1), the leader's objective goals appear as:

$$c_{11L}X_1 + c_{12L}X_2 + d_{1L}^- - d_{1L}^+ = T_{1L},$$
$$c_{11U}X_1 + c_{12U}X_2 + d_{1U}^- - d_{1U}^+ = T_{1U}, \qquad \text{... (4.4)}$$

Similarly, from the expression in (4.2), the follower's objective goals take the form:

$$c_{21L}X_1 + c_{22L}X_2 + d_{2L}^- - d_{2L}^+ = T_{2L},$$
$$c_{21U}X_1 + c_{22U}X_2 + d_{2U}^- - d_{2U}^+ = T_{2U}, \qquad \text{... (4.5)}$$

where d_{kL}^-, d_{kL}^+ and d_{kU}^-, d_{kU}^+ (≥ 0), $k = 1, 2$ represent the under- and over- deviational variables, respectively, associated with the respective goals. Again, the expression in (4.3) can be transformed to the conventional goals as

$$X_1 + \mathbf{d}_L - \mathbf{d}_L = X_1^P$$
$$X_1 + \mathbf{d}_U - \mathbf{d}_U = X_1^{lb} \qquad (4.6)$$

where, $\mathbf{d}_L$, $\mathbf{d}_L$ (≥ 0) and $\mathbf{d}_U$, $\mathbf{d}_U'$ (≥ 0) represents the vectors of under- and overdeviational variables, and dimension of them depends on the dimension of X_1.

Now, in the possibility approach in IP (9) method, the aim of the DMs is to minimize the possible regrets for under-deviations from T_{kL} and over-deviations from T_{kU} $(k = 1, 2)$ and also the regrets for vectors of under- and over-deviations from X_1^P and X_1^{lb} respectively, defined for the leader's control vector X_1 to reach a solution for achieving the aspired goal levels within their respective target intervals specified in the decision situation.

As a matter of fact, minimization of the deviational variables d_{kL} and d_{kU}, $(k = 1, 2)$ and the vectors of deviational variables d_L and d_U in the context of minimizing the total possible regrets for goal achievement is taken into account, and then the designed objective function of the proposed GP model is termed as the 'regret function'.

4.2 Construction of Regret Function

The development of regret function, for minimization of the deviational variables can be made in different ways (9) depending on the needs and desires of the DMs in the decision making context.

Here, from the optimistic point of view of the DMs, both the aspects of GP, *minsum* GP in (4) for minimizing the sum of the weighted unwanted deviational variables as well as *minmax* GP in (19) for minimizing the maximum of the deviations, are simultaneously taken into account as a convex combination of them to reach a satisfactory solution for goal achievement within the specified target intervals of the defined goals of the problem.

Now, for model simplification let it be assumed that n_1 $(n_1 < n)$ be the number of decision variables involved with the control vector X_1.

The regret function appears as:

$$\text{Minimize } Z = \lambda\left[\sum_{i=1}^{n_1+2}\left(w_{iL}^{-}\, d_{iL}^{-} + w_{iU}^{+}\, d_{iU}^{+}\right)\right] + (1-\lambda)\left[\max_{i \in n_1+2}\left(d_{iL}^{-} + d_{iU}^{+}\right)\right] \quad \text{... (4.7)}$$

where Z represents the regret function for goal achievement; d_{iL}^{-}, d_{iU}^{+} (≥ 0) $(i = 1, 2, ..., n_1 + 2)$, are successively renamed for d_{kL}^{-}, d_{kU}^{+} $(k = 1, 2)$ and n_1 components of each of the d_L^{-}, d_U^{+}, and

where w_{iL}^{-}, w_{iU}^{+} (≥ 0) with $\sum_{i=1}^{n_1+2}\left(w_{iL}^{-} + w_{iU}^{+}\right) = 1$ are the numerical weights of importance of achieving the goals within their respective target intervals, and $0 < \lambda < 1$.

Now, let

$$\max_{i \in n_1+2}\left[\left(d_{iL}^{-} + d_{iU}^{+}\right)\right] = d \quad \text{... (4.8)}$$

Then, the executable GP model of the problem can be presented as:

Find $X(X_1, X_2)$ so as to:

$$\text{Minimize } Z = \lambda\left[\sum_{i=1}^{n_1+2}\left(w_{iL}^{-}\, d_{iL}^{-} + w_{iU}^{+}\, d_{iU}^{+}\right)\right] + (1-\lambda)\, d$$

and satisfy the goal expressions in (4.4), (4.5) and (4.6)

subject to

$$\left(d_{iL}^{-} + d_{iU}^{+}\right) \le d, \quad i = 1, 2, \ldots, n_1 + 2 \qquad \ldots (4.9)$$

and the given system constraints in (2.3).

Now, for the developed GP model of the proposed problem, the task of both the DMs is to search the solution to satisfy the goal levels to the extent possible by evaluating the defined regret function for overall benefit of the organization in the decision making context.

Now, since GA is a goal satisficer in (7) rather than optimizer, the proposed GA scheme can be employed here to minimize the regret function Z and thereby to reach a satisfactory decision for distribution of proper decision powers to the DMs.

Here, the fitness function appears as:

$$\text{eval}\,(E_v) = (Z)_v$$

$$\left\{\lambda\left[\sum_{i=1}^{n_1+2}\left(w_{iL}^{-}\, d_{iL}^{-} + w_{iU}^{+}\, d_{iU}^{+}\right)\right] + (1-\lambda)\, d\right\}_v, \qquad \ldots (4.10)$$

where, v represents a chromosome.

Here, the best chromosome E^* with highest fitness score at a generation is determined by

$$E^* = \min_{v}\left[\text{eval}\left(E_v\right)\right]$$

To illustrate the approach, a numerical example is solved in the Section 5.

5. AN ILLUSTRATIVE EXAMPLE

The BLPP with interval coefficients can be considered as:

Find $X(x_1, x_2)$ so as to:

$$\max_{x_1} F_1(x_1, x_2) = (1, 3)\,x_1 - (1, 2)\,x_2 \quad \text{(leader's problem)}$$

where, for given x_1 , x_2 solves

$$\max_{x_2} F_2(x_1, x_2) = (1, 2)\,x_1 + (1, 3)\,x_2$$

(follower's problem)

subject to

$$\begin{aligned} &3x_1 - 5x_2 \leq 15,\ 3x_1 - x_2 \leq 21, \\ &3x_1 + x_2 \geq 27,\ 3x_1 + 4x_2 \leq 45, \\ &x_1 + 3x_2 \leq 30,\ x_1, x_2 \geq 0. \end{aligned} \qquad \text{... (5.1)}$$

Now, the use of different genetic parameters and their values adopted in the solution search process are defined as follows.

The input parameters:

- Max_gen; //number of generations
- Pop_size; // population size
- pc; //probability of crossover
- pm; //probability of mutation

The programming language C is used in the process of coding the evaluation program. The environment of execution is Intel Pentium IV with 2.66 GHz. Clock-pulse and 1 GB RAM.

The, chromosome length = 30 is considered with a view to searching solution in the domain of feasible solution set (S) defined in the decision situation. The population size as in the standard GA method in (7) is taken 100.

The number of generations = 300 is initially taken to conduct the experiment. The different experiments with the different values of p_c $(0 < p_c < 1)$ and p_m $(0 < p_m < 1)$, in the ranges $(0.7 \leq p_c \leq 0.9)$ and $(0.03 \leq p_m \leq 0.8)$ are made in the proposed GA scheme. It is found that $p_c = 0.8$ and $p_m = 0.08$ are successful in the decision search process.

Now, using the interval arithmetic rule and employing the GA scheme, the individual best and least solutions of the leader are obtained as

$$(x_1, x_2, T_{1U}) = (8, 3; 13) \text{ and } (x_1, x_2, T_{1L}) = (7, 6; 8),$$

respectively.

Similarly, the best and least solutions of the follower are obtained as

$$(x_1, x_2, T_{2U}) = (7, 6; 19) \text{ and } (x_1, x_2, T_{1L}) = (8, 3; 14),$$

respectively.

Then, following the procedure, the leader's problem with the target interval can be obtained as:

$$(1, 3)\, x_1 - (1, 2)\, x_2 = (8, 13)$$

Similarly, the follower's problem is obtained as:

$$(1, 2)\, x_1 + (1, 3)\, x_2 = (14, 19)$$

Now, the decision situation, the leader feels that the achieved value 8 of x_1 can be relaxed upto 7.5 but not beyond of it. So

$$x_1^p = 7.5 \left(x_1^{ll} < 7.5 < x_1^{lb}\right)$$

is considered as the lower bound of x_1 of its target interval.

Then, the control variable x_1 of the leader with interval coefficients and target interval can be presented as:

$$(1, 1)\, x_1 = (7.5, 8)$$

Now, by following the procedure, the linear objective goals are obtained as:

$$\begin{aligned}
x_1 - x_2 + d_{1L}^- - d_{1L}^+ &= 8 \\
3x_1 - 2x_2 + d_{1U}^- - d_{1U}^+ &= 13 \\
x_1 + x_2 + d_{2L}^- - d_{2L}^+ &= 14 \\
2x_1 + 3x_2 + d_{2U}^- - d_{2U}^+ &= 19 \\
x_1 + d_{3L}^- - d_{3L}^+ &= 7.5 \\
x_1 + d_{3U}^- - d_{3U}^+ &= 8
\end{aligned} \qquad \text{... (5.2)}$$

where, $d_{iL}^-, d_{iL}^+, d_{iU}^-, d_{iU}^+ \geq 0, \quad i = 1, 2, 3.$

Now, in the decision making context, the deviational variables to be minimized in the regret function depend on the needs and desires of the DMs.

In the present decision situation, the executable GP model with the regret function is constructed as:

Find $X(x_1, x_2)$ so as to

$$\text{Minimize } Z = \lambda \sum_{i=1}^{3} \left(w_{iL}^{-} d_{iL}^{-} + w_{iU}^{+} d_{iU}^{+} \right) + (1-\lambda) d$$

subject to

$$\sum_{i=1}^{3} d_{iL}^{-} \le d,$$

$$\sum_{i=1}^{3} d_{iU}^{-} \le d, \qquad \text{... (5.3)}$$

and the goal constraints in (5.2) and the system constraints in (5.1).

Now, assigning equal weights, each value of which $= \left(\frac{1}{6}\right)$, to all the goals for their achievement, and introducing $\lambda = 0.5$, the problem is solved by employing the GA scheme with fitness function Z defined in (5.3) and the same parameter values considered previously.

The resultant decision is obtained as

$$(x_1, x_2) = (8, 3), \text{ with } (F_1, F_2) = (13, 14).$$

The result shows that a satisfactory decision is reached in the decision making environment.

Note: The BLPP with crisp coefficients in the fuzzy decision environment is considered by Pal and Moitra (14) is

$$\max_{X_1} F_1(x_1, x_2) = 2x_1 - x_2 \qquad \text{(leader's problem)}$$

where, for given x_1, x_2 solves

$$\max_{X_2} F_2(x_1, x_2) = x_1 + 2x_2 \qquad \text{(follower's problem)}$$

subject to the given system constraints in (5.1).

In the FGP formulation, if the best solutions of leader and follower are considered as the fuzzy aspiration levels of the respective objectives and their least solutions as the lower tolerance limits, then the solution of the problem in the framework of FGP is obtained as

$$(x_1, x_2) = (8.6, 4.8) \text{ with } (F_1, F_2) = (12.4, 18.2).$$

A comparison shows that a better solution is achieved here in terms of balancing the decision powers to the DMs of both the hierarchical levels. Now, it is to be followed that if the IP technique is used to the problem, then the solution is found as $(x_1, x_2) = (8, 3)$ with $(F_1, F_2) = (13, 14)$.

A further comparison shows that the solution under the present approach is a superior one from the view point of satisfaction of both the DMs in the hierarchical decision making environment.

6. CONCLUSION

In this paper, how the GA method can be efficiently employed to solve BLPPs is presented. In future study, the proposed method can be extended to solve MLPPs as well as the other decentralized planning problems from the view point of their potential use to real-life problems.

The proposed approach can be extended to solve chance constrained hierarchical decision problems, which may be a problem for future research.

However, it is expected that the approach presented here may lead to further research on real-life hierarchical decision problems in the inexact decision making environment.

ACKNOWLEDGEMENT

The authors are thankful to the anonymous referee and Prof R.N. Bhaumik, Conference Chair ICRFSC 2009, for their valuable suggestions in improving the quality of the paper.

REFERENCES

[1] Alcazar, J., Moroto, C. and Ruiz, R., "Solving the Multi-Mode Resource Constrained Project Scheduling Problems with Genetic Algorithms", *J. of the Operational Research Society*, **54**, pp. 614–626, 2003.

[2] Bialas, W.F. and Carwan, M.H., "On Twolevel Optimization", *IEEE Transactions on Automatic Control*, **27**, pp. 211–214, 1982.

[3] Bialas, W.F. and Carwan, M.H., "Two-Level Linear Programming". Management Sciences, **30**, pp. 1004–1020, 1984.

[4] Biswas, A. and Pal, B.B., "Application of Fuzzy Goal Programming Technique to Land use Planning in Agricultural System", Omega, **33**, pp. 391–398, 2005.

[5] Biswas, A. and Pal, B.B., "Fuzzy Goal Programming Procedure for Multiobjective Bilevel Programming Problems", *International Journal of Computer, Mathematical Sciences and Appl.*, **1**, No. 1, pp. 87–98, 2007.

[6] Candler, W. and Townsley, R., "A Linear Two-Level Programming Problem", *Computers and Operations Research*, **9**, pp. 59–76, 1982.

[7] Goldberg, D.E., "Genetic Algorithms in Search, Optimization, and Machine Learning", Addison-Wesley, Reading, MA, 1989.

[8] Holland, J.H., "Genetic Algorithms and Optimal Allocations of Trials", *SIAM Journal of Computing*, **2 (2)**, pp. 88–105, 1973.

[9] Inuiguchi, M. and Kume, Y., "Goal Programming Problems with Interval Coefficients and Target Intervals", *Euro. J. Operational Res.*, **52**, pp. 345–361, 1991.

[10] Lai, Y.J., "Hierarchical Optimization. A Satisfactory Solution, Fuzzy Sets and Systems", **77**, pp. 321–335, 1996.

[11] Michalewicz, Z., "Numerical Optimization and Constraints", *in Proc. of 6th International Conference on Genetic Algorithms*, pp. 151–158, 1995.

[12] Moore, R.E., "Interval Analysis", Prentice Hall, New Jersey, 1996.

[13] Oliviera, C., and Anunes, C.H., "Multiple Objective Linear Programming Models with Interval Coefficients–An Illustrated Overview", *Euro. J. Operational Res.*, **118**, pp. 1434–1463, 2007.

[14] Pal, B.B., and Moitra, B.N., "A Fuzzy Goal Programming Approach for Solving Bilevel Programming Problems", *Lecture Notes in Artificial Intelligence (LNAI)*, **2275**, pp. 91–98, 2002.

[15] Pal, B.B., and Moitra, B.N., "A Goal Programming Procedure for Solving Problems with Multiple Fuzzy Goals Using Dynamic Programming", *European Journal of Operational Research*, **144**, pp. 480–491, 2003.

[16] Pal, B.B., and Gupta, S., "A Goal Programming Approach for Solving Interval Valued Multiobjective Fractional Programming Problems Using Genetic Algorithm", *Proceedings of IEEE 10th Colloquium and International Conference on Industrial and Information Systems 2008 (ICIIS 2008), (apeared in the IEEE explore)*, **440**, pp. 1–6, 2008.

[17] Pal, B.B., and Sen, S., "A Goal Programming Procedure for Solving Interval Valued Multiobjective Fractional Programming Problems", *Proceedings of 16th International Conference on Advanced Computing and Communication (ADCOM 2008)*, 297, pp. 1–6, 2008.

[18] Powell, D., and Skolnick, M., "Using Genetic Algorithms in Engineering Design Optimization with Non-Linear Constraints", *in the Proc. of 5th International conference on Genetic Algorithms*, pp. 424–430, 1993.

[19] Romero, C., "Handbook of Critical Issues in Goal Programming", Pergamon Publ. Corp., 1991.

[20] Shih, H.S., Lai, Y.J., and Lee, E.S., "Fuzzy Approaches for Multi-Level Programming Problems", *Computers and Operations Research*, **23**, pp. 73–91, 1996.

[21] Shih, H.S. and Lee, S., "Compensatory Fuzzy Multiple Level Decision Making, Fuzzy Sets and Systems", **14**, pp. 71 - 87, 2000.

[22] Simon, H.A., "Administrative Behavior", Fress Press, New York 1957.

[23] Steuer, R.E., "Algorithm for Linear Programming Problems with Interval Objectives Function Coefficients", *Management Science*, **26**, pp. 333 - 348, 1981

[24] Tong, S., "Interval Number and Fuzzy Number Linear Programming, Fuzzy Sets and Systems", **66**, pp. 301 - 306, 1994.

[25] Wen, U.P. and Hsu, S.T., "Efficient Solutions for the Linear Bilevel Programming Problem", *European Journal of Operational Research*, **62**, pp. 354 - 362, 1991

39

A STOCHASTIC SIMULATION BASED GENETIC ALGORITHM FOR SOLVING CHANCE CONSTRAINED FUZZY GOAL PROGRAMMING PROBLEMS

Bijay Baran Pal[1]*, Somsubhra Gupta[2] and Mousumi Kumar[3]

[1]*Department of Mathematics, University of Kalyani, Kalyani-741235, W.B., India, E-mail: bbpal18@hotmail.com*

[2]*Department of Information Technology JIS College of Engineering, Kalyani-741235, W.B., India, E-mail: gsomsubhra@gmail.com*

[3]*Department of Mathematics, Alipurduar College, Alipurduar-736122, Jalpaiguri W.B., India, E-mail: roysankha@yahoo.co.in*

***Abstract**: In this article, a fuzzy goal programming (FGP) procedure for solving certain structured chance constrained programming (CCP) problem is presented.*

In the proposed approach, achievement of the membership functions (MF) defined for the fuzzily described chance constrained goal with specified probability as the aspiration levels to the highest degree (unity) to the extent possible is considered.

In the model formulation of the problem, instead of converting the probabilistic MF to their deterministic equivalent, the stochastic simulation technique under the goal satisficing philosophy (12) is introduced to reach a feasible decision at each iterative step in the process of solving the problem.

In the solution process, a genetic algorithm (GA) scheme with use of the proposed simulation approach is adopted to reach a most satisfactory decision in the uncertain decision making environment.

A numerical example is solved to illustrate the potential use of the approach over the conventional fuzzy programming (FP) approaches.

Keywords: *Chance Constrained Programming, Fuzzy Goal Programming, Genetic Algorithm, Goal Programming, Stochastic Fuzzy Goal Programming.*

1. INTRODUCTION

In the real-life decision situations, the decision maker (DM) is generally faced with two types of uncertainty (probabilistically and fuzzily defined) in setting the parameter values to the problems. In the probabilistically uncertain decision situation, the parameters involved with a problem are random in nature with certain probability distributions.

For instance in agricultural planning situation, the crops-production depends on adequate supply of water which very much depends on rainfall as well as availability of skilled agricultural labourers which are always uncertain in nature.

On the other hand, in a fuzzily described uncertain decision situation, the parameter values are imprecise in nature due to vagueness of human judgments.

Stochastic programming (SP) as an optimization tool based on the probability theory has been developed in various ways in [2, 22]. In the field of SP, the CCP in [1, 17] has appeared as a prominent tool to solve problems with random parameter values. The CCP has been studied extensively in the past.

The uses of GAs [10] as well as FGP approaches in (4) to multiobjective CCP problems have also been studied in [15, 20, 23] in the past.

In the conventional SP approaches in [11], the objectives and also the system constraints with stochastic parameters are converted into the deterministic equivalent in the process of solving problems in uncertain environment. However, in most of the previous studies in [21, 23], certain types of probability distributions of independent continuous random parameters involved with the problem are considered in an exclusive way.

But, in the real-world decision situation, it is found that the discrete and continuous both types of random variables are involved with the problem. Again, randomness in both the coefficients and resource vector occurs simultaneously in some decision making situations. In such a case, the conventional deterministic approaches become computationally expensive and complex.

To overcome the above difficulty, the stochastic simulation technique [19] has appeared as one of the prominent tools to solve problems in the domain of chance constrained decision making. The use of such techniques to solve dependent chance programming (DCP) problems with single and multiplicity of objectives in different conventional programming environments have been studied extensively by Liu [5], Liu et. al. [24] in the past. In the field of DCP, the use of possibility theory introduced by Zadeh [16], has been extensively studied by Dubois and Prade [9], and it has been widely used [6] to solve DCP problems. The GA based solution approach with the use of stochastic simulation to fuzzily described DCP problems has been studied by Liu and Iwamura [8] in the past.

However, consideration of both the aspects of randomness and fuzziness to multiobjective decision making (MODM) problems is a great challenge to the researchers and extensive study in this area is still at an early stage. Again, the use of stochastic simulation to the MODM problems involving both the discrete and continuous random variables is yet to be widely circulated in the literature.

In this article, a certain structure of a stochastic fuzzy goal programming (SFGP) problems with multiplicity of chance constrained objective goals for achievement of probabilistically defined aspiration levels is considered, where both the discrete and continuous random variables with different probability distributions are taken into account.

In the model formulation the achievement of the highest degree (unity) to the extent possible of each of the probabilistically defined MFs of the fuzzy objective goals on this basis of the weights of importance is considered.

In the decision process, a GA based solution search process with the use of stochastic simulation is adopted to reach a feasible and satisfactory decision in the decision making context.

The potential use of the approach is illustrated by a numerical example.

2. PROBLEM FORMULATION

Similar to the conventional GP in [14] and FGP in [3], the chance constrained FGP formulation may be considered solely with stochastic fuzzy goals in [18] or a set of structural crisp constraints in [23]. Again, imposition of certain restriction on the decision vector in [7] may also be considered there depending on the decision making situation.

In the present SFGP model formulation, the stochastic fuzzy goals with certain restriction on the vector of decision variables are taken into consideration.

The generic form of a chance constrained FGP problem can be presented as:

Find $X (x_1, x_2, \ldots, x_n)$ so as to satisfy

$$\text{Prob}\,[f_i(X) \leq b_i] \gtrsim P_i,\ i = 1, 2, \ldots m \qquad \ldots (2.1)$$

and $0 \leq X \leq U$,

where $X (\geq 0)$ is the vector of decision variables, U denotes the vector of upper bounds on X.

Prob stands for probability, Prob (.) indicates the stochastically defined ith goal and p_i $(0 < p_i < 1)$ is the fuzzily described probability of satisficing the ith goal, and where the symbol $\gtrsim$ is the fuzzification of $\geq$ in the sense of Zimmermann [13].

Now, the goal expression in (2.1) can be explicitly presented as [18]:

$$\text{Prob}\left(\sum_{j=1}^{n} c_{ij}\, x_j \leq b_i\right) \gtrsim p_i, \quad i = 1, 2, \ldots, m$$

$$\text{with } 0 \leq x_j \leq u_j, \qquad j = 1, 2, \ldots, n \qquad \ldots (2.2)$$

In the decision situation, since the DM is a satisficer rather than optimizer in a fuzzy decision making context, certain probability q_i $(q_i < p_i)$ as lower tolerance limit for defining the range of satisfaction of the DM with the solution achievement to a certain degree is considered in the model formulation of the problem.

Here, the three different situations for DM's satisfaction are presented as followed.

(*i*) The DM is fully satisfied, if

$$\text{Prob}\left(\sum_{j=1}^{n} c_{ij}\, x_j \le b_i\right) \ge p_i, \qquad i = 1, 2, \ldots, m$$

(*ii*) The DM is almost satisfied, if

$$q_i \le \text{Prob}\left(\sum_{j=1}^{n} c_{ij}\, x_j \le b_i\right) < p_i, \quad i = 1, 2, \ldots, m$$

(*iii*) The DM is dissatisfied, if

$$\text{Prob}\left(\sum_{j=1}^{n} c_{ij}\, x_j \le b_i\right) < q_i, \quad i = 1, 2, \ldots, m$$

Now, for the above three given conditions, the MF $\mu_i(X)$ for measuring the degree of achievement of the *i*th goal can be constructed as follows.

2.1 Construction of Membership Function

The MF can be presented as

$$\mu_i(X) = \begin{cases} 1, & \text{if } \text{Prob}\left(\sum_{j=1}^{n} c_{ij} x_j \le b_i\right) \ge p_i \\ \dfrac{\text{Prob}\left(\sum_{j=1}^{n} c_{ij} x_j \le b_i\right) - q_i}{p_i - q_i} & \text{if } q_i \le \text{Prob}\left(\sum_{j=1}^{n} c_{ij} x_j \le b_i\right) < p_i \\ 0, & \text{if } \text{Prob}\left(\sum_{j=1}^{n} c_{ij} x_j \le b_i\right) < q_i \end{cases} \quad \ldots (2.3)$$

Now, in the fuzzy decision situation, the aim of the DM is to maximize the membership value of each of the defined MFs in (2.3). Since a number of objectives are involved and the objectives often conflict each other regarding achievement of their respective aspired goal levels, the weighted *minsum* FGP approach [3] is addressed to formulate the achievement function to achieve the membership goals on the basis of their weights of importance in decision making environment.

3. FGP MODEL FORMULATION

The *minsum* FGP model [3] can be presented as

$$\text{Maximize: } Z = \sum_{i=1}^{m} w_i \mu_i$$

Subject to

$$\mu_i \leq \frac{\text{Prob}\left(\sum_{j=1}^{n} c_{ij} x_j \leq b_i\right) - q_i}{p_i - q_i},$$

$$0 \leq \mu_i \leq 1, \qquad 0 \leq x_j \leq u_j, \qquad \text{... (3.1)}$$

where $w_i (> 0)$ $(i = 1, 2, ..., m)$ is the numerical weights of importance of maximizing μ_i, and where w_i is determined as [4]:

$$w_i = \frac{1}{p_i - q_i}, \quad i = 1, 2, ..., m$$

Now, the stochastic simulation for achieving the solution by satisfying the chance constraint goals in (3.1) is described in the following Section.

3.1 Stochastic Simulation for Chance Constrained Goal

The chance constraints in (3.1) can be expressed as

$$\text{Prob}\left(\sum_{j=1}^{n} c_{ij} x_j \leq b_i\right) \geq \mu_i P_i + (1 - \mu_i) q_i$$

$$i = 1, 2, \ldots, m \qquad \text{... (3.2)}$$

It is to be noted here that the right-hand side value is actually the convex combination of the optimistic value p_i and the pessimistic value q_i. For which, the satisfaction of the DM with the chance constraints lies in the tolerance range specified in the decision situation. Now, the simulation process is defined as follows.

Let, $$g_i(X, r_i) = \left(\sum_{j=1}^{n} c_{ij} x_j - b_i \right),$$

where r_i is the $(n + 1)$ dimensional random vector of variables,

$$r_i = (c_{i1}, c_{i2}, \ldots, c_{in}; b_i) \quad i = 1, 2, \ldots, m$$

Then, the problem in (3.2) takes the form

$$\text{Prob}\left[g_i(X, r_i) \le 0\right] \ge \mu_i P_i + (1 - \mu_i) q_i$$

$$i = 1, 2, \ldots, m \qquad \ldots (3.3)$$

In the process of running the problem, the mN independent random vectors is considered as

$$r_i^{(k)} = \left[c_{i1}^{(k)}, c_{i2}^{(k)}, c_{in}^{(k)}, c_i^{(k)}\right]$$

$$k = 1, 2, \ldots, N;$$

$$i = 1, 2, \ldots, m.$$

Then, let N' be the number of occasions on which the condition in (3.3) is satisfied.

Here, from the basic concept of the probability theory, the condition satisfying probability is defined by $P_i = N'/\text{mN}$.

The simulation process is summarized in the following steps:

3.2 Simulation Algorithm

Step 1: $N' = 0$

Step 2: Generate the random number according to the prescribed probability measure Prob.

Step 3: Compute the value of the probability measuring function defined in (3.3).

Step 4: (*i*) Set $N' = N' + 1$; (*ii*) Repeat the steps 2 to 3, mN times.

Step 5: Compute P_i: If $P_i \geq \mu_i\, p_i + (1 - \mu_i)\, q_i$, $i = 1, 2, ..., m$ then the defined chance constraints in (3.3) are satisfied, and go to the Step 6. Otherwise, repeat the Steps 2 to 5.

Step 6: Return the value.

Now, the GA scheme based on the proposed simulation process for obtaining the best feasible value of (to the objective function) and thereby obtaining the satisfactory solution is presented in the following Section.

3.3 GA Scheme for Stochastic Simulation

The GA method used in solution search process is described as follows:

Step 1: Representation and initialization: Let $V = (x_1, x_2, ..., x_n)$ be the binary coded representation of a chromosome in the population. The population size is defined by p_size and p_size chromosomes are initialized randomly in the search domain. A uniform random number corresponding to each x_j ($j = 1, 2 ..., n$) is generated within $(0, u_j)$, where u_j is the upper bound of x_j specified in the decision making context.

Step 2: Feasibility checking: The feasibility of a solution is checked by employing the simulation technique defined in the Section B.

Step 3: Evaluation function: The evaluation function defined by eval (V). It is used to assign a probability of fitness to a chromosome V so that its chance of being selected is proportional to its fitness value relative to the other chromosomes in the population. This implies that a chromosome with higher fitness value have a better chance to take part in the reproduction process for producing offspring. The roulette-wheel selection (10) scheme is considered for parent selection in a population $V_1, V_2 \ldots V_{p\_size}$

Here, since the objective is to maximize the achievement, a chromosome with higher score is considered a fitter one.

Now, let a parameter a $\in (0, 1)$ in the genetic system be given.

Then the rank based evaluation function can be defined as (21):

$$eval\ (V_l) = a\,(1 - a)^{l-1},\ l = 1, 2, ..., \text{p\_size}. \quad ... (3.4)$$

Here, it is to be observed that the evaluation function with the best score corresponds to $l = 1$ and the worst score corresponds to l = p_size.

Step 4: Selection: The selection of a chromosome as a parent for producing offspring is described as follows (21):

(*a*) Compute the cumulative probability q_1 for each chromosome V_l as follows:

Set $\quad q_0 = 0,$

$$q_l = \sum_{k=1}^{l} \text{eval}\ (V_k), \quad l = 1, 2, ..., \text{p\_size} \quad ... (3.5)$$

(*b*) Generate a random number r in $(0, q_{\text{p\_size}})$

(c) Select the lth chromosome V_l $(1 \le l \le \text{p\_size})$ such that. $q_{l-1} < r \le q_l$.

(*d*) Repeat the steps (*b*) and (*c*) p_size times and obtain p_size chromosomes.

Step 5: Crossover: The probability of crossover is defined in the genetic system as p_c. Then (p_c, p_size) is the expected number of chromosomes, which undergo the crossover operation. A random number r is generated from the interval [0, 1]. A chromosome V_l is selected as a parent provided $r < p_c$. The process is repeated p_size times and (p_c. p_size) parents are selected for crossover operation as follows:

Let V_1' and V_2', be a pair of chromosomes, which take part in the crossover operation. Then, an integer n' is randomly generate from the interval (1, m) and the genes are exchanged after the n' position of the chromosomes V_1' and V_2' and then two offspring are produced. If the offspring are feasible, then the parents are replaced by them, otherwise the crossover operation is repeated.

Step 6: Mutation: The probability of mutation of the genetic system is defined by p_m. Then (p_m.p_size) is the expected number of chromosomes, which undergo the mutation operations. A random number r is generated from the interval (0, 1). A chromosome V_l is selected as a parent for mutation if $r < p_m$. The process is repeated p_size times, and (p_m.p_size) parents are selected for mutation operation. Then an integer n' is generated as the mutation position from the interval $(1, m)$ and mutation is made in bit-by bit basis. If the new chromosome is infeasible then the process is repeated until it becomes feasible.

Step 7: Termination: The execution of the whole process terminates when the best decision is reached at a certain generation number in the genetic search process.

The potential use of the approach is illustrated by the following example in the Section 4.

4. A NUMERICAL EXAMPLE

The following problem with three chance constraints fuzzy goals is considered:

Find (x_1, x_2, x_3) so as to: satisfy

G1: $\text{Prob}\left(c_{11}x_1 + c_{12}x_2 + c_{13}x_3 \le b_1\right) \gtrsim 0.85$

G2: $\text{Prob}\left(c_{21}x_1 + c_{22}x_2 + c_{23}x_3 \le b_2\right) \gtrsim 0.75$

G3: $\text{Prob}\left(c_{31}x_1 + c_{32}x_2 + c_{33}x_3 \le b_3\right) \gtrsim 0.9$

$$0 \le x_1 \le 5,\ 0 \le x_2 \le 20,\ 0 \le x_3 \le 2, \quad \ldots (4.1)$$

where c_{ij}, b_i $(i, j = 1, 2, 3)$ are independent random variables, in which the distributions followed by the random variables are defined as follows:

(*i*) c_{11}, c_{12}, c_{13} and b_1 are with geometric distribution having the parameter values 0.95, 0.80, 0.9 and 0.01, respectively.

(*ii*) c_{21}, c_{22}, c_{23} and b_2 are exponentially distributed with mean values 0.05, 0.20, 0.02 and 20.0, respectively.

(*iii*) c_{31}, c_{32}, c_{33} and b_3 follow negative binomial distribution with parameter values (1, 0.5), (2, 0.8), (5, 0.6), and (5, 0.1) respectively.

Now, in the decision situation, the pessimistic values $q_1 = 0.7$, $q_2 = 0.6$ and $q_3 = 0.7$ are specified by the DM for the successive goals of the problem.

Then, following the procedure, the chance constraints set in (4.1) for satisfaction of the goal achievement of the DM takes the form:

$$0.7 \le \text{Prob}\,(c_{11}\,x_1 + c_{12}\,x_2 + c_{13}\,x_3 \le b_1) \le 0.85$$

$$0.6 \le \text{Prob}\,(c_{21}\,x_1 + c_{22}\,x_2 + c_{23}\,x_3 \le b_2) \le 0.75$$

$$0.7 \le \text{Prob}\,(c_{31}\,x_1 + c_{32}\,x_2 + c_{33}\,x_3 \le b_3) \le 0.9$$

Then, the FGP formulation with the chance constraints defined for the membership functions are obtained as:

Maximize Z: $\dfrac{\mu_1}{0.15} + \dfrac{\mu_2}{0.15} + \dfrac{\mu_3}{0.2}$ so as to

Satisfy:

$$\mu_1 \le \frac{\text{Prob}\left(c_{11}\,x_1 + c_{12}\,x_2 + c_{13}\,x_3 \le b_1\right) - 0.7}{0.15}$$

$$\mu_2 \le \frac{\text{Prob}\,(c_{21}\,x_1 + c_{22}\,x_2 + c_{23}\,x_3 \le b_2) - 0.6}{0.15}$$

$$\mu_3 \le \frac{\text{Prob}\left(c_{31}\,x_1 + c_{32}\,x_2 + c_{33}\,x_3 \le b_2\right) - 0.7}{0.2}$$

$$0 \le x_1 \le 5, \quad 0 \le x_2 \le 20, \quad 0 \le x_3 \le 2$$

Now, the proposed stochastic simulation-based GA, developed using Programming Language C with hardware support of Dell Power Edge R900 Server having 2 GB RAM is used to solve the problem.

In the genetic search process, the following parameter values are taken into consideration

- population size is 100,
- probability of crossover is 0.8,
- probability of mutation is 0.08.

The optimal decision is obtained with 1000 cycles of simulation run and after 1000 generation number.

The achieved solution is $(x_{1,}\ x_{2,}\ x_3) = (1.5287,\ 10.5663,\ 0.2836)$ with $Z = 18.34$ and with the membership values

$$\mu_1 = 1,\ \mu_2 = 1,\ \mu_{13} = 1$$

Note 1: It is to be noted that the membership values of the defined goals of the problem are achieved exactly in the decision situation. The occurrence of such a situation is due to the specified different parameter values involved with the problem. However, in a practical decision situation, estimation of the parameter values highly depends on the decision making environment. Again, parametric analysis on the parameter values can be made there for making proper decision in the decision making context. However, the result exhibits the potential use of the approach for solving SFGP problem.

Note 2: If the conventional max-min fuzzy operator (13) is used in the process of solving the problem, then the obtained solution is found as

$$(x_1, x_2, x_3) = (3.6493, 2.9072, 1.8368)$$

A comparison shows that a better solution is achieved here under the proposed approach.

5. CONCLUSION

In this article, a certain structure of GP problems under randomness and fuzziness is considered. Here, a new FGP methodology based on GA and stochastic simulation technique is discussed to solve SFGP problems. The major advantage of the approach is that randomness in both the technological coefficients and the resource vector of the fuzzy goal constraints can be considered in the framework of the proposed model. Further, if the crisp (or fuzzy) system constraints are to be taken into account in the model formulation, then they can easily be incorporated within the framework of the proposed model without involving any computational load.

An extension of the approach for non-linear fuzzy SFGP problem may be a problem in future studies. Finally, it is hoped that the approach presented here may open up a new look into the way of solving real-world chance constrained FGP problems in the current uncertain MODM arena.

ACKNOWLEDGEMENT

The authors are thankful to the anonymous Referees and the Conference Chair, ICRFSC2009 for their helpful suggestions and comments which have lead to improve the quality of presentation of the paper.

REFERENCES

[1] A. Charness and W.W. Coopers, "Chance Constrained Programming", *Management Science*, **6**, pp. 73–79, 1959.

[2] A. Pre'kopa, "Contribution to the Theory of Stochastic Programming", *Mathematical Prog.*, **4**, pp. 202–221, 1973.

[3] B.B. Pal, B.N. Moitra, "A Fuzzy Goal Programming Procedure for Solving Bilevel Programming Problems", *Lecture Notes in Artificial Intelligence*, **2275**, Heidelberg, Springer Verlag pp. 91–98, 2002.

[4] B.B. Pal, B.N. Moitra, U. Maulik, "A Goal Programming Procedure for Fuzzy Multiobjective Linear Fractional Programming Problem", *Fuzzy Sets and Systems*, **139**, pp. 395–405, 2003.

[5] B. Liu, "Dependent Chance Programming in Fuzzy Environment", *Fuzzy Sets and Systems*, **109**, pp. 97–106, 2000.

[6] B. Liu, "Fuzzy Random Dependent Chance Programming", *IEEE transaction on Fuzzy Systems*, **9**, pp. 721–726, 2001.

[7] B. Liu, "Theory and Practice of Uncertain Programming", Physica-Verlag, Heidelberg, 2002.

[8] B. Liu, and K. Iwamura, "A Note on Chance Constrained Programming with Fuzzy Coefficients", *Fuzzy Sets and Systems*, **100**, Nos. 1–3, pp. 229–233, 1998.

[9] D. Dubois, and H. Prade, "Combining Fuzzy Imprecision with Probabilistic Uncertainty in Decision Making", *Lecture Notes*, Springer-Verlag, 1988.

[10] D.E. Goldberg, "Genetic Algorithm in Search, Optimization and Machine Learning", Addison-Wesley, Reading, MA, 1989.

[11] G.B. Contini, "Stochastic Approach to Goal Programming", *Operations Research*, **16** pp. 576–586, 1968.

[12] H.A. Simon, "Administrative Behavior", Fress Press, New York, 1957.

[13] H.J. Zimmermann, "Fuzzy Programming and Linear Programming with Several Objective Functions", *Fuzzy Sets and Systems*, **1**, pp. 45–55, 1978.

[14] J.P Ignizio, "Goal Programming and Extensions", D.C. Health, Lexington, MA, 1976.

[15] K. Iwamura, and B. Liu, "A Genetic Algorithm for Chance Constrained Programming", *Journal of Information and Optimization Sciences*, **17 (2)**, pp. 409–422, 1996.

[16] L.A. Zadeh, "Fuzzy Sets a Basis for Theory of Possibility", *Fuzzy Sets and Systems*, **1**, pp. 3–28, 1978.

[17] L.B. Miller and H. Wagner, "Chance-Constrained Programming with Joint Constraints". *Operations Research*, **13**, pp. 930–945, 1965.

[18] M.G. Iskander, "On Solving Stochastic Fuzzy Goal Programming Using a Suggested Interactive Approach", *The Journal of Fuzzy Mathematics*, **9 (2)**, pp. 437–446, 2001.

[19] P. Kall, P. and S.W. Wallace, "Stochastic Programming". John Wiley and Sons, 1996.

[20] R.H. Mohamed, "A Chance Constrained Fuzzy Goal Program", *Fuzzy Sets and Systems*, **47**, pp. 183–186, 1992.

[21] R.K. Jana and M. P. Biswal, "A Stochastic Simulation Based Genetic Algorithm for Chance Constraint Programming Problems with Continuous Random Variables". *International Journal of Computer Mathematics*, **81 (9)**, pp. 1069–1076, 2004.

[22] S. Kataoka, "A Stochastic Programming Model", *Econometrica*, **31**, pp. 181–196, 1963.

[23] W. Mohammed, "Chance Constrained Fuzzy Goal Programming with Right-Hand Side Uniform Random Variable Coefficients", *Fuzzy Sets and Systems*, **109** pp. 107–110, 2000.

[24] Y.K. Liu, and B.K. Liu, "Fuzzy Random Programming with Equilibrium Chance Consraint", *Information Sciences*, **170**, pp. 363–395, 2000.

40

A LINEAR FUZZY GOAL PROGRAMMING METHOD FOR SOLVING FRACTIONAL BILEVEL PROGRAMMING PROBLEMS

Bijay Baran Pal[1*], Shyamal Sen[2] and Mousumi Kumar[3]

[1]*Department of Mathematics, University of Kalyani*
Kalyani-741235, W.B., India.
E-mail: bbpal18@hotmail.com

[2]*Department of Mathematics, BKC College*
Kolkata-700108, W.B., India.
E-mail: ssenk@yahoo.co.in

[3]*Department of Mathematics, Alipurduar College*
Alipurduar Court-736122, W.B., India.
E-mail: roysankha@yahoo.co.in

Abstract: *This paper presents a fuzzy goal programming (FGP) procedure for modeling and solving fractional bilevel programming problems by using a linearization approach in a hierarchical decision making organization.*

In the proposed approach, the individual best and worst solutions of the two decision makers (DMs) located at the two hierarchical levels are determined first. Then, the tolerance membership functions (MFs) of the fuzzily defined fractional objective goals are constructed for measuring the degree of satisfaction of the DMs regarding achievement of their objectives and the degree of optimality of the decision vector controlled by the upper level DM.

In the FGP model formulation, the linear transformation approach presented by Pal et. al. [10] is introduced to the defined fractional membership goals to solve the problem by using the minsum FGP [11] methodology.

In the solution process, achievement of the highest membership value (unity) of the membership goals defined for the MFs of the fuzzy goals to the extent possible by minimizing the under-deviational variables associated with the membership goals on the basis of weights of importance of achieving the fuzzy goals is taken into consideration.

A numerical example is solved and the solution is compared with the solution of FGP approach [14] studied previously.

Keywords: *Goal programming, Fuzzy programming, Bilevel programming, Fractional programming, Linear transformation, Fractional Bilevel Programming Problem.*

1. INTRODUCTION

Bilevel programming (BLP) is a special case of multilevel programming (MLP) in [1] in the area of mathematical programming (MP) with two level structures in a hierarchical decision making organization. In BLP, two decision makers (DMs) are involved and they are located at two different hierarchical levels with the interest of optimizing the individual objective functions in the decision making environment.

The concept of BLP was first introduced by Candler and Townsely in [4]. The field of BLP was then studied deeply form the point of view of its potential use to different real-life problems like agricultural planning, economic system, financial planning, warfare and network design, and specially for resolving the conflict in a hierarchical decision making situation.

During 1980s, several solution approaches in [2, 4] have been investigated for solving BLP problems (BLPPs) from the view point of their potential use to hierarchical decision system. It may mentioned here that most of the classical approaches developed so far in the past are not computationally efficient and sometimes they leads to a paradox that the lower level DM dominates the decision of the upper level DM. Consequently, a decision deadlock arises in most of the hierarchical decision situation. To overcome such a difficulty, the fuzzy set theoretic approach in [5] to the hierarchical decision problems has been introduced by Lai in [7] in 1996. Here, the basic concept of the fuzzy programming (FP) approaches to hierarchical decision problems studied in [8] previously is the same, where repeated evaluation of the problem by considering different values of membership functions, not the highest degree (unity), are needed to reach a satisfactory solution in the decision making context. Here, to overcome the computational load, FGP approach in [13] to BLPPs has been studied by Pal and Moitra in [8, 9, 11, 12], and other in the past.

Now, in a real-life hierarchical decision situation, it is to be mention that fractional objectives are frequently involved to the decision problems in the context of making a balance of trade and optimization of benefit in the planning environment. For instance, the problems for optimizing the profit-cost, inventory-sales ratios, and also the interest of taking other different ratios arise in the decision making situation.

The FGP approach to fractional BLPPs (FBLPPs) has been studied by Pal et. al. in [3] in the recent past. But, the extensive study on large scale hierarchical decision problems is at an early stage.

In this paper, a FGP procedure with the use of a linearization technique for solving BLPP having fractional objectives is presented. In the model formulation of the problem, first the membership goals of the membership functions defined for the fuzzily described objectives are introduced for measuring the degree of achievement of the objectives in terms of the membership values in the decision making context.

In the solution process, minimization of the under-deviational variables of the fuzzy goals from the aspired goal level (unity) on the basis of weights of importance is taken into account to reach a satisfactory decision for distribution of decision powers to the DMs in the decision making environment.

2. FRACTIONAL BLPP FORMULATION

In the hierarchical decision making environment, it is assumed that the DMs are motivated to cooperate each other and each tries to optimize his/her own benefit paying considerable attention to the preference of the other.

Then, let x_1 and x_2 be the vectors of variables controlled by the upper level DM (the leader) and lower level DM (the follower), respectively.

Then, the general form of a FBLPP can be presented as:

$$\underset{x_1}{\text{Max}}\ F_1\ (x_1, x_2) = \frac{c_{11}x_1 + c_{12}x_2 + \alpha_1}{d_{11}x_1 + d_{12}\ x_2 + \beta_1}$$

(Leader's problem) ... (2.1)

and, for given x_1, x_2 solves

$$\underset{x_2}{\text{Max}}\ F_2\ (x_1, x_2) = \frac{c_{21}x_1 + c_{22}x_2 + \alpha_1}{d_{21}x_1 + d_{22}\ x_2 + \beta_1}$$

(Follower's problem) ... (2.2)

subject to

$$(x_1, x_2) \in S = \{(x_1, x_2) \mid A_1\ x_1 + A_2\ x_2 \le b, x_1, x_2 \ge 0\}$$

... (2.3)

where b, c_{ij}, d_{ij} (i, j = 1, 2) are the constant vectors, A_1 and A_2 are constant matrices; α_i , β_i (i = 1, 2) are constants.

It is customary to assume that

$(d_{i1}\ x_1 + d_{i2}\ x_2 + \beta_i) > 0$, ($i$ = 1, 2), to avoid any infeasibility of making decision. Also, it is assumed that S ($\neq \Phi$) is bounded.

Now, the FGP formulation of the problem is described in the Section 3.

3. FUZZY PROGRAMMING FORMULATION

To formulate the FGP model of the BLPP in an imprecise environment, the fuzzy aspiration levels of the objectives of both the DMs and the decision vector x_1 controlled by the leader are determined first.

Then, the defined fuzzy goals are characterized by the membership functions to measure the degree of goal achievement in terms of membership values.

3.1. Construction of Fuzzy Goals

Let $\left(x_1^{\ell_1}, x_2^{\ell_1}; F_1^1\right)$ and $\left(x_1^{\ell_2}, x_2^{\ell_2}; F_1^2\right)$ be the independent best and worst solutions, respectively, of the leader,

where $$F_1^1 = \max_{(x_1, x_2) \in S} F_1\ (x_1, x_2),$$

and $$F_1^2 = \min_{(x_1, x_2) \in S} F_1\ (x_1, x_2).$$

Then, let $\left(x_1^{f_1}, x_2^{f_1}; F_2^1\right)$ and $\left(x_1^{f_2}, x_2^{f_2}; F_2^2\right)$ be the independent best and worst solutions, respectively, of the follower when calculated in isolation,

where $$F_2^1 = \max_{(x_1, x_2) \in S} F_2(x_1, x_2)$$

and $$F_2^2 = \min_{(x_1, x_2) \in S} F_2(x_1, x_2)$$

Then, the fuzzy goals of the leader and follower can be defined as $F_1 \gtrsim F_1^1$, $F_2 \gtrsim F_2^1$ respectively.

Again, the fuzzy goal of the control vector x_1 is obtained as $x_1 \gtrsim x_1^{\ell_1}$.

Now, in the fuzzy decision making context, the lower tolerance limits of the leader and follower can be defined as $F_1^2 (F_1^2 < F_1^1)$ and $F_2^2 (F_2^2 < F_2^1)$, respectively.

Further, since the leader has a higher power of making decision, a certain relaxation of $X_1^{\ell_1}$ as a lower tolerance limit should be given for searching of a better decision by the follower.

Let $x_1^r \left(x_1^{\ell_2} < x_1^r < x_1^{\ell_1}\right)$ be the lower tolerance limit of $X_1^{\ell_1}$.

Then, characterization of membership functions of the defined fuzzy goals is presented in the following Section 3.2.

3.2. Characterization of Membership Function

The tolerance membership function for the defined fuzzy goals can be expressed as [11]:

$$\mu_{F_1}\left[F_1(x_1, x_2)\right] = \begin{cases} 1, & \text{if } F_1(x_1, x_2) \geq F_1^1 \\ \dfrac{F_1(x_1, x_2) - F_1^2}{F_1^1 - F_1^2}, & \text{if } F_1^2 \leq F_1(x_1, x_2) < F_1^1 \\ 0, & \text{if } F_1(x_1, x_2) < F_1^2 \end{cases} \quad \text{... (3.1)}$$

$$\mu_{F_2}\left[F_2(x_1, x_2)\right] = \begin{cases} 1, & \text{if } F_2(x_1, x_2) \geq F_2^1 \\ \dfrac{F_2(x_1, x_2) - F_2^2}{F_2^1 - F_2^2}, & \text{if } F_2^2 \leq F_2(x_1, x_2) < F_2^1 \\ 0, & \text{if } F_2(x_1, x_2) < F_2^2 \end{cases} \quad \text{... (3.2)}$$

$$\mu_{x_1}(x_1) = \begin{cases} 1, & \text{if } x_1 \geq x_1^{\ell\,\prime} \\ \dfrac{x_1 - x_1^r}{x_1^{\ell\,\prime} - x_1^r}, & \text{if } x_1^r \leq x_1 < x_1^{\ell\,\prime} \\ 0, & \text{if } x_1 < x_1^r \end{cases} \qquad \text{... (3.3)}$$

Now, the FGP model formulation of the problem is presented in the Section 4.

4. FGP MODEL FORMULATION

In a fuzzy decision making situation, the aim of each of the DMs is to achieve the highest membership values of each of the membership functions defined for his/her in the decision making environment.

In FGP formulation of the problem, the membership functions are transformed into membership goals by assigning the highest membership values (unity) as the aspiration level and introducing under- and over-deviational variables to each of them.

Then, the membership goals can be successively presented as:

$$\frac{F_1(x_1, x_2) - F_1^2}{F_1^1 - F_1^2} + \rho_{11} - \rho_{12} = 1 \qquad \text{... (4.1)}$$

$$\frac{F_2(x_1, x_2) - F_2^2}{F_2^1 - F_2^2} + \rho_{21} - \rho_{22} = 1 \qquad \text{... (4.2)}$$

$$\frac{x_1 - x_1^r}{x_1^{\ell\,\prime} - x_1^r} + \rho_{31} - \rho_{32} = I \qquad \text{... (4.3)}$$

where $\rho_{k1}, \rho_{k2} \geq 0$, $(k = 1, 2)$ represent the under- and over-deviational variables, respectively, associated with the kth membership goal, and where $\rho_{31}, \rho_{32} \geq 0$ represent the vectors of under- and over-deviational variables associated with the membership goals defined for the decision vector x_1, and I is a column vector with all elements equal to 1 and the dimension of it depends on the dimension of the decision vector x_1.

Now, it is to be observed that the membership goals in (4.1) and (4.2) are fractional in form. To avoid the inherent computational complexity involved with the fractional goals in [6], the linear transformation technique in [10] is used to solve the problem by employ the linear FGP methodology.

4.1 Linear Transformation of Membership Goals

Using the expressions of F_1 (.) and F_2 (.) in (2.1) and (2.2), the membership goals in (4.1) and (4.2) can be represented in a unified form in vector notation as:

$$T_k\left[\frac{C_k X+\alpha_k}{D_k X+\alpha_k}-F_k^2\right]+\rho_{k1}-\rho_{k2}=1, k=1,2. \quad \text{... (4.4)}$$

where $C_k = [c_{k1}, c_{k2}]$ and $D_k = [d_{k1}, d_{k2}]$ are the row vectors, and $X = [x_1, x_2]$ is a column vector and,

$$T_k = \frac{1}{F_k^1 - F_k^2}, k = 1, 2$$

Then, the simplified form of the expression in (4.4) can be obtained as:

$$\left[\frac{C'_k X+\alpha'_k}{D_k X+\beta_k}\right]+\rho_{k1}-\rho_{k2}=1, \quad \text{... (4.5)}$$

where $$C'_k = T_k\left(C_k - F_k^2 D_k\right) \text{ and}$$

$$\alpha'_k = T_k\left(\alpha_k - F_k^2\beta_k\right), \quad k = 1, 2.$$

Now, letting

$$\frac{1}{D_k X+\beta_k} = y_{k0}\ (>0), \quad k=1,2 \quad \text{... (4.6)}$$

the expression in (4.5) takes the form

$$\left[(C'_k X+\alpha'_k)\, y_{k0}\right]+\rho_{k1}-\rho_{k2}=1$$

or, $$\left[C'_k\,(y_{k0} X)+\alpha'_k\, y_{k0}\right]+\rho_{k1}-\rho_{k2}=1 \quad \text{... (4.7)}$$

Then, in the linearization process, a new vector of variables Y_k is defined such that

$$Y_k = y_{k0} X \quad \text{... (4.8)}$$

where, the dimension of Y_k is the same as that of X.

Now, for consideration of such a variable transformation, the expression in (4.7) appears as:

$$\left(C_k' Y_k + \alpha_k' y_{0k}\right) + \rho_{k1} - \rho_{k2} = 1, \quad k = 1, 2 \quad \text{... (4.9)}$$

Here, it is to be followed that the expression of the new vector of variables Y_k in (4.8) can be presented in an explicit form as

$$y_{kj} = y_{k0} \cdot x_j, \; j = 1, 2, ..., n; \; k = 1, 2. \quad \text{... (4.10)}$$

Then, from the component-wise relationships of the expression in (4.10), the ratios of the elements of Y_k can be derived as:

$$\frac{y_{kj}}{y_{1j}} = \frac{y_{k0}}{y_{10}} = \gamma_k \text{ (say)}, \; j = 1, 2, ..., n; \; k = 1, 2 \quad \text{... (4.11)}$$

Using the relations in (4.11), to the expression (4.8), Y_k can be obtained as:

$$Y_k = \gamma_k Y_k, \text{ with } y_{k0} = \gamma_k y_{10} \quad \text{... (4.12)}$$

Now, using the expression in (4.12), the membership goal expressions in (4.9) in linear form appear as:

$$\left(C_k' Y_1 + \alpha_k' y_{10}\right) + \eta_{k1} - \eta_{k2} - \delta_k = 0, \quad k = 1, 2 \quad \text{... (4.13)}$$

where $\eta_{k1}, \eta_{k2} \geq 0$ represent the under- and over-deviational variables, respectively, and where $\eta_{k1} = \delta_k \rho_{k1}$, $\eta_{k2} = \delta_k \rho_{k2}$ and

$$\delta_k = \frac{1}{\gamma_k}; \quad k = 1, 2.$$

Further, letting

$$\frac{1}{x_1^{t_1} - x_1^r} = y_{30} \quad \text{... (4.14)}$$

the goal expressions in (4.3) can be expressed as

$$Y_3 - y_{10} x_1^r + \eta_{31} - \eta_{32} = \delta_3 I \quad \text{... (4.15)}$$

where $\quad Y_3 = y_{10} x_1, \quad \eta_{31} = \delta_3 \rho_{31}, \quad \eta_{32} = \delta_3 \rho_{32},$

$$\delta_3 = \frac{1}{\gamma_3}$$

and dimension of Y_2 depends on the dimension of x_1.

Again, in the sequel of linearization process, the expressions in (4.6) and (4.14) take the form:

$$D_k Y_1 + \beta_k y_{10} - \delta_k = 0, \quad k = 1, 2 \quad \text{... (4.16)}$$

$$y_{10}\left(x_1^{t_1} - x_1^r\right) = \delta_3\, \boldsymbol{i} \qquad \text{... (4.17)}$$

Further, for the consideration of the change of variables, the constraints set in (2.2) can be represented as:

$$AY_1 \le y_{10}b, \qquad \text{... (4.18)}$$

where, A is the vector of the matrices A_1 and A_2.

Now, the most widely used and the simplest version of GP (i.e. *minsum* GP) [11] can be used here to formulate the model of the problem.

The executable FGP model of the problem appears as:

Find $Y\left(y_{10}, y_{11}, y_{12}, ..., y_{1n}\right)$ so as to

$$\text{Minimize } Z = \sum_{k=1}^{2}\left(w_{k1}\,\eta_{k1}\right) + \boldsymbol{w}_{31}\,\boldsymbol{\eta}_{31} \qquad \text{... (4.19)}$$

and satisfy the membership goals in (4.3) and (4.13), subject to the constraint sets in (4.16), (4.17) and (4.18),

where Z represents the goal achievement function, $w_{k1}\ (>0)$, $k = 1$, 2, are the numerical weights associated with η_{k1}, and represent the relative importance of achieving the aspired levels of the goals, and $\boldsymbol{w}_{31}$ (> 0) is the vector of numerical weights associated with $\boldsymbol{\eta}_{31}$, and they are determined as (11):

$$\begin{cases} w_{k1} = \dfrac{1}{F_k^1 - F_k^2}, \text{ for the } \mu_k \text{ defined in (3.1) and (3.2)}, \quad k = 1, 2 \\ \text{and } \boldsymbol{w}_{31} = \dfrac{1}{\boldsymbol{x}_1^{t_1} - \boldsymbol{x}_1^r}, \text{ for the } \mu_k \text{ defined in (3.3)} \end{cases}$$

The FGP model in (4.19) provides the most satisfactory solution for achievement of the aspired goal levels of the membership goals to the extent possible on the basis of the needs and desires of DMs in the decision making environment.

The following numerical example is solved to illustrate the approach.

5. NUMERICAL EXAMPLE

The illustrative example studied by Pal et. al [9] is considered.

The FBLPP is of the form:

$$\underset{x_1}{\text{Maximize}}\ F_1(x_1, x_2) = \frac{2x_1 + x_2}{2x_1 + 3x_2 + 1} \qquad \text{(Leader's problem)}$$

and, for given x_1, x_2 solves

$$\underset{x_2}{\text{Maximize}}\ F_2(x_1, x_2) = \frac{x_1 + 2x_2}{x_1 + x_2 + 1} \qquad \text{(Follower's problem)}$$

subject to

$$\begin{aligned} -x_1 + 2x_2 &\le 3, \\ 2x_1 - x_2 &\le 3 \\ x_1 + 2x_2 &\ge 3, \\ x_1, x_2 &\ge 0 \end{aligned}$$

Now, following the procedure, the best and worst solutions of the leader are obtained as

$$\left(x_1, x_2; F_1^1\right) = \left(2, 1; \frac{5}{8}\right), \text{ and}$$

$$\left(x_1, x_2; F_1^2\right) = \left(1, 2; \frac{4}{9}\right), \text{ respectively.}$$

Similarly, the solution of the follower are obtained as

$$\left(x_1, x_2; F_2^{f_1}\right) = \left(3, 3; \frac{9}{7}\right), \text{ and}$$

$$\left(x_1, x_2; F_2^2\right) = (2, 1; 1), \text{ respectively.}$$

Then, the fuzzy goals can be defined as

$$F_1 \gtrsim \frac{5}{8}, \quad F_2 \gtrsim \frac{9}{7}, \quad \text{and } x_1 \gtrsim 2.$$

The lower tolerance limits for goal achievement of F_1 and F_2 are found as $\frac{4}{9}$ and 1, respectively. Again, the lower tolerance limit of x_1 is reasonably taken as 1.

Then, the membership goals for the defined membership functions can be obtained as:

$$\mu_1 : \frac{8}{13} \cdot \frac{(10x_1 - 3x_2 - 4)}{(2x_1 + 3x_2 + 1)} + \rho_{11} - \rho_{12} = 1$$

$$\mu_2\colon \frac{7}{2}\cdot\frac{(x_2-1)}{(x_1+x_2+1)}+\rho_{21}-\rho_{22}=1$$

$$\mu_3\colon x_1-1+\rho_{31}-\rho_{32}=1$$

Now, following the linearization process, the linear FGP model of the problem can be obtained as:

$$\text{Minimize } Z=\frac{72}{13}\eta_{11}+\frac{7}{2}\eta_{21}+\eta_{31} \text{ so as to satisfy}$$

$$8\left(y_{11}-3y_{12}\,4y_{10}\right)+\eta_{11}-\eta_{12}=1$$

$$7\left(y_{12}-y_{10}\right)+\eta_{21}-\eta_{21}=\delta_2$$

$$y_{11}-y_{10}+\eta_{31}-\eta_{32}=\delta_3$$

subject to

$$13\left(2y_{11}+3y_{12}+y_{10}\right)=1$$

$$2\left(y_{11}+3y_{12}+y_{10}\right)=\delta_2$$

$$y_{10}=\delta_3$$

and

$$-y_{11}+2y_{12}\le 3y_{10}$$

$$2y_{11}-y_{12}\le 3y_{10}$$

$$y_{11}+2y_{12}\ge 3y_{10}$$

where, $\delta_1,\delta_2,\delta_3,y_{10}>0,\ y_{11},y_{12}\ge 0$.

Then, using the software the LINGO (ver 6.0), the solution is obtained as

$$y_{10}=0.009625385,\ y_{12}=0.009625385, \text{ and } y_{11}=0.01923077.$$

The resultant solution of the problem is then obtained as:

$$(x_1,x_2) = (2,1), \text{ with } (F_1,F_2) = (0.625,1)$$

The result shows that a satisfactory decision is reached here in the decision making environment.

Note: If the problem is solved by using the additive FGP model in (14), where minimization of the achievement function (Z) without incorporation of defined weights and subject to $\mu_i \le 1$ the associated MF expressions and $\mu_i \le 1$ is taken into account, then without linearization of the defined fractional membership goals, the solution of the problem is found as

$$(x_1,x_2)=(2,2.5), \text{ with } (F_1,F_2)=(0.52,1.27).$$

A comparison shows that a better solution is achieved here under the proposed approach in terms of satisfaction of both the DMs regarding goal achievement in the order of hierarchy of execution of their decision powers in the decision making environment.

6. CONCLUSION

In this article, a linearization approach to FGP formulation of a FBLPP is presented. The proposed approach can easily be extended to solve multiobjective BLPPs with fractional criteria as well as other bilevel decentralized planning problems.

An extension of the approach to fractional quadratic BLPPs may be a problem for future study.

Finally, it is expected that the FGP approach presented here for solving FBLPPs can contribute to future study on multiobjective hierarchical decision making problems.

ACKNOWLEDGEMENT

The authors are thankful to the anonymous referee and Prof. R.N. Bhaumik, Conference Chair ICRFSC 2009, for their helpful suggestions in improving the quality of presentation of the paper.

REFERENCES

[1] Bard, J.F. and Falk, J.E., "An Explicit Solution to the Multi-Level Programming Problem", *Computers and Operations Research*, **9**, pp. 77 - 100, 1982.

[2] Bialas, W.F. and Karwan, M.H., "Two-Level Linear Programming", *Management Sciences*, **30**, pp. 1004 – 1020, 1984.

[3] Biswas, A. and Pal, B.B., "Fuzzy Goal Programming Procedure for Multiobjective Bilevel Programming Problems", *International Journal of Computer, Mathematical Sciences and Applications*, **1**, No. 1, pp. 87 – 98, 2007.

[4] Candler, W. and Townsley, R., "A Linear Two-Level Programming Problem", *Computer and Operations Research*, 9, pp. 59 – 76, 1982.

[5] Hannan, E.L., "Linear Programming with Multiple Fuzzy Goals", *Fuzzy Sets and Systems*, **6**, pp. 235 – 248, 1981.

[6] Hannan, E.L.: "On an Interpretation of Fractional Objectives in Goal Programming as Related to Papers by Awer Buch al. and Hannan", *Management Science*, **27**, No. 7, pp. 847 – 848, 1981.

[7] Lai, Y.J., "Hierarchical Optimization: A Satisfactory Solution", *Fuzzy Sets and Systems*, **77**, pp. 321–335, 1996.

[8] Moitra, B.N., Pal, B.B., "A Fuzzy Goal Programming Approach for Solving Bilevel Programming Problems", *in: N.R. Pal, M. Sugeno (Eds.), Advances in Soft Computing–Afss 2002, Springer-Verlag, Berlin, Heidelberg, New York*, No. **2275**, pp. 91–98, 2002.

[9] Pal, B.B. and Moitra, B.N., "Goal Programming Approach to Fuzzy Programming with Multiple Objectives", *in: Proceedings of the National Conference on "Networking of Machines, Microprocessors, IT and HRD-Need of the Nation in the Next Millennium"*, pp. B-100-B-109, 1999.

[10] Pal, B.B., Sen, S., "A Goal Programming Procedure for Solving Interval Valued Multiobjective Fraction Programming Problem", *Proceedings of the 16th International Conference on Advance Computing and Communication (ADCOM), (corresponding in IEEE explore)*, pp. 297–302, 2008.

[11] Pal, B.B., Moitra, B.N., Maulik, U., "A Goal Programming Procedure for Fuzzy Multiobjective Linear Fractional Programming Problem", *Fuzzy Sets and Systems*, **139**, No. 2, pp. 395–405, 2003.

[12] Pal, B.B., Moitra, B.N., "A Fuzzy Goal Programming Procedure for Solving Quadratic Bilevel Programming Problems", *International Journal of Intelligent Systems*, **18**, No. 5, pp. 529–540, 2003.

[13] Pal, B.B. and Moitra, B.N. (2003), "A Goal Programming Procedure for Solving Problems with Multiple Fuzzy Goals Using Dynamic Programming", *European Journal of Operational Research*, **144**, No. 3, pp. 480–491.

[14] Tiwari, R.N., Dharmar S. and Rao J.R., "Fuzzy Goal Programming an Additive Model", *Fuzzy Sets and Systems*, **24**, pp. 27–34, 1987.

41

A GOAL PROGRAMMING METHOD FOR SOLVING ACADEMIC PERSONNEL PLANNING PROBLEMS WITH INTERVAL-VALUED RESOURCE GOALS IN UNIVERSITY MANAGEMENT SYSTEM USING GENETIC ALGORITHM

Bijay Baran Pal[1*], Somsubhra Gupta[2], Papun Biswas[3] and Anirban Mukhopadhyay[4]

[1]Department of Mathematics, University of Kalyani, Kalyani-741235, W.B., India, E-mail: bbpal18@hotmail.com*

[2]Department of Information Technology, JIS College of Engineering, Kalyani-741235, W.B., India, E-mail: gsomsubhra@gmail.com

[3]Department of Electrical Engineering, Narula Institute of Technology, Kolkata-700109, W.B., India, E-mail: papunbiswas@yahoo.com

[4]Department of Computer Science and Engineering, University of Kalyani, Kalyani-741235, W.B., India, Email: anirban@klyuniv.ac.in

Abstract: *This paper presents how the genetic algorithm (GA) method can be efficiently used to the goal programming (GP) formulation of academic resource planning problems with fractional goals and target intervals in university management system.*

A case example of the University of Kalyani, West Bengal (WB), India is solved to demonstrate the proposed model. The model solution is compared with the existing staff allocation pattern as well as the fuzzy goal programming (FGP) solution studied [11] previously.

Keywords: *Genetic Algorithm, Goal Programming, Fractional Goal, Interval Programming, University Management System.*

1. INTRODUCTION

The interest in higher education in new study areas has increased a lot in the recent years from the view point of academic interest and job prospect to different activity areas. As a matter of fact, initiative for opening of new academic departments has been taken into account by most of the higher academic institutions. But, most of the institutions are faced with a lot of problems for proper allocation of academic and other staff in the departments due to budget limititions.

Since the mid-1960s, different mathematical programming (MP) models studied for university management was surveyed by Schroeder in [19].

Now, since most of the academic planning problems are multiobjective in nature, the conventional goal programming (GP) approach in [7, 17], as one of the most promising tool for multiobjective decision analysis, has been introduced to the field of academic planning problems in [9, 20, 12, 13].

Again, now in most of the real-life decision situations, it is to be observed hat fuzzy model parameters are introduced in different ways in the field of fuzzy programming (FP) in [21] for solving different multiobjective decision making problems in [5, 10].

The use of fuzzy goal programming (FGP) in [10] to academic resource planning problems has been studied [11] in the past.

In the real-world decision making situations, GAs based on natural selection and population genetics [6], have also appeared as robust computational tools in the field of optimization problems [4]. In the context of solving staff allocation problems in university system, the GA based FGP approach has been studied [13] in the recent past.

It is to be observed here that the assigning of fuzzy goal values to the objectives may not always be possible in practice due to imprecise in nature of human judgments. As a consequence, decision trouble for goal achievement within the permissible tolerance ranges may arise in a practical decision situation.

To overcome the above difficulty, interval programming (IP) approaches in [8] have appeared as prominent tools for solving decision problems with interval parameter values instead of fuzzy description of them.

The GA approach to IP problems with fractional criteria has been reported [14] in the recent past.

However, the use of IP approach as well as GA based solution approach to academic planning problems is at an early stage.

In the present academic resource planning context, the certain ratios of teaching staff, non-teaching staff and total number of students which are inherently associated with the problem and which have the characteristic of fractional programming are described in the form of interval-valued to make a satisfactory decision in the inexact decision environment.

In the model formulation of the problem, first the interval-valued goals are converted into the standard goals in the framework of GP by using interval arithmetic technique [8]. In the executable GP model, the fractional goals of the problem in the ratio forms are directly introduced to the genetic search process of the GA method, instead of using traditional linearization technique [18] used for fractional goals in the mathematical programming approaches.

A case example is solved to illustrate the potential use of the approach.

2. PROBLEM FORMULATION

In the decision situation, the two types of goals, interval-valued goals and crisp goals are involved with the problem.

The goals are defined as follows :

2.1 Definition of Interval-Valued Goals

The goals in fractional from appear as

$$\text{Optimize } Z_k(X) = \frac{\left[a_k^L, a_k^U\right]X + \left[\alpha_k^L, \alpha_k^U\right]}{\left[b_k^L, b_k^U\right]X + \left[\beta_k^L, \beta_k^U\right]} = \left[t_k^L, t_k^U\right],$$

$$k = 1, 2 \ldots K_1 \qquad \ldots (2.1)$$

It is customary to assume that $\left[b_k^L, b_k^U\right]X + \left[\beta_k^L, \beta_k^U\right] > 0$ to preserve the feasibility of the solution.

The goals in linear form can be presented as

$$\text{Optimize } Z_k(X) = \left[a_k^L, a_k^U\right]X + \left[\alpha_k^L, \alpha_k^U\right] = \left[t_k^L, t_k^U\right],$$
$$k = (K_1 + 1), \ldots, K_2 \qquad \ldots (2.2)$$

In the above two forms of the goals, X represents the vector of decision variables associated with the kth objective $Z_k(.)$, $\left[a_k^L, a_k^U\right]$ and $\left[b_k^L, b_k^U\right]$ are the vectors of interval-valued coefficients, $\left[t_k^L, t_k^U\right]$ indicates the vector of target intervals, and $\left[\alpha_k^L, \alpha_k^U\right]$ and $\left[\beta_k^L, \beta_k^U\right]$ are the interval valued constant vectors associated with the respective goals, where L and U stands for the lower- and upper-bounds respectively.

2.2 Definition of Crisp Goals

The goals with crisp coefficients and right hand side values appear as

$$\text{Optimize: } Z_k(X) \binom{\geq}{\leq} t_k;$$
$$k = (K_2 + 1), (K_2 + 2), \ldots, K \qquad \ldots (2.3)$$

where t_k represents the crisp target value of the kth objective.

Now, in the model formulation of the problem, the interval valued goals are to be converted into the conventional goals.

2.3 Transformation of Interval-Valued Goals

Using interval arithmetic technique [8], the goal Expressions in (2.1) and (2.2) can be explicitly rewritten as follows.

The fractional interval goals in (2.1) appear as

$$\left(\frac{\sum_{j=1}^{n} a_{kj}^L x_j + \alpha_k^L}{\sum_{j=1}^{n} b_{kj}^U x_j + \beta_k^U}, \frac{\sum_{j=1}^{n} a_{kj}^U x_j + \alpha_k^U}{\sum_{j=1}^{n} b_{kj}^L x_j + \beta_k^L} \right) = \left[t_k^L, t_k^U\right],$$

$$k = 1, 2, \ldots, K_1 \qquad \ldots (2.4)$$

The linear interval goals in (2) take the form

$$\left[\sum_{j=1}^{n} a_{kj}^{L} x_j + \alpha_k^L, \sum_{j=1}^{n} a_{kj}^{U} x_j + \alpha_k^U \right] = \left[t_k^L, t_k^U \right],$$

$$k = (K_1 + 1), (K_1 + 2), \ldots, K_2 \qquad \ldots (2.5)$$

Then, to formulate the GP model of the problem, the goals defined in (2.3), (2.4) and (2.5) are to be transformed into the flexible goals by introducing under- and over- deviational variables to each of them.

2.4 Construction of Flexible Goals

The standard flexible goal expressions of the goals in (2.4) appear as

$$\frac{\sum_{j=1}^{n} a_{kj}^{U} x_j + \alpha_k^U}{\sum_{j=1}^{n} a_{kj}^{L} x_j + \beta_k^L} + d_{kL}^{-} - d_{kL}^{+} = t_k^L \qquad \ldots (2.6)$$

$$\frac{\sum_{j=1}^{n} a_{kj}^{L} x_j + \alpha_k^L}{\sum_{j=1}^{n} b_{kj}^{U} x_j + \beta_k^U} + d_{kU}^{-} - d_{kU}^{+} = t_k^U, \qquad \ldots (2.7)$$

$$k = 1, 2, \ldots, K_1$$

Similarly the flexible goal Expressions in (2.5) take the form

$$\sum_{j=1}^{n} a_{kj}^{U} x_j + \alpha_k^U + d_{kL}^{-} - d_{kL}^{+} = t_k^L, \qquad \ldots (2.8)$$

$$\sum_{j=1}^{n} a_{kj}^{L} x_j + \alpha_k^L + d_{kU}^{-} - d_{kU}^{+} = t_k^U, \qquad \ldots (2.9)$$

$$k = (K_1 + 1), (K_1 + 2), \ldots, K_2$$

where $\left(d_{kL}^{-}, d_{kU}^{-}\right)$ and $\left(d_{kL}^{+}, d_{kU}^{+}\right)$ with $d_{kL}^{-} . d_{kL}^{+} = 0$ and $d_{kU}^{-} . d_{kU}^{+} = 0$, represents the sets of under- and over-deviational variables, respectively, associated with the respective goals.

Again the goal expression in (2.3) in the explicit form appears as

$$Z_k(X) + d_k^- - d_k^+ = t_k, \quad k = (K_2 + 1), ..., K \qquad ... (2.10)$$

where $d_k^-, d_k^+ \geq 0$ with $d_k^- . d_k^+ = 0$ represent the under- and over-deviational variables, respectively, associated with the *k*th goal, $k = (K_2 + 1), ..., K$.

Now, the GA scheme adopted in the solution search process is presented in the following Section 3.

3. DESIGN OF GA SCHEME

The genetic parameters and the core function adopted in the search process are presented in the following algorithmic steps:

Step 1: Representation and Initialization: Let V denote the binary coded representation of chromosome in a population, where $V = (x_1, x_2, ..., x_n)$. The population size is defined by pop_size, and pop_size chromosomes are randomly initialized in its search domain.

Step 2: Fitness function: The fitness [6] score of each individual is evaluated by the defined objective function. The fitness function can be presented as follows:

$$\text{eval } (V_\ell) = (Z)_\ell, \quad \ell = 1, 2, ..., \text{pop\_size}$$

where Z represents the objective function to be evaluated.

Then, in the search process, the best chromosome with the highest fitness value at each generation is determined as

$$V^* = \min \{\text{eval } (V_\ell) | \ \ell = 1, 2, ..., \text{pop\_size}\},$$

which depends on function searching of the best value of an objective.

Step 3: Selection: The selection process defines the choosing of individuals (the parent solutions) from the population for the mating purpose. The Roulette wheel scheme [4] is used for parent selection. In this scheme, the selection of two individuals is made on the basis of their successive highest fitness scores.

Step 4: Crossover: The parameter P_c is defined as the probability of crossover. The conventional arithmetic crossover operator (single-point crossover) of a genetic system is applied here in the sense that the resulting offspring always satisfy the given system constraints set. In the crossover mechanism, a chromosome (solution individual) is selected as a parent for producing offspring, if for a defined random number $r \in [0, 1]$, $r < P_c$ is satisfied.

Here, for the two selected parents X_1 and $X_2 \in S$ (the feasible search space), the arithmetic crossover yields the offspring X_1^1 and X_2^1 as

$$X_1^1 = \alpha X_1 + (1 - \alpha) X_2,$$
$$X_2^1 = (1 - \alpha) X_1 + \alpha X_2, \quad 0 < \alpha < 1,$$

where X_1^1 and $X_2^1 \in S$.

Step 5: Mutation: The parameter P_m, as in the conventional GA scheme, is defined as the probability of mutation. The mutation operation is performed on a bit-by-bit basis, where for a random number $r \in [0, 1]$, a chromosome is selected for mutation provided that $r < P_m$.

Step 6: Termination: The execution of whole process terminates when the fitted chromosome is reported at certain generation number in the solution search process.

Now, the different types of parameters and variables involved with the problem are defined in the following Section 3.1.

3.1 Parameters and Variables

3.1.1 Definition of Parameters

The two types of parameters, crisp and interval-valued, are defined as follows:

$[FTS]_{ijt}$ = Minimum number of full-time teaching staff (FTS) required in the department i, rank j at the time period t.

$[F_{it}^L, F_{it}^U]$ = The specified interval in which the total number of FTS should be employed to the department i for smooth running the department at the time period t.

$[S]_{it}$ = Total number of students (TS) in the department i, at the time period t.

$[FS]_{ijt}$ = Annual (average) salary of a FTS in the department i, rank j at the time period t.

$[NS]_{it}$ = Annual (average) salary of a non-teaching staff (NTS) in the department i at the time period t.

$[PS]_{it}$ = Annual remuneration for a part-time teaching staff (PTS) in the department i at the time period t.

$[q_i^L, q_i^U]$ = Target interval to maintain the ratio of PTS and FTS in the department i.

$[r_i^L, r_i^U]$ = Target interval to maintain the ratio of NTS and Total Teaching Staff (TTS) (FTS and PTS) in the department i.

$[s_i^L, s_i^U]$ = Target interval to maintain the ratio of TS and TTS in the department i.

$[b_{it}^L, b_{it}^U]$ = Target interval for the Pay-roll budget allocated to the department i at the time period t.

3.1.2 Definition of Decision Variables

The following decision variables are involved in the proposed model.

f_{ijt} = Number of FTS to be employed in the department i, rank j at the time period t for smooth functioning of the academic activities of the department.

N_{it} = Number of NTS required in the department i at the time period t.

p_{it} = Number of PTS required in the department i at the time period t.

Now, the crisp goals and interval-valued target goals of the model are described in the following Section 4.

4. DESCRIPTION OF CRISP GOALS AND INTERVAL-VALUED GOALS

4.1 FTS Goals

Both the crisp goals and interval-valued goals are involved regarding employment of FTS in a department.

4.1.1 Crisp Goals

For smooth functioning of academic activities, university should always have to try to provide a minimum number of FTS at each rank to each of the departments at the time period t.

The goals with crisp targets levels can be presented as:

$$f_{ijt} + d_p^- - d_p^+ = [FTS]_{ijt},$$
$$i = 1, 2, ..., I; \quad j = 1, 2, ..., J; \quad p = 1, 2, ..., IJ$$

4.1.2 Interval-Valued Goals

For potential academic performance, a total number FTS within a specified interval must be employed to each department.

The goals with target intervals appear as

$$\sum_{j=1}^{J} f_{ijt} = \left[F_{it}^L, F_{it}^U\right], \quad i = 1, 2, ..., I$$

4.2 Ratio Goals

For enrichment of academic performance at each of the departments, certain ratios in the respective specified target intervals need to be satisfied.

The ratio goals are defined as follows.

4.2.1 PTS-FTS Ratio Goals

The goal expressions with target-intervals appear as

$$\frac{P_{it}}{\sum_{j=1}^{n} f_{ijt}} = \left(q_i^L, q_i^U\right), \quad i = 1, 2, ..., I$$

4.2.2 NTS-TTS Ratio Goals

The goal expressions with the target intervals take the form

$$\frac{N_{it}}{\left(\sum_{j=1}^{n} f_{ijt} + P_{it}\right)} = \left[r_i^L, r_{it}^U\right], \quad i = 1, 2, ..., I$$

4.2.3 TS-TTS Ratio Goals

The goal expression can be presented as

$$\frac{[S]_{it}}{\left(\sum_{j=1}^{n} f_{ijt} + P_{it}\right)} = \left[s_i^L, s_i^U\right], \quad i = 1, 2, ..., I$$

4.3 Budget Goal

Due to limitation of total pay-roll budget to run the university, the pay-roll budget for each of the departments is described as interval valued.

The budget goal expression appears as

$$\sum_{j=1}^{J} [FS]_{ijt} \cdot f_{ijt} + [NS_{it}] . N_{it} + [PS_{it}] P_{it} = \left[b_{it}^L, b_{it}^U\right]$$

$$i = 1, 2, ..., I$$

Now, conversation of the interval-valued goals to the standard form of goals in the framework of GP and the model solution of the problem are presented via a case example in the Section 5.

5. AN ILLUSTRATIVE EXAMPLE: A CASE STUDY

The academic resource allocation problem of the University of Kalyani, W.B., India is considered to expound the potential use of the proposed approach. The resource allocation problems of the six new departments: Microbiology (MB), Computer Science & Engg. (CSE), Master of Business Administration (MBA), Geography (GEO), Molecular Biology and Bio-Technology (MB and BT) and Physiology (PHY) established in the recent past are considered.

The data for the proposed model were collected from the Budget allocation programme published in [1] by the University of Kalyani (in the financial year 2008-2009), Kalyani, W.B., India.

The pay-roll budget allocation made to each of the six departments in the financial year 2008-2009 is considered as the lower bound (b_{i1}^L) and the estimated budget allocation for the financial year 2009-2010 is considered as the upper bound (b_{i1}^U) for each of the department i. Then, the target interval-valued budget for the departments, are presented in the Table 1.

Table 1: *Data Description for Interval-Valued Pay-Roll Budget*

Department	*MB (1)* $[b_{1r}^L, b_{1t}^U]$	*CSE (2)* $[b_{2r}^L, b_{2t}^U]$	*MBA (3)* $[b_{3r}^L, b_{3t}^U]$	*GEO (4)* $[b_{4r}^L, b_{4t}^U]$	*MB and BT (5)* $[b_{5r}^L, b_{5t}^U]$	*PHY (6)* $[b_{6r}^L, b_{6t}^U]$
Pay-roll Budget in interval (in Rs. lac)	[11.41, 15.45]	[26, 31.10]	[16.11, 18.53]	[14.95, 16.95]	[12.86, 15.94]	[11.78, 14.44]

The data for the annual average salaries of FTS, NTS and PTS are described in the Table 2.

Table 2: *Annual Average Salaries of FTS, NTS and Cost for PTS*

Rank	*Professor (1)*	*Reader (2)*	*Lecturer (3)*	*PTS*	*NTS*
Salary/ Remuneration (Rs. Lack)	3.30	2.40	2.00	0.705	0.48

Note: In the UGC (India) system, Reader is equivalent to Associate Professor and lecturer is equivalent to Assistant Professor.

The target intervals for the different ratio goals are presented in the Table 3.

Table 3: *Target Intervals for the Ratio Goals*

Department	*Number of students*	*PTS-FTS* $[q_i^l, q_i^U]$	*NTS-TTS* $[r_i^L, r_i^U]$	*TS-TTS* $[s_i^L, s_i^U]$
1	32	[0.25, 0.35]	[0.38, 0.4]	[0.18, 0.22]
2	110	[0.25, 0.35]	[0.38, 0.4]	[0.03, 0.064]
3	50	[0.25, 0.35]	[0.33, 0.35]	[0.10, 0.14]
4	50	[0.25, 0.35]	[0.45, 0.5]	[0.10, 0.14]
5	24	[0.25, 0.35]	[0.38, 0.4]	[0.25, 0.292]
6	30	[0.25, 0.35]	[0.38, 0.4]	[0.20, 0.233]

Again, as suggested by the UGC and other funding authorities (Govt. of India), the target intervals for the total FTS of each department are given in the Table 4:

Table 4: *Target Intervals for FTS*

Department	*1*	*2*	*3*	*4*	*5*	*6*
Target intervals for total FTS $[F_{i1}^L, F_{i1}^U]$	[6, 8]	[8, 10]	[8, 10]	[6, 8]	[6, 8]	[6, 8]

Now, using the data tables and the other necessary collected data, the crisp and interval goals are constructed in the Section 5.1. Here, since the data for the running period ($t = 1$) is considered, the time specification (t) is omitted for simplicity during presentation of the case model.

5.1 Construction of Model Goals

5.1.1 Crisp Goals

FTS Goal: It is always desired that a minimum number of FTS of each of the ranks in each department should be employed by the university.

The goals expressions by introducing aspired levels appear as:

$$f_{11} + d_1^- - d_1^+ = 1,\quad f_{12} + d_2^- - d_2^+ = 2,\quad f_{13} + d_3^- - d_3^+ = 3,$$
$$f_{21} + d_4^- - d_4^+ = 2,\quad f_{22} + d_5^- - d_5^+ = 2,\quad f_{23} + d_6^- - d_6^+ = 4,$$
$$f_{31} + d_7^- - d_7^+ = 1,\quad f_{32} + d_8^- - d_8^+ = 2,\quad f_{33} + d_9^- - d_9^+ = 2,$$
$$f_{41} + d_{10}^- - d_{10}^+ = 2,\quad f_{42} + d_{11}^- - d_{11}^+ = 1,\quad f_{43} + d_{12}^- - d_{12}^+ = 2,$$
$$f_{61} + d_{16}^- - d_{16}^+ = 1,\quad f_{62} + d_{17}^- - d_{17}^+ = 3,\quad f_{63} + d_{18}^- - d_{18}^+ = 1,$$

5.1.2 Target Interval Associated Goals

FTS goals: Using the data in the Table 4, and following the procedure, the goals for the total FTS for each of the departments are described as follows:

$$f_{11} + f_{12} + f_{13} + d_{1L}^- - d_{1L}^+ = 6,\quad f_{11} + f_{12} + f_{13} + d_{1U}^- - d_{1U}^+ = 8,$$
$$f_{21} + f_{22} + f_{23} + d_{2L}^- - d_{2L}^+ = 8,\quad f_{21} + f_{22} + f_{23} + d_{2U}^- - d_{2U}^+ = 10,$$
$$f_{31} + f_{32} + f_{33} + d_{3L}^- - d_{3L}^+ = 8,\quad f_{31} + f_{32} + f_{33} + d_{3U}^- - d_{3U}^+ = 10,$$
$$f_{41} + f_{42} + f_{43} + d_{4L}^- - d_{1L}^+ = 6,\quad f_{41} + f_{42} + f_{43} + d_{4U}^- - d_{4U}^+ = 8,$$
$$f_{51} + f_{52} + f_{53} + d_{5L}^- - d_{5L}^+ = 6,\quad f_{51} + f_{52} + f_{53} + d_{5U}^- - d_{5U}^+ = 8,$$
$$f_{61} + f_{62} + f_{63} + d_{6L}^- - d_{6L}^+ = 6,\quad f_{61} + f_{62} + f_{63} + d_{6U}^- - d_{6U}^+ = 8,$$

Budget Goals: Using the Tables 1 and 2, budget goals can be presented as

$$3.30\, f_{11} + 2.40\, f_{12} + 2\, f_{13} + 0.705\, N_1 + 0.48\, p_1 + d_{7L}^- - d_{7L}^+ = 11.41$$

$$3.30\, f_{11} + 2.40\, f_{12} + 2\, f_{13} + 0.705\, N_1 + 0.48\, p_1 + d_{7U}^- - d_{7U}^+ = 15.45$$

$$3.30\, f_{21} + 2.40\, f_{22} + 2\, f_{23} + 0.705\, N_2 + 0.48\, p_2 + d_{8L}^- - d_{8L}^+ = 26.00$$

$$3.30\, f_{21} + 2.40\, f_{22} + 2\, f_{23} + 0.705\, N_2 + 0.48\, p_2 + d_{8U}^- - d_{8U}^+ = 27.00$$

$$3.30\, f_{31} + 2.40\, f_{32} + 2\, f_{33} + 0.705\, N_3 + 0.48\, p_3 + d_{9L}^- - d_{9L}^+ = 16.11$$

$$3.30\, f_{31} + 2.40\, f_{32} + 2\, f_{33} + 0.705\, N_3 + 0.48\, p_3 + d_{9U}^- - d_{9U}^+ = 18.53$$

$$3.30\, f_{41} + 2.40\, f_{42} + 2\, f_{43} + 0.705\, N_4 + 0.48\, p_4 + d_{10L}^- - d_{10L}^+ = 14.95$$

$$3.30\, f_{41} + 2.40\, f_{42} + 2\, f_{43} + 0.705\, N_4 + 0.48\, p_4 + d_{10U}^- - d_{10U}^+ = 16.95$$

$$3.30\, f_{51} + 2.40\, f_{52} + 2\, f_{53} + 0.705\, N_5 + 0.48\, p_5 + d_{11L}^- - d_{11L}^+ = 12.86$$

$$3.30\, f_{51} + 2.40\, f_{52} + 2\, f_{53} + 0.705\, N_5 + 0.48\, p_5 + d_{11U}^- - d_{11U}^+ = 15.94$$

$$3.30\, f_{61} + 2.40\, f_{62} + 2\, f_{63} + 0.705\, N_6 + 0.48\, p_6 + d_{12L}^- - d_{12L}^+ = 11.78$$

$$3.30\, f_{61} + 2.40\, f_{62} + 2\, f_{63} + 0.705\, N_6 + 0.48\, p_6 + d_{12U}^- - d_{12U}^+ = 14.44$$

Then using the data in the Table 3, the different ratio goals can be obtained as follows.

PTS-FTS Ratio Goals: The PTS and FTS ratio objective goals are obtained as

$$\frac{p_1}{f_{11} + f_{12} + f_{13}} + d_{13L}^- - d_{13L}^+ = 0.25,$$

$$\frac{p_1}{f_{11} + f_{12} + f_{13}} + d_{13U}^- - d_{13U}^+ = 0.35$$

$$\frac{p_2}{f_{21} + f_{22} + f_{23}} + d_{14L}^- - d_{14L}^+ = 0.25,$$

$$\frac{p_2}{f_{21} + f_{22} + f_{23}} + d_{14U}^- - d_{14U}^+ = 0.35$$

$$\frac{p_3}{f_{31} + f_{32} + f_{33}} + d_{15L}^- - d_{15L}^+ = 0.25,$$

$$\frac{p_3}{f_{31}+f_{32}+f_{33}}+d_{15U}^{-}-d_{15U}^{+}=0.35,$$

$$\frac{p_4}{f_{41}+f_{42}+f_{43}}+d_{16L}^{-}-d_{16L}^{+}=0.25,$$

$$\frac{p_4}{f_{41}+f_{42}+f_{43}}+d_{16U}^{-}-d_{16U}^{+}=0.35,$$

$$\frac{p_5}{f_{51}+f_{52}+f_{53}}+d_{17L}^{-}-d_{17L}^{+}=0.25,$$

$$\frac{p_5}{f_{51}+f_{52}+f_{53}}+d_{17U}^{-}-d_{17U}^{+}=0.35$$

$$\frac{p_6}{f_{61}+f_{62}+f_{63}}+d_{18L}^{-}-d_{18L}^{+}=0.25,$$

$$\frac{p_6}{f_{61}+f_{62}+f_{63}}+d_{18U}^{-}-d_{18U}^{+}=0.35$$

NTS-TTS Ratio Goals: The goal expressions take the form

$$\frac{N_1}{\left[\left(f_{11}+f_{12}+f_{13}\right)+p_1\right]}+d_{19L}^{-}-d_{19L}^{+}=0.38$$

$$\frac{N_1}{\left[\left(f_{11}+f_{12}+f_{13}\right)+p_1\right]}+d_{19U}^{-}-d_{19U}^{+}=0.40$$

$$\frac{N_2}{\left[\left(f_{21}+f_{22}+f_{23}\right)+p_2\right]}+d_{20L}^{-}-d_{20L}^{+}=0.38$$

$$\frac{N_2}{\left[\left(f_{21}+f_{22}+f_{23}\right)+p_2\right]}+d_{20U}^{-}-d_{20U}^{+}=0.40$$

$$\frac{N_3}{\left[\left(f_{31}+f_{32}+f_{33}\right)+p_3\right]}+d_{21L}^{-}-d_{21L}^{+}=0.30$$

$$\frac{N_3}{\left[\left(f_{31}+f_{32}+f_{33}\right)+p_3\right]}+d_{21U}^{-}-d_{21U}^{+}=0.35$$

$$\frac{N_4}{\left[\left(f_{41}+f_{42}+f_{43}\right)+p_4\right]}+d_{22L}^{-}-d_{22L}^{+}=0.10$$

$$\frac{N_4}{\left[\left(f_{41}+f_{42}+f_{43}\right)+p_4\right]}+d_{22U}^{-}-d_{22U}^{+}=0.14$$

$$\frac{N_5}{\left[\left(f_{51}+f_{52}+f_{53}\right)+p_5\right]}+d_{23L}^{-}-d_{23L}^{+}=0.10$$

$$\frac{N_5}{\left[\left(f_{51}+f_{52}+f_{53}\right)+p_5\right]}+d_{23U}^{-}-d_{23U}^{+}=0.14$$

$$\frac{N_6}{\left[\left(f_{61}+f_{62}+f_{63}\right)+p_6\right]}+d_{24L}^{-}-d_{24L}^{+}=0.20$$

$$\frac{N_6}{\left[\left(f_{61}+f_{62}+f_{63}\right)+p_6\right]}+d_{24U}^{-}-d_{24U}^{+}=0.233$$

TS-TTS Ratio Goals: The goal expression can be obtained as

$$\frac{1}{\left[\left(f_{11}+f_{12}+f_{13}\right)+p_1\right]}+d_{25L}^{-}-d_{25L}^{+}=0.18$$

$$\frac{1}{\left[\left(f_{11}+f_{12}+f_{13}\right)+p_1\right]}+d_{25U}^{-}-d_{25U}^{+}=0.22$$

$$\frac{1}{\left[\left(f_{21}+f_{22}+f_{23}\right)+p_2\right]}+d_{26L}^{-}-d_{26L}^{+}=0.03$$

$$\frac{1}{\left[\left(f_{21}+f_{22}+f_{23}\right)+p_2\right]}+d_{26U}^{-}-d_{26U}^{+}=0.064$$

$$\frac{1}{\left[\left(f_{31}+f_{32}+f_{33}\right)+p_3\right]}+d_{27L}^{-}-d_{27L}^{+}=0.25$$

$$\frac{1}{\left[\left(f_{31}+f_{32}+f_{33}\right)+p_3\right]}+d_{27U}^{-}-d_{27U}^{+}=0.292$$

$$\frac{1}{\left[\left(f_{41}+f_{42}+f_{43}\right)+p_4\right]}+d_{28L}^{-}-d_{28L}^{+}=0.10$$

$$\frac{1}{\left[\left(f_{41}+f_{42}+f_{43}\right)+p_4\right]}+d_{28U}^{-}-d_{28U}^{+}=0.14$$

$$\frac{1}{\left[\left(f_{51}+f_{52}+f_{53}\right)+p_5\right]}+d_{29L}^{-}-d_{29L}^{+}=0.10$$

$$\frac{1}{\left[\left(f_{51}+f_{52}+f_{53}\right)+p_5\right]}+d_{29U}^{-}-d_{29U}^{+}=0.14$$

$$\frac{1}{\left[\left(f_{61}+f_{62}+f_{63}\right)+p_6\right]}+d_{30L}^{-}-d_{30L}^{+}=0.20$$

$$\frac{1}{\left[\left(f_{61}+f_{62}+f_{63}\right)+p_6\right]}+d_{30U}^{-}-d_{30U}^{+}=0.233$$

Now, in a decision situation, the aim of the DM is to achieve the objective goal values by means of minimizing the associated deviational variables in the achievement function of the GP formulation of the problem to the extent possible in the decision making environment. In the GP model formulation of the present problem, both the aspects of GP [*minsum* GP [7] and *minmax* GP [16]] are taken together as a convex combination of them to reach the satisfactory decision within the specified ranges for achievement of the interval-valued goals. Here, the term 'regret function', instead of goal achievement function in GP, is used in the notion of minimizing the possible regrets towards achieving the goal values within the defined intervals from the optimistic point of view of the decision maker. The regret function is formulated by taking two priority levels (P_1 and P_2) under the preemptive priority based GP [16].

The regret function appears as:

Find $\left\{f_{ij}, N_i, p_i \mid i=1,2,\ldots,6;\ j=1,2,3,4\right\}$ so as to

Minimize $Z =$

$$\left[\begin{array}{c}\left(d_1^{-}+d_2^{-}+\ldots.+d_{18}^{-}\right)+\\ \left\{\lambda_1\left[\sum_{i=1}^{6} w_{i1}\left(d_{iL}^{-}+d_{iU}^{+}\right)+\sum_{i=13}^{30} w_{i1}\left(d_{iL}^{-}+d_{iU}^{+}\right)\right]+\right\}\\ \left(1-\lambda_1\right)V_1, P_2\left[\begin{array}{c}\lambda_2\sum_{i=7}^{12} w_{i2}\left(d_{iL}^{-}+d_{iU}^{+}\right)\\ +\left(1-\lambda_2\right)V_2\end{array}\right]\end{array}\right] \quad \ldots(5.1)$$

And satisfy $d_{iL}^{-} + d_{iU}^{+} \leq V_1$, $i = 1, 2, ..., 6$, and $i = 13, 14, ..., 30$, and $d_{iL}^{-} + d_{iU}^{+} \leq V_2$, $i = 7, 8, ..., 12$, $\lambda_1, \lambda_2 \in (0, 1)$. $d_{iL}^{-}, d_{iU}^{+} \geq 0$ with $d_{iL}^{-} . d_{iU}^{+} = 0$, $i = 1, 2, ..., 30$.

Where $$V_1 = \text{Max}\left\{\sum_{i=1}^{6} w_{i1}\left(d_{iL}^{-} + d_{iU}^{+}\right) + \sum_{i=13}^{30} w_{i1}\left(d_{iL}^{-} + d_{iU}^{+}\right)\right\}$$

and $$V_2 = \text{Max}\left\{\sum_{i=7}^{12} w_{i2}\left(d_{iL}^{-} + d_{iU}^{+}\right)\right\},$$

and $\left(\sum_{i=1}^{6} w_{i1} + \sum_{i=13}^{30} w_{i2}\right) = 1$ and $\sum_{i=7}^{12} w_{i2} = 1$, and satisfy the goal expressions described in the decision making context.

Now, to solve the problem by using the proposed GA scheme defined in the Section 3, the regret function (Z) becomes the fitness function and it takes the form

$$\text{Eval}\ (V_l) = (Z)_{lr} = (.)$$

where $(Z)_{lr}$ represents the achievement function Z for lth chromosome at the rth priority level ($l = 1, 2, ...,$ pop_size; $r = 1, 2$), and where (.) designates the corresponding functional value of Z defined in (5.1). Here, the highest fitness score at a generation is determined as

$$E^* = \{\min\ \{\text{eval}\ (E_l);\ l = 1, 2.\ \text{Pop-size}\}\}$$

In the execution process, the following GA parameters are used.

Probability of crossover $P_c = 0.8$, Probability of mutation $P_m = 0.08$, Population size =100, Chromosome length = 30

Again, in the context of goal achievement, equal weights to all the goals, i.e. $w_{i1} = w_{i2} = \frac{1}{30}$ with $\lambda_1 = \lambda_2 = 0.5$ are given in the decision making environment.

The program for the GA scheme is designed in the Programming Language C^{++}. The execution is done in a Intel Pentium IV with 2.66 GHz. Clock-pulse and 1GB RAM. The optimal solution is reached after 200 generations.

The model solution is presented in the Table 5.

Table 5: *Staff Allocation Structure Under the Proposed Model*

Department	*MB*	*CSE*	*MBA*	*GEO*	*MB and BT*	*PHY*
Professor	1	2	1	1	1	1
Reader	1	2	2	1	2	3
Lecturer	4	6	6	4	3	2
PTS	2	3	3	2	1	2
NTS	3	5	4	4	3	3

The existing staff allocation structure is given in the Table 6.

Table 6: *Existing Staff Allocation Structure (2008-09)*

Department	*MB*	*CSE*	*MBA*	*GEO*	*MB and BT*	*PHY*
Professor	1	1	0	0	0	0
Reader	2	3	2	1	1	2
Lecturer	3	2	1	1	2	1
NTS	2	7	2	4	2	3

A comparison of the model solution with the existing allocation structure reflects that a satisfactory solution is achieved here under the proposed approach in terms of the needs and desires of the departments.

Note: If the fuzzy version of the problem for goal satisficing criteria is considered which is obtained by using the mid-point arithmetic in IP to be defined interval parameter sets where minimization of only the under-deviational variables of membership goals with highest membership value (unity) defined for the membership function of the fuzzy objectives is taken into account, and if the linearization approach to the academic planning problem studied by Pal and Sen [11] is used, then the obtained solution for staff allocation is displayed in Table 7.

Table 7: *Staff Allocation Structure Under Linear FGP Model*

Department	*MB*	*CSE*	*MBA*	*GEO*	*MB and BT*	*PHY*
Professor	1	2	1	1	2	1
Reader	1	1	1	1	1	1
Lecturer	6	9	4	3	1	2
PTS	2	3	2	1	1	1
NTS	4	6	3	3	2	2

Here, again a comparison with the previous two results shows that the solution under the proposed approach is a superior one from the viewpoint of achieving a better staff allocation structure for smooth functioning of the academic activities of the university.

6. CONCLUSION

The main advantage of the GA based FGP solution is that the computational complexity for involvement of fractional goals [18] and the computational load with the use of the traditional linearization approach [15] can easily be overcome here by using the proposed method. If other type of goals (linear/nonlinear) is to be involved within the streamline proposed problem for enhancement of academic performance, then they can easily be incorporated with the framework of the model.

However, it is hoped that the method presented here can lead to contribute future research on academic planning for enrichment of higher education in university system.

ACKNOWLEDGEMENT

The authors are thankful to the anonymous referee and conference chair of ICRFSC 2009, for their valuable suggestions to improve the presentation of the paper.

REFERENCES

[1] BUDGET ESTIMATES (both Revenue and Development). "THE FINANCIAL YEAR (2009-2010) AND Revised Estimates (2008-2009)", University of Kalyani, W.B (India), Publishes by the Finance Officer, Kalyani University Press, 2008.

[2] Davis, L.D., ed. 1991, "Handbook of Genetic Algorithms", Van Nostrand Reinhold.

[3] Deb., K. 2002, "Multi-objective Optimization Using Evolutionary Algorithms", John Wiley and Sons, Ltd.

[4] Goldberg, D.E. 1989, "Genetic Algorithm in Search, Optimization and Machine Learning", Addison-Wesley, Reading. MA.

[5] Hannan, E.L. "Linear Programming with Multiple Fuzzy Goals", *Fuzzy Sets and Systems*, **6**, pp. 235–248, 1981.

[6] Holland, J.H. "Genetic Algorithms and Optimal Allocations of Trials", *SIAM Journal of Computing*, **2**, No. 2, pp. 88–105.

[7] Ignizio, J.P. 1976, "Goal Programming and Extensions", Lexington Books. Lexington MA.

[8] Inuiguchi, M. and Kume, Y. "Goal Programming Problems with Interval Coefficients and Target Intervals". *European Journal of Operational Research*, **52**, pp. 345–360, 1991.

[9] Kwak, N.K. and Lee, C. "A Multicriteria Decision Making Approach to University Resource Allocations and Information Infrastructure Planning", *European Journal of Operational Research*, **110**, pp. 234–242, 1998.

[10] Narasimhan, R. "On Fuzzy Goal Programming–Some Comments", *Decision Sciences*, **11**, pp. 532–538, 1980.

[11] Pal, B.B. and Sen, S. "A Linear Goal Programming Procedure for Academic Personnel Management Problems in University System", *In Proceedings of IEEE 10th Colloquium and International Conference on Industrial and Information Systems (ICIIS 2008)*, **437**, pp. 1–7, 2008.

[12] Pal, B.B. and Basu, I., "A Long Range Resource Planning in University Management via Goal Programming with Penalty Functions", "Multiple Criteria Decision Making Through Goal Programming", *Ph.D Thesis, University of Kalyani,* India, pp. 77–96, 1997.

[13] Pal, B.B. and Gupta, S. "An Application of Genetic Algorithm Method to Fuzzy Goal Programming Model for Academic Personnel Planning Problems in University Management System", *In Proceedings of 7th All India People's Technology Congress (AIPTC)*, pp. 49–58, 2009.

[14] Pal, B.B. and Gupta, S. "A Goal Programming Approach for Solving Interval Valued Multiobjective Fractional Programming Problems by using Genetic Algorithm" *In Proceedings of IEEE 10th colloquium and International Conference on Industrial and Information Science 2008 (ICIIS 2008)*. 2008.

[15] Pal. B.B. and Sen, S. "A Goal Programming Procedure for Solving Interval Valued Multiobjective Fractional Programming Problems", *In Proceedings of Sixteenth Int. Conf. on Advanced Computing and Communication 2008, (ADCOM 2008)*, **297**, pp. 1–6, 2008.

[16] Romero, C. 1991. "*Handbook of Critical Issues in Goal Programming*", Pergamon Publ. Corp.

[17] Romero, C. "A Survey of Generalized Goal Programming", *European Journal of Operational Research*, **25**, No. 2, pp. 183–191, 1986.

[18] Schaible, S. "Fractional Programming, Applications and Algorithms", *European Journal of Operational Research*, **7**, pp. 111–120, 1981.

[19] Schroeder, R.G. "A Survey of Management Science in University Operations", *Management Science*, **19**, No. 8, pp. 895–906, 1973.

[20] Schroeder, R.G. "Resource Planning in University Management by Goal Programming, *Operations Research*, **22**, pp. 700–710, 1974.

[21] Zimmermann, H.J. "Fuzzy Programming and Linear Programming with Several Objective Functions", *Fuzzy Sets and Systems*, **1**, pp. 45–55, 1978.

42

CLASSIFICATION OF FUSED FACE IMAGES USING MULTILAYER PERCEPTRON NEURAL NETWORK

Debotosh Bhattacharjee[2], Mrinal Kanti Bhowmik[1], Mita Nasipuri[2], Dipak Kumar Basu[2] and Mahantapas Kundu[2]

[1]*Department of Computer Science and Engineering, Tripura University, Suryamaninagar-799130, Tripura, India*

[2]*Department of Computer Science and Engineering, Jadavpur University, Kolkata-700032, India*

Abstract: *This paper presents a concept of image pixel fusion of visual and thermal faces, which can significantly improve the overall performance of a face recognition system. Several factors affect face recognition performance including pose variations, facial expression changes, occlusions, and most importantly illumination changes. So, image pixel fusion of thermal and visual images is a solution to overcome the drawbacks present in the individual thermal and visual face images. Fused images are projected into eigenspace and finally classified using a multi-layer perceptron. In the experiments we have used Object Tracking and Classification Beyond Visible Spectrum (OTCBVS) database benchmark thermal and visual face images. Experimental results show that the proposed approach significantly improves the verification and identification performance and the success rate is 95.07%. The main objective of employing fusion is to produce a fused image that provides the most detailed and reliable information. Fusion of multiple images together produces a more efficient representation of the image.*

Keywords: *Image fusion, Thermal infrared images, Eigenspace projection, Multilayer Perceptron, Backpropagation learning, Face recognition, Classification.*

1. INTRODUCTION

Human face recognition is a challenging task and its domain of application is very vast, covering different areas like security systems, defense applications, and intelligent machines. It involves different image processing issues like face detection, recognition and feature extraction (1) (2) (3). Recently, fusion of face images is gaining acceptance as a superior biometric in case of face recognition. Data fusion is a method which combines different types of data gathered by the simultaneous use of several sensing modalities to generate a new type of data. Various perceptual mechanisms integrate these senses to produce the internal representation of the sensed environment. The integration tends to be synergistic in the scene that information inferred from the process cannot be obtained from any proper subset of the sense modalities. This property of synergism is one that should be sought for when implementing multisensor integration for machine perception. The principal motivation for the fusion approach is to exploit such synergism in the technique for combined interpretation of images obtained from multiple sensors.

So far, research work on fusion has been carried out for years. And the obtained fusion methods can be classified into two categories as stated in (11). One is about weakly coupled fusion methods, and the other is about strongly coupled fusion methods. In the first category of fusion methods, fusion of data produced by sensory modules does not affect the operation of the modules. On the contrary, for strongly coupled fusion methods, the modules producing the data to be fused are being affected in some way by other information from other modules.

For the detailed review on current advances in visual and thermal face recognition refer to (12). Simple image fusion in spatial domain is discussed in (13), where face recognition is used for testing the fusion of face database images. Both image fusion and decision fusion is employed in (14) to improve the accuracy of the face recognition system.

Recently, researchers have investigated the use of fusion of thermal infrared and visual face images for person identification to tackle the drawbacks of individual thermal and visual images (7) (8) (9) (10) (11). In this paper, we present a novel approach to the problem of face recognition that realizes the full potential of the fusion of thermal IR band and visual band images. In this work at first thermal and visual face images are combined together and create the fused image of corresponding thermal and visual face images, after that using these transformed fused images eigenfaces are computed and finally those eigenfaces thus found are classified using a multilayer perceptron.

The organization of the rest of this paper is as follows. In Section 2, the overview of the system is discussed, in Section 3 experimental results and discussions are given. Finally, section IV concludes this work.

2. THE SYSTEM OVERVIEW

Here we present a technique for human face recognition. In this work we have used Object Tracking and Classification Beyond Visible Spectrum (OTCBVS) database benchmark thermal and visual face images. Every thermal face image and visual face image is first combined and converted into fused image. These transformed images are separated into two groups namely training set and testing set. The eigenspace is computed using training images. All the training and testing images are projected into the created eigenspace and named as fused eigenfaces. Once these conversions are done the next task is to use a classifier to classify them. A multilayer perceptron is used for this purpose. The block diagram of the system is given in figure 1. In this figure dotted line indicates feedback from different steps to their previous steps to improve the efficiency of the system, e.g. if the classification results are not satisfactory some adjustment in the internal parameters may be done. In case of eigenspace projection number of eigenvectors, used to create eigenspace, may be increased or decreased to improve better representation of eigenfaces in the reduced space in order to achieve higher accuracy in recognition of faces.

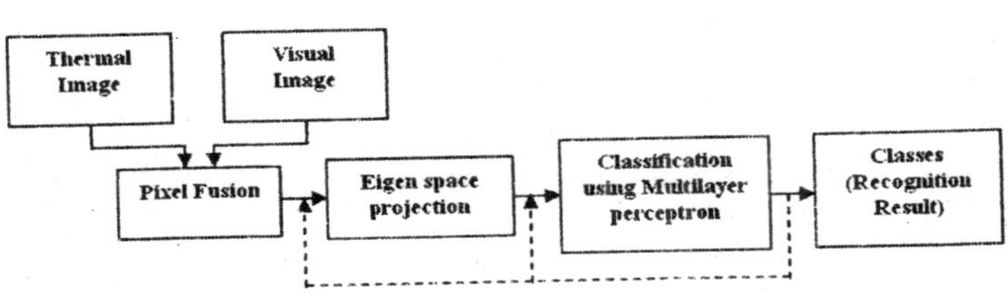

Fig. 1: *Block Diagram of the System Presented Here*

2.1 Thermal Infrared Face Images

Thermal infrared face images are formed as a map of the major blood vessels present in the face. Therefore, a face recognition system designed based on thermal infrared face images cannot be evaded or fooled by forgery, or disguise, as can occur using the visible spectrum for facial recognition. Compared to visual face-recognition systems this recognition system will be less vulnerable to varying conditions, such as head angle, expression, or lighting.

2.2 Image Fusion Technique

The task of interpreting images, either visual images alone or thermal images alone, is an unconstraint problem. The thermal image can at best yield estimates of surface temperature that in general, is not specific in distinguishing between object classes. The features extracted from visual intensity images also lack the specificity required for uniquely determining the identity of the imaged object. The interpretation of each type of image thus leads to ambiguous inferences about the nature of the objects in the scene. The use of thermal data gathered by an infrared camera, along with the visual image, is seen as a way of resolving some of these ambiguities. In the other hand, thermal images are obtained by sensing radiation in the infrared spectrum. The radiation sensed is either emitted by an object at a non-zero absolute temperature, or reflected by it. The mechanisms that produce thermal images are different from those that produce visual images. Thermal image produced by an object's surface can be interpreted to identify these mechanisms. Thus thermal images can provide information about the object being imaged which is not available from a visual image (11).

A great deal of effort has been expended on automated scene analysis using visual images, and some work has been done in recognizing objects in a scene using infrared images. However, there has been little effort on interpreting thermal images of outdoor scenes based on a study of the mechanism that gives rise to the differences in the thermal behavior of object surfaces in the scene. Also, nor has been any effort been made to integrate information extracted from the two modalities of imaging.

In our method the process of image fusion is where pixel data of 70% of visual image and 30% of thermal image of same class or same image is brought together into a common operating image or now commonly referred to as a Common Relevant Operating Picture (CROP) (10). This implies an additional degree of filtering and intelligence applied to the pixel streams to present pertinent information to the user. So image pixel fusion has the capacity to enable seamless working in a heterogeneous work environment with more complex data. For accurate and effective face recognition we require more informative images. Image by one source (i.e. thermal) may lack some information which might be available in images by other source (i.e. visual). So if it becomes possible to combine the features of both the images viz. visual and thermal face images then efficient, robust, and accurate face recognition can be developed.

We describe below in detail the fusion scheme considered in this work. We assume that each face is represented by a pair of images, one in the IR spectrum and one in the visible spectrum. Both images have been combined prior to fusion to ensure similar ranges of values.

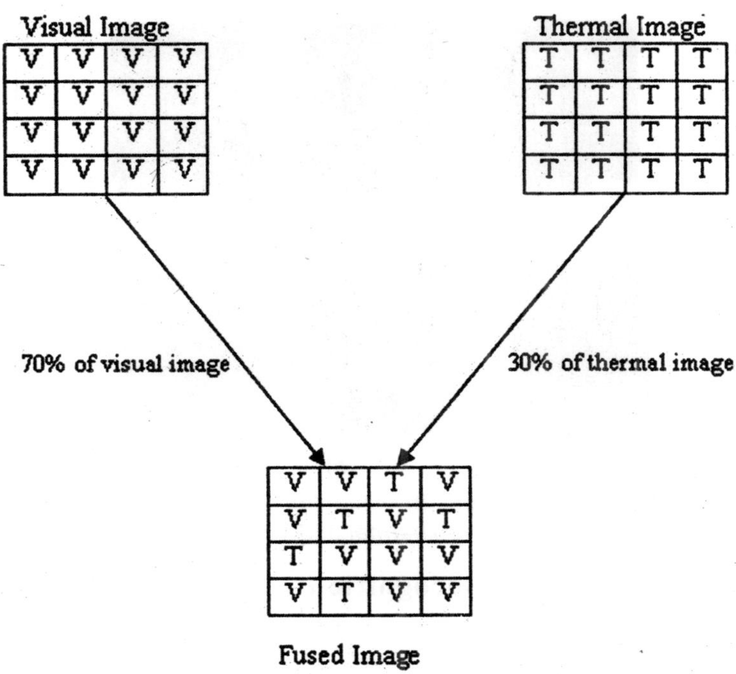

Fig. 2: *Fusion Technique*

We fused visual and thermal images. Ideally, the fusion of common pixels can be done by pixel-wise weighted summation of visual and thermal images.

$$F(x, y) = a(x, y)\, V(x, y) + b(x, y)\, T(x, y) \text{ (12)} \qquad \text{... (1)}$$

where $F(x, y)$ is a fused output of a visual image, $V(x, y)$, and a thermal image, $T(x, y)$, while $a(x, y)$ and $b(x, y)$ represent the weighting factors for visual and thermal images respectively. In this work, we have considered $a(x, y) = 0.70$ and $b(x, y) = 0.30$

(a)

(b)

(c)

Fig. 3: *(a) Thermal Images, (b) Visual Images, (c) Fused Images of Corresponding Thermal and Visual Images*

2.3 Eigenfaces For Recognition

In mathematical terms, we wish to find principal components (4) (5) (6) of the distribution of faces, or the eigenvectors of the covariance matrix of the set of face images. These eigenvectors can be thought of as set of features which together characterize the variations between face images. Each image location contributes more or less to each eigenvector, so that we can display the eigenvector as sort of ghostly face which we call an eigenface. Each face image in the training set can be presented exactly in terms of a linear combination of the eigenfaces. The number of a possible eigenfaces is equal to the number of face images in the training set. However the faces can also be approximated using only the "best" eigenfaces-those that have the largest eigenvalues, and which therefore account for the most variance within the set face images. The best U eigenfaces constitute a U-dimensional subspace, which may be called as "face space" of all possible images. Identifying images through eigenspace projection takes three basic steps. First the eigenspace must be created using training images. After that all those training images are projected into the eigenspace and call them eigenfaces. Train a classifier using these eigenfaces. Finally, the test images are identified by projecting them into the eigenspace and classifying them by the trained classifier.

2.4 ANN Using Backpropagation With Momentum

Neural networks, with their remarkable ability to derive meaning from complicated or imprecise data, can be used to extract patterns and detect trends that are too complex to be noticed by either humans or other computer techniques. A trained neural network can be thought of as an "expert" in the category of information it has been given to analyze. The Back propagation learning algorithm is one of the most historical developments in Neural Networks. It has reawakened the scientific and engineering community to the modeling and processing of many quantitative phenomena using neural networks. This learning algorithm is applied to multilayer feed forward networks consisting of processing elements with continuous differentiable activation functions. Such networks associated with the back propagation learning algorithm are also called back propagation networks.

3. EXPERIMENTAL RESULTS AND DISCUSSIONS

This work has been simulated using MATLAB 7. For comparison of results experiments are conducted for fused images. A thorough system performance investigation, which covers all conditions of human face recognition, has been conducted. They are face recognition under (*i*) variations in size, (*ii*) variations in lighting conditions, (*iii*) variations in facial expressions, (*iv*) variations in pose.

We first analyze the performance of our algorithm using OTCBVS database which is a standard benchmark thermal and visual face images for face recognition technologies.

3.1 OTCBVS Database

Our experiments were perform on the face database which is Object Tracking and Classification Beyond Visible spectrum (OTCBVS) benchmark database contains a set of thermal and visual face images. There are 700 images of visual and 700 thermal images of 16 different persons. For some subject, the images were taken at different times which contain quite a high degree of variability in lighting, facial expression (open/closed eyes, smiling/non smiling, etc.) pose (Up right, frontal position, etc.) and facial details (Glasses/no Glasses). All the images were taken against a dark homogeneous background with the subjects in and upright, fontal position, with tolerance for some tilting and rotation of up to 20 degree. The variation in scale is up to about 10% all the images in the database.

3.2 Classification of Fused Eigenfaces Using Multilayer Perceptron

Out of total 700 thermal and visual images 400 images are taken out of which 200 are thermal images and 200 are visual images. Combining these thermal and visual images we get 200 fused images. 100 of these images are used as training set and rest 100 images are taken as testing images. The training set contains 10 classes which mean that each class has 10 images. Now 5 images from one particular class (which are not used as a training image) and 5 more images of the other classes are taken from the testing set. According to this process for all the 10 classes we get the result which is shown in the following graph of Figure 4. In this work a multilayer neural network with back propagation has been used. The learning algorithm error back propagation with momentum is used here.

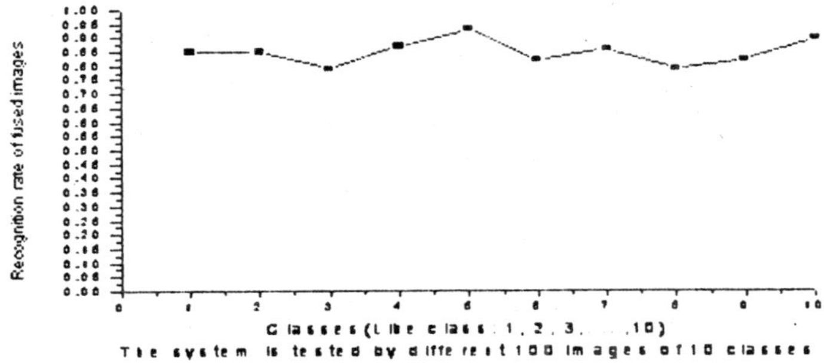

Fig. 4: *Study of Recognition Rate of Fused Images*

In order to access the effectiveness of the fused images we compare the fused face image recognition. Results obtained after applying the same procedure for fused image of thermal and visual images are shown in Fig. 4. Here, experiments are conducted for different number of images during testing.

Table 1: *Comparison of Different Fusion Algorithms*

Fusion scheme	*Reference*	*Combining*	*Weights [IITV/IR (Image intensified TV sensor with infrared images)]*
ADD (Adaptive Fusion)	(15)	None	70%/30%
DWT1 (Discrete Wavelet Transform)	(21, 16)	Weighted average	70%/30%
LAP1 (Laplacian Pyramid)	(17, 16)	Weighted average	70%/30%
FSD1 (Filter Subtract Decimate)	(18, 16)	Weighted average	70%/30%
GRAD1 (Gradient Pyramid)	(19, 16)	Weighted average	70%/30%
MORPH1 (Morphological Pyramid)	(20, 16)	Weighted average	70%/30%
SiDWT1 (Shift Invariant Discrete Wavelet Transform) Discrete Wavelet Transform)	(22, 16)	Weighted average	70%/30%

Table Contd.

Table 1 Contd.

DWT2 (Discrete Wavelet Transform)	(21, 16)	Choosing Maximum	None
LAP2 (Laplacian Pyramid)	(17, 16)	Choosing Maximum	None
FSD2 (Filter Subtract Decimate)	(18, 16)	Choosing Maximum	None
GRAD2 (Gradient Pyramid)	(19, 16)	Choosing Maximum	None
MORPH2 (Morphological Pyramid)	(20, 16)	Choosing Maximum	None
SiDWT2 (Shift Invariant Discrete Wavelet Transform)	(22, 16)	Choosing Maximum	None
Present Method Average	Paper	Weighted	70%/30%

These algorithms (in Table 1) were tested for fusing low-light images produced by an image-intensified TV (IITV) sensor with infrared images produced by a Forward Looking Infrared (FLIR) camera, after registering these images as well as possible. Based upon the preliminary survey, we determined the most promising fusion approaches to be those given in Table 1. It is noted that the alternative of the combining method listed in Table 1 is employed at the lowest frequency band for the pyramid transform-based fusion, and low-low band for DWT and SiDWT fusion. The combining method for other frequency bands is "choosing maximum" in our tests. In the process of determining the methods in Table 1, we eliminated many algorithms. Some of them performed very poorly. The alternative of either the "choosing maximum" or the "weighted average" combining method at the lowest or low-low frequency band generates different MDB (17) fusion results with respect to contrast and brightness.

4. CONCLUSION

In this paper we have presented a human face recognition technique using fusion of pixel level fused image of thermal face and visual face results with varying illumination, facial expression, pose, and facial details. After the fusion of images as weighted sum, the fused images are projected into eigenspace. Those fused eigenfaces are classified using multilayer perceptron. Eigenspace is constituted by the images belong to the training set of the MLP. The efficiency of our scheme has been demonstrated on Object Tracking and Classification Beyond Visible spectrum (OTCBVS) benchmark database and recognition rate obtained is 95.07%.

5. ACKNOWLEDGMENT

Authors are thankful to the project entitled "Development of Techniques for Human Face Based Online Authentication System Phase-I" sponsored by Department of Information Technology under the Ministry of Communications and Information Technology, New Delhi-110003, Government of India Vide No. 12(14)/08-ESD, Dated 27/01/2009 at the Department of Computer Science & Engineering, Tripura University-799130, Tripura (West), India for providing the necessary infrastructural facilities for carrying out this work.

REFERENCES

[1] W. Zhao, R. Chellappa, A. Rosenfeld, and P. Phillips, " Face Recognition: A literature survey", *ACM Computer Survey*, **35**, No. 4, pp. 399–458, December 2003.

[2] P. Jonathan Phillips, Alvin Martin, C.L.Wilson, Mark Przybocki, "An Introduction to Evaluating Biometric System", *National Institute of Standards and Technology*, page 56–62, February, 2000.

[3] A. Samal and P. A. Iyenger, "Automatic Recognition and Analysis of Human Faces and Facial Expressions: A Survey", *Pattern Recognition*, **25**, No. 1, 1992.

[4] M. Turk and A. Pentland, "Eigenfaces for Recognition", *Journal of Cognitive Neuro-science*, March 1991, **3**, No. 1, pp. 71–86.

[5] L. Sirovich and M. Kirby, "A Low-dimensional Procedure for the Characterization of Human Faces," *J. Opt. Soc. Amer. A*, **4 (3)**, pp. 519–524, 1987.

[6] Gottumukkal R., Asari V.K. (2004). "An Improved Face Recognition Technique Based on Modular PCA Approach", *Pattern Recognition Letters*, **25**, 429–436.

[7] S. Singh, A. Gyaourva, G. Bebis and I. Pavlidis, "Infrared and Visible Image Fusion for Face Recognition", *Proc. SPIE*, **5404**, pp. 585–596, Aug. 2004.

[8] Hanif, M., Ali, U., "Optimized Isual and Thermal Image Fusion for Efficient Face Recognition", *2006 9th International Conference on Information Fusion, proceedings of IEEE*, 10-13 July, 2006, pp. 1–6.

[9] Chen, Y. Blum, R.S., "Experimental Tests of Image Fusion for Night Vision", *2005 8th International Conference on Information Fusion, proceedings of IEEE*, **1**, pp. 8.

[10] David Hughes, "Sinking in a Sea of Pixels—The Case for Pixel Fusion", *2006 Silicon Graphics, Inc., http://www.sgi.com/pdfs.*

[11] Zhang Yin, Andrew A. Malcolm, "Thermal and Visual Image Processing and Fusion", *http://www.simtech.a-star.edu.sg/Research/Technical Reports.*

[12] Jingu Heo, "Fusion of Visual and Thermal Face Recognition Techniques: A Comparative Study", *The University of Tennessee, Knoxville, October 2003, http://imaging.utk.edu/publications/papers/dissertation.*

[13] Taj J.A., Ali U., Qureshi R.J., Khan S.A., "Fusion of Visual and Thermal Images for Efficient Face Recognition using Gabor Filter", *accepted for publication in Proc. of The 4th ACS/IEEE International Conference on Computer Systems and Appiications*, March 8–11, 2006, Dubasi/Sharjah, UAE.

[14] Ali U., Taj J.A. "Gabor Filter based Efficient Thermal and Visual Face Recognition Fusion Architecture", *accepted for publication in the fourth International Conference on Active Media Technology*, June 7–9, 2006 Brisbane, Australia.

[15] E.J. Bender, C.E. Reese, and G. S. van derWal, "Comparison of Additive Image Fusion Versus Feature-level Image Fusion Techniques for Enhanced Night Driving", *Proceedings of SPIE, Vacuum and Solid State Photoelectronic Imagers, Detectors, and Systems*, 4796, 2002.

[16] Z. Zhang and R.S. Blum, "A Categorization of Multiscale-Decomposition-Based Image Fusion Schemes with a Performance Study for a Digital Camera Application", *Proceedings of IEEE*, **87 (8)**, 1315–1326, 1999.

[17] Peter J. Burt and Edward H. Adelson, "The Laplacian Parymid as a Compact Image Code", *IEEE Transaction on Communications*, **31 (4)**, 532–540, 1983.

[18] C. H. Anderson, "A Filter-Subtract-Decimate Hierarchical Pyramid Signal Analyzing and Synthesizing Technique", *United States Patent,* **4, 718, 104**, Washington, D.C., 1987.

[19] Peter J. Burt, "A Gradient Pyramid Basis for Patternselective Image Fusion. Society for information Display", *Digest of Technical Papers*, pp. 467–547, 1992.

[20] A. Toet, "A Morphological Pyramidal Image Decomposition", *Pattern Recognition Letters*, **9**, 255–261, 1989.

[21] T. Huntsberger and B. Jawerth, "Wavelet Based Sensor Fusion", *In Proc. SPIE*, **2059**, pp. 488–498, 1993.

[22] O. Rockinger, "Image Sequence Fusion Using a Shift Invariant Wavelet Transform", *Proceedings of the International Conference on Image Processing*, pp. 288–291, 1997.

43

ON ANOTHER DECOMPOSITION OF FUZZY AUTOMATA

S.P. Tiwari

Department of Applied Mathematics Indian School of Mines University Dhanbad-826004, India,
E-mail: sptiwarimaths@gmail.com

Abstract: *This paper presents a non-unique decomposition of a fuzzy automaton (possibly with infinite state-set) as a union of its f-primaries. Also, it is shown that the concepts of primaries and f-primaries of a fuzzy automaton are same in the case of compact fuzzy automaton.*

1. INTRODUCTION

The decompositions of finite and infinite automata have been studied in [1, 3, 4, 7]. Awell-known result in this context about (crisp) finite automata, is the so called "Primary Decomposition Theorem". It says that such automata can be so decomposed via their primaries. Its counterpart for fuzzy automata, having finite state-sets, has already been proved in [5]. It turns out that this theorem may not hold good for arbitrary fuzzy automata having infinite state-sets. However, in [8], it is shown that a counterpart of this result holds for fuzzy automata, having even infinite state-sets, provided their state-set topologies are compact [c.f. [6] for a similar observation]. Continuing the study of decomposition theory of fuzzy automata, in [9], a unique decom – position of an arbitrary fuzzy automaton has been discussed. The present work principally motivated from [2], is about another

* The author acknowledge with thanks the support recieved through a research grant No. MRP(AM)/Acad/06/2008, provided by Indian School of Mines University, Dhanbad

decomposition of an arbitrary fuzzy automaton in terms of its *f*-primaries.

2. PRELIMINARIES

In this section we collect some results, which are useful in the next section. We start from the following concept of fuzzy automata which resembles the concept of fuzzy machines, as given in [5].

Definition 2.1: A fuzzy automaton is a triple $M = (Q, X, \delta)$, where Q is a nonempty set (of states of M), X is a monoid (the input monoid of M), whose identity shall be denoted as e, and δ is a fuzzy subset of $Q \times X \times Q$, i.e. a map $\delta: Q \times X \times Q \to (0, 1)$, such that $\forall q, p \in Q$, $\forall x, y \in X$,

$$\delta(q, e, p) = \begin{cases} 1 & \text{if } q = p \\ 0 & \text{if } q \neq p \end{cases}$$

and $\quad (q, xy, p) = \vee [(q, x, r) \wedge \delta(r, y, p) : r \in Q]$.

Remark 2.1: As in (6), we take the view that a (crisp) automaton is a triple (Q, X, δ), where Q is a (possibly infinite) set, X is a monoid with identity e, and $\delta : Q \times X \to Q$ is a function such that $\delta(q, e) = q$ and $\delta(q, xy) = \delta[\delta(q, x), y]$, $\forall q \in Q$, $\forall x, y \in X$. By identifying δ with the function $\hat{\delta} : Q \times X \times Q \to (0, 1)$, given by $\hat{\delta}(q, x, p) = 1$ or 0 according as $\delta(q, x) = p$ or $\delta(q, x) \neq p$, we see that the concept of a fuzzy automaton generlizes that of a (crisp) automaton.

Definition 2.2 [(8)]: Given a fuzzy automaton (Q, X, δ) and $A \subseteq Q$, the *source* and the *successor* of A are respectively the sets

$$\sigma_Q(A) = [q \in Q : (q, x, p) > 0, \text{ for some } (x, p) \in X \times A], \text{ and}$$
$$s_Q(A) = [p \in Q : (q, x, p) > 0, \text{ for some } (x, q) \in X \times A].$$

We shall frequently write $\sigma_Q(A)$ and $s_Q(A)$ as just $\in (A)$ and $s(A)$, and $\sigma[(q)]$ and $s[(q)]$ as just $\sigma(q)$ and $s(q)$.

Proposition 2.1 [(8)]: Let (Q, X, δ) be a fuzzy automaton.

(*a*) The source and the successor, viewed as functions $\sigma : 2^Q \to 2^Q$ and $s : 2^Q \to 2^Q$, turn out to be Kuratowski closure operators on Q, inducing two topologies on Q (which we shall respectively denote as $T(Q)$ and $T^*(Q)$).

(*b*) The topologies $T(Q)$ and $T^*(Q)$ are dual in the sense that each $A \subseteq Q$ is $T(Q)$-open if and only if A is $T^*(Q)$-closed.

Definition 2.3 [(5)]: A fuzzy automaton (R, X, λ) is called a subautomaton of a fuzzy automaton (Q, X, δ) if $R \subseteq Q$, $s_Q(R) = R$, and $\delta_{/R \times X \times R} = \lambda$.

Proposition 2.2 [(8)]: Let $M = (Q, X, \delta)$ and $N = (R, Y, \lambda)$ be fuzzy automata. Then N is subautomaton of M if and only if R is $T(Q)$-open.

Now, we recall the concept of primaries, which play a key role in the study of decomposition of fuzzy automata.

Definition 2.4 [(5)]: Let (Q, X, δ) be a fuzzy automaton. Then a subset $R \subseteq Q$ is called

(*i*) *genetically closed* if $\exists P \subseteq R$ such that $\sigma(P) \subseteq s(P)$ and $s(P) = R$;

(*ii*) a *primary* subset of Q if R is a nonempty minimal genetically closed subset of Q.

Definition 2.5 [(8)]: A primary of a fuzzy automaton $M = (Q, X, \delta)$ is a subautomaton of M whose state-set is a primary subset of Q.

In (8), the primaries of a fuzzy automata have been characterized in terms of the topology on its state-set, by using the concept of *core*, which is reproduced below.

Definition 2.6: The *core* of any subset R of the state-set Q of a fuzzy automaton is the set

$$\mu(R) = \{q \in R : (q) \subseteq R\}.$$

We shall frequently write $\mu[(q)]$ as just $\mu(q)$. Before stating the next proposition, we first recall the following definition from Willard (10).

Definition 2.7: A closed subset of topological space is called *regular closed* if it is equal to the closure of its interior

For example, the closures of open subsets are always regular closed.

Proposition 2.3 [(8)]: *A* closed subset R of Q is genetically closed if and only if R is a $T^*(Q)$-regular closed subset of Q.

Proposition 2.4 [(8)]: The state-set of a primary of a fuzzy automaton (Q, X, δ) is a minimal $T(Q)$-regular closed subset of Q.

Proposition 2.5 [(8)]: Let (Q, X, δ) be a fuzzy automaton and $p \in Q$. Then $s\ (p)$ is a primary of Q if and only if $p \in \mu\ [s\ (p)]$.

We close this section by recalling the definition of a compact fuzzy automaton and some of the associated results from (8).

Definition 2.8: A fuzzy automaton (Q, X, δ) is called *compact* if the topology $T\ (Q)$ is compact.

Proposition 2.6: A fuzzy automaton (Q, X, δ) is compact if and only if there exists a finite subset Q' of Q such that $s\ (Q') = Q$, or equivalently, if and only if $\sigma\ (Q)$ is finite.

Proposition 2.7: A primary of a compact fuzzy automaton is a maximal singly generated subautomaton.

3. *f*-PRIMARIES AND DECOMPOSITION OF A FUZZY AUTOMATON

In this section, chiefly inspired from (2), we have given the concept of *f*-primaries of a fuzzy automaton and shown that each fuzzy automaton can be expressed, *though non-uniquely*, as a union of its '*f*-primaries'. In this section, unless mentioned otherwise, Q is assumed to be equipped with the topology $T^*\ (Q)$.

Definition 3.1: Let $M = (Q, X, \delta)$ be a fuzzy automaton. Then a subset R of Q is called finitely closed if for each finite subset T of R there exists $q \in Q$ such that $T \subseteq s\ (q)$.

Remark 3.1: For each $q \in Q$, $s\ (q)$ is obviously a finitely closed subset of Q.

Lemma 3.1: Let (R, X, δ) be a primary of a fuzzy automaton (Q, X, δ). Then $s\ (\sigma\ (p)) = R$, $\forall p \in \mu\ (R)$.

Proof: Let $p \in \mu\ (R)$. Then $\sigma\ (p) \subseteq R$ and so $s\ \{\sigma\ (p) \subseteq R$ [as $s\ (R) = R$]. But as $s\ [\sigma\ (p)]$ is a regular closed subset (since the closures of open subsets are always regular closed) and R is a minimal regular closed subset of Q, $s\ (\sigma\ (p)) = R$.

Lemma 3.2: Let (R, X, λ) be a primary of a fuzzy automaton (Q, X, δ). Then for every finite subset T of R, $T \subseteq s\ (r)$, for some $r \in R$.

Proof: We prove this lemma by induction on the number of elements in T. The result is obvious if $|T| = 1$. Now assume the result to be true for all T having $k - 1$ elements. Consider some $T \subseteq R$ having k elements. Pick any $p \in T$. By induction hypothesis $\exists q \in R$ such that $T - \{p\} \subseteq s(q)$. Thus $T \subseteq s(q) \cup \{p\} \subseteq s(\{q, p\})$. Put $S = \{q, p\}$ and let $m \in \mu(R)$. Then by Lemma 3.1, $R = s(\sigma(m))$. Note that $q \in R \Rightarrow q \in s(\sigma(m)) \Rightarrow (m', x, q) > 0$, for some $(m', x) \in \sigma(m) \times X$. Now $m' \in \sigma(m) \Rightarrow \sigma(m') \subseteq \sigma(m) \subseteq R$ (as $m \in \mu(R)$) $\Rightarrow m' \in \mu(R) \Rightarrow s(\sigma(m')) = R$ (by Lemma 3.1). Consequently, $p \in R \Rightarrow p \in s(\sigma(m')) \Rightarrow \delta(r, y, p) > 0$, for some $(r, y) \in \sigma(m') \times X$. From $r \in \sigma(m')$, we get $m' \in s(r)$, whereby $s(m') \subseteq s(r)$. This, together with the facts that $q \in s(m')$ and $p \in s(r)$, gives $q, p \in s(r)$. Hence $S = \{q, p\} \subseteq s(r)$. Thus $s(\{q, p\}) \subseteq s(r)$. Hence $T \subseteq s(r)$.

Proposition 3.1: Let (R, X, λ) be a primary of a fuzzy automaton (Q, X, δ). Then R is a maximal finitely closed subset of Q.

Proof: Lemma 3.2 already shows that R must be finitely closed. So only the maximality of R needs to be shown. Let S be another finitely closed subset of Q such that $R \subseteq S$. Let $q \in S$ and $p \in \mu(R)$. Since $\mu(R) \subseteq R \subseteq S$, $p \in S$. Thus $p, q \in S$. As S is a finitely closed subset of Q, there exists $r \in Q$ such that $(p, q) \subseteq s(r)$, whereby $p, q \in s(r)$. Also, as $p \in \mu(R)$, $\sigma(p) \subseteq R$. Now $p \in s(r) \Rightarrow r \in \sigma(p)$. Hence $r \in R$, whereby $s(r) \subseteq s(R) = R$. But then $q \in R$ (as $q \in s(r)$), showing that $S \subseteq R$. Hence $S = R$, proving that R is a maximal finitely closed subset of Q.

Remark 3.2: The converse of the above theorem is not true, even for a (crisp) automaton, as the following counter-example shows.

Counter-example 3.1: Let $M = (Q, X, \delta)$ be the (crisp) automaton, for which Q is the set of integers, $X = (0, 1, 2, ...)$, and $\delta : Q \times X \to Q$ is given by $\delta(n, r) = n - r$, $\forall n \in Q$, $\forall x \in X$.

We note that $\forall m, n \in Q$, $n \in \sigma(m)$ if and only if $n \geq m$. First, we show that M is the only primary of itself.

Let A be a nonempty subset of Q. Then two cases arise.

Case 1: *When A is finite:* In this case $\mu(A) = \phi$, as there does not exist any $n \in A$ such that $\sigma(n) \subseteq A$ [since $\forall n \in Q$, $\sigma(n)$ is infinite]. Thus $s[\mu(A)] = \phi$, whereby A is not a regular closed subset of Q.

Case 2: *When A is infinite:* In this case, if $\mu(A) = \phi$, then A cannot be regular closed as indicated above. So let $\mu(A) \neq \phi$ and $n \in \mu(A)$. Then $n \in A$ and $\sigma(n) \subseteq A$. Also, $\forall m \in \sigma(n)$, $\sigma(m) \subseteq \sigma(\sigma(n)) = \sigma$

$(n) \subseteq A$. Thus $m \in \mu(A) \subseteq s(\mu(A))$, $\forall m \in \sigma(n)$, i.e. $m \in s(\mu(A))$, $\forall m \geq n$. Next, as $n \in \mu(A)$, we have $s(n) \subseteq s(\mu(A))$, whereby $k \in s(\mu(A))$, $\forall k \in s(n)$, i.e. $k \in s(\mu(A))$, $\forall k \leq n$. Thus we have shown that $q \in s(\sigma(A))$, $\forall q \geq n$ and $\forall q \leq n$, i.e. $q \in s[\mu(A)]$, $\forall q \in Q$, showing that $s[\mu(A)] = Q$.

Hence $\forall A \subseteq Q$, $s(\mu(A)) \neq A$, if $A \neq Q$. Thus Q is the only regular closed subset of itself. Let $R = (1, 2, 3, ...)$ be a subset of Q. Then obviously it is finitely closed. But, R cannot be regular closed, as Q is the only regular closed subset of itself. Thus a finitely closed subset is not neccesarily regular closed, and hence not a primary.

Proposition 3.2: Let $M = (Q, X, \delta)$ be a fuzzy automaton. Then each finitely closed subset of Q is $T(Q)$-open.

Proof: Let R be a finitely closed subset of Q. We have to show that $s(R) = R$. Let T be a finite subset of $s(R)$. For each $q \in T$, we can find some $p \in R$ with $q \in s(p)$. The set S of all such p's must be finite as T is finite. Also, $S \subseteq R$. Obviously, $T \subseteq s(S)$. Now as R is finitely closed, $\exists t \in Q$ such that $S \subseteq s(t)$. Hence $s(S) \subseteq s(t)$, whereby $T \subseteq s(t)$. Thus for each finite subset T of $s(R)$, $\exists t \in Q$ such that $T \subseteq s(t)$. Hence $s(R)$ is a finitely closed subset of Q. But as R is a maximal finitely closed set and $R \subseteq s(R)$, we find that $s(R) = R$.

Definition 3.2: An *f-primary* of a fuzzy automaton $M = (Q, X, \delta)$ is a subautomaton of M whose state-set is a maximal finitely closed subset of Q.

Proposition 3.3: Let $M = (Q, X, \delta)$ be a fuzzy automaton. Then every primary of M is an f-primary of M.

Proof: Follows trivially from Proposition 3.1 and Definition 3.2.

Remark 3.3: From Remark 3.2, it is clear that the converse of above proposition is not true.

Proposition 3.4: Let $M = (Q, X, \delta)$ be a fuzzy automaton and $p \in Q$. Then $s(p)$ is a primary of M if and only if $s(p)$ is an f-primary of M.

Proof: In view of Proposition 3.3, we only need to prove that if $s(p)$ is an f-primary then it is a primary. So let $s(p)$ be an f-primary of M. In view of Proposition 2.5, it suffices to show that $p \in \mu[s(p)]$, or that $\sigma(p) \subseteq s(p)$. Let $q \in \sigma(p)$. Then $p \in s(q)$, whereby $s(p) \subseteq$

$s(q)$. As $s(q)$ is finitely closed and $s(p)$, being an f-primary, is maximal finitely closed, $s(q) = s(p)$. Thus $q \in s(p)$, showing that $\sigma(p) \subseteq s(p)$.

Proposition 3.5: Let $M = (Q, X, \delta)$ be a compact fuzzy autumaton and $N = (R, X, \lambda)$ be its subautomaton. Then N is an f-primary of M if and only if N is a primary of M.

Proof: Once again, we need to prove only the sufficiency. Let N be an f-primary of M. Then due to the compactness of M, there exists a finite subset R' of R such that $s(R') = R$ (cf. Proposition 2.6). Also, as R is finitely closed, $\exists q \in Q$ such that $R' \subseteq s(q)$, whereby $s(R') \subseteq s(q)$. Thus $R \subseteq s(q)$. But as $s(q)$ is finitely closed and R is maximal finitely closed (as N is an f-primary), $s(q) = R$. Hence N is a primary of M (cf. Proposition 2.7).

Theorem 3.1: Every fuzzy automaton can be expressed non-uniquely as the union of its f-primaries.

Proof: From Remark 3.1, for each $q \in Q$, $s(q)$ is a finitely closed subset of Q, whereby each $q \in Q$ will be contained in some finitely closed subset of Q. Let $q \in Q$ and let $\mathcal{F}_q = \{R \subseteq Q : R$ is finitely closed and $q \in R\}$. $\mathcal{F}_q$ is nonempty as, at least, it contains $s(q)$ (cf. Remark 3.1). Let $R_1 \subseteq R_2 \subseteq R_3 \subseteq \ldots$ be any chain in $\mathcal{F}_q$. Then $\cup_{j \in J} R_j$ is also a finitly closed subset of Q, because for any finite $T \subseteq \cup_{j \in J} R_j$, for some $j \in J$, whereby $\exists p \in Q$ such that $T \subseteq s(p)$. Thus $\cup_{j \in J} R_j$ is an upper bound in $\mathcal{F}_q$ of the chain $R_1 \subseteq R_2 \subseteq R_3 \subseteq \ldots$. Therefore by Zorn's lemma, $\mathcal{F}_q$ has a maximal element. Thus for each $q \in Q$, there exists a maximal finitely closed subset containing it, i.e. an f-primary of Q containing q. Also, $\cup_{p \in Q} s(p) = Q$. Hence every fuzzy automaton can be expressed as the union of its f-primaries. To see the non-uniqueness of the union in the statement of last result, consider the following counter-example.

Counter-example 3.2: Let $M = (Q, X, \delta)$ be the (crisp) automaton, where Q is the set of integers, $X = (0, 1, 2, \ldots)$, and $\delta : Q \times X \to Q$ is given by

$$\delta(n, x) = \begin{cases} n + x & \text{if } n > 0 \\ -|n + x| & \text{if } n = 0 \\ n & \text{if } n < 0, x = 0 \\ |n + x| & \text{if } n < 0, x \neq 0 \end{cases}$$

$\forall n \in Q, \forall x \in X.$

We note that each $s(-i)$, $i \in N$ (the set of natural numbers) and $s(0)$ are f-primaries of M. Here Q can be expressed in two ways, given by $Q = \cup_{i \in N} s(-i)$ and $Q = [\cup_{i \in N} s(-i)] \cup s(0)$.

REFERENCES

[1] Z. Bavel, "Structure and Transition Preserving Functions of Finite Automata", *Jour. Assoc. Comput. Mach.*, **15**, (1968), 135–158.

[2] Z. Bavel and J.W. Thomas, "On the Decomposability of Monadic Algebras and Automata", *in Proc. of the 8th Annual Symp. on Switching and Automata Theory*, (1967), 322–335.

[3] W.M.L. Holcombe, "Algebraic Automata Theory", Cambridge University Press, 1987.

[4] K. Krohn and J. Rhodes, "Algebraic Theory of Machines I. Prime Decomposition Theorems for Finite Semigroups and Machines", *Trans. Amer. Math. Soc.*, **116**, (1965), 450–464.

[5] J.N. Mordeson and D.S. Malik, "Fuzzy Automata and Languages: Theory and Applications", *Chapman and Hall/CRC*, 2002.

[6] W. Shukla and A.K. Srivastava, "A Topology for Automata: A Note, Information and Control, **32**, (1976), 163–168.

[7] A.K. Srivastava and W. Shukla, "A Topology for Automata II", *Internat. Jour. Math. and Math. Sc.*, **9**, (1986), 425–428.

[8] A.K. Srivastava and S.P. Tiwari, "A Topology for Fuzzy Automata", *in: Proc. 2002 AFSS Internat. Conf. on Fuzzy Systems, Lecture Notes in Artificial Intelligence, Springer-Verlag*, **2275**, (2002), 485–490.

[9] S.P. Tiwari and A.K. Srivastava, "On a Decomposition of Fuzzy Automata", *Fuzzy Sets and Systems*, **151**, (2005), 503–511.

[10] S. Willard, "General Topology", Addison-Wesley, 1972.

44

A FUZZY GOAL PROGRAMMING APPROACH TO CHANCE CONSTRAINED MULTIOBJECTIVE DECISION MAKING PROBLEMS

Bijay Baran Pal[1*] and Shyamal Sen[2]

*[1]Department of Mathematics, University of Kalyani
Kalyani-741235, West Bengal, India.
E-mail: bbpal18@hotmail.com*

*[2]Department of Mathematics, BKC College
Kolkata-700108, West Bengal, India.
E-mail: ssenk@yahoo.co.in*

***Abstract:** In this article, a fuzzy goal programming (FGP) procedure for solving multiobjective decision making (MODM) problems having the chance constraints is presented.*

In the proposed approach, first the chance constraints are converted into its deterministic equivalent in the decision making environment. Then, the individual best and worst solutions of each of the objectives are determined to represent the fuzzy version of the problem.

In the model formulation of the problem, the concept of tolerance membership functions for measuring the degree of satisfactions of the fuzzily described objectives of the decision maker (DM) are defined.

In the decision process, the FGP solution approach is used for achievement of the highest membership value (unity) of each of the defined membership goals to the extent possible by minimizing the under-deviational variables on the basis of weights of importance of achieving the defined fuzzy goals and thereby obtaining the most satisfactory solution in the decision making situation.

A numerical example is solved and the model solution is compared with the solution of the conventional fuzzy programming (FP) approach studied (14) previously.

***Keywords:** Chance Constrained Programming, Fuzzy Programming, Fuzzy Goal Programming, Goal Programming, Stochastic Programming.*

1. INTRODUCTION

In the real-world decision situations, the DMs are often faced with the problem of inexact human judgments as well as inherent uncertainty in the parameters involved with the problem.

The two types of prominent approaches for solving such problems are stochastic programming (SP) which deals with probabilistically uncertain data and FP which deals with imprecisely described data.

The field of SP based on the theory of probability, initially introduced by Charnes and Cooper [2], as the chance constrained programming (CCP) has been studied [8] extensively and applied to various real-life problems in [1].

On the other hand, the FP based on the theory of fuzzy sets, initially introduced by Zadeh [13], has been studied [14] deeply during the mid '70s to '80s of the last century from the point of view of its potential use to different real-life problems [11] with imprecisely defined data.

Now, in the real-life decision situations, it has been realized that both the probabilistic and fuzzy data are frequently involved, and both the aspects of SP and FP need be taken into account in actual practice of modeling and solving the problems and thereby taking proper decisions. However, consideration of both the aspects in a problem creates a great challenge for developing efficient solutions method to deal with both the stochastic and fuzzy terms in actual practice.

The constructive modeling aspects on programming problems under randomness and fuzziness was first studied by Luhandjula [6] in 1983. The methodological development of fuzzy stochastic programming (FSP) approaches for solving linear programming (LP) problems has been surveyed by Luhandjula [7] in 2006.

Now, since most of the real world problems are multiobjective in nature, FGP [9, 12] in the area of fuzzy MODM [14] and as extension of conventional goal programming (GP) [4] for multiobjective decision analysis has been studied deeply and applied to real-life problems [10]. The FSP approaches to multiobjective decision problems have been studied [3] in the past. The CCP, as a prominent tool in the field of SP, for solving FP problems with multiplicity of objectives has also been studied [5] in the recent past. But, the extensive study in this area is still at an early stage.

Again, although FP approaches to CCP problems have been investigated (3, 8) in the last few years, it is too early to deep study in the area of FSP from the view point of its potential use of real-life problems. Further, the use of FGP method to MODM problems with chance constraints is yet to circulate in the literature.

Now, it may be mentioned here that the fuzzification of objectives/ constraints as well as randomization of various parameters with different probability distributions depends on the decision making environment.

In this paper, for simplicity, fuzzy MODM problems with only the random resource vector of the chance constraints with known probability distribution is taken into consideration. Such types of problems are frequently arises in real decision situation. For instance, there may exist an agricultural production planning problem to optimize the total production of different seasonal crops which dependent very much on the adequate supply of water resource. But, water supply to different seasons from various sources is very scarce, because it is highly dependent on the amount rainfall received during the cropping plan period.

In the model formulation of the problem, the use of mean and variance as conventionally made in solving CCP problems is taken into account to transform the defined chance constraints to their equivalent crisp system constraints. Then, the membership function of the fuzzy goal objectives obtained by introducing fuzzy aspiration level and tolerance limit to each of them by which correspond to the best and worst decisions subject to the crisply defined system constraints.

In the solution process, achievement of the highest value (unity) of the membership goals defined for the membership function to the extent possible on the basis of their weights of importance is considered by minimizing the associated under-deviational variables to reach a most satisfactory decision in the decision making environment.

To illustrate the proposed approach, a numerical example is solved.

2. PROBLEM FORMULATION

The general format of a chance constrained MODM problems can be stated as:

Find $X(x_1, x_2, \ldots, x_n)$ so as to

$$\text{Maximize } Z_k(X), \quad k \in K_1 \text{ and}$$
$$\text{Minimize } Z_k(X), \quad k \in K_2 \text{ subject to}$$

$$x \in S \left\{ x \in R^n \mid \Pr\left[Ax \begin{pmatrix} \geq \\ \leq \end{pmatrix} b \right] \geq p, x \geq 0, b \in R^m \right\} \quad \ldots (2.1)$$

where Pr stands for probabilistically defined constraints, A is a coefficient matrix and b is a resource vector and p $(0 < p < 1)$ is the vector of satisficing probability levels defined for the random occurrence of the parameter values in the constraints set. Again, it is assumed that the feasible region S is nonempty $(S \neq \Phi)$, and where $K_1 \bigcup K_2 = \{1, 2, \ldots, K\}$, $K_1 \bigcap K_2 = \Phi$.

In the present decision situation, it is assumed that the vector b is a vector of independent continuous normally distributed random variable. Then, the conversion to crisp equivalent of the chance constraints in (2.1) is described in the following Section 2.1

2.1 Deterministic Equivalent of Chance Constraints

The chance constraints set in (1) with $\geq$ type restriction can be explicitly presented as

$$Pr\left[\sum_{j=1}^{n} a_{ij}\, x_j \geq b_i \right] \geq p_i \quad \ldots (2.2)$$

Let $E(b_i)$ and $\mathrm{Var}(b_i)$ be the mean and variance of the random variable b_i with the characteristic of normal distribution.

Now, the following notion of distribution function for a random variable is defined in the sequel of finding the crisp equivalent of the constraints in (2.2). Let $F_i(y)$ be the distribution function of the ith random variable b_i. Then, since $F_i(y)$ is a monotonically non-decreasing function, the value of the corresponding variable is determined as

$$F_i^{-1}(\varepsilon) = \{\text{Max } y/P_r(b_i \le y) \le \varepsilon\},\ 0 \le \varepsilon \le 1 \qquad \dots (2.3)$$

Now, the defined constraints set in (2.2) can be expressed as

$$Pr\left[\sum_{j=1}^{n} a_{ij}x_j \ge b_i\right] = Pr\left[\frac{b_i - E(b_i)}{\sqrt{[\text{var}(b_i)]}} \le \frac{\sum_{j=1}^{n} a_{ij}x_j - E(b_i)}{\sqrt{[\text{var}(b_i)]}}\right] \qquad \dots (2.4)$$

Then, using the prescribed probability levels p_i, the expression in (2.4) takes the form

$$Pr\left[\frac{b_i - E(b_i)}{\sqrt{[\text{var}(b_i)]}} \le \frac{\sum_{j=1}^{n} a_{ij}x_j - E(b_i)}{\sqrt{[\text{var}(b_i)]}}\right] \ge p_i$$

$$i = 1, 2, \dots, m \qquad \dots (2.5)$$

Now, using the notion of distribution function defined in (2.3) to the expression in (2.5), the deterministic equivalent of the constraints in (2.2) with ≥ type restriction can be obtained as a set of linear system constraints as

$$\sum_{j=1}^{n} a_{ij}x_j \ge E(b_i) + F_i^{-1}(p_i)\sqrt{[\text{var}(b_i)]}$$

$$i = 1, 2, \dots, m \qquad \dots (2.6)$$

Proceeding in an analogous way, the deterministic equivalent of the set of chance constraints with ≤ type can be obtained as

$$\sum_{j=1}^{n} a_{ij}\, x_j \le E(b_i) + F_i^{-1}(1 - p_i)\sqrt{[\text{var}(b_i)]},$$

$$i = (m + 1), (m + 2), ..., N. \quad ... (2.7)$$

Now, the FP formulation of the problem in (2.1) is presented in the Section 3.

3. FP PROBLEM FORMULATION

In the field of FP, an imprecise aspiration level is assigned to each of the objectives and the tolerance limit for achievement of each of them is taken into consideration.

Let (X^*_{Bk}, Z^*_{Bk}) and (X^*_{Wk}, Z^*_{Wk}) be the best and worst solutions for maximization of the Kth objective Z_k, $k \in K_1$, subject to the constraints in (2.6) and (2.7) where B and W are used to indicate the best and worst solutions, respectively. Again, let (X^*_{Bk}, Z^*_{Bk}) and (X^*_{Wk}, Z^*_{Wk}) be the best and worst solutions, respectively, for minimization of the objective Z_k, $k \in K_2$.

Then, the fuzzy goals of the problem appear as

$$Z_k \gtrsim Z^*_{Bk},\ k \in K_1$$

and

$$Z_k \lesssim Z^*_{Bk},\ k \in K_2 \quad ... (3.1)$$

with the lower and upper tolerance limits Z^*_{Wk} $(k \in K_1)$ and Z^*_{Wk} $(k \in K_2)$, respectively, where $\gtrsim$ and $\lesssim$ indicates the fuzziness of the restrictions $\geq$ and $\leq$, respectively.

Now, in the field of FP, the fuzzy goals are characterized by their respective membership functions.

3.1 Characterization of Membership Function

Let $\mu_k(X)$ be the membership function representation of the kth fuzzy goal.

Then, for $\gtrsim$ type of restriction, $\mu_k(X)$ takes the form

$$\mu_k(X) = \begin{cases} 1, & \text{if } Z_k(X) \geq Z^*_{Bk} \\ \dfrac{Z_k(X) - Z^*_{Wk}}{t_{1K}}, & \text{if } Z^*_{Wk}(X) \leq Z_k(X) < Z^*_{Bk} \\ 0, & \text{if } Z_k(X) < Z^*_{Wk} \end{cases} \quad ... (3.2)$$

where $t_{1k} = (Z^*_{Bk} - Z^*_{Wk})$ is the tolerance range for achievement of the kth fuzzy goal, $k \in K_1$.

Similarly, for $\lesssim$ type of restriction, $m_k(X)$ appear as

$$\mu_k(X) = \begin{cases} 1, & \text{if } Z_k(X) \le Z^*_{Bk} \\ \dfrac{Z^*_{Wk} - Z_k(X)}{t_{2K}}, & \text{if } Z^*_{Bk}(X) \le Z_k(X) \le Z^*_{Wk} \\ 0, & \text{if } Z_k(X) > Z^*_{Wk} \end{cases} \quad ...(3.3)$$

where $t_{2k} = (Z^*_{Wk} - Z^*_{Bk})$ is the tolerance range for achievement of the kth fuzzy goal, $k \in K_2$.

Now, in the field of FP, achieving of highest membership value (unity) of each of the objectives is highly desired by the DM in a decision situation. In such a case, the FGP (9) as a robust tool for fuzzy multiobjective decision analysis is taken into consideration in the decision making environment.

4. FGP MODEL FORMULATION

In FGP formulation, the defined membership functions are converted into the membership goals by introducing under-and over-deviational variables and assigning unity as the aspiration level to each of them. In the 'goal achievement function', minimization of the under-deviational variables of the goals on the basis of weights of importance of achieving the goals are taken into consideration. Such a version FGP is called the *minsum* FGP (9) and which is the simplest and widely used approach to multiobjective decision making problem. Then, executable *minsum* FGP formulation appears as

Find X so as to

$$\text{Minimize } Z = \sum_{k=1}^{K} w_k^- d_k^-$$

and satisfy

$$\frac{Z_k(X) - Z^*_{Wk}}{t_{1k}} + d_k^- - d_k^+ = 1, \quad k \in K_1$$

$$\frac{Z^*_{Wk} - Z_k(X)}{t_{2k}} + d_k^- - d_k^+ = 1,$$

$$k \in K_2 \qquad \dots (4.1)$$

subject to the system constraints in (2.6) and (2.7), where Z represents the fuzzy goal achievement function, d_k^- and d_k^+ (≥ 0) represent the under- and over-deviational variables associated with the kth membership goal, (> 0), $k = 1, 2, \dots, K$ represent the numerical weights associated with the respective under deviational variables, and where w_k^- ($k = 1, 2, \dots K$) are determined as (9).

$$w_k^- = \begin{cases} \dfrac{1}{t_{1k}}, & \text{for the } \mu_k \text{ defined in (3.2)} \\ \dfrac{1}{t_{2k}}, & \text{for the } \mu_k \text{ defined in (3.3)} \end{cases}$$

To illustrate the proposed approach, the following numerical example is solved in the Section 5.

5. AN ILLUSTRATIVE EXAMPLE

The following example with three objectives and three probability constraints with normally distributed random resource vector is considered.

Find $X(x_1, x_2, x_3)$ so as to

$$\text{Maximize } Z_1 = 6x_1 + 2x_2 + 3x_3$$
$$\text{Maximize } Z_2 = 3x_1 + 5x_2 + 4x_3$$
$$\text{Maximize } Z_3 = 2x_1 + 7x_2 + x_3$$

subject to

$$Pr\left(x_1 + x_2 + x_3 \geq b_1\right) \geq 0.95$$
$$Pr\left(5x_1 + x_2 + 6x_3 \leq b_1\right) \geq 0.10$$
$$Pr\ \left(2x_1 + 3x_2 + x_3 \leq b_2\right) \geq 0.15$$
$$x_1, x_2, x_3 \geq 0 \qquad \dots (5.1)$$

where b_1, b_2, b_3 are independent normal random variables with the following means and variances

$E(b_1) = 2.5$, Var $(b_1) = 2$; $E(b_2) = 8$, Var $(b_2) = 9$;
$E(b_3) = 10$, Var $(b_3) = 12$

Now using the values of $E(b_i)$ and var (b_i) $(i = 1, 2, 3)$, and following the procedure, the deterministic equivalent of the successive constraints in (5.1) are determined as

$$x_1 + x_2 + x_3 \geq 4.826$$

$$5x_1 + x_2 + 6x_3 \leq 11.855$$

$$2x_1 + 3x_2 + x_3 \leq 13.592 \qquad \text{... (5.2)}$$

Then, the best and worst decision of the first objective subject to the constraints in (5.2) is obtained as

$(x_1, x_2, x_3; Z^*_{B1}) = (1.69, 3.40, 0; 16.94)$ and

$(x_1, x_2, x_3; Z^*_{W1}) = (0, 4.38, 0.44; 10.09)$, respectively.

Similarly, the best and worst decisions of the second and third objectives are successively found as

$(x_1, x_2, x_3; Z^*_{B2}) = (0, 0.49, 1.29; 25.70)$, and

$(x_1, x_2, x_3; Z^*_{W2}) = (1.76, 3.06, 0; 20.61)$, respectively, for the second objective Z_2,

$(x_1, x_2, x_3; Z^*_{B3}) = (0, 4.38, 0.44; 31.12)$, and

$(x_1, x_2, x_3; Z^*_{W3}) = (1.75, 3.07, 0; 24.99)$, respectively, for the third objective Z_3.

Now, based on the above numerical values, the membership functions of the defined fuzzy goals can be constructed by following (3.2) and (3.3).

Then, following the procedure, the resulting FGP model is obtained as

Find $X(x_1, x_2, x_3)$ so as to

$$\text{Minimize } Z = 0.146d_1^- + 0.196d_2^- + 0.163d_3^-$$

and satisfy:

$$0.146(6x_1 + 2x_2 + 3x_3 - 10.09) + d_1^- - d_1^+ = 1$$

$$0.196(3x_1 + 5x_2 + 4x_3 - 20.61) + d_2^- - d_2^+ = 1$$

$$0.163(2x_1 + 7x_2 + x_3 - 24.99) + d_3^- - d_3^+ = 1$$

$$x_1, x_2, x_3 \geq 0 \qquad \text{... (5.3)}$$

subject to the constraints set in (5.2).

The resultant solution of the problem in (5.3) obtained by using the software LINGO (var. 6.0) is

$$(x_1, x_2, x_3) = (0.68, 3.81, 0.76) \text{ with}$$

$$(z_1, z_2, z_3) = (14.05, 24.21, 28.86).$$

The result shows that a most satisfactory decision is achieved here in the decision making environment.

Note 1: If the problem is solved by using the conventional FP approach studied in (14) previously, then the solution is found as

$$(x_1, x_2, x_3) = (0.30, 4.04, 0.86) \text{ with}$$

$$(z_1, z_2, z_3) = (14.27, 24.05, 28.73).$$

A comparison shows that a better decision is achieved here under the proposed approach from the view point of obtaining reasonably balanced decision in the decision making environment.

6. CONCLUSION

The main advantage of the proposed approach is that a compromise decision for achievement of the aspired goal levels of the objectives defined individually for each of them can be made on the basis of their weights of importance and the admissible tolerance values of the aspired goal levels of the objectives.

Here, under the proposed approach, the other defined random parameters can easily be accommodated within the framework of the proposed model.

If the coefficients of the fuzzy objective goals are also considered random, then that may be a problem for future study. An extension of the approach to nonlinear MODM problems involving both the fuzziness and randomness may be one of the current research problems. However, it is hoped that the approach presented here can contribute to future study in the field of real-life multiobjective decision making problems in the current uncertain environment.

ACKNOWLEDGEMENT

The authors are thankful to the anonymous referee and the Conference Chair of ICRFSC 2009 for their valuable suggestions to improve the quality of the presentation of the paper.

REFERENCES

[1] Bravo, M. and Ganzalez I., "Applying Stochastic Goal Programming: A Case Study on Water use Planning", *European Journal of Operational Research*, **196**, pp. 1123–1129, 2009.

[2] Charnes, A. and Cooper, W.W., "Chance-Constrained Programming", *Management Science*, **6**, pp. 73–79, 1959.

[3] Hulsurkar, S., Biswal, M.P. and Sinha, S.B., "Fuzzy Programming Approach to Multi-objective Stochastic Linear Programming Problems", *Fuzzy Sets and Systems*, **88**, pp. 173–181, 1997.

[4] Ignizio, J.P., "Goal Programming and Extensions", Lexington, Massachusetts, D.C. Health, 1976.

[5] Iskander, M.G., "A Fuzzy Weighted Additive Approach for Stochastic Fuzzy Goal Programming", *Applied Mathematics and Computation*, **154**, Issue no. 3, pp. 543–553, 2004.

[6] Luhandjula, M.K., "Linear Programming Under Randomness and Fuzziness", *Fuzzy Sets and Systems*, **18**, pp. 45–55, 1983.

[7] Luhandjula, M.K., "Fuzzy Stochastic Linear Programming: Survey and Future Research Directions", *European Journal of Operational Research*, **174**, pp. 1353–1367, 2006.

[8] Liu, B., "Dependent Chance Programming in Fuzzy Environments", *Fuzzy Sets and Systems*, **109**, pp. 97–106, 2000.

[9] Pal, B.B., Moitra, B.N. and Maulik, U., "A Goal Programming Procedure for Fuzzy Multiobjective Linear Fractional Programming Problem", *Fuzzy Sets and Systems*, **139**, pp. 395–405, 2003.

[10] Pal, B.B. and Sen, S. "A Linear Goal Programming Procedure for Academic Personnel Management Problems in University System", *The Proceedings of IEEE 10th Colloquium and third International Conference on Industrial and Information Systems (ICIIS 2008)*, paper ID. 437, pp. 1–7, 2008.

[11] Slowinski, R., "A Multicriteria Fuzzy Linear Programming Method for Water Supply System Development Planning", *Fuzzy Sets and Systems*, **19**, pp. 217–237, 1986.

[12] Tiwari, R.N., Dharmar S. and Rao J.R., "Fuzzy Goal Programming–An Additive Model", *Fuzzy Sets and Systems*, **24**, pp. 27–34, 1987.

[13] Zadeh, L.A., "The Concept of Linguistic Variable and its Application to Approximate Reasoning", *Part I, II, III, Information Sciences*, **8**, pp. 199–224, pp. 301–352, 1975.

[14] Zimmermann, H.J., "Fuzzy Programming and Linear Programming with Several Objective Functions", *Fuzzy Sets and Systems*, **1**, pp. 45–55, 1978.

45

ON THE PROPERTIES OF ROUGH APPROXIMATIONS AND TYPES OF CLASSIFICATIONS

B.K. Tripathy[1], D. Mohanty[2], and M.S. Prakash Kumar Ch.[1]

[1]School of Computing Sciences, VIT University, Vellore-632 014, Tamil Nadu, India.

[2]Department of Mathematics, Simanta Mahavidyalaya, Jharpokharia, Orissa-757 006, India.

Abstract: *The study of rough definability of classifications was initiated by Busse ([3]). Classifications are of great interest as in the process of learning from examples, rules are derived from classifications generated by single decisions. It was established by Busse that not all concepts of rough definability of sets can be extended to classifications. The four propositions established in [3] were extended to obtain necessary and sufficient type results in [12], from which several other results could be derived. The concept of types of classifications is analyzed in detail in [12] and it is shown that the five cases of types considered by Busse are the only basic ones and the other six possibilities reduce to the five cases of Busse either directly or transitively. An elementary study of the structure of classifications with respect to types of elements was initiated in [12]. In this paper, we prove a general theorem which provides a complete picture of the types of elements in a classification. We shall provide an algorithm to generate types of classifications, given the number of elements in the classification. The theorem of this paper establishes the claim that the notion of complement in bi-valued and multivalued logic are different. Finally, we analyze generation of rules as derived from the theorem.*

Keywords: *Rough set, classification, R-definable, R-lower approximation, R-upper approximation, Type 1, Type 2, Type 3 and Type 4.*

1. INTRODUCTION

Most of our traditional tools for formal modeling, reasoning and computing are crisp, deterministic and precise in character. Real situations are very often not crisp and deterministic and they cannot be described precisely. For a complete description of a real system often one would require by far more detailed data than a human being could ever recognize simultaneously, process and understand. This observation led to the extension of the concept of crisp sets so as to model imprecise data which can enhance their modeling power.

Two such models are the concept of Fuzzy sets introduced by Zadeh [13] in 1965 and the concept of Rough sets introduced by Pawlak [5] in 1982. Rough set has been found to be an excellent tool to capture indiscernibility of objects. The basic assumption of rough set theory is that human knowledge about a universe depends upon their capability to classify its objects. Classifications of a universe and equivalence relations defined on it are known to be interchangeable notions. So, for mathematical reasons equivalence relations were considered by Pawlak to define rough sets. A rough set is represented by a pair of crisp sets, called the lower and upper approximations of the set. The lower approximation of a rough set comprises of those elements of the universe, which can be said to belong to it definitely with the available knowledge. The upper approximation on the other hand comprises of those elements which are possibly in the set with respect to the available information. The concept of rough sets was primarily concerned with the study of intelligent systems characterized by insufficient and incomplete information.

2. DEFINITIONS AND PROPERTIES

Let U be a universe of discourse and R be an equivalence relation over U. By U/R we denote the family of all equivalence classes of R, referred to as categories or concepts of R and the equivalence class of an element $x \in U$ is denoted by $[x]_R$.

Definition 2.1: *By a knowledge base*, we understand a relational system $K = (U, \mathfrak{R})$, where U is as above and $\mathfrak{R}$ is a family of equivalence relations over U.

Definition 2.2: For any subset $P(\neq \phi) \subset \mathfrak{R}$, the intersection of all equivalence relations in P is denoted by $IND\ (P)$ and is called the *indiscernibility relation over P*. We define

$$IND\,(K) = \{IND\,(P)\colon\ P(\neq \phi) \subset \Re\}.$$

Definition 2.3: Let $X \subseteq U$ and R be in *IND* (*K*). The sets $\underline{R}X$ and $\overline{R}X$ are called the *R-lower* and *R-upper approximations* of X respectively and are defined as follows:

$$\underline{R}X = \{x \in U\colon [x]_R \subseteq X\},$$

$$\overline{R}X = \{x \in U\colon [x]_R \cap X \neq \phi\}.$$

Definition 2.4: For $X \subseteq U$, the *R-boundary of X* is denoted by $BN_R\,(X)$ and is defined as

$$BN_R\,(X) = \overline{R}X - \underline{R}X.$$

Definition 2.5: A set $X \subseteq U$ is said to be *rough with respect to R* if and only if $\underline{R}X \neq \overline{R}X$; that is, $BN_R\,(X) \neq \phi$.

X is said to be *R-definable* if and only if $\underline{R}X = \overline{R}X$ or $BN_R\,(X) = \phi$.

It may be noted that *R*-definable sets are crisp sets with respect to *R*. Many properties of the lower and upper approximations of rough sets, union of rough sets and intersection of rough sets have been obtained [2].

Definition 2.6: Suppose $x \in U$ and $X \subseteq U$. We say *x is certainly in X* with respect to *R* if and only if $x \in \underline{R}X$ and *x is possibly in X* with respect to *R* if and only if $x \in \overline{R}X$.

3. APPROXIMATION OF CLASSIFICATIONS

Classifications of universes play central roles in basic rough set theory. We define below a classification formally.

Definition 3.1: Let $F = \{X_1, X_2 \ldots X_n\}$ be a family of non empty sets defined over U. We say that F is a *classification of U* if and only if $X_i \cap X_j = \phi$ for $i \neq j$ and $\bigcup_{i=1}^{n} X_i = U$.

Definition 3.2: Let F be as above and R be an equivalence relation over U. Then $\underline{R}F$ and $\overline{R}F$ denote respectively the *R-lower* and *R-upper approximations* of the family F and are defined as

$$\underline{R}F = \{\underline{R}X_1, \underline{R}X_2, \ldots\ldots\ldots\ldots\ldots, \underline{R}X_n\} \text{ and}$$

$$\bar{R}F = \{\bar{R}X_1, \bar{R}X_2, \ldots\ldots\ldots\ldots \bar{R}X_n\}.$$

Throughout this paper we assume that F is a classification of U and R is an equivalence relation over U.

Grzymala-Busse [3] introduced and established some properties of approximation of classifications. These results are irreversible by nature ([6], p. 24). Also, these results establish that the two concepts; approximation of sets and approximation of families of sets (or classifications) are two different issues and that the equivalence classes of approximate classifications cannot be arbitrary sets. It has further been observed that if we have positive example of each category in the approximate classification then we must also have negative examples of each category. In [12] the results of Busse have been extended to obtain two necessary and sufficient type theorems. Several corollaries including the four theorems of Busse [3] could be obtained from these theorems. We avoid quoting these results here.

4. SOME PROPERTIES OF CLASSIFICATIONS

In this section we shall establish some properties of measures of uncertainty and discuss in detail on properties of classifications with two elements and three elements.

4.1 Measures of Uncertainty

We denote the number of elements in a set A by card (A).

4.1.1 *Classification*

We shall discuss the measures of uncertainty for F being a classification of U.

Definition 4.1.1: Let $F = \{X_1, X_2, \ldots, X_n\}$ be a classification of U and R be an equivalence relation on U. Then we denote the *accuracy of approximation of F by R* by $\beta_R(F)$ and define it as

$$\beta_R(F) = \left(\sum_{i=1}^{n} card(\underline{R}X_i)\right) / \left(\sum_{i=1}^{n} card(\bar{R}X_i)\right)$$

Definition 4.1.2: Let F and R be as above. Then we denote the *quality of approximation of F by R* is denoted by $\gamma_R(F)$ and define it as

$$\gamma_R(F) = (\sum_{i=1}^{n} card\,(\underline{R}X_i))/card\,(U).$$

The accuracy of classification expresses the percentage of possible correct decision when classifying objects employing the knowledge of *R*. The quality of a classification expresses the percentage of objects which can be correctly classified to classes of *F* employing knowledge of *R*.

Let R_1 and R_2 be any two equivalence relations on *U* and F_1 and F_2 be the corresponding classifications of *U*.

Property 4.1.1: For any *R*-definable classification *F* in *U*, $\beta_R(F) = \gamma_R(F) = 1$.

Hence, if a classification *F* is *R*-definable then it is totally independent on *R*. Here we note that even the converse of the above property is true.

Property 4.1.2: If $\beta_R(F) = \gamma_R(F) = 1$ then *F* is *R*-definable in *U*.

Proof: If $\beta_R\,(F) = 1$ *then*

$\sum_{i=1}^{n} card\,(\underline{R}X_i) = \sum_{i=1}^{n} card\,(\bar{R}X_i)$. As $card\,(\underline{R}X_i) \leq card\,(\bar{R}X_i)$, it follows that $\underline{R}X_i = \bar{R}X_i$. So, *F* is *R*-definable in *U*.

The following relation between $\beta_R\,(F)$ and $\gamma_R\,(F)$ was established in [8].

Property 4.1.3: For any classification *F* in *U* and an equivalence relation *R* on *U*, $0 \leq \beta_R\,(F) \leq \gamma_R\,(F) \leq 1$.

A cover is a generalization of a classification in the sense that the elements in it are not necessarily pair wise disjoint. The following property for covers is trivial.

Property 4.1.4: If *F* is a covering of *U* then $0 \leq \beta_R\,(F) \leq 1$ and $0 \leq \gamma_R(F) \leq n$.

4.1.2 Types of Rough Sets

Types of rough sets describe their topological properties. In practical applications of rough sets we need both the accuracy measure and the types. A rough set can be classified into the following four types [5]:

Type 1: If $\underline{R}X \neq \phi$ and $\overline{R}X \neq U$, then we say that *X is roughly R-definable*.

Type 2: If $\underline{R}X = \phi$ and $\overline{R}X \neq U$, then we say that *X* is *internally R-undefinable*.

Type 3: If $\underline{R}X \neq \phi$ and $\overline{R}X = U$, then we say that *X is externally R-undefinable*.

Type 4: If $\underline{R}X = \phi$ and $\overline{R}X = U$, then we say that *X* is *totally R-undefinable*.

4.2 Types of Classifications

In this section we present Types of classifications and their rough definability as stated by Busse [3]. It is well known that classifications are of great interest in the process of learning from examples and deriving rules from classifications generated by single decisions.

Definition 4.2.1: Let *R* be an equivalence relation on *U*. A classification $F = \{X_1, X_2, ..., X_n\}$ of *U* will be called roughly *R-definable, weak in U* if and only if there exists a number $i \in N_n$ such that $\underline{R}X_i \neq \phi$.

Definition 4.2.2: Let *R* and *F* be as above. Then *F* is said to be *roughly R-definable strong* in *U* (or of *Type 1*) if and only if $\underline{R}X_i \neq \phi$ for each $i \in N_n$.

Definition 4.2.3: Let *R* and *F* be as above. Then *F* is said to be *internally R-undefinable, weak* in *U* if and only if $\underline{R}X_i = \phi$ for each $i \in N_n$ and there exists $j \in N_n$ such that $\overline{R}X_j \neq U$.

Definition 4.2.4: Let *R* and *F* be as above. Then *F* is said to be internally *R*-undefinable, *strong* in *U* (or of *Type 2*) if and only if $\underline{R}X_i = \phi$ and $\overline{R}X_i \neq U$ for each $i \in N_n$.

It has been observed by Busse ([3]) that no *externally R-undefinable* set *X* exists in *U*. So, extension of the notion of Type 3 rough sets to classifications is not possible.

Definition 4.2.5: Let R and F be as above. Then F is said to be *totally R-undefinable* in U (or of *Type 4*) if and only if $\underline{R}X_i = \phi$ and $\bar{R}X_i = U$ for each $i \in N_n$.

The possible types of classifications have been analyzed in [11] and it has been obtained that there are eleven possibilities. Out of these eleven possibilities only five have been considered by Busse [3]. The rest of the cases have been shown to reduce either directly or transitively to one of these cases [11]. This provides justification for only five cases being considered in [3].

4.3 Types of Classifications

It has been observed [6] that approximation of sets and approximation of families of sets are two different issues and equivalence classes of approximate classification cannot be arbitrary sets, although they are strongly related. The concepts of complements in case of sets and in case of classifications are different, which may lead to new concepts of negation in the case of binary and multivalued logic.

In the next section we shall analyze the structure and properties of classifications having 2 elements and classifications having 3 elements.

4.3.1 Classifications with two or three Elements

To establish the results of this section and next section, we shall use the following table on Types of union of rough sets of different Types established by Tripathy and Mitra [10]. For brevity of notation, we use *T-i* for Type-*i*, *i* = 1, 2, 3, 4.

$\cup$	*T*-1	*T*-2	*T*-3	*T*-3
T-1	*T*-1/*T*-3	*T*-1/*T*-3	*T*-3	*T*-3
T-2	*T*-1/*T*-3	*T*-1/*T*-2/*T*-3/*T*-4	*T*-3	*T*-3/*T*-4
T-3	*T*-3	*T*-3	*T*-3	*T*-3
T-4	*T*-3	*T*-3/*T*-4	*T*-3	*T*-3/*T*-4

The following two results have been established in [8] for classifications having only two elements or three elements.

Theorem 4.3.1: A classification with two elements is roughly *R*-definable weak or of *T*-1 or of *T*-4 only.

Theorem 4.3.2: Taking X as above, the only feasible combinations of Types of X_1, X_2 and X_3 are {T-1, T-1, T-1}, {T-1, T-1, T-2}, {T-1, T-2, T-2}, {T-2, T-2, T-2}, {T-2, T-2, T-4}, {T-2, T-4, T-4}, {T-3, T-2, T-2} and {T-4, T-4, T-4}.

4.4 The General Result on Types of Classifications

In this section we prove a result which characterizes the general case of a classification having 'n' elements. We shall be using the above table on types of union of elements of different types.

Theorem 4.4.1: Let $\{X_1, X_2, ... X_n\}$ be a classification of U. Then for $n \geq 3$, the number of possibilities in terms of types of elements $X_i, i = 1, 2, ... n$ is 2 (n + 1).

Proof: When n = 3, the result is established above. For simplicity we write i for T-i, i = 1, 2, 3, 4. So, the different cases when n = 3 are 111, 112, 122, 222, 224, 244, 444 and 223. We can classify them into 4 groups. Group-1 has elements having at least one type-1and rest type-2, Group-2 has elements having all type-2, Group-3 has elements with at least one type-4 and rest type-2 and Group-4 has elements having one type-3 and rest type-2. This grouping of elements shall be used in the next step of the proof.

Let us consider a classification with 4 elements. We have to add an element of any type to elements of the elements of the above four groups. To elements of Group-1 we cannot add type-3 or type-4 elements as from the union table it follows that union of type-3 or type-4 with any other type elements leads to type- 3 or type-4 elements and complement of a type-1 element is a type-1 element. We can add type-1 or type-2 elements. But due to symmetry of union of elements we get only one additional element in this group. So, the elements are 1111, 1112, 1122 and 1222. To the elements of Group-2, we can add any type of element. Out of these, adding type-1 has been taken care in Group-1. Adding type-4 will be taken care in Group-3 and adding type-3 will be taken care in Group-4. So, we get the only element in this Group as 2222.

To Group-3 elements we can not add type-1 elements as union of type-4 elements with any other type elements results in type-3 or type-4 elements and none of these is complement of type-1. We can not add type-3 element as union of type-3 with any other type provides type-3 elements only and it is not the complement of type-4. So, we can add only type-2or type-4 elements. Using symmetric property of union of different types of elements we get the elements of Group-3 as 2224, 2244, 2444 and 4444.

Finally to Group-4 elements we can add only type-2 elements, which are the complement of type-3, which is the union of any other type element with type-3. So, we get 2223. Hence, the total number of elements for classifications with 4 elements is 10 = 2 (4 + 1). So, the result is true in this case also.

Next, suppose the result is true for classifications with n elements. At this stage, $|Group-1| = n$, $|Group-2| = 1$, $|Group-3| = n$ and $|Group-4| = 1$. Applying the same logic as we have done in finding the combination of types for classifications with four elements from those of classifications with three elements, we see that the number of elements in different groups changes as; $|Group-1| = n+1$, $|Group-2| = 1$, $|Group-3| = n+1$ and $|Group-4| = 1$. So, the total number of types of combinations in a classification with $(n + 1)$ elements is $(n + 1) + 1 + (n + 1) + 1 = 2 (n + 2) = 2 \{(n + 1) + 1\}$. Hence, the result is true in this case.

Thus, by mathematical induction the result is true for all values of n.

4.4.1 Structure of Classifications

From Theorem 4.4.1, we derive a complete picture of the structure of classifications.

Property 4.4.1.1: In a classification with n elements, there are exactly n cases which are Roughly R-definable weak.

Property 4.4.1.2: In a classification with n elements, there is exactly one case which is roughly R-definable strong.

Property 4.4.1.3: In a classification with n elements, there are exactly $(n - 1)$ cases which are Internally R-undefinable weak.

Property 4.4.1.4: In a classification with n elements, there is exactly one case which is R-undefinable strong.

Property 4.4.1.5: In a classification with n elements, there are exactly one case which is totally R-undefinable.

Note 4.4.1.1: As mentioned earlier, out of the 11 possible cases of types for classification only 5 are significant and the other types reduce to these 5 cases directly or transitively. The above properties, which are consequences of Theorem 4.4.1, establish this fact.

Note 4.4.1.2: Also, we find that the structure of complements in case of sets as established in Theorem 4.3.1 is different from the structure of complements in case of classifications as established in Theorem 4.3.2 and Theorem 4.4.1.

4.4.2 An Algorithm

In this section, we provide an algorithm to generate all the possible combinations of types of elements for a given number 'n' of elements in the classification.

Algorithm:

Input: $n \geq 3$

Step 1 (Initialization): $G1$ (3) = {111, 112}, $G2$ (3) = {222}, $G3$ (3) = {224, 244, 444} and $G4$ (3) = {223}.

Step 2: **If** ($n \neq 3$) **then**

For $i = 3$ **to** n **Do**

$G1\ (i + 1) = \{x_j : x \in G1\ (i)\ and\ j = 1\ or\ 2\}$,

$G2\ (i + 1) = \{x_2 : x \in G2\ (i)\}$

$G3\ (i + 1) = \{x_j : x \in G3\ (i)\ and\ j = 2\ or\ 4\}$

$G4\ (i + 1) = \{x_3 : x \in G4\ (i)\}$.

(Here xk means the string x concatenated with the value k.)

Step 3: $G1 = G1\ (n)$, $G2 = G2\ (n)$, $G3 = G3\ (n)$ and $G4 = G4\ (n)$.

Output: $G1$, $G2$, $G3$ and $G4$.

Complexity: The complexity of the above algorithm is of $\boldsymbol{O}\ (\boldsymbol{n})$.

5. RULE GENERATION

Suppose we are given a pair $S = (U, A)$ of non-empty, finite sets U and A, where U is the universe of objects, and A a set consisting of attributes, that is functions $a: U \rightarrow V_a$, where V_a is the set of values of attribute a, called the domain of a. The pair $S = (U, A)$ is called an information system. Any information system can be represented by a data table with rows labeled by objects and columns labeled by attributes. Any pair (x, a), where $x \in U$ and $a \in A$ defines the table entry consisting of the value $a(x)$.

Any subset B of A determines a binary relation $I(B)$ on U, called an indiscernibility relation, defined by $x\, I(B)\, y$ if and only if $a(x) = a(y)$ for every $a \in B$. Clearly $I(B)$ for all B is an equivalence relation. It can be noted that taking $\mathfrak{R} = \{I(a): a \in A\}$ we get a knowledge base defined above and $I(B) = IND(P)$ where $P = \{I(a): a \in B\}$.

Sometimes we distinguish in an information system $S = (U, A)$ a partition of A into two classes $C, D \subseteq A$ of attributes, called condition and decision (action) attributes, respectively. The tuple $S = (U, C, D)$ is called a *decision system*.

By *rules* on information systems we mean conditional statements that specify actions under conditions. Two sets of rules are induced from the lower and upper approximations: certain and possible. Rules computed from lower approximations are certain rules and rules computed from upper approximations are possible rules. Certain rules are categorical and may be further employed using classical logic. Possible rules may be further processed using either classical logic or any theory to deal with uncertainty. The rule induction system, producing certain and possible rule, is called LERS (Learning from examples using rough Sets). System LERS tests the input data for consistency. In the case of inconsistent data the system computes the lower and upper approximations of each concept. The user has a choice to use the machine learning approach or the knowledge acquisition approach. In the first case, the system induces a single minimal discriminant description for each concept. In the second case, the system induces all the rules, in the minimal form of each that can be induced from the input data. In both cases, the user has a choice between local or global approach. In the local approach, a subsystem LEM2 (learning from examples Module, version 2) computes local coverings of attribute-value pairs [1]. Local coverings are constructed from minimal complexes. The minimal complex contains attribute-value pairs, selected on the basis of their relevancy to the concept. In the case of a tie, the next criterion is the maximum of conditional probability of the concept given in the attribute-value pair. In the global approach, each concept is represented by a substitutional partition with two blocks: the first block is the concept; the second block is its complement. A minimal subset of the set of all attributes, such that the substitutional partition depends upon it, is called a global covering. A global covering may be selected on the basis of lower boundaries [2].

5.1 Properties and Observations

The following properties on generation of rules were obtained in [12].

5.1.1 Properties

(P1): A necessary and sufficient condition for classification C to induce certain rules is $\underline{C}\chi \neq \phi$.

(P2): The number of certain rules is equal to the number of nonempty boundaries in the classification.

By boundaries in the classification we mean the boundaries of each of the elements in the classification taken together.

(P3): The number of possible rules is equal to the number of non-empty boundaries in the classification.

Note 5.2.1 Observations: It is worth noting that 9 observations were made in [11] on generation from information systems.

Now, using properties 4.4.1.1 to 4.4.1.5, we can shed some light on the number of rules of different kinds, which can be obtained for information systems.

6. CONCLUSION

In basic rough set theory we are interested in classifications. Busse [3] has introduced and established some properties of approximations of classifications by the way obtaining a method for generation of rules using them. In [12] the theorems of Busse are generalized to obtain results of necessary and sufficient type. Also, the structures of classifications with two and three elements were obtained. In this paper, we established a theorem, which provides the structure of a classification with respect to type of elements. This result establishes that the concept of complement in case of sets and classifications are different. We provided an algorithm to compute the different combinations of types of elements for a classification, given the number of elements. Few elementary results on the properties of measures of accuracy of approximations are established. Some estimation on the number of rules of different kinds can be obtained from the properties of types of classifications obtained in Section 4.4.

REFERENCES

[1] Chan C.C. and Grzymala-Busse J., "On the Attribute Redundancy and the Learning Programs ID3", PRISM and LEM.

[2] Chan C.C. and Grzymala-Busse J., "Rough-set Boundaries as a Tool for Learning Rules from Examples", *Proc. of the ISMIS-89, 4th Int. Symp. on Methodologies for intelligent Systems*, (1989), pp. 281–288.

[3] Grzymala-Busse J., "Knowledge Acquisition Under Uncertainty–A Rough Set Approach", *Journal of Intelligent and Robotics Systems*, **1**, (1988), pp. 3–16.

[4] Grzymala-Busse J., "LERS-A System for Learning from Examples Based on Rough Sets", *In: Intelligent Decision Support-Handbook of Applications and Advances of the Rough Sets Theory*, (Ed: R. Slowinski), Kluwer Academic Publishers, (1991).

[5] Pawlak Z., "Rough Sets", *International Journal of Information and Computer Science*, (1982), pp. 341–356.

[6] Pawlak Z., "Rough Sets, Theoretical Aspects of Reasoning About Data", Kluwer academic Publishers. (1991).

[7] Pawlak Z. and Skowron A., "Rudiments of Rough Sets", *Information Sciences*, **177**, (2007), pp. 3–27.

[8] Pawlak Z. and Skowron A., "Rough Sets: Some extensions", *Information Sciences*, **177 (1)**, (2007), pp. 28–40.

[9] Pawlak Z. and Skowron A., "Rough Sets and Boolean Reasoning", *Information Sciences*, **177 (1)**, (2007), pp. 41–73.

[10] Tripathy B.K. and Mitra A. "Topological Properties of Rough Sets and their Applications", *Accepted for Publication in the International Journal of Granular Computing, Rough Sets and Intelligent Systems (IJGCRSIS)*, (2009).

[11] Tripathy B.K., Ojha J. and Mohanty D. "On Rough Definability and Types of Approximation of Classifications", *IEEE-International conference*, Patiala, 5–7 March, (2009).

[12] Tripathy B.K., Ojha J. and Mohanty D. "On Rough Approximations of Classifications and Properties of Decision Systems". *To be presented at the 1st IIMA International Conference on Advanced Data Analysis, Business Analytics and Intelligence*, June 6-7, Ahmedabad (India), (2009).

[13] Zadeh L.A., "Fuzzy Sets", *Information and Control*, **8**, (1965), pp. 338–353.

46

RECONSTRUCTED EFFECTIVE OBJECTIVE FUNCTION OF FUZZY C-MEANS USING PROPERTIES OF KERNEL FUNCTION

S.R. Kannan

Abstract: *Effective objective function of fuzzy c-means clustering algorithm using mathematical method is an important one in finding a meaningful map in the real world data. In this paper, a new objective function of fuzzy c-means was introduced by completely reconstructing the basic distance function using Kernel Function and Gaussian Functions. At initial stage it introduces a new kernel function from Gaussian function by incorporating the properties of kernel function and then it constructs an effective objective function for fuzzy c-means using the new kernel function for data analyzing. This paper calls the proposed new objective function of fuzzy c-means as New Kernel Induced Objective Function of Fuzzy C-Means. The advantage of our new proposed objective function is that it is capable in clustering high dimensional data set. Also the results have been achieved with less iteration using the proposed algorithm*

Keywords: *Fuzzy C-Means, Objective Function, Kernel Function, Distance Function, Data Clustering.*

1. INTRODUCTION

Clustering is a process for grouping objects in such a way that samples of the same group are more similar to one another than samples belonging to different groups. Many clustering strategies have been used, such as the supervised clustering and unsupervised clustering scheme, each of which has its own special characteristics. Mainly unsupervised clustering methods have been introduced for grouping objects, since it does not require any label of data.

Unsupervised clustering has basically divided into hard clustering and soft fuzzy clustering methods. The conventional hard clustering method restricts each point of the data set to exclusively just one cluster. As a consequence, with this approach the segmentation results are often very crisp, i.e. each object of the data belongs to exactly just one class. However, in many real world situations, make this hard (crisp) clustering a difficult task. Due to the fuzzy clustering (4) was proposed, which produced the idea of partial membership of belonging described by a membership function. Fuzzy clustering as a soft clustering method has been widely studied and successfully applied in data for real world problems (13, 22, 23, 24). But main disadvantages of fuzzy clustering algorithm are its need for a large amount of time to converge and it is more sensitive to the noise and outliers in the data.

This paper realized the underlined drawbacks, and it tries to rectify the drawbacks by introducing new objective function of fuzzy *c*-means with the help of Gaussian function, Lagrangian multipliers method, Kernel functions.

2. BACKGROUND

2.2 Fuzzy Clustering

The idea of basic fuzzy clustering called Fuzzy *C*-Means (FCM) has invented by Bezdek. It provides a method that shows how to group data points that populate some multidimensional space into a specific number of different clusters. *Fuzzy c-means* (FCM) is a data clustering technique wherein each data point belongs to a cluster to some degree that is specified by a membership grade.

The FCM algorithm took several names before FCM. These include Fuzzy ISODATA and Fuzzy *ck*-Means. The idea of using fuzzy set theory for clustering is credited to Ruspini [Ruspini, 1969 (25); Ruspini, 1970 (26)]. Dunn is credited with the first specific formulation of FCM, [Dunn, 1973 (10)], but its generalisation and current framing is credited to Bezdek, (Bezdek, 1981) (4). A collection of influential papers in the development of fuzzy clustering methods can be found in [Bezdek and Pal, 1992 (5)]. The FCM objective function and its generalisations are the most heavily studied fuzzy model in Pattern Recognition. FCM to be a clustering algorithm that provides a fuzzy partition of the input data set. However, there is an infinite range of possible fuzzy partitions. Therefore, an optimisation model or objective function must be devised to search for the optimal partition according to the chosen objective function. FCM is, thus, first and foremost an objective function. The way that most researchers have solved the optimisation problem has been through an iterative locally optimal technique, called the FCM algorithm. This is not the only way to solve the FCM objective function, for example, in [AlSultan and Selim, 1993 (2)] it is solved by the Simulated Annealing optimisation technique; in the problem is reformulated and general optimisation methods are suggested for its solution; in [Al-Sultan and Fedjki, 1997 (3)] it is solved by a combinatorial optimisation technique called Tabu Search; in (Hall *et al.*,) (6) it is solved by the *genetic algorithm* which is an optimisation technique based on evolutionary computation; and in [Runkler and Bezdek, 1999 (24)] it is solved within an alternate optimisation framework. In fact, it is not impossible that an exact solution to the problem may be formulated.

2.3 Fuzzy C-Mean Algorithm

Fuzzy *c*-means (FCM) is a method of clustering which allows one piece of data to belong to two or more clusters. This method (developed by Dunn in 1977 and improved by Bezdek in 1981) is frequently used in pattern recognition. It is based on minimization of the following objective function:

$$J_m = \sum_{i=1}^{N} \sum_{j=1}^{C} u_{ij}^m \left\| x_i - c_j \right\|^2, \ 1 \leq m < \infty$$

where m is any real number greater than 1, u_{ij} is the degree of membership of x_i in the cluster j, x_i is the ith of d-dimensional measured data, c_j is the d-dimension center of the cluster, and $\|*\|$ is any norm expressing the similarity between any measured data and the center.

Fuzzy partitioning is carried out through an iterative optimization of the objective function shown above, with the update of membership u_{ij} and the cluster centers c_j by:

$$u_{ij} = \frac{1}{\sum_{k=1}^{C}\left(\frac{\| x_i - c_j \|}{\| x_i - c_k \|}\right)^{\frac{2}{m-1}}}, \quad c_j = \frac{\sum_{i=1}^{N} u_{ij}^m - x_i}{\sum_{i=1}^{N} u_{ij}^m}$$

This iteration will stop when $\max_{ij}\left[\left| u_{ij}^{(k+1)} - u_{ij}^{(k)} \right|\right] < \varepsilon$ where ε is a termination criterion between 0 and 1, whereas k are the iteration steps. This procedure converges to a local minimum or a saddle point of J_m.

3. RECONSTRUCTED EFFECTIVE OBJECTIVE FUNCTION OF FUZZY C-MEANS

The objective function of basic Fuzzy C-Means function for finding appropriate structure in data is given by (Bezdek)

This paper proposes *Reconstructed Effective Objective Function of Fuzzy C-Means* by introducing Kernel function that allows the clustering of objects to be more reasonable.

The modified proposed objective function with kernel function is given by

$$J(U, V) = \sum_{i=1}^{n} \sum_{k=1}^{c} u_{ik}^m \| \varphi(x_i) - \varphi(v_k) \|^2 \quad \text{... (1)}$$

Where ψ stands as map. And the distance function can be expressed using in product space.

To obtain *Kernel induced Fuzzy C-Means based Gaussian Function the distance function can be modified as*

$$\| \varphi(x_i) - \varphi(v_k) \|^2 = k(x_i, x_i) + k(v_k, v_k) - 2k(x_i, v_k) \quad \text{... (2)}$$

where $i = 1, 2, ..., n$ and $k = 1, 2, ..., c$.

The Eq. (2) can be modified as

$$\| \varphi(x_i) - \varphi(v_k) \|^2 = 2[\beta + 1 - k(x_i, v_k)] \quad \text{... (3)}$$

Using (3) the new Kernel induced Fuzzy C-Means based Gaussian Function given by

$$J(U, V) = 2 \sum_{i=1}^{n} \sum_{k=1}^{c} u_{ik}^{m} \; [\beta + 1 - k(x_i, v_k)] \qquad \ldots (4)$$

Solving the constrained objective function in (4) we obtain

$$u_{1c} = \left\{ \frac{\lambda_1}{2m\left[\beta + 1 - k\left(x_1, v_c\right)\right]} \right\}^{1/m-1} \qquad \ldots (5)$$

Since our proposed partition matrix in an objective function satisfies the following conditions

$$\sum_{k=1}^{c} u_{ik} = 1, \quad \text{for } 1 = k = n$$

We obtain

$$\left(\frac{\lambda_1}{2m}\right)^{\frac{1}{m-1}} = \frac{1}{\left\{ \sum_{j=1}^{c} \left[\frac{1}{\beta + 1 - k\left(x_1, v_j\right)} \right]^{1/m-1} \right\}} \qquad \ldots (6)$$

Substituting (6) in 5, we have

$$u_{1c} = \frac{\left[\frac{1}{\beta + 1 - k\left(x_1, v_c\right)} \right]^{\frac{1}{m-1}}}{\sum_{j=1}^{c} \left[\frac{1}{\beta + 1 - k\left(x_1, v_1\right)} \right]^{\frac{1}{m-1}}}$$

In similar way we obtain

$$u_{2c} = \frac{\left[\frac{1}{\beta + 1 - k\left(x_2, v_c\right)} \right]^{\frac{1}{m-1}}}{\sum_{j=1}^{c} \left[\frac{1}{\beta + 1 - k\left(x_2, v_j\right)} \right]^{\frac{1}{m-1}}}$$

$$u_{nc} = \frac{\left[\dfrac{1}{\beta + 1 - k\left(x_n, v_c\right)}\right]^{\frac{1}{m-1}}}{\sum_{j=1}^{c}\left[\dfrac{1}{\beta + 1 - k\left(x_n, v_j\right)}\right]^{\frac{1}{m-1}}}$$

In general, we have

$$u_{ik} = \frac{\left[\dfrac{1}{\beta + 1 - k\left(x_i, v_k\right)}\right]^{\frac{1}{m-1}}}{\sum_{j=1}^{c}\left[\dfrac{1}{\beta + 1 - k\left(x_i, v_j\right)}\right]^{\frac{1}{m-1}}} \qquad \text{... (7)}$$

Solving Eq. (4) this obtains membership grades for objects in data for finding meaningful groups.

$$v_1 = \frac{\sum_{i=1}^{n} u_{i1}^{m}\left[k\left(k_i, v_1\right) - \beta\right] x_i}{\sum_{i=1}^{n} u_{i1}^{m}\left[k\left(k_i, v_1\right) - \beta\right]}$$

$$v_2 = \frac{\sum_{i=1}^{n} u_{i2}^{m}\left[k\left(x_i, v_2\right) - \beta\right] x_i}{\sum_{i=1}^{n} u_{i2}^{m}\left[k\left(x_i, v_2\right) - \beta\right]}$$

$$v_c = \frac{\sum_{i=1}^{n} u_{ic}^{m} \left[k\left(x_i, v_c\right) - \beta \right] x_i}{\sum_{i=1}^{n} u_{ic}^{m} \left[k\left(x_i, v_c\right) - \beta \right]} \qquad \text{... (8)}$$

In general, we have

$$v_k = \frac{\sum_{i=1}^{n} u_{ik}^{m} \left[k\left(x_i, v_k\right) - \beta \right] x_i}{\sum_{i=1}^{n} u_{ik}^{m} \left[k\left(x_i, v_k\right) - \beta \right]} \qquad \text{... (9)}$$

4. RESULTS AND DISCUSSIONS

In this section, we present some experimental results on random data. For the purpose of comparison this paper use existed fuzzy *c*-means for the same random data.

The first artificial image is created by random data given (in Table 1) in Fig. 1, another two are the brain real MR slices.

Table 1: *Random Data*

S. no	*Data*		*S. no*	*Data*	
	x_1	x_2		x_1	x_1
1	1.8	2	11	12	4
2	2	2.2	12	11.5	3.5
3	2	1.8	13	12.5	3.5
4	2	3.5	14	21	10
5	8.8	3	15	21	11
6	9	3.2	16	20.5	10.5
7	9	2.8	17	21.5	10.5
8	9.2	3	18	2	4
9	7	2.8	19	19	20
10	12	3	20	11	12

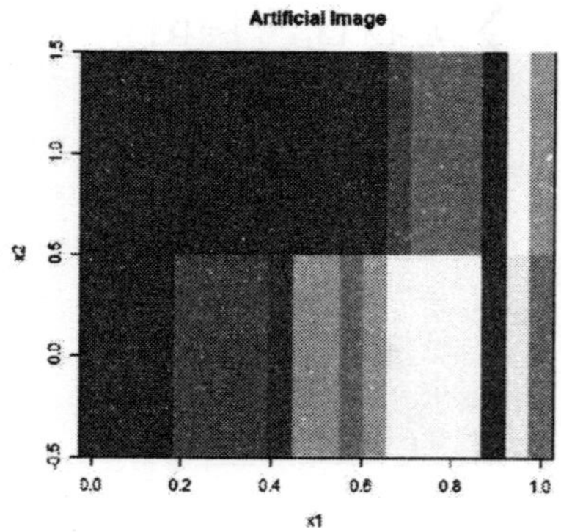

Fig. 1: *Image by Random Data*

The artificial image is shown in Fig. 1, contains 20 objects taken for the random data, each object of random data is taking different color to construct an artificial image.

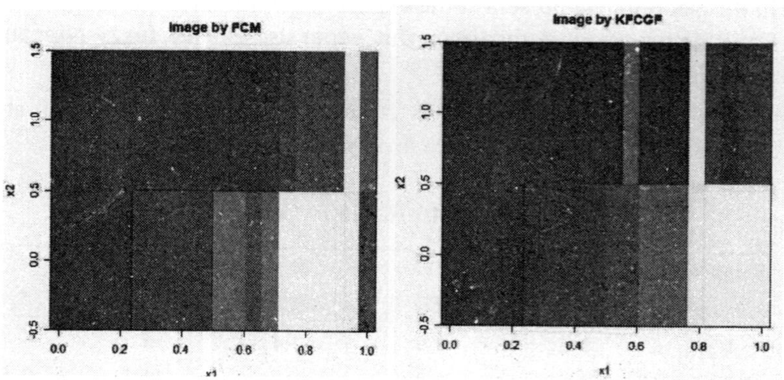

Fig. 2: *(a) Segmented Image by FCM, (b) Segmented Image by Effective Objective Fuzzy C-Means*

Figures 2 (*a*), 2 (*b*), shows the segmentation results of existed FCM, and Effective Objective Fuzzy *C*-Means respectively. Those objects must be allocated to, say, 2 classes.

FCM can not separate the two classes completely, but Effective Objective Fuzzy *C*-Means well succeed in clustering the data into two classes as shown in Figs 2 (*a*) and (*b*).

Table 2: *Final Memberships of two Clusters of FCM and Effective Objective Fuzzy C-Means Method and Object Allocation*

S. no	*Data* x_1	x_2	*Membership for cluster 1*	*Membership for cluster 2*	*Appropriate clusters*
1	1.8	2	0.986186	0.013814	1
2	2	2.2	0.992940	0.007060	1
3	2	1.8	0.982537	0.017463	1
4	2	3.5	0.989281	0.010719	1
5	8.8	3	0.132239	0.867761	2
6	9	3.2	0.103746	0.896254	2
7	9	2.8	0.117448	0.882552	2
s8	9.2	3	0.090034	0.909966	2
9	7	2.8	0.437255	0.562745	2
10	12	3	0.026485	0.973515	2
11	12	4	0.016826	0.983174	2
12	11.5	3.5	0.006171	0.993829	2
13	12.5	3.5	0.034943	0.965057	2
14	21	10	0.432943	0.567057	2
15	21	11	0.442378	0.557622	2
16	20.5	10.5	0.430682	0.569318	2
17	21.5	10.5	0.444227	0.555773	2
18	2	4	0.972505	0.027495	1
19	19	20	0.491037	0.508963	2
20	11	12	0.379073	0.620927	2

From the resulted images using FCM and Effective Objective Fuzzy *C*-Means and Object allocation Table 2, we note that *FCM* is much less fragmented than the original image and the incorporation of spatial constraints into the classification has somewhat the disadvantage of blurring of some fine details, but Effective Objective Fuzzy *C*-Means had better result than FCM, as shown in Fig. 2 (b).

5. CONCLUSION

This paper proposed an effective objective function for fuzzy *c*-means algorithm and applied it artificial image generated by random data.

The algorithm was formulated by using properties of Kernel function with basic objective function of the FCM algorithm to have proper effective data analyzing method. This paper shown the performance of effective objective function of fuzzy *c*-means algorithm with artificial image and it compared the results with conventional FCM method. It is clear from our comparison that effective objective function for fuzzy *c*-means is better results than FCM.

REFERENCES

[1] Ahmed M.N., Yamany S.M., Mohamed N., Farag A.A., and Moriarty T.A., "Modified Fuzzy *C*-Means Algorithm for Bias Field Estimation and Segmentation of MRI Data". *IEEE Trans. on Medical Imaging,* (2002), **21**, 193–199.

[2] Al-Sultan K.S., and Selim S.Z. "A Global Algorithm for the Fuzzy Clustering Problem", *Pattern Recognition,* (1993), **26 (9)**, 1357–1361.

[3] Al-Sultan K.S., and Fedjki C.A., "A Tabu Search-Based Algorithm for the Fuzzy Clustering Problem". *Pattern Recognition (1997),* **30 (12)**, 2023–2030.

[4] Bezdek J.C., "Pattern Recognition with Fuzzy Objective Function Algorithms, New York, Plenum Press, 1981.

[5] Bezdek J.C. and Pal S.K., eds. "Fuzzy Models for Pattern Recognition". New York, IEEE Press 1992.

[6] Bezdek J.C., Hall L.O., and Clarke L.P., "Review of MR Image Segmentation Techniques Using Pattern Recognition", *Med. Phys.,* (1993), **20**, 1033–1048.

[7] Brand M.E., Bohan T.P., and Kramer L.A. "Estimation of CSF and Gray Matter Volumes in Hydrocephalic Children Using Fuzzy Clustering of MR Images", *Comput. Med. Imaging Graph.* (1994), **18**, 25–34.

[8] Carvalho B.M., Gau J.C., Herman G.T., and Kong Y. "Algorithms for Fuzzy Segmentation", *Pattern Analysis and Applications,* (1999), **2 (1)**, 73–81.

[9] Clark M.C., Hall L.O., Goldgof D.B., Velthuizen R., Murtagh F.R., Silbiger M.S., "Automatic Tumor-Segmentation Using Knowledge-Based Techniques". *IEEE Transactions on Medical Imaging,* (1998), **117**, 187–201.

[10] Dunn J.C. "A Fuzzy Relative of the ISODATA Process and its use in Detecting Compact Well-Separated Clusters". *Journal of Cybernetics,* (1973), **3**, 32–57.

[11] Fletcher-Heath L.M., Hall L.O., Goldgof D.B., Murtagh F.R., "Automatic Segmentation of Non-Enhancing Brain Tumors in Magnetic Resonance Images". *Artifical Intelligence in Medicine,* (2001), **21**, 43–63.

[12] Gering D.T., Grimson W.E.L., Kikinis R., 2002, "Recognizing Deviations from Normalcy for Brain Tumor Segmentation". In: Dohi T., Kikinis R. (Eds.), *Medical Image Computing and Computer - Assisted Intervention MICCAI,* Springer (2002), **2488**, 388–395.

[13] Hall L.O., Bensaid A.M., Clarke L.P., Velthuizen R.P., Silbiger M.S., and Bezdek J.C.A., "Comparison of Neural Network and Fuzzy Clustering Techniques in Segmenting Magnetic Resonance Images of the Brain". *IEEE Trans. Neural Networks,* **(1992)**, **3 (5)**, 672–682.

[14] Jing-Hao Xue, Aleksandra Pizurica, Wilfried Philips, Etienne Kerre, Rik Vande Walle, Ignace Lemahieu, "An Integrated Method of Adaptive Enhancement for Unsupervised Segmentation of MRI Brain Images", *Pattern Recognition Letters,* (2003), **24 (15)**, 2549–2560.

[15] Krishnan N., Nelson Kennedy Babu C.V., Joseph Rajapandian V., and Richard Devaraj N., "A Fuzzy Image Segmentation using Feedforward Neural Networks with Supervised Learning". *Proceedings of the International Conference on Cognition and Recognition,* 396–402.

[16] Kwon M.J., Han Y.J., Shin I.H., and Park H.W., "Hierarchical Fuzzy Segmentation of Brain MR Images". *Int. J. Imaging Systems and Technology,* (2003), **13**, 115–125.

[17] Li X., Li L., Lu H., Chen D., and Liang Z, "Inhomogeneity Correction for Magnetic Resonance Images with Fuzzy *C*-Mean Algorithm". *Proc. SPIE,* (2003), **5032**, 995–1005.

[18] Li C.L., Goldgof D.B., and Hall L.O. "Knowledge-Based Classification and Tissue Labeling of MR Images of Human Brain". *IEEE Trans. Med. Imag.,* 1993, **12 (4)**, 740–750.

[19] Liew A.W.C., Leung S.H., and Lau W.H., "Fuzzy Image Clustering Incorporating Spatial Continuity". *IEE Proc. Visual Image Signal Process,* (2000), **147**, 185–192.

[20] Matthew C. Clark, Lowrance O. Hall, Dmitry B. Goldgof, Laurence P. Clarke, Robert P., Velthuizen and Martin, S. Silbigger, "MRI Segmentation Using Fuzzy Clustering Techniques", *IEEE Engineering in Medicine and Biology,* (1994), **13 (5)**, 730–742.

[21] Moussaoui A., Benmahammed K., Ferahta N and Chen V., "A New MR Brain Image Segmentation Using an Optimal Semisupervised Fuzzy *C*-Means and pdf Estimation". *Electronic Letters on Computer Vision and Image Analysis,* (2005), **5 (4)** 1–11.

[22] Noordam J.C., van den Broek W.H.A.M., and Buydens L.M.C., "Geometrically Guided Fuzzy *C*-Means Clustering for Multivariate Image

Segmentation". *Proc. Int. Conf. on Pattern Recognition*, (2000). **1**, 462–465.

[23] Pham D.L., and Prince J.L. "Adaptive Fuzzy Segmentation of Magnetic Resonance Images". *IEEE Trans. Medical Imaging*, (1999). **18**, 737–752.

[24] Runkler T.A., Bezdek J.C., "Alternating Cluster Estimation: A New Tool for Clustering and Function Approximation". *IEEE Trans. on Fuzzy Systems*, (1999), **7 (4)**, 377–393.

[25] Ruspini E.H., "A New Approach to Clustering". *Information and Control*, (1969). **15 (1)**, 22–32.

[26] Ruspini E.H., "Numerical Methods for Fuzzy Clustering". *Inform. Sci.*, (1970), **2**, 319–350.

47

GENE IDENTIFICATION USING DISTRIBUTION OF CODONS–A NEW APPROACH

Shashi Bajaj and Pradip Kumar Sen

Department of Mathematics, Jadavpur University, India

Abstract: *From the beginning, differentiating coding portions from non-coding portions in DNA sequences has been a main research topic for the researchers in bio-informatics. In this context, the role of statistical properties of the frequencies of nucleotides has been considered by the authors in a previous paper. Codons are being considered here in a similar manner for the same species Drosophila Melanogaster. The distribution of sixty four codons (all possible triplets of the four nucleotides A, C, G, T) is considered first. These are divided into twenty two groups according as the twenty different amino acids plus one start and three stop codons. Finally the frequencies of seven such codon groups are used to distinguish coding and non-coding sequences. In this connection, we deduce a criterion which gives more then 70% accuracy for all chromosomes.*

Keywords: *DNA, Coding and non-coding sequences, Compositional bias, Codon, Amino acid, Discriminating statistic.*

1. INTRODUCTION

The purpose of this paper is to develop methods related to gene identification based on differences between INTRONs and EXONs. In a DNA sequence only the coding portion, i.e. EXONs can take part in carrying a trait. But 90% of whole sequences are INTRONs, which were previously considered as garbage. Recent studies indicate otherwise, though the exact function of INTRONs is not clear. A significant amount of research work has been done to differentiate the coding and non coding parts of DNA sequences. Some of the standard techniques are Hidden Markov Models, Neural Networks (GRAIL), the basic codon model of Goldman and Yang (1994) [5], fixed-effect codon Model [6], Grammars, Homology, etc. ([7], [8]).

Our aim here is to study the difference between coding and non coding sequences of a genome using statistical analysis. This type of approach, though it makes obvious sense, is relatively uncommon ([2], [3], [4]).

The method considered in this paper comes under compositional bias, which in its simplest form, considers distribution of bases [1]. Here we started with sixty four codons which are all possible triplets of four nucleotides 'A', 'C', 'G', 'T'. As start (ATG) and stop (TAA, TAG, TGA) codons have special functions, we next do the same analysis for sixty codons excluding start and stop codons. Each codon encodes one of the twenty amino acids used in the synthesis of proteins, each amino acids being encoded by one or more codons. Thus the codons form twenty different groups. When we do the same analysis using these groups, it gives better results. With some numerical considerations, we find that there are seven groups among the twenty which may be mostly involved in identifying a sequence as coding or non-coding. So, finally we form another data set containing eight codon groups, seven as mentioned above and one combining the remaining thirteen groups. The same analysis gives a much improved result. In what follows, we describe the step by step analysis.

2. DATA

For this analysis we have used data from GenBank. The considered species is Drosophila Melanogaster (Fruit Fly). We have chosen the species, as its genomic sequence is considered reliable and is commonly investigated by researchers.

There are four chromosomes of this genome: CHR2, CHR3, CHR4 and CHRX. The first two are further divided into two parts 2L, 2R and 3L, 3R respectively. GenBank provides the data in various formats, but we have taken the "gbk" format only which is the most suitable for our work.

3. PRELIMINARY PROCESSING

First we have divided the genomic sequences of each chrcmosome into five parts, to be stored in five files, the first four of which are similar to the original "gbk" genomic sequences file, but without the ORIGIN portion for the whole base sequence. Two files contain coding sequence, further classified into two types:

(*i*) CDS A (the continuous coding sequences),

(*ii*) CDS B (the coding sequences interrupted by INTRON A's).

The next two files contain non-coding sequences. They are also of two types:

(*i*) INTRON A (the non-coding sequences embedded in CDS B's),

(*ii*) INTRON B (these is also INTRON but not embedded in the CDS B's).

The fifth file is formed by the ORIGIN part of the original file.

With each of the first four files using the fifth, we create fasta format files. These fasta files are our basis for the next part of the analysis. The first step of our analysis uses the relative frequencies of the sixty four codons in the four types of sequences. In the next step, we obtain chi-square type squared distances. Finally the values are used to create frequency distribution tables and frequency curves to draw inference.

4. DIFFERENTIATING CRITERION

For a more detailed analysis of the difference among the four types of sequences we use a distribution of a statistic similar to a chi-square statistic for goodness of fit tests. The standard chi-square statistic is

$$\sum_i \frac{(Observed\ Value - Expected\ Value)^2}{Expected\ Value}$$

$$= \sum_i \frac{(nP_i - n\pi_i)^2}{n\pi_i}$$

Where, P_i = Observed Proportion, π_i = Theoretical Proportion, n = Sample Size.

To avoid the huge scaling effect of the large sequence size n we use

$$d^2 = \sum_{i=1}^{r} (P_i - \pi_i)^2$$

instead of the above. Here r is 64, 60, 20 and 8 in the different stages of the analysis. We are going to use it as a discriminating statistic. We can consider this statistic as the squared Euclidian distance between the distributions $\{P_i\}$ and $\{\pi_i\}$.

We use five different choices for the choice of $\{\pi_i, i = 1, \ldots, r\}$: overall distributions for CDS A, CDS B, INTRON A, INTRON B, or all four types combined.

Next, we construct the frequency distribution of d^2 for each of the five cases. Figure 1 shows the results only for Chromosome 2L using CDS A as standard. The horizontal axis represents $500d^2$ for easy scaling. We make the following observations from the analysis:

1. There is a difference between coding and non-coding sequences as expected.
2. An interesting point is that the two coding and the two non-coding types show difference among themselves.
3. The CDS A graph has the most fluctuation among all the graphs.
4. The behaviour is similar for all the chromosomes, except for Chromosome 4 which is slightly different from the others.
5. The best difference is expected by using as base the combined distribution, which is supposed to be a balance. However the best results are obtained by using CDS A as base for all chromosomes, though it is very similar and marginally better than combined.

Our final target is to determine a critical value M such that

(*a*) Percentage of coding sequences with $d^2 < M$ is high and Percentage of non-coding sequences with $d^2 > M$ is low, or

(*b*) Percentage of coding sequences with $d^2 < M$ is low and Percentage of non-coding sequences with $d^2 > M$ is high.

For this we compare the cumulative relative frequencies for coding and non-coding and look for the maximum difference. We do this for each chromosome for each of the five cases C_1 to C_5. The results are shown in Table 1.

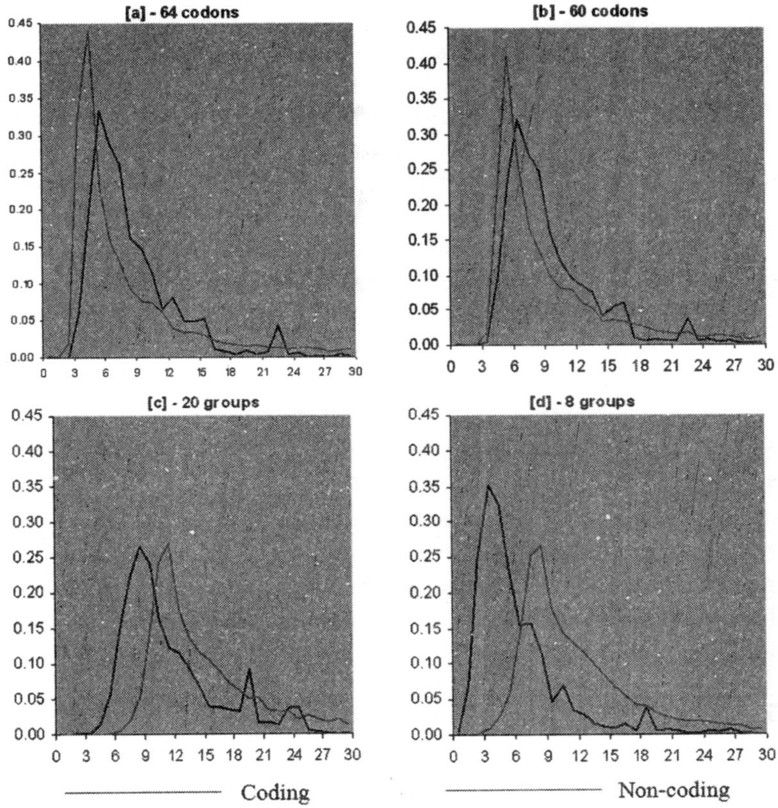

Fig. 1: *Distribution of 500d² using CDS A as Standard*

Table 1: *Critical Value of 500d² for Differentiating Coding from Non-coding Sequences*

C	*CDS A (C_1)*		*CDS B (C_2)*		*INTRON A (C_3)*		*INTRON B (C_4)*		*COMBINED (C_5)*	
H	*Coding*	*Non-Coding*	*Coding*	*Non-Coding*	*Coding*	*Non-Coding*	*Coding*	*Non-Coding*	*Coding*	*Non-Coding*
R	***Critical Value***		***Critical Value***		***Critical Value***		***Critical Value***		***Critical Value***	
2	0.7663	0.1194	0.7301	0.0701	0.6337	0.1104	0.5728	0.0563	0.5969	0.051
L	**6.5**		**6.5**		**6.5**		**5.5**		**5.5**	

Table Contd.

Table 1 Contd.

2	0.7405	0.0904	0.7325	0.0876	0.6287	0.1525	0.5655	0.0956	0.7163	0.2087
R	**6.5**		**6.5**		**6.5**		**5.5**		**6.5**	
3	0.7285	0.0839	0.7218	0.0734	0.6123	0.1035	0.5349	0.0528	0.5714	0.0499
L	**6.5**		**6.5**		**6.5**		**5.5**		**5.5**	
3	0.7378	0.0849	0.7107	0.0727	0.6111	0.1237	0.6805	0.2321	0.6936	0.1893
R	**6.5**		**6.5**		**6.5**		**5.5**		**6.5**	
4	0.7511	0.01	0.905	0.0698	0.819	0.0399	0.8462	0.0399	0.8824	0.0764
	8.5		**7.5**		**7.5**		**7.5**		**7.5**	
X	0.735	0.0912	0.7255	0.091	0.6244	0.1797	0.5739	0.1134	0.5989	0.0939
	6.5		**6.5**		**6.5**		**5.5**		**5.5**	

In the above table if we overlook the chromosome 4 data then from the remaining we may easily decide on **6.5** as the critical value. Actually the critical value is $d^2 = 6.5/500 = 0.0013$. Chromosome 4 always gives some abnormal results because of much fewer sequences than all the other chromosomes. If we consider CDS A as base then it may be concluded that almost 70% of coding sequences are below the critical point and 90% of non-coding sequences are above that point.

5. CONCLUSION

Our original aim was to develop a method to classify a nucleotide sequence as coding or non-coding. The method developed can be summarized as follows:

For the given sequence calculate $\boldsymbol{d^2}$ *with CDS A as base. If* $\boldsymbol{d^2} < 0.0013$ *the sequence is classified as coding and if* $\boldsymbol{d^2} > 0.0013$ *it is classified as non-coding.*

Thus it is a powerful distinguishing criterion with more than 70% accuracy. We expect that the same type of criterion will apply to other species. In two other species considered by the authors, the critical value needs to be modified.

ACKNOWLEDGEMENTS

We are grateful to the University Grants Commission for financial assistance under the Major Research Projects scheme. We also thank the Department of Mathematics, Jadavpur University for infrastructural support.

REFERENCES

[1] Bajaj S. and Sen P.K., (2009), "Differentiating Coding from Non-coding–A Statistical Analysis of DNA Sequence". InfoBiz (Accepted for publication).

[2] Borodovsky M., E. Koonin, and K. Rudd, (1994), "New Genes in Old Sequences: A Strategy for Finding Genes in a Bacterial Genome". *Trends in Biochemical Sciences*, **19**, 309–313.

[3] Burge C. and Karlin S., (1997), "Prediction of Complete Gene Structures in Human Genomic DNA". *J. Mol Biol*, **268**, 78–94.

[4] Fickett J.W., (1982), "Recognition of Protein Coding Regions in DNA Sequences", *Nucleic Acids Res*, **10**, 5303–5318.

[5] Le Bao, Hong Gu, Katherine A. Dunn, and Joseph P. Bielawski, "Likelihood-Based Clustering (LiBaC) for Codon Models, a Method for Grouping Sites According to Similarities in the Underlying Process of Evolution". *Molecular Biology and Evolution*, **25**, 1995–2007.

[6] Le Bao, Hong Gu, Katherine A. Dunn and Joseph P. Bielawski, "Methods for Selecting Fixed-effect Models for Heterogeneous Codon Evolution, with Comments on their Application to Gene and Genome Data". *BMC Evolutionary Biology*, 2007, **7**, 1471–2148.

[7] Mathe C., Sagot MF, Schiex T., Rouze P., (2002), "Current Methods of Gene Prediction, their Strengths and Weaknesses", *Nucleic Acids Reasearch*, **30**, 4103–4117.

[8] Pachter L., (1998), "Introduction to Computational Molecular Biology–Gene Recognition", *MIT Lecture notes*.

48

CHARACTERIZATION OF PRIME IDEALS IN Γ-SEMIGROUPS THROUGH FUZZY IDEALS

S. Coumaressane

Department of Mathematics, Perunthalaivar Kamarajar Government Arts College, Kalitheerthalkuppam, Affiliated by Pondicherry University, Puducherry-605107, India

E-mail: coumaressane_s@yahoo.com

Abstract: *The concept of fuzzy prime ideal in Γ-semigroups is introduced and characterize prime ideals in Γ-semigroups through fuzzification. We have shown that image of a fuzzy prime ideal in Γ-semigroups contains only two elements. The notion of fuzzy m-system in Γ-semigroups is also introduced and related results are obtained. We have shown that if μ is a fuzzy prime ideal, 1 – μ is a fuzzy m-system in Γ-semigroups.*

1. INTRODUCTION

In 1965, Zadeh ([14]) introduced the concept of a fuzzy subset and studied its properties on the parallel lines to set theory. In 1971, Rosenfeld ([9]) defined the fuzzy subgroup and gave some of its properties. Rosenfeld's definition of a fuzzy group is a turning point for pure mathematicians. Since then, the study of fuzzy algebraic structure has been pursued in many directions such as groups, rings, modules, vector spaces and so on. In 1981, Das ([1]) explained the interrelationship between the fuzzy subgroups and its t-level subsets. Sen ([11]) introduced Γ-semigroups in 1981. Sen and Saha ([10], [12]) introduced Γ-semigroups different from the first definition of Γ-semigroups in the sense of Sen ([11]).

In this paper we introduced the notions of fuzzy left (right, two-sided) ideal, fuzzy prime ideal and fuzzy *m*-system of Γ-semigroups. We have shown that the image of a fuzzy prime ideal of Γ-semigroup contains two elements. By an example, we have shown that level subset μ_t is an *m*-system, but μ is not a fuzzy *m*-system in Γ-semigroup. However, we found a condition under which μ is a fuzzy *m*-system. Finally, we have shown that if μ is a fuzzy prime ideal, then $1 - \mu$ is a fuzzy *m*-system in Γ-semigroup.

2. PRELIMINARY NOTES

We would like to reproduce some definitions and results proposed by the pioneers in this field earlier for the sake of completeness.

Definition 2.1: Let *M* and Γ be any two nonempty sets. *M* is called a Γ-semigroup if for all $a, b, c \in M$ and $x, y \in \Gamma$;

1. There is a mapping $M \times \Gamma \times M \to M$ defined by $(a, x, b) \to a\,x\,b$.
2. There is a mapping $\Gamma \times M \times \Gamma \to \Gamma$ defined by $(x, a, y) \to x\,a\,y$.
3. $(axb)\,yc = a\,(xby)\,c = ax\,(byc)$.

Notation 2.2: For subsets *A*, *B* of a Γ-semigroup *M*, let

$$A\Gamma B = \{a\gamma b \mid a \in A,\ b \in B \text{ and } \gamma \in \Gamma\}.$$

Definition 2.3: Let *M* be a Γ-semigroup and *A* a nonempty subset of *M*. *A* is called a *right* (*resp. left*) *ideal* of *M* if $A\Gamma M \subseteq A$ (resp. $M\Gamma A \subseteq A$).

A subset *A* of *M* is called an *ideal* of *M* if it is a right and a left ideal of *M*.

From now on, throughout this paper *M* will denote Γ-semigroup unless otherwise specified.

Definition 2.4: A mapping $\mu : M \to [0, 1]$ is called a *fuzzy subset* of *M*.

A fuzzy subset $\mu: M \to [0, 1]$ is nonempty if μ is not the constant map which assumes the value 0. For any two fuzzy subsets λ and μ of *M*, $\lambda \subseteq \mu$ means that $\lambda\,(a) \leq \mu\,(a)$ for all $a \in M$. $1 - \mu$ is a fuzzy subset of *M* defined by $(1 - \mu)\,(x) = 1 - \mu\,(x)$ for all $x \in M$. The characteristic function of Γ-semigroup *M* is denoted by ***M*** and its subset *A* is denoted by $\mathcal{X}_A$. If μ is a fuzzy subset of *M*, then the image of μ denoted by $Im(\mu) = \{\mu(m) \mid m \in M\}$ and $|Im(\mu)|$ denotes the cardinality of $Im(\mu)$. We denote by $I(a)$ the ideal of M generated by $a \in M$. We can show that $I(a) = a \cup M\Gamma a \cup a\Gamma M \cup M\Gamma a\Gamma M$. Hereafter we consider only nonempty fuzzy subsets of M.

Definition 2.5: Let μ be any fuzzy subset of M. For $t \in [0, 1]$, the set $\mu_t = \{x \in M \mid \mu(x) \geq t\}$ is called a *level subset* of μ.

Definition 2.6: Let μ and λ be any two fuzzy subsets of M. Then $\mu \cap \lambda$, $\mu \cup \lambda$ and $\mu * \lambda$ are fuzzy subsets of M defined by

$$(\mu \cap \lambda)(x) = \min\{\mu(x), \lambda(x)\},$$

$$(\mu \cup \lambda)(x) = \max\{\mu(x), \lambda(x)\}, \text{ and}$$

$$(\mu * \lambda)(z)$$

$$= \begin{cases} \sup\limits_{z = x\gamma y} \{\min\{\mu(x), \lambda(y)\}\} & \text{if } z \text{ is expressed as } z = x\gamma y, \\ & \text{where } x, y, z \in M \text{ and } \gamma \in \Gamma. \\ 0 & \text{otherwise.} \end{cases}$$

Definition 2.7: For any $x \in M$ and $t \in [0, 1]$, define the *fuzzy point* x_t as

$$x_t(y) = \begin{cases} t & \text{if } y = x, \\ 0 & \text{if } y \neq x. \end{cases}$$

If x_t is a fuzzy point and μ is any fuzzy subset of M such that $x_t \subseteq \mu$, then we write $x_t \in \mu$. Note that $x_t \in \mu$ if and only if $x \in \mu_t$ where μ_t is a level subset of μ.

For any fuzzy subset μ of M, it is obvious that $\mu = \bigcup\limits_{a_t \in \mu} a_t$

The following lemma can be shown immediately.

Lemma 2.8: For any nonempty subsets A and B of M,

1. $\mathcal{X}_A * \mathcal{X}_B = \mathcal{X}_{A\Gamma B}$;
2. $\mathcal{X}_A \cap \mathcal{X}_B = \mathcal{X}_{A \cap B}$.

Definition 2.9: A fuzzy subset μ of M is called a fuzzy left (resp. right) ideal of M if $\mu(m\gamma a) \geq \mu(a)$ $(\mu(a\gamma m) \geq \mu(a))$ for all $m, a \in M$ and $\gamma \in \Gamma$.

If μ is both fuzzy left and right ideal of M, then μ is called a *fuzzy ideal* of M.

Lemma 2.10: [2] A fuzzy subset λ of M is a fuzzy left (right) ideal of M if and only if $\boldsymbol{M} * \lambda \subseteq \lambda$ ($\lambda * \boldsymbol{M} \subseteq \lambda$).

Lemma 2.11: [2] Let μ be a fuzzy subset of M. μ is a fuzzy left (resp. right) ideal of M if and only if each level subset μ_t, $t \in Im(\mu)$, is a left (resp. right) ideal in M.

Theorem 2.12: [2] Let A be a nonempty subset in M and μ be a fuzzy subset of M defined by

$$\mu(x) = \begin{cases} s & \text{if } x \in A, \\ t & \text{otherwise} \end{cases}$$

for all $x \in M$ and $s, t \in [0, 1]$ with $s > t$. μ is a fuzzy left (resp. right) ideal of M if and only if A is a left (resp. right) ideal in M. Moreover

$$M_\mu = \{x \in M \mid \mu(x) = s\} = A.$$

Corollary 2.13: Let A be a subset in M. A is a left (resp. right) ideal in M if and only if $\mathcal{X}_A$ is a fuzzy left (resp. right) ideal of M.

Theorem 2.14: [2] Let m be a fixed element in M. If μ is a fuzzy left (resp. right) ideal of M, then $\mu^m = \{x \in M \mid \mu(x) \geq \mu(m)\} = \mu_t$, where $t = \mu(m)$, is a left (resp. right) ideal in M.

Corollary 2.15: Let μ be a fuzzy left (resp. right) ideal of M. Then the set M_μ is a left (resp. right) ideal in M.

The following results can be seen easily.

Lemma 2.16: Let μ be a fuzzy ideal of M. If $x_r \subseteq \mu$ for any fuzzy point x_r, then for all fuzzy ideals λ of M,

1. $x_r * \lambda \subseteq \mu$;
2. $\lambda * x_r \subseteq \mu$.

Lemma 2.17: Let θ and σ be any two fuzzy subsets of M. If $\theta \subseteq \mu$ and $\sigma \subseteq \lambda$, then $\theta * \sigma \subseteq \mu * \lambda$ and $\sigma * \theta \subseteq \lambda * \mu$ for any fuzzy subsets μ and λ.

Corollary 2.18: Let a_r and b_s be any two fuzzy points of M such that $a_r \in \mu$ and $b_s \in \lambda$, then $a_r * b_s \subseteq \mu * \lambda$ and $b_s * a_r \subseteq \lambda * \mu$, for any fuzzy subsets μ and λ of M.

3. FUZZY PRIME IDEALS OF Γ-SEMIGROUPS

Definition 3.1: An ideal T of M is called *prime* if for any ideals A and B in M such that $A\Gamma B \subseteq T$ implies $A \subseteq T$ or $B \subseteq T$.

Theorem 3.2: Let T be an ideal of M. Then the following are equivalent:

1. T is prime.
2. For any $a, b \in M$ such that $a\Gamma M\Gamma b \subseteq T$, we have $a \in T$ or $b \in T$.
3. For any $a, b \in M$ such that $I(a)\,\Gamma\, I(b) \subseteq T$, we have $a \in T$ or $b \in T$.
4. For any two right ideals A and B in M such that $A\Gamma B \subseteq T$, we have $A \subseteq T$ or $B \subseteq T$.
5. For any two left ideals A and B in M such that $A\Gamma B \subseteq T$, we have $A \subseteq T$ or $B \subseteq T$.

Definition 3.3: A fuzzy ideal μ of M is called *prime* if μ is not a constant function and for any two fuzzy ideals λ, θ of M, $\lambda * \theta \subseteq \mu$ implies that either $\lambda \subseteq \mu$ or $\theta \subseteq \mu$.

The following results one can easily obtain.

Theorem 3.4: Let I be an ideal in M. I is a prime ideal of M if and only if $\mathcal{X}I$ is a fuzzy prime ideal of M.

Theorem 3.5: Let μ is a fuzzy ideal of M. If μ is a fuzzy prime ideal of M, then $M_\mu = \{x \in M \mid \mu(x) = s\}$, where $s = \sup_{y \in M} \{\mu(y)\}$, is a prime ideal of M.

Lemma 3.6: If μ is a fuzzy ideal of M and $a \in M$, then $\mu(x) \geq \mu(a)$ for all $x \in I(a)$.

Lemma 3.7: Let I be a prime ideal of M and $\alpha \in [0, 1)$. If μ is a fuzzy subset of M defined by

$$\mu(x) = \begin{cases} 1 & \text{if } x \in I, \\ \alpha & \text{otherwise,} \end{cases}$$

then μ is a fuzzy prime ideal of M.

The proofs of following three theorems are similar to rings theory [7] with the corresponding changes for Γ-semigroups.

Theorem 3.8: Let μ be a fuzzy subset of M. If μ is a fuzzy prime ideal of M, then $|I_m(\mu)| = 2$.

Proof: Since μ is not constant, $|Im(\mu)| \geq 2$. Suppose that $|Im(\mu)| \geq 3$. Let $\sup\limits_{y \in M} \mu(y) = s$ and $k = g.l.b\{\mu(x) | x \in M\}$. Then there exist $t, m \in Im(\mu)$ such that $t \leq m \leq s$ and $t \geq k$. Let λ and θ be fuzzy subsets of M such that $\lambda(x) = \frac{1}{2}(t + m)$ for all $x \in M$ and $\theta(x) = k$ if $x \notin \mu_m = \{x \in M | \mu(x) \geq m\}$, $\theta(x) = s$ if $\in \mu_m$. Clearly λ is a fuzzy ideal of M. We show that θ is a fuzzy ideal of M. Let $x, y \in M$ and $\gamma \in \Gamma$. If $x, y \in \mu_m$, then $x\gamma y \in \mu_m$ and $\theta(x\gamma y) = s = \min\{\theta(x), \theta(y)\}$. If $x \in \mu_m$ and $y \notin \mu_m$, then $x\gamma y \in \mu_m$. $\theta(x\gamma y) = s \geq k = \min\{\theta(x), \theta(y)\}$. If $x, y \notin \mu_m$, then $\theta(x) = \theta(y) = k$. Hence $\theta(x\gamma y) \geq \min\{\theta(x), \theta(y)\}$. Thus $\theta(x\gamma y) \geq \min\{\theta(x), \theta(y)\}$ for all $x, y \in M$ and $\gamma \in \Gamma$. Hence θ is a fuzzy ideal of M. We now claim that $\lambda * \theta \subseteq \mu$. Let $x \in M$. Consider the following cases.

(*i*) $x \in \mu_m$. Then $\mu(x) \geq m$ and

$$(\lambda * \theta)(x) = \sup_{x = u\gamma v}\{\min\{\lambda(u), \theta(v)\}\} \leq \frac{1}{2}(t + m) < m \leq \mu(x)$$

(*ii*) $x \notin \mu_m$. Then for any $u, v \in M$ and $\gamma \in \Gamma$ such that $x = u\gamma v$, $u \notin \mu_m$ and $v \notin \mu_m$. Then $\theta(v) = k$. Hence $(\lambda * \theta)(x) = \sup\limits_{x = u\gamma v}\{\min\{\lambda(u), \theta(v)\}\} = k \leq \mu(x)$.

Thus in any case, $(\lambda * \theta)(x) \leq \mu(x)$. Hence $\lambda * \theta \subseteq \mu$. Now there exists $u \in M$ such that $\mu(u) = t$. Then $\lambda(u) = \frac{1}{2}(t + m) > t = \mu(u)$. Hence $\lambda \not\subseteq \mu$. Also there exists $x \in M$ such that $\mu(x) = m$. Then $x \in \mu_m$ and thus $\theta(x) = s > m = \mu(x)$. Hence $\theta \not\subseteq \mu$. Thus neither $\lambda \subseteq \mu$ nor $\theta \subseteq \mu$. This shows that μ is not a fuzzy prime ideal of M, which is a contradiction to the hypothesis. Hence $|Im(\mu)| = 2$.

Theorem 3.9: Let μ be a fuzzy subset of M. If μ is a fuzzy prime ideal of M, then $s = 1$, where $s = \sup\limits_{y \in M} \mu(y)$.

Theorem 3.10: Let μ be a fuzzy subset of M. If $|Im(\mu)| = 2$, $s = \sup\limits_{y \in M} \mu(y)$ and $M_\mu = \{x \in M | \mu(x) = 1\}$ is a prime ideal of M, then μ is a fuzzy prime ideal of M.

The following Theorem 3.11 is an immediate consequence of Lemma 3.7, Theorems 3.9 and 3.10.

Theorem 3.11: Let μ be any fuzzy subset of M. μ is a fuzzy prime ideal of M if and only if $Im(\mu) = \{1, \alpha\}$ where $\alpha \in [0, 1)$ and M_μ is a prime ideal of M.

4. FUZZY *m*-SYSTEMS IN Γ-SEMIGROUPS

In this section we introduce *m*-system and fuzzy *m*-system in M.

Definition 4.1: A subset A of M is called *m*-system if for any $a, b \in M$ there exist $\gamma_1, \gamma_2 \in \Gamma$ and $x \in M$ such that $a\gamma_1 x\gamma_2 b \in A$.

Lemma 4.2: Let P be an ideal of M. P is prime if and only if $M \backslash P$ is an *m*-system in M.

Definition 4.3: A fuzzy subset μ of M is called fuzzy *m*-system if for any $t, s \in [0, 1)$ and $a, b \in M$, $\mu(a) > t$, $\mu(b) > s$ implies there exist $\gamma_1, \gamma_2 \in \Gamma$ and $x \in M$ such that $\mu(a\gamma_1 x\gamma_2 b) > \max\{t, s\}$.

Example 4.4: Consider $(Z_6, \otimes_6)$ where $\otimes_6$ denotes multiplication modulo 6. Composition of any two elements is taken as multiplication modulo 6. Take $M = \Gamma = Z_6$. Then Z_6 is a Γ-semigroup. Define $\mu : Z_6 \to [0, 1]$ by $\mu(x) = 1$ if $x = 1, 3$; $\mu(x) = 0$ if $x = 0, 2, 4, 5$. For any $t, s \in [0, 1]$, $\mu(1) > t$ and $\mu(3) > s$. If $\gamma_1 = \gamma_2 = 3$ and if $x = 3 \in M$, then $\mu(1 \cdot \gamma_1 \cdot x \cdot \gamma_2 \cdot 3) = \mu(3) > \max\{t, s\}$. Thus μ is a fuzzy *m*-system in M.

Theorem 4.5: Let A be a subset of M. A is an *m*-system in M if and only if $\mathcal{X}_A$ is a fuzzy *m*-system in M.

Remark 4.6: Let μ be a fuzzy subset in M. μ holds a property like ideal, bi-ideal, quasi-ideal, etc. if and only if its level subset μ_t also satisfies the same property in M. However, μ is a fuzzy subset in M such that the level subset μ_t is an *m*-system in M, $t \in Im(\mu)$, which does not imply μ is a fuzzy *m*-system of M as the following example shows.

Example 4.7: Consider $(Z_6, \otimes_6)$ where $\otimes_6$ denotes multiplication modulo 6. Define $\mu : Z_6 \to [0, 1]$ by $\mu(x) = 1$ if $x = 1$; $\mu(x) = 0.5$ if $x = 3$ and $\mu(x) = 0$ if $x = 0, 2, 4, 5$. Take $M = \Gamma = Z_6$. For any $t \in [0, 1]$, we have $\mu_t = \{1\}$ or $\{1, 3\}$ or $\{0, 1, 2, 3, 4, 5\}$. Hence μ_t is an *m*-system in M for all t. But μ is not a fuzzy *m*-system in M, since $\mu(1)$

> 0.9 μ (3) > 0.4 but there is no $x \in Z_6$ and $\gamma_1, \gamma_2 \in \Gamma$ such that $\mu(1\gamma_1 x\gamma_2 3) > \max\{0.9, 0.4\}$.

However we have the following Theorem.

Theorem 4.8: Let μ be a fuzzy subset in M with $a \in I(x)$ implies $\mu(a) \geq \mu(x)$. μ is a fuzzy m-system in M if and only if $\mu^r = \{x \in M| \mu(x) > r\}$ is an m-system in M.

Proof: Let μ be a fuzzy m-system in M. Let $x, y \in \mu^r$, for some $r \in [0, 1)$. This implies that $\mu(x) > r$ and $\mu(y) > r$. Since μ is a fuzzy m-system in M, there exist $\gamma_1, \gamma_2 \in \Gamma$ and $z \in M$ such that $\mu(x\gamma_1 z\gamma_2 y) > r$ implies $x\gamma_1 z\gamma_2 y \in \mu^r$. Thus μ^r is an m-system in M.

Conversely, let us assume that μ^r is an m-system in M for all $r \in [0, 1)$. Suppose that $\mu(x) > r$ and $\mu(y) > s$, for some $r, s \in [0, 1)$ and $x, y \in M$. If $r = s$, the result is immediate. Without loss of generality let us assume that $s > r$. $\mu(x) > r$ and $\mu(y) > s > r$. Since μ^r is an m-system in M, then there exist $\gamma_1, \gamma_2 \in \Gamma$ and $z \in M$ such that $x\gamma_1 z\gamma_2 y \in \mu^r$. Now $x\gamma_1 z\gamma_2 y \in I(y)$ and hence $\mu(x\gamma_1 z\gamma_2 y) > \mu(y) > s$. Thus μ is a fuzzy m-system in M.

Theorem 4.9: Let μ be a fuzzy ideal of M. μ is a fuzzy prime ideal of M if and only if $1 - \mu$ is a fuzzy m-system of M.

Proof: Let us assume that μ is a fuzzy prime ideal of M. For any $t, s \in [0, 1)$, suppose there exist $a, b \in M$ such that $(1 - \mu)(a) > t$ and $(1 - \mu)(b) > s$. Hence $\mu(a) < 1 - t$ and $\mu(b) < 1 - s$. Since μ is a fuzzy prime ideal, $Im(\mu) = \{1, \alpha\}, \alpha \in [0, 1)$. Thus $\alpha < 1 - t, \alpha < 1 - s$ and $\mu(a) = \mu(b) = \alpha$. Let $P = \{x \in M|\mu(x) = 1\}$. Then by Theorem 3.5, P is a prime ideal in M and $a, b \notin P$. This implies $a, b \in M \backslash P$ which is an m-system in M. Thus there exist $x \in M$ and $\gamma_1, \gamma_2 \in \Gamma$ such that $a\gamma_1 x\gamma_2 b \in M \backslash P$ and we have $\mu(a\gamma_1 x\gamma_2 \text{b}) = \alpha$. Now $\mu(a\gamma_1 x\gamma_2 b) = \alpha < \min\{(1 - t), (1 - s)\} = 1 - \max\{t, s\}$. Now $\max\{t, s\} < 1 - \mu(\text{a}\gamma_1 x\gamma_2 b) = (1 - \mu)(a\gamma_1 x\gamma_2 b)$.

Conversely, let us assume that $1 - \mu$ is a fuzzy m-system in M. Let μ_1, μ_2 be two fuzzy ideals such that $\mu_1 * \mu_2 \subseteq \mu$. Suppose that $\mu_1 \not\subseteq \mu$ and $\mu_2 \not\subseteq \mu$. Now $\mu_1 = \bigcup_{a_p \in \mu_1} a_p$ and $\mu_2 = \bigcup_{b_q \in \mu_2} b_q$. Then there exist $a_s \in \mu_1$ and $b_t \in \mu_2$, $s, t \in [0, 1)$, such that $\mu(a) < s$ and $\mu(b) < t$. This implies that $1 - \mu(a) > 1 - s$, $1 - \mu(b) > 1 - t$. Since $1 - \mu$ is a fuzzy m-system in M, there exist $x \in M$ and $\gamma_1, \gamma_2 \in \Gamma$ such that $(1 - \mu)$

$(a\gamma_1 x\gamma_2 b) > \max \{(1-s), (1-t)\} = 1 - \min \{s, t\}$. Thus $\mu\,(a\gamma_1 x\gamma_2 b) < \min \{s, t\}$ and $(a\gamma_1 x\gamma_2 b)_{\min \{s, t\}} \notin \mu$. Now by Corollary 2.18, we have $(a\gamma_1 x\gamma_2 b)_{\min \{s, t\}} \subseteq a_s * b_t \subseteq \mu_1 * \mu_2 \subseteq \mu$, which is a contradiction to $(a\gamma_1 x\gamma_2 b)_{\min \{s, t\}} \notin \mu$. Hence μ is a fuzzy prime ideal of M.

REFERENCES

[1] P.S. Das, "Fuzzy Groups and Level Subgroups", *Journal of Mathematical Analysis and Applications*, **84**, (1981), No. 1, 264–269.

[2] P. Dheena and S. Coumaressane, "Characterization of Regular Γ-semigroups Through Fuzzy Ideals", *Iranian Journal of Fuzzy Systems*, **4**, (2007), No. 2, 57–68.

[3] R. Kumar, "Fuzzy Algebra Volume 1", *University of Delhi Publication Division*, Delhi, 1993.

[4] Y.I. Kwon and S.K. Lee, "The Weakly Semi-prime Ideals of *po*–Γ-semigroups", *Kangweon–Kyungki Mathematical Journal*, **5**, (1997), No. 2, 135–139.

[5] ...., "On the Left Regular *po*–Γ-semigroups", *Kangweon–Kyungki Mathematical Journal*, **6**, (1998), No. 2, 149–154.

[6] W.J. Liu, "Fuzzy Invariant Subgroups and Fuzzy Ideals", *Fuzzy Sets and Systems*, **8**, (1982), No. 2, 133–139.

[7] D.S. Malik and John N. Mordeson, "Fuzzy Prime Ideals of a Ring", *Fuzzy Sets and Systems*, **37**, (1990), No. 1, 93–98.

[8] T.K. Mukherjee and M.K. Sen, "On Fuzzy Ideals of a Ring 1", *Fuzzy Sets and Systems*, **21**, (1987), No. 1, 99–104.

[9] A. Rosenfeld, "Fuzzy Groups", *Journal of Mathematical Analysis and Applications*, **35**, (1971), No. 3, 512–517.

[10] N.K. Saha, "On Γ-semigroup 2", *Bulletin of the Calcutta Mathematical Society*, **79**, (1987), No. 6, 331–335.

[11] M.K. Sen, "On Γ-semigroups", *Proceedings of the International Conference on Algebra and it's Application*, Decker Publication, New York, **301**, (1981).

[12] M.K. Sen and N.K. Saha, "On Γ-semigroup 1", *Bulletin of the Calcutta Mathematical Society*, **78**, (1986), No. 3, 180–186.

[13] U.M. Swamy and K.L.N. Swamy, "Fuzzy Prime Ideals of Rings", *Journal of Mathematical Analysis and Applications*, **134**, (1988), No. 1, 94–103.

[14] L.A. Zadeh, "Fuzzy Sets", *Information and Control*, **8**, (1965), 338–353.